CU00683744

FIELD GUIDE TO THE WILD FLOWERS
of the
WESTERN
MEDITERRANEAN

Puglia, Italy

Cabo de Gata, Spain

Cap de Formentor, Mallorca

FIELD GUIDE TO THE WILD FLOWERS
of the
WESTERN
MEDITERRANEAN

Kew Publishing
Royal Botanic Gardens, Kew

The author has asserted his right to be identified as the author of this work in accordance with the
Copyright, Designs and Patents Act 1988

First published in 2016 by Royal Botanic Gardens, Kew
Royal Botanic gardens, Kew, Richmond, Surrey, TW9 3AB, UK
www.kew.org

ISBN 9781842466162
eISBN 9781842466179

British Library Cataloguing in Publication Data
A catalogue record for this book is available from the British Library.

Distributed on behalf of the Royal Botanic Gardens, Kew, in North America by the
University of Chicago Press, 1427 East 60th Street, Chicago, IL 60637, USA.

Design: Christine Beard
Page layout: Nick Otway
Editor: Sharon Whitehead

Printed and bound in Italy by Printer Trento

For information or to purchase all Kew titles please visit shop.kew.org/kewbooksonline or email
publishing@kew.org

Kew's mission is to be the global resource in plant and fungal knowledge and the world's leading
botanic garden.

Kew receives about half of its running costs from Government through the Department for Environment,
Food and Rural Affairs (Defra). All other funding needed to support Kew's vital work comes from members,
foundations, donors and commercial activities, including book sales.

All proceeds go to support Kew's work in saving the world's plants for life.

Acknowledgements

This field guide would not have been possible without the significant contribution from numerous friends and colleagues to whom I am very grateful. In particular, I thank Simon Hiscock for his inspiration and for the many happy years of teaching undergraduate field courses in Portugal, along with Alexandra Allen and Andy Bailey. I am also very grateful for the expertise, helpful suggestions and significant contributions made by Finn Rasmussen, Antonio Pujadas, Gianniantonio Domina, Tim Rich and Serge D. Muller during the preparation of the manuscript. Thanks to Salvatore Cozzolino, Paolo Grunanger, Richard Bateman and Pete Hollingsworth for their expertise on specific plant groups. Many of the photographs from Sicily were provided by Gianniantonio Domina, the rarer North African species by Serge D. Muller, and orchid species by Niels Faurholdt, all of which are valuable contributions. Species described in this book were predominantly authenticated with *Flora Europaea*. *Flora Iberica*, although incomplete during the compilation of this book, was also a valuable identification resource.

CONTENTS

PLACES TO SEE WILD FLOWERS IN
THE WESTERN MEDITERRANEAN   13

WILD FLOWER HABITATS 31

SPECIES DESCRIPTIONS 40

Cabo de Gata, Almeria

Introduction

This book is a guide to the flowers of the western Mediterranean floristic region: an area encompassing southern Europe from the Portuguese Algarve in the west to Italy in the east, the islands (including the Balearic Islands, Corsica and Sardinia), and North Africa from Morocco in the west to Tunisia in the east. The western Mediterranean Basin is characterised by a climate of hot dry summers and mild wet winters and has a different flora to the eastern Mediterranean and Asia Minor, which warrant a guide of their own.

The Mediterranean regions of Morocco, Algeria and Tunisia are covered, however the other phyto-geographic zones of northern Africa, for example the montane and desert zones, each have their own specific flora. Similarly, many of the alpine species that grow in the western Mediterranean and the Macaronesian flora have also been excluded; these are covered in other guides.

Mastic (*Pistacia lentiscus*) is one of the most common plants in the western Mediterranean from Portugal and North Africa in the west, to Italy in the east.

The Mediterranean flora is one of the richest in the world, and the number of plant species in the western Mediterranean alone exceeds 10,000. This book focuses on the most common and conspicuous species that occur in the area (such as the ubiquitous mastic, *Pistacia lentiscus*), but also describes local rarities and endemics that are of interest to a particular area (for example *Linaria algarviana*, a rarity found only in southern Portugal). Non-native species that are naturalised are also included, because they form a significant contribution to the flora, and because it is not always obvious whether a

Linaria algarviana is a rare endemic restricted to just a few coastal shale cliffs in the Algarve.

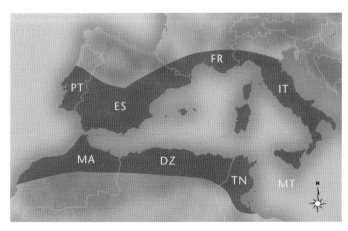

Geographical coverage of this book, indicated in purple.

ES: Spain

DZ: Algeria

FR: France

IT: Italy

PT: Portugal

MA: Morocco

MT: Malta

TN: Tunisia

plant is native or not. Some plant families are covered in more detail than others. This is because some families are of wider interest or more conspicuous. For example a greater proportion of orchids (Orchidaceae) are described than of grasses (Poaceae).

HOW TO USE THIS BOOK

This flora is a visual field guide with colour photographs for quick and easy identification, which are sequenced in order of plant relatedness. To identify a plant using this book, the reader's approach will typically be as follows:

1. Scan through the photographs for a plant that looks most similar to the species in question. To identify the relevant section of the book quickly, the reader may first need to identify the family to which the plant belongs to (see *How to identify plant families*, p.11).

2. Use the plant descriptions associated with the photographs to verify the identification.

3. Check the descriptions for closely related species which may appear to be similar on first glance. These are described in the same section, with the key distinguishing characteristics highlighted.

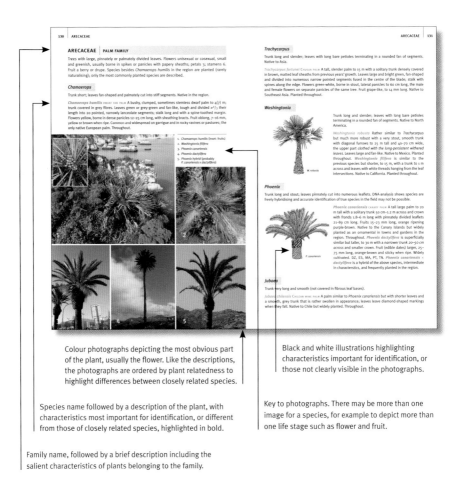

Colour photographs depicting the most obvious part of the plant, usually the flower. Like the descriptions, the photographs are ordered by plant relatedness to highlight differences between closely related species.

Black and white illustrations highlighting characteristics important for identification, or those not clearly visible in the photographs.

Species name followed by a description of the plant, with characteristics most important for identification, or different from those of closely related species, highlighted in bold.

Key to photographs. There may be more than one image for a species, for example to depict more than one life stage such as flower and fruit.

Family name, followed by a brief description including the salient characteristics of plants belonging to the family.

Classification based on plant relatedness has been chosen because plants in a family tend to share common characteristics (see *How plants are classified in this book*, p.5). Grouping plants in ways other than by relatedness, for instance by flower colour, can be misleading. This is because the flowers of closely related species are often a very different colour, and may vary markedly even within a population, for example the flowers of *Bartsia trixago*, which range from bright yellow to pale pink.

KEYS

Exhaustive identification keys, which are used in traditional floras, have been avoided because they demand a level of detail that is more than most readers want and because they can be lead to an incorrect identification if all the species in a given family are not included. Instead, larger groups of similar species have been sub-divided into groups (labelled A, B, C etc.) which share distinctive characteristics. The differences between closely related taxa are highlighted in the plant descriptions.

PLANT DESCRIPTIONS

The plant descriptions have predominantly been written in a standardised way for consistency and for quick and easy comparison amongst species. Each description includes details on dimensions, habitat and distribution (see below). However it is not useful to describe all species in the exactly same way. For example it might be useful to highlight a particular set of characteristics (such as filament hairiness) which are important in distinguishing species in some groups but not in others. An excessive use of technical terms has been avoided where possible. However to describe species accurately, the use of some technical terms is necessary, therefore a list of definitions is included in the Glossary.

Name

Each plant species is listed by its scientific name, a naming convention made up of two parts: the first indicates the genus to which the species belongs, and the second indicates the species within the genus. Some plant descriptions also include subspecies. These normally differ subtly from the typical form of the species, but are not distinct enough to be described at the species rank. A detailed account of how the plant descriptions are classified is given below (see *How plants are classified in this book*, p.5).

Dimensions

The dimensions described will of course vary under different environmental conditions and with the age of the plant, but are comparative between species and act as an approximate guide.

Habitat

The habitat is described because this is often useful for identification. For example species of sea lavender (*Limonium* spp.) are generally found growing in saline environments, and are therefore unlikely to be encountered in, for instance, oak woodland.

Flowering time

The flowering time is not normally given in the descriptions because the exact flowering time varies very much from year to year depending on the weather, and also with the location and elevation. The majority of species flower in the spring or early summer in the western Mediterranean, and for those which flower in the autumn or winter, the flowering time is highlighted in the description.

Distribution

The precise distribution for many of the species described is simply not known. It must be emphasised that the distributions of species in this book are based on plant records, on the literature and on the author's experience and are an approximate guide only. It is therefore possible that a species may be encountered outside the distribution given in this guide. Most of the species described are widespread, but rarities and endemics are also described if they are of particular interest or are likely to be encountered. The countries a species is likely to occur in are abbreviated according to international convention: DZ (Algeria), ES (Spain), FR (France), IT (Italy), PT (Portugal), MA (Morocco), MT (Malta) and TN (Tunisia). For less widespread species or island endemics, an indication of distribution is also given for the larger islands (defined as: the Balearic Islands, Corsica, Sardinia and Sicily).

ILLUSTRATIONS AND PHOTOGRAPHS

Due to the comprehensive coverage, it has not been possible to include photographs of all species. Therefore photographs have been selected to illustrate the most common and widespread species from all the major families and genera, as well as some exotic species, rarities and endemics that are of particular interest. Names of plants for which photographs are included are indicated in blue bold. The majority of photographs depict the flowering stage which is usually the most useful for identification. Black and white illustrations are given to complement the plant descriptions and to illustrate characteristics that are useful for identification, particularly for closely related groups of plants.

Cistanche phelypaea

Wild flower Classification and Identification

HOW PLANTS ARE CLASSIFIED IN THIS BOOK

Plant classification is the assignment of plants into a hierarchy of groups according to their relatedness. All the plants described in this book are assigned to a hierarchy that begins with 4 groups: the gymnosperms (non-flowering seed plants), the basal angiosperms (an ancient lineage of plants including water lilies, magnolias and laurels), the monocots (herbaceous perennials with parallel-veined leaves such as orchids, grasses and lilies) and the eudicots (representing the majority of plants, trees and shrubs, typically characterised by net-veined leaves). Within these broad divisions is a further division of orders (ending in '- ales') which are made up of related families, which in turn are made up of genera. For example the olive (*Olea europaea*) is a species that belongs to the genus *Olea* which is in the family Oleaceae, which is part of the order Lamiales, a division of the eudicots. A brief description of each family and genus is given because members within each of these groups tend to share common characteristics that are useful for identification.

The classification described above has traditionally been based on morphology (the appearance of the plant). It is now accepted that genetic analysis is a more accurate means of classifying plants, and is the basis of the globally accepted classification of flowering plants, a system called the 'Angiosperm Phylogeny Group' (APG) classification. In order to show the relatedness of a given group of plants (see *How to use this book*, p.2), the sequence of this flora is based on this APG system of classification. For this reason, the assignment of species to a family in this book will sometimes differ from that of traditional floras, in which plants were classified according to their morphology rather than their genetic relatedness. For example many species traditionally ascribed to the figwort family (Scrophulariaceae), such as the snapdragons (*Antirrhinum*), are now known to belong to the plantain family (Plantaginaceae).

TAXONOMY

As described above, following the advent of DNA-based analyses, traditional plant family classifications have been revised extensively. This is also true of plant species names. As more information becomes available about relatedness of species, their names are revised. This can lead to confusion where traditionally used names are still widely used, or where botanists do not agree on the status of a species, subspecies or variety. Therefore plant names will vary from one guide to another, and over time. The naming convention used in this guide predominantly follows *Flora Iberica*, which is the most up-to-date flora for many species in the region. However this guide deviates from *Flora Iberica* where more contemporary information is available, where there does not appear to be a regional consensus for a given species or subspecies, or for species that occur outside this flora's coverage. To avoid confusion, synonyms are included for species that are widely known under an alternative or obsolete

A Plant Family Tree

FAMILY

A group (name usually ending '–eae') of related genera. Every family has a number of defining characteristics useful for identification (see *How to Identify Plant Families*, p.11).

ORCHIDACEAE

GENUS (plural: genera)

A group of related species. All species in a given genus share the same common generic name (the first word in a name made up of two parts). Species in the same genus share many characteristics and are sometimes difficult to tell apart.

Ophrys *Anacamptis*

SPECIES

A unit within a genus. Each species has a unique specific epithet (the second word in a name made up of two parts). Every species has uniquely defining traits necessary for identification.

Ophrys speculum *Anacamptis pyramidalis*

SUBSPECIES

Some species are further divided into subspecies. These normally differ subtly from the typical form of the species, but are not distinct enough to be described at the species rank.

Ophrys speculum
subsp. *lusitanica*

Ophrys speculum
subsp. *speculum*

name, indicated by '(Syn.)' in the descriptions. For very complex groups, it would not be possible to list all described taxa. For example the common dandelion (*Taraxacum officinale*) comprises a myriad of microspecies, but for the purpose of this book, only one description is given. Subspecies are included where they are geographically important or particularly distinct. Whilst it is conventional to include all subspecies that occur in the region covered, it would be beyond the scope of this guide to do so, particularly for very variable species for which numerous subspecies have been described.

THE SPECIES SELECTED

It would be impossible to include every species that grows in the western Mediterranean in one field guide. Therefore the ca. 2,400 species described in this book have been selected using the following three lenses:

Geography: The species most likely to be encountered across all the countries covered are included, as well as local rarities of interest to a particular area, for example the islands. Fewer species are described from areas less frequented or well-known, such as remote parts of Mediterranean Algeria and Tunisia, than for example the Algarve or the Balearic Islands.

Habitat: Species have been selected from a range of habitats (see *Wild flower habitats*, p.31). The bulk of species are from garrigue and maquis, which are the predominant habitats in the region. Those from other habitats are often characteristic of other eco-regions. For example many species in deciduous forests are also common in temperate Northern Europe; similarly some from the Atlantic coast of Morocco (the Atlantic littoral) are shared with Macaronesia, the Sahara, and the Algero-Tunisian coast. One or two of the species described are not strictly Mediterranean but included for interest, for example the rare mountain-dwelling orchid *Dactylorhiza durandii* from the Moroccan Rif.

Taxonomy: All the main plant families are included. However some families are covered in more detail than others. This is because some families are disproportionately large or abundant (see *How to identify plant families*, p.11), or because they are of wider interest or more conspicuous, for example the orchid family (Orchidaceae).

IDENTIFYING PLANTS IN THE FIELD

Even though plant classification is now based on DNA-based analyses, identification in the field still relies on plant descriptions such as those included in this flora. A grasp of the basic anatomy of plants, including leaf type and flower structure (see below) is therefore essential for wild flower identification. The habit, form and appearance of a plant will often all be indicators of the family to which it belongs (see *How to identify plant families*, p.11). How flowers are arranged is important. For example an umbel (an umbrella-shaped inflorescence) is characteristic of the carrot family (Apiaceae), whereas a capitulum (a sunflower-like inflorescence made up of many tiny flowers or florets) is typical of the daisy family (Asteraceae). The use of a hand lens is recommended to look closely at smaller features of the plant, which may be useful for identification. For taxonomically challenging groups such as the broomrapes (*Orobanche*), for example, the degree of hairiness of the filaments is an important diagnostic, whereas the shape and form of the fruit is more important in closely related species of medick (*Medicago*).

Some common flower types seen in the west Mediterranean flora: (A) Flower with an actinomorphic perianth of petals and sepals (*Cistus*); (B) Inflorescence a capitulum, typically seen in Asteraceae (*Centaurea*); (C) Inflorescence an umbel (*Cachrys*); (D) Inflorescence composed of spikelets, typically seen in Poaceae (*Aegilops*); (E) Flower a zygomorphic perianth of tepals (*Gladiolus*); (F) Inflorescence a lax raceme (*Leopoldia*); (G) Inflorescence a corymb (*Iberis*); (H) Inflorescence a dense, spike-like raceme (*Bartsia*).

Some common leaf types seen in the west Mediterranean flora: (A) Ovate-elliptic (*Cistus*); (B) Round and shallowly palmately lobed (*Ficus*); (C) Clasping (perfoliate) (*Lonicera*); (D) Round and cordate (heart-shaped) at the base (*Cyclamen*); (E) Crowded in opposite files (*Corema*); (F) Trifoliate (*Bituminaria*); (G) Pinnately divided into leaflets (*Erophaca*); (H) Whorled (*Rubia*); (I) Lanceolate (*Plantago*); (J) Arrow-shaped (*Arum*); (K) Linear (*Asphodelus*); (L) Deeply pinnately lobed and toothed (*Sonchus*).

FLOWER ANATOMY SIMPLIFIED

Flower anatomy is a very important tool for wild flower identification. Flowers are broadly classed as being either regular, or actinomorphic (radially symmetrical), or zygomorphic (bilaterally symmetrical), a trend that is often consistent within a family. Other important features include the relative dimensions, colour and hairiness of flower parts such as the filaments and anthers (collectively referred to as the stamens), the ovary, style and stigma (collectively referred to as the carpels), and the sepals (which make up the calyx) and petals (which make up the corolla) which together are referred to as the perianth. The main floral structures are shown below.

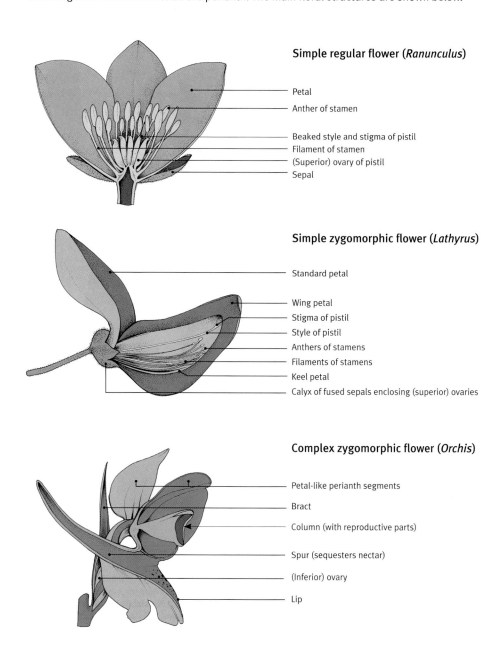

Simple regular flower (*Ranunculus*)

— Petal
— Anther of stamen

— Beaked style and stigma of pistil
— Filament of stamen
— (Superior) ovary of pistil
— Sepal

Simple zygomorphic flower (*Lathyrus*)

— Standard petal

— Wing petal
— Stigma of pistil
— Style of pistil
— Anthers of stamens
— Filaments of stamens
— Keel petal
— Calyx of fused sepals enclosing (superior) ovaries

Complex zygomorphic flower (*Orchis*)

— Petal-like perianth segments

— Bract

— Column (with reproductive parts)

— Spur (sequesters nectar)

— (Inferior) ovary

— Lip

HOW TO IDENTIFY PLANT FAMILIES

An understanding of the salient features of the main plant families will enable the reader to navigate identification even in a seemingly bewildering array of wild flowers. Furthermore, related families often share characteristics, so even if a plant does not belong to any of the families described below, this list should guide the reader to the relevant section of the book. Eight important families are particularly abundant, and contribute to about half the entire flora of the region. These are listed in order of importance below, in addition to a further five families which contain fewer species, but are nevertheless common in many plant communities.

1. **ASTERACEAE daisy family.** Individual flowers minute, tubular and 5-lobed or flat and strap-like, and aggregated into daisy-like, dandelion-like or thistle-like flower-heads (capitula) surrounded by bracts. Fruit single-seeded, often with a 'parachute' (pappus) attached.

2. **LEGUMINOSAE (Fabaceae) pea family.** Leaves alternate, often pinnate, with stipules. Calyx 5-toothed, sometimes 2-lipped; flowers zygomorphic with 5 petals, the upper a standard, the laterals as 2 wings and the lower fused into a keel; stamens 10. Fruit often pea-pod-like, splitting when ripe, sometimes coiled.

3. **POACEAE grass family.** Leaves alternate, usually linear, often in 2 ranks and sheathing the stem. Flowers complicated: usually cosexual with 1, 3 or 6 stamens and a pistil enclosed in 2 bracts, the lemma and palea, the whole making a floret; florets arranged in spikelets with empty bracts (glumes) at the base; styles normally 2. Fruit grain-like (an achene).

4. **CARYOPHYLLACEAE pink family.** Leaves opposite. Sepals 5, separate or fused into a tube. Petals 5, separate but sometimes spilt. Fruit a capsule with 6, 8 or 10 teeth.

5. **LAMIACEAE mint family.** Plants often glandular and aromatic with square stems, leaves opposite. Flowers borne in (often) congested whorls towards the stem apex. Calyx with 5 teeth, often 2-lipped; corolla normally 2-lipped; stamens 2 or 4. Fruit made up of nutlets hidden at the calyx base.

6. **BRASSICACEAE cabbage family.** Leaves usually opposite. Flowers often white or yellow, with 4 separate sepals; petals 4, separate and forming a cross. Fruit typically a 2-parted pod.

7. **APIACEAE carrot family.** Leaves alternate, simple or 2-4 times divided (feathery). Flowers tiny, petals 5, often uneven, borne in umbrella-like heads (umbels); stamens normally 5. Fruit 2-parted, often flattened.

8. **LILIACEAE lily family.** Traditionally a single family now split into several closely related families characterised by leaves with parallel veins, often strap-like. Tepals (perianth segments) petal-like, separate or fused; stamens 6; ovary superior. Fruit a 3-parted capsule or berry.

9. **ORCHIDACEAE orchid family.** Leaves in rosettes or alternate, narrow and untoothed. Flowers in upright spikes or racemes; sepals 3, petals 3, the laterals spreading or united with one or all sepals, the lower forming a variably shaped lip; pollen in a paired sticky mass (pollinia); ovary inferior. Fruit a capsule.

10. **PLANTAGINACEAE plantain family.** Now including many species formerly in the Scrophulariaceae. Leaves alternate or opposite. Flowers borne in cymes or racemes; sepals 4–5, petals 4–5, often fused into a corolla tube, sometimes 2-lipped; stamens 2, 4 or 5. Fruit a capsule. The Orobanchaceae is a related family of parasites many of which lack green pigment (chlorophyll).

11. **AMARANTHACEAE amaranth family.** Now including many species formerly in the Chenopodiaceae. Leaves alternate or opposite. Flowers regular, borne in cymes or panicles, usually with 5 (1-8) tepals, often fused; stamens 1–5; ovary superior. Flowers often enlarging in fruit; fruit 1-seeded.

12. **BORAGINACEAE Borage family.** Leaves normally alternate and bristly. Flowers borne in coiled cymes; calyx 5-toothed; petals 5, fused into a variably long tube, often pink, white or blue; stamens 5. Fruit made up of 5 nutlets.

13. **CISTACEAE Rock rose family.** A well-represented family in the region. Leaves normally opposite, often with stipules. Flowers cosexual and regular, often showy; sepals 3 or 5, petals 5; stamens numerous. Fruit a 1–10-valved capsule.

Places to see wild flowers in the western Mediterranean

A selection of locations from across the western Mediterranean is described here, each with details of its unique flora including rarities or plants of particular interest. Many of the places described have been identified as 'Important Plant Areas', or are nature reserves or national parks. The best time to visit to see wild flowers is also indicated, however this varies with the local weather from year to year.

IBERIAN PENINSULA

PORTUGAL

The Algarve

The Algarve region of southern Portugal has an Atlantic outlook and is not geographically part of the Mediterranean Basin, but has a Mediterranean climate and flora. The region covers an area of approximately 5,500 square kilometres stretching about 135 km from the windward Atlantic coast in the west, to the more

Map of the mainland Iberian Peninsula. Areas of particular interest for wild flowers described in this book are indicated in purple.

sheltered Spanish border region in the east. The low-lying coastal belt lies 40 km from the sea and rises gradually northwards through a region of rolling hills into a somewhat discontinuous mountain range which forms a natural boundary with the neighbouring province of Alentejo. Due

Coastal dune systems in the Algarve support rarities including *Ionopsidium acaule* (left) and *Linaria ficalhoana* (right).

to this varied geology, the region has a rich flora, including sand dune plant communities with rarities such as the tiny endemics, *Linaria ficalhoana* and *Ionopsidium acaule* which are legally protected, as well as parasitic *Cistanche phelypaea* which grows in impressive stands in the saltmarshes at Faro and in the west. Garrigue and maquis (known locally as 'Matos') communities are dominated by gum rock rose (*Cistus ladanifer*), grey-leaved rock rose (*C. albidus*), Phoenicean juniper (*Juniperus phoenicea*), French lavender (*Lavandula stoechas*) and mastic (*Pistacia lentiscus*) as well as a number of geophytes including various bee orchids *Ophrys lutea*, *O. speculum*, *O. bombyliflora* and *O. fusca*. Olive (*Olea europaea*), carob (*Ceratonia siliqua*) and orange (*Citrus* x *sinensis*) cultivation are common, in addition to extensive cork oak (*Quercus suber*) woods in the north.

Insectivorous *Drosophyllum lusitanicum* has suffered a dramatic reduction in the Algarve, and now occurs at just a handful of sites.

Of particular interest is the protected South West Alentejo and Cape St. Vincent National Park (Parque Nacional de Sudoeste Alentejano e Costa Vicentina), which contains one of the richest floras in Europe, and a number of endemics including the yellow-flowered crucifers *Biscutella sempervirens* subsp. *vicentina* and *Diplotaxis siifolia* subsp. *vicentina*, and the tiny purple annual *Linaria algarviana*. Other rarities in the Algarve region are the poorly known, parasitic *Cynomorium coccineum* of

The Algarve's Cape St. Vincent National Park juts into the Atlantic, and boasts numerous endemics, including the rare *Silene rothmaleri* (top inset), thought to be extinct until recently, which grows on just a few unstable shale slopes. Cliff-tops of the cape are a riot of colour in late spring after the winter rains have passed (bottom inset).

which just one or two thriving sites remain, and the insectivorous *Drosophyllum lusitanicum* which is precarious and sporadic the region. Despite the protected status of much of the region, the flora is very much under threat from development. The best time to visit is between mid-March and late April; much of the landscape at lower elevations is dry and baron by mid-summer.

SPAIN

Andalucía

The flora of Andalucía is both well-documented and extremely rich, characterised by approximately 4,000 vascular plants of which over 10% are endemic. Andalucía is characterised by a similar climate, geology and flora to northwest Africa, indeed the region is separated from Morocco by just 12 km at the narrowest point of the Gibraltar Strait. In particular, the Aljibe Mountains of Andalucía have a flora similar to that of higher elevations of the Tanger area of Morocco, as do the Spanish Serranía de Ronda and Axarquía to the Moroccan western Rif and Taguist, and the coastal sea belts of both countries. It is, of course, well beyond the scope of this book to describe the entire flora of these particularly diverse floristic regions, therefore with a

The pink flowers of grey-leaved rock rose (*Cistus albidus*) are a common sight on rocky outcrops and maquis along Andalucía's southern coastline in spring.

Much of Andalucía's landscape is dry and baron, and dominated by drought-tolerant, slerophyllous shrubs such as *Genista umbellata*.

few exceptions, a focus is placed on the typically Mediterranean flora found at lower elevations. Much of the coastline of Andalucía is developed, but the protected coastal areas still maintain a very rich flora. Rocky slopes are typically dominated by *Genista umbellata*, grey-leaved rock rose (*Cistus albidus*), Phoenicean juniper (*Juniperus phoenicea*), mastic (*Pistacia lentiscus*), and grasses such as *Macrochloa tenacissima*. Tall, purple-pink spikes of the parasitic broomrape *Orobanche latisquama* among their host plant, rosemary (*Rosmarinus officinalis*), as well as clumps of the rare endemic *Limonium insigne* sprout from rocky outcrops, and the area has a very rich diversity of geophytes in early spring. There are two national parks in Andalucía: the Doñana and the Sierra Nevada.

Doñana, Andalucía

The Doñana National Park lies within in the provinces of Huelva and Seville and has long been established as one of the most important wetland refuges for wildlife in Europe. It covers 540 km² of which 135 km² are protected. The park includes habitats including maquis, scrub and woodland, as well up to 3,000 water bodies in wet years, including lagoons, marshes and streams. There are sand dune and salt marsh plant communities in Las Marismas and the Guadalquivir River Delta region. Maquis at higher elevations is dominated by French lavender (*Lavandula stoechas*), *Helichrysum italicum* and *Cistus libanotis*. Damper depressions are dominated by Dorset heath (*Erica ciliaris*), ling (*Calluna vulgaris*) and dwarf gorse (*Ulex minor*), often fringed with patches of *Halimium halimifolium* and *Stauracanthus genistoides*. Ponds and flooded meadows are dominated by the grass *Cynodon dactylon*, by the alien exotic *Cotula coronopifolia*, and by the delicate lesser water-plantain (*Baldellia ranunculoides*), and by coral necklace (*Illecebrum verticillatum*), Pond water-crowfoot (*Ranunculus peltatus*) dominates in deeper waters.

Sierra Nevada, Andalucía

The Sierra Nevada is a mountain range in the provinces of Granada and Almería. Much of the flora of the park is typically montane (not a focus of this book) and includes a number of rarities and endemics, including the daffodil *Narcissus nevadensis*, as well as other geophytes such as *Crocus nudiflorus*. The montane vegetation is dominated by stands of holm oak (*Quercus ilex*), sweet chestnut (*Castanea sativa*) and Pyrenean oak (*Q. pyrenaica*). Below 2,000 m, the vegetation is more typically Mediterranean, and dominated by junipers (*Juniperus communis*, *J. oxycedrus*), pines and oaks (*Pinus sylvestris*, *Q. faginea*) and *Crataegus monogyna*. On the lower slopes, olives (*Olea europaea*), butcher's broom (*Ruscus aculeatus*) and mastic (*Pistacia lentiscus*) are common.

Tabernas, Andalucía

The Tabernas desert, located in Spain's south-eastern province of Almería, is one of the only true semi-deserts in Europe, and is superficially similar to the absolute deserts of Arabia, North Africa and North America. The area is isolated from Mediterranean fronts, and annual rainfall can fall short of 200 mm per year, giving rise to a very barren landscape with little vegetation, that nevertheless hosts a number of rare xerophytes. Rarities include the sea lavender *Limonium insigne* and the conspicuous parasitic *Cynomorium coccineum*, which elsewhere in the Mediterranean region is normally found on dry, crumbling sea cliffs. Prickly pears (*Opuntia maxima*) and oleander (*Nerium oleander*) are a common and prominent feature in the area.

Cap de Creus, Catalonia

The Cap de Creus is a rocky headland that juts out into the Mediterranean from the northeast Spanish mainland region of Catalonia, about 25 km from the French border. The area is a natural park encompassing 190 km² of bleak, rocky garrigue that is subject to strong coastal winds and salt spray, largely devoid of trees, and is rather reminiscent of the southern Portugal's jugged and windswept Cape St. Vincent. Common species in the park include mastic (*Pistacia lentiscus*), prickly juniper (*Juniperus oxycedrus*), tree heather (*Erica arborea*), spiny legumes such as *Calicotome spinosa*, rock roses (*Cistus salviifolius, C. monspeliensis* and *C. albidus*), and a myriad of halophytic plants that grow on coastal cliffs. Mauve spikes of parasitic broomrape *Orobanche amethystea*, which grow on the roots of *Eryngium campestre*, are not uncommon along roadsides in the park, and ivy broomrape (*O. hederae*) grows at the base of the walls of the monastery. The park is best visited in April to May.

The Cap de Creus on Spain's eastern coast boasts rich wind-swept coastal garrigue plant communities. Parasitic ivy broomrape (*Orobanche hederae*) (right) grows in sheltered spots, such as at the base of old walls.

The Balearic Islands

The Balearic Islands (the largest of which comprise Mallorca, Menorca and Ibiza), as well as being a popular holiday destination are an interesting area to observe wild flowers. They host a diverse range of species typical of Mediterranean maquis, aquatic, oak forest, coastal littoral and montane habitats, and a high number of island endemics, particularly on the mountain slopes and rocky coasts.

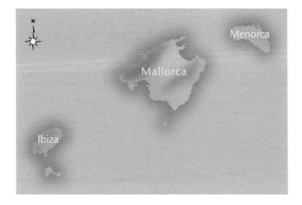

The Tabernas desert is home to plants which are tolerant of extreme drought. Rarities include the sea lavender (*Limonium insigne*) (left above) and the bizarre parasite *Cynomorium coccineum* (left below).

Mallorca, Balearic Islands

Mallorca's Cap de Formentor is a rocky peninsula, rising to 384 m above sea level. The area boasts breath-taking views and is a botanist's paradise, with numerous endemics and rarities on the exposed slopes, including spiny shrubs (such as *Astragalus balearicus*, *Launaea cervicornis* and *Teucrium balearicum*), the spectacular dead horse arum (*Helicodiceros muscivorus*), delicate *Lotus tetraphyllus* in rocky crevices, and the intriguing, autumn-flowering *Arum pictum* in wooded areas. The Parc Natural de S'Albufera is a predominantly wetland national park in the north of the island between Alcúdia and Ca'n Picafort. It contains a range of habitats including lagoons, dune belts and flood plains. Bull rushes (*Typha latifolia*), tamarisk (*Tamarix africana*) and elm (*Ulmus minor*) are common in damp and water-logged fresh water areas, whilst brackish areas support halophytic species such as *Salicornia* spp. and *Arthrocnemum* spp. Geophytes such as the tassel hyacinth (*Leopoldia comosa*) and orchid species, including the small-flowered tongue orchid (*Serapias parviflora*), *Orchis robusta*, *O. fragrans* and the bee orchid (*Ophrys apifera*), are also frequent in the park and surrounding area. The best time to visit is in late spring to early summer.

Mallorca's Cap de Formentor is a botanist's paradise, home to numerous rarities, including autumn-flowering *Arum pictum* (inset) which grows among pine trees in more sheltered spots.

Menorca, Balearic Islands

Menorca's Necropolis de Cala Morell is a coastal site that is afforded protection from development by its historic interest. Rocky slopes and stony ground in the site and the surrounding coastline are dominated by rock samphire (*Crithmum maritimum*), hairy sea heath (*Frankenia hirsuta*), *Lotus cytisoides*, *Anthemis maritima* and *Helichrysum stoechas*. Further inland the most prevalent species are sedges such as *Schoenus nigricans*, annuals including *Reichardia picroides* and

Menorca's Necropolis de Cala Morell has an interesting mosaic of coastal garrigue plant communities.

shrubs such as Phoenicean juniper (*Juniperus phoenicea*), rosemary (*Rosmarinus officinalis*), and lavender cotton (*Santolina chamaecyparissus*). This is one of only a few places to see the rare, yellow, endemic broomrape *Orobanche iammonensis* which sprouts from rocky outrcrops. The best time to visit is between March and May.

FRANCE

Most of the south of France is of interest to the botanist even though its Mediterranean coastline is relatively short. The list below contains a handful of floristically rich places in the region.

The Luberon Massif

The Luberon in the south of France comprises three mountain ranges and covers an area of 600 km². It has a particularly rich flora with an estimated 1,500 plant species. The area has a mild climate, which is influenced by the Alps and the Mediterranean, and a varied geology, with

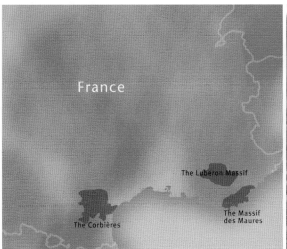

France has a relatively short Mediterranean coastline, which is nevertheless floristically rich, particularly in orchids, such as the lizard orchid (*Himantoglossum hircinum*) (right).

habitats including limestone garrigue, oak woodland, mountain slopes and pine woods, as well as a range of disturbed habitats including agricultural land, vineyards, olive groves and lavender fields. Orchids are common in the area, particularly lizard orchids (*Himantoglossum hircinum*), pyramidal orchids (*Anacamptis pyramidalis*) and helleborines (*Epipactis muelleri*), as well as Bertoloni's bee orchid (*Ophrys bertolonii*). Limestone maquis and garrigue are dominated by rock roses (*Cistus albidus*, *C. laurifolius*) and *Helichrysum stoechas*. Other plants of interest in the area include the striking violet limodore (*Limodorum abortivum*) and *Gladiolus italicus*.

The Massif des Maures

The Massif des Maures lies to the north of Saint-Tropez and comprises a series of acidic rock ridges and hill forests flanked by the Gapeau and Argens valleys to the north and the Mediterranean Sea to the south. The area is characterised by scrub, woods, grassland, heaths and maquis. The massif is home to large stands of pines (*Pinus halepensis*, *P. pinaster*), cork oak (*Quercus suber*), and other typically Mediterranean species, including mastic (*Pistacia lentiscus*), myrtle (*Myrtus communis*), *Phillyrea angustifolia*, *Smilax aspera* and rosemary (*Rosmarinus officinalis*).

The Corbières

The Corbières comprise a limestone mountain range in the Languedoc-Roussillon in southwest France, bordering the Mediterranean Sea to the east. The flora of the Corbières includes both Mediterranean and Pyrenean elements.

Corsica

Corsica lies southeast of the French mainland and is one of the largest Mediterranean islands, with 1,000 km of coastline and an extensive mountain range with diverse habitats, including rugged granite cliffs and gorges, pine and oak woods, as well as coastal maquis. Corsica has a rich and varied flora, comprising approximately 3,000 species and a number of rarities and endemics restricted to the island, or shared with neighbouring Sardinia. The island is particularly rich in

Left: Lavender (*Lavandula* x *intermedia*) is a common crop in the Provence region of southern France.

Right: *Serapias neglecta*, which occurs locally on Corsica, is one of a plethora of rare tongue orchids which occur in the western Mediterranean. PHOTO: NIELS FAURHOLDT

rare and endemic geophytes including the tongue orchids *Serapias nurrica* and *S. neglecta*, as well as the delicate *Romulea corsica*, *Leucojum longifolium* and *Crocus minimus*. The valley of the Gorge de la Restonica is home to the abundant winter-flowering, lilac crocus *C. corsicus*, whilst the *Quercus* forests of the western coast's Spelunca Gorge host the striking, white-flowered *Pancratium illyricum* as well as the unusual violet limodore (*Limodorum abortivum*). Like most islands in the Mediterranean, Corsica's flora is threatened by development and by the introduction of invasive alien species, particularly in lowland areas.

ITALY

La Maremma

The Maremma region extends from southwestern Tuscany (Maremma Livornese & Maremma Grossetana) to northern Lazio on the western coast of Italy, and hosts a range of habitats including volcanic outcrops, abandoned agricultural coastal plains, garrigue, pine forest, swamps, saltmarshes and coastal dunes. The diverse range in geology, climate and habitat contribute to a diverse flora, which includes an estimated 25% of the entire Italian flora. Dune systems are dominated by *Juniperus oxycedrus* and by smaller perennials including *Teucrium polium*. Stands of umbrella pine (*Pinus pinea*) create an impressive backdrop to the coast. Species of particular interest include the aquatic, carnivorous bladderwort *Utricularia australis* and various orchids (*Orchis palustris*, *Himantoglossum*

Map of Italy (green) including mainland Italy, Sardinia and Sicily. Areas of particular interest for wild flowers covered here are indicated in purple.

Extensive stands of umbrella pines (*Pinus pinea*) are common in the Maremma, in which clearings are home to a myriad of perennials, such as the pink-flowered *Ajuga iva* (inset).

robertianum, *Cephalanthera longifolia*, *Epipactis muelleri*, *Limodorum abortivum*, *Ophrys fuciflora* and *O. insectifera* to name just a few). Large stands of giant fennel (*Ferula communis*) are common on roadsides on the approach to the park from the north, and make an impressive feature. Rarities include an endemic crocus (*Crocus etruscus*) which is locally common in the broadleaf woodlands of Massa Marittima. The area is best visited from late April to June.

Vesuvius

Mount Vesuvius is a volcanic national park located about 9 km east of Naples on the coastal Campania plain, and comprises two geological formations: the Vesuvius cone and the Somma caldera. The arid, volcanic slopes of Vesuvius are characterized by typical Mediterranean maquis and pine forests, dominated by shrubs such as *Genista tinctoria* and home to numerous geophytes including a range of orchid species. By contrast, the slopes of the Somma caldera are much wetter and more densely vegetated with deciduous woodland dominated largely by chestnut (*Castanea sativa*) and oak (*Quercus* spp.).

Puglia

Puglia is the southeastern-most region of Italy, bordering the Adriatic Sea in the east and the Ionian Sea to the southeast. The south of Puglia is dominated by olive groves and by drought-tolerant plants, many of which are more characteristic of the eastern Mediterranean. Rarities include the endemic, critically endangered *Arum apulum*, and Adriatic-dwelling *Limonium cancellatum*. The region includes Mount Gargano, as well as an archipelago of islands and an extensive range of ancient deciduous hill forest. Habitats are rich and varied, including beach and oak forests, pine forests, rocky coastland, wetlands and limestone outcrops. The park is renowned for its extraordinary diversity of orchid species: boasting nearly 70 species, the area has been described as the 'orchid capital of Europe'. Orchid species in the Gargano include the giant orchid (*Himantoglossum robertianum*), lady orchid (*Orchis purpurea*) and man orchid (*Orchis*

Much of the landscape and plants in southern Puglia are more characteristic of the eastern Mediterranean such as the Adriatic *Limonium cancellatum* (top right). The critically endangered *Arum apulum* (bottom right) is not found outside the region.

anthropophorum), as well as a multitude of bee orchids (including *Ophrys lutea*, the endemic *O. fuciflora* subsp. *apulica*, *O. fusca* and *O. bertolonii*) and tongue orchids such as *Serapias lingua*, *S. cordigera*, *S. vomeracea* and *S. parviflora*. The best time to visit is from April to May.

Sicily

Sicily is the largest of the Mediterranean islands. Its size and extremely varied geology have given rise to a very rich and varied flora. Habitats are very diverse, and include limestone on the mountains around Palermo and Trapani in the west, the deciduous beech forests on the clay substrates of the Nebrodi mountains, the carbonate substrates of the northern range, and the Etna volcano which rises 3,300 m above sea level. Each of these areas hosts a unique flora, with elements typical of the western Mediterranean, as well as floristic links with the eastern Mediterranean and North African floras. A range of rarities and over 20 endemic species are found in the Aeolian island archipelago as a whole. Rarities include *Limonium lopadusanum* and *Brassica macrocarpa*, a relative of the common wild cabbage, which are both restricted to just a few coasts in the Sicilian archipelago, and the endangered *Calendula maritima* which occurs only on the mainland between Marsala and Monte Cofano and neighboring islets.

Antirrhinum siculum (left), endemic to Sicily and Malta, and *Sarcopoterium spinosum* (right), a shrub more typical of the eastern Mediterranean, are both common on Sicily.

Sardinia

Sardinia is the second largest of the Mediterranean islands after Sicily, and like neighbouring Corsica, boasts one of the richest floras in the Mediterranean with an estimated 2,300 species, of which nearly 350 are endemic. Important habitats on the island include the rocky coastline, which stretches for 1,800 km², the mountain massifs of granite, dolomite limestone and sandstone, as well as numerous wetland habitats and alluvial plains, such as Campidano in the southwest and the Nurra in the northwest.

6,000 km² of the island is protected (about 25% of the island's surface area), and the island has three national parks: the Asinara National Park, the Arcipelago di La Maddalena National Park, and the Gennargentu National Park. Sardinia also has a further 10 regional parks and numerous reserves. The Asinara National Park is dominated by maquis, predominantly mastic (*Pistacia lentiscus*) and olive (*Olea europaea*). Other common species in the park include *Convolvulus althaeoides*, viper's bugloss (*Echium vulgare*) and *Galactites tomentosa*. The rare, endemic, spiny shrub *Centaurea horrida* occurs on rocky outcrops. The Arcipelago di La Maddalena National Park is situated in the northeast of the island, and includes a multitude of islands, islets, and rock formations. The park has a typically Mediterranean flora, dominated by Phoenicean Juniper (*Juniperus phoenicea*), strawberry tree (*Arbutus unedo*) and mastic (*Pistacia lentiscus*). The Gennargentu National Park is located on the east coast of Sardinia, and includes garrigue, mountain and coastal habitats. Rarities in the park include *Crocus minimus*, the peony *Paeonia mascula*, *Romulea requienii* and *Ornithogalum collinum*.

NORTHERN AFRICA

The North African flora is very rich and diverse because of the region's markedly varied geography, which encompasses a mosaic of typically Mediterranean habitats, including numerous islands and islets, deciduous and evergreen forests, maquis, garrigue, pasture, wetland, coastal cliffs and dunes, and semi-desert. This book mainly describes floristic elements from the Mediterranean regions of Morocco, Algeria and Tunisia, which are predominantly similar to those of the European western Mediterranean, but which also have a high degree of species endemism. It is beyond the scope of this book to describe the other significant ecoregions of North Africa, including the Atlantic littoral, Saharan region and mountain ranges, which are not typically Mediterranean in their species composition (though a few exceptions are described in this guide, where they overlap with Mediterranean plant communities or for interest).

The flora of Mediterranean North Africa is not as well documented as that of the European territory. The distribution and status of many species is improperly known, particularly in more remote areas, and deserves further attention to allow effective setting of conservation priorities, especially in light of the significant threat from agricultural intensification including over-grazing and deforestation in the region. Indeed many of the endemic species have been discovered only recently, emphasising the need for further attention.

Often, the most diverse and accessible places to visit to see the North African flora are the national parks, of which there are 10 in Morocco, 10 in Algeria (many of which are coastal), and 8 in Tunisia. A small number of these have been identified as sites of particular interest for their plant species diversity and are described below. The best time to visit is generally between March and April, but varies very much within the region.

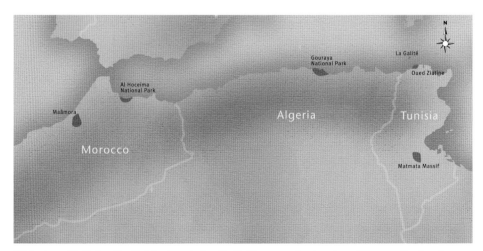

This book specifically describes floristic elements from the Mediterranean-climate regions of Morocco, Algeria and Tunisia, and does not, on the whole, cover the Saharan region, or mountain ranges which make up much of North Africa's land area (and which are not strictly floristically Mediterranean). North Africa is vast, and only a few places of botanical interest are covered here.

PHOTOGRAPH: GUY ANDERSON

Cistanche species such as *C. violacea* (left) and *C. phelypaea* (right) are spectacular parasitic plants which are widespread in deserts and dunes across North Africa. These plants lack green pigment and true leaves, owing to their parasitic habit.

MOROCCO

Morocco covers a total area of over 710,000 km², and boasts 160 sites of biological and ecological interest. In fact the flora of Morocco is one of the most diverse in the Mediterranean region, with an estimated 4,000 vascular plant species of which over 20% are endemic. This high floristic diversity is a consequence of Morocco's varied geology and climate, which is influenced both by the Atlantic and the Sahara. This book focuses on the flora of the Mediterranean region and covers the low elevation areas north of a line extending from Sale-Rabat to the more arid Oujda and the Algerian frontier. This area characterised by a similar climate, geology and flora to the European western Mediterranean, and in particular to the Andalucía region of southern Spain, from which it is separated by the Gibraltar Strait, just 12 km wide at its narrowest point. Consistent with that of the other regions in the western Mediterranean, the Moroccan flora is characterised by over 100 families, of which the daisy, pea and grass families (Asteraceae, Leguminosae and Poaceae, respectively) are the most species-rich. Plant Mediterranean communities are dominated by arar tree (*Tetraclinis articulata*) and deciduous oaks (*Quercus canariensis* and *Q. pyrenaica*). Other common species include Spanish juniper (*Juniperus thurifera*), Phoenician juniper (*J. phoenicea*), prickly juniper (*J. oxycedrus*), carob (*Ceratonia siliqua*), olive (*Olea europaea*) and mastic (*Pistacia lentiscus*). The spectacular parasite *Cistanche violacea* is not found in Europe, but is frequent in parts of North Africa, and occurs in both desert and Mediterranean sand dune communities. Moroccan rarities include the endemic sea holly *Eryngium atlanticum* and succulent *Kleinia anteuphorbium*.

Maâmora

The Maâmora is one of the more extensive stands of cork oak (*Quercus suber*) in the region, situated in northwest Morocco, between Rabat and Kenitra in the west and the Beht river in the east, covering an area of approximately 700 km², and ranging in altitude from 30–230 m. Common plant species in the park include dwarf fan palms (*Chamaerops humilis*), sage-leaved rock rose (*Cistus salviifolius*), giant fennel (*Ferula communis*), *Hyparrhenia hirta* and French lavender (*Lavandula stoechas*). Common tree species in the park include the North African endemic maritime pine (*Pinus pinaster* subsp. *hamiltonii*), and introduced *Eucalyptus camaldulensis*.

Al Hoceima National Park

The Al Hoceima National Park lies close to the town of Al Hoceima in northern Morocco. Covering an area of 470 km², it is the largest protected area of Mediterranean coastline in Morocco, and includes limestone halophytic plant communities along the cliffs, in addition to common species in the park such as kermes oak (*Quercus coccifera*), carob tree (*Ceratonia siliqua*), mastic (*Pistacia lentiscus*), olive (*Olea europaea*), grey-leaved rock rose (*Cistus albidus*), pomegranate (*Punica granatum*), *Consolida regalis*, heaths (*Erica* spp.), as well as a range of typical Mediterranean geophyte species.

The Moroccan Wetlands

The Mediterranean wetlands of Morocco are of particular interest. These comprise a range of lakes, lagoons, marshes and peat fens, for example the Moulouya river mouth, Sebkha of Bou Areg, the Tahaddarte river, the Ghomara coast, and the lagoon and barrage of Smir. These wetland complexes are extremely rich, largely unknown, and seriously threatened by human activities. A number of species are of interest in the Moroccan wetlands, including the delicate coral necklace (*Illecebrum verticillatum*) which scrambles over flooded sandy pastures, *Limoniastrum monopetalum* which occurs on saline alluvial sands, tamarisk (*Tamarix africana*) which is common in wet fresh and brackish habitats, the unusual yellow-flowered insectivorous bladderwort *Utriculara australis*, as well as a range of regional rarities including Dorset heath (*Erica ciliaris*). These wetland habitats are very much under threat from drainage for development and cultivation of *Eucalyptus* spp.

PHOTOGRAPH: SERGE D.MULLER

PHOTOGRAPH: SERGE D.MULLER

Species associated with Moroccan garrigue communities include French Lavender (*Lavandula stoechas*), the rare endemic sea holly *Eryngium atlanticum* (top right) and succulent *Kleinia anteuphorbium* (bottom right).

Species associated with Moroccan wetland communities include tamarisk (*Tamarix africana*) (left) and coral necklace (*Illecebrum verticillatum*) (right).

ALGERIA

Algeria covers a vast area of more than 2,000,000 km², and is the largest of all countries with a Mediterranean coastline. Like Morocco, Algeria has a diverse range of floristic regions characterised by Mediterranean, montane and desert plant communities. The country is of considerable botanical interest, and includes 21 Important Plant Areas in the north and 3 coastal national parks, El Kala National Park, the Gouraya National Park and the Taza National Park, all of which have a noteworthy diversity of plant species and include a number of rarities. Algeria shares many species with neighbouring Morocco and Tunisia, but is also home to more than 200 endemic species. The predominant Algerian plant communities occupy: coastal dunes and limestone cliffs; sclerophyllous forests dominated by kermes oak (*Quercus coccifera*), holm oak (*Q. ilex*) and cork oak (*Q. suber*); deciduous forest dominated by *Q. canariensis*; evergreen forests dominated by pines (*Pinus halepensis* and *P. pinaster*); maquis dominated by mastic (*Pistacia lentiscus*), rosemary (*Rosmarinus officinalis*) and *Calicotome spinosa*; as well as wetland habitats including marshes, flooded meadows, lakes, and peat fens where the rare North African endemic prostrate daisy (*Bellis prostrata*) occurs.

Gouraya National Park

The Gouraya National Park is a UNESCO-recognised reserve that lies in the northeast of Algeria Béjaïa Province, and includes a mountainous limestone massif which dominates the north west of the town of Béjaïa, a small calcareous massif in the western zone, and a wetland, Lake Mézaïa. Of particular interest are the coastal limestone cliffs and the rocky fields, both of which are home to many rare and endemic plant species. The tree spurge (*Euphorbia dendroides*) is a prominent species in the park on rocky slopes and ledges. Limestone cliffs here are home to halophytes such as yellow sea aster (*Pallenis maritima*), *Lotus cytisoides* and sweet Alison (*Lobularia maritima*). Maquis is dominated by kermes oak (*Quercus coccifera*), *Phillyrea latifolia*,

PHOTOGRAPH: SERGE D. MULLER

The rare *Bellis prostrata* (left) is found in Algerian wetland communities, whilst myrtle (*Myrtus communis*) (right) is common on coastal garrigue, as it is elsewhere in the western Mediterranean.

Rhamnus lycioides, *Prasium majus*, *Asparagus albus*, dwarf fan palm (*Chamaerops humilis*) and *Sedum sediforme*. Forests are characterised by Aleppo pine (*Pinus halepensis*), olive trees (*Olea europea*), carob (*Ceratonia siliqua*), myrtle (*Myrtus communis*) and *Pistacia terebinthus*. Bee orchids occur in the park (*Ophrys apifera*, *O. speculum* and *O. tenthredinifera*), as do other geophytes including *Gladiolus communis* which occurs on calcareous outcrops. Species that are common in wetter habitats, such as flooded areas and river banks, include herb Robert (*Geranium robertianum*), black bryony (*Dioscorea communis*), Italian arum (*Arum italicum*) and narrow-leaved ash (*Fraxinus angustifolia*).

The Algerian Islands

The Algerian coastline has many islands and islets, which are isolated and uninhabited, and are of particular interest because they show floristic similarity with both continental Africa and the larger western Mediterranean islands, including Sicily, Sardinia and Corsica. They are not particularly species-rich owing to their small size, however due to their geographical isolation and lack of habitation, these islands are a refuge for rarities. The Rechgoun and Habibas Archipelago in the northwest of Algeria is an island complex of volcanic origin with a degree of plant species endemism shared with both neighbouring African and European continents. The Kabylian-Numidian small islands on the north coast of Algeria are home to populations of *Pancratium foetidum* and *Allium commutatum* (the latter otherwise more north-central in its distribution). The islands are best visited in March to April, but on the whole are not easily accessible, and are described here for their noteworthy interest.

TUNISIA

Tunisia is the smallest country in North Africa by land area, covering about 165,000 km². It is a Maghreb country bordered by Algeria to the west, Libya to the southeast and the Mediterranean Sea to the north. The climate ranges from arid-saharan in the south to humid-Mediterranean in the northern mountain ranges. The Mediterranean region of Tunisia is more humid than the Mediterranean parts of Morocco and Algeria, and the flora is less diverse with a lower degree of endemism. The predominant native plant communities are typical of Mediterranean maquis and forest, comprising sclerophyllous forests dominated by kermes oak (*Quercus coccifera*) and cork oak (*Q. suber*), deciduous forest dominated by *Q. canariensis*, evergreen forests dominated by pines (*Pinus halepensis* and *P. pinaster*), maquis dominated by olive (*Olea europaea*) and mastic (*Pistacia lentiscus*), and steppic formations dominated by Phoenicean juniper (*Juniperus phoenicea*), *Artemisia* spp. and grasses such as *Macrochloa tenacissima*. Wetland species include the very rare Tunisian dock (*Rumex tunetanus*), an endemic of northern Tunisia, which is close to extinction and is found at only one site in the Mogods.

La Galite Archipelago

La Galite Archipelago is a mainly granite-based island complex lying 40 km from the continent with many rarities that are more characteristic of Corsica and Sardinia and continental Europe than of continental North Africa, including *Allium commutatum*, *Ononis minutissima*, and interestingly, *Serapias nurrica*, a rare tongue orchid previously believed to be restricted to Sardinia, Corsica, Menorca, Sicily and mainland Italy.

Matmata Massif

The Matmata massif, near the east coast of Tunisia, includes Toujane, an area recognised as an Important Plant Area. Toujane is characterised by an arid Mediterranean climate and a typical scrubland plant community, dominated by Phoenician juniper (*Juniperus phoenicea*) and grasses such as *Macrochloa tenacissima*, as well as by *Teucrium polium* and the rare *Rosmarinus eriocalyx*.

Oued Ziatine

The Oued Ziatine is a river basin recognised as an Important Plant Area that runs to the Mediterranean at Cap Serrat, in the north of Tunisia. This watercourse is characterised by a humid Mediterranean climate, and is largely forested with alder (*Alnus glutinosa*). It is home to rarities including the annual *Solenopsis bicolor* and the small daisy *Bellis prostrata*, a species restricted to the coastal wetlands and coastal plains of northeast Algeria and Tunisia (and recently also encountered in Morocco).

Tunisian dock (*Rumex tunetanus*) (left) is a rare endemic of northern Tunisia on the brink of extinction. *Macrochloa tenacissima* (above) is a typical Mediterranean grass common in Tunisia, as well as much of the western Mediterranean.

THE MALTESE ARCHIPELAGO

The Maltese archipelago lies in the central Mediterranean approximately 80 km south of Sicily and 280 km east of Tunisia, and comprises Malta, Gozo and Comino, as well as numerous minor islands that are uninhabited. The largest island, Malta, has an area of 316 km² and is one of the world's smallest and most densely populated countries. Due to Malta's small size, dense population, and paucity of permanent freshwater and montane habitats, its flora is less rich than that of other Mediterranean countries, however its geographic isolation has contributed to a degree of endemism. The principle habitats on the archipelago include woodland dominated by holm oak (*Quercus Ilex*) and Aleppo pine (*Pinus halepensis*), sclerophyllous maquis and garrigue dominated by olive (*Olea europaea*), mastic (*Pistacia lentiscus*), yellow germander (*Teucrium flavum*) and carob (*Ceratonia siliqua*), and degraded plant communities dominated by grasses including *Hyparrhenia hirta* and many annuals in the carrot and daisy families (Apiaceae and Asteraceae, respectively). Rarities include an endemic, Maltese everlasting (*Helichrysum melitense*), which grows only on the western cliffs of Gozo. Other native geophytes on the islands include *Ophrys sphegodes*, the sea daffodil (*Pancratium maritimum*), *Crocus longiflorus*, *Ornithogalum arabicum* and *Iris pseudopumila*.

The Maltese archipelago has a typically Mediterranean flora, with woodland plant communities dominated by Aleppo pine (*Pinus halepensis*) (below) like much of the region.

Wild flower habitats

GARRIGUE AND MAQUIS

The western Mediterranean Basin's climate supports a rich shrub-dominated vegetation at lower altitudes. These shrubs are typically tough and spiny with hairy or waxy leaves, a growth form known as sclerophyllous ('hard leaved'), referring to the presence of sclereids (cells with thickened cell walls), which are an adaptation to retain moisture. The sclerophyllous habit has evolved independently in unrelated plant families across the globe in response to arid conditions, and the floral landscapes of all five Mediterranean climate regions, including the Mediterranean Basin, South Africa, California, Australia and Chile, appear superficially similar. In the western Mediterranean Basin this spiny, evergreen scrub vegetation on deep, acid soils is known variously as maquis (French), mattoral (Spanish) or macchia (Italian). Garrigue refers to the low-lying form of maquis associated with shallower soils, typically on coastal limestone where weather fronts contribute to a distinctly more stunted growth habit than that of maquis, even though the dominant species are often the same.

Prickly juniper (*Juniperus oxycedrus*) in northern Spain. A dominant plant in many garrigue and maquis plant communities.

Plants associated with maquis and garrigue are resinous and aromatic. These habitats are dominated by kermes oak (*Quercus coccifera*), juniper (*Juniperus* spp.), lavender (*Lavandula* spp.), rosemary (*Rosmarinus officinalis*), thyme (*Thymus* spp.) and rock roses (*Cistus* spp.) as well as various shrubs in the pea family (Leguminosae). Maquis on deeper, acidic substrates is dominated by heathers (*Erica* spp., *Calluna* spp.), broom (*Cytisus scoparius*) and strawberry trees (*Arbutus unedo*), interspersed with the more lime-tolerant species listed above. The understoreys are rich in geophytes such as daffodils (*Narcissus* spp.), alliums (*Allium* spp.), irises (Iridaceae) such as the barbary nut iris (*Moraea sisyrinchium*) and a myriad of orchids (Orchidaceae), all of which survive the hot, dry summers in the form of underground corms, bulbs and tubers. A range of spring-flowering grasses (Poaceae) and annuals of the pink, pea and borage families (Caryophyllaceae, Leguminosae, Boraginaceae) are common in more disturbed areas, and lie dormant during the summer months in the form of seeds.

Limestone formations in southern Portugal: the bedrock of garrigue throughout the western Mediterranean, and an important habitat for geophytes such as the barbary nut iris (*Moraea sisyrinchium*) (inset).

In the western Mediterranean Basin this spiny, evergreen scrub vegetation on deep, acid soils at lower altitudes is known as maquis. Typical plants that grow in this habitat include the holm oak (*Quercus ilex*) (top right) and rosemary (*Rosmarinus officinalis*) (bottom right).

Low-lying, weather-beaten vegetation associated with shallow coastal limestone is known as garrigue. Plants that grow in this habitat are similar to those of maquis, but often more stunted. Examples include the native dwarf fan palm *Chamaerops humilis* (top left) and the gum rock rose (*Cistus ladanifer*) (bottom left).

AGRICULTURAL LAND

The landscape of the western Mediterranean Basin was at one time dominated by evergreen oak forest, the majority of which was cleared for agriculture during Neolithic times. The predominant crops in the region include wheat, rice, pulses, vegetables, citrus fruits, and grapes. In addition crops introduced in antiquity which do not require intensive farming have widely naturalised and are an important part of the flora, for example the carob, olive, fig and almond.

The carob (*Ceratonia siliqua*) is an evergreen tree belonging to the pea family (Leguminosae) that has long been cultivated as a source of animal feed and as a cocoa substitute. Interestingly, the flower clusters sprout directly from the branches of the tree — a process known as cauliflory. The olive (*Olea europaea*) is of major agricultural importance throughout the Mediterranean Basin as a source of olive oil. The ancestry of the olive is not known because truly wild forms have been lost through millennia of cultivation, however several geographically distinct subspecies have been recognised in addition to the numerous cultivars that are farmed. Traditional, low-intensity small-holding olive farms and neglected olive groves are very important habitats for hundreds of species of annuals, small perennials and geophytes including orchids. Sadly, however, much of the olive farming practice in the western Mediterranean is now intensively managed, which has led to a marked decline in such species. Fig (*Ficus carica*) and almond (*Prunus dulcis*) groves and low-input vineyards are also important reserves for wild flowers in the region when managed traditionally, and host a similar myriad of species to olive groves.

Much of the Mediterranean Basin is under threat from agricultural intensification. For example, cork oak (*Quercus suber*) forests which sustain unique and important deciduous woodland ecosystems are increasingly replaced by stands of fast-growing *Eucalyptus* and pine (*Pinus*) forests, which are cultivated for pulpwood, because they realise greater profits in today's market.

Crops introduced in antiquity in the region, which do not require intensive farming, and have widely naturalised and are an important part of the flora: the carob (left), fig (top middle) olive (bottom middle) and the almond (right).

COASTAL HABITATS

The western Mediterranean coastline hosts an extraordinarily rich flora. Sheltered coves flanked by garrigue-covered slopes, exposed limestone and shale cliffs and undeveloped sand dune systems support very rich maritime plant communities throughout the region, particularly in protected areas. All coastal habitats are exposed to a high salinity, and the plants that grow there (halophytes) are specifically adapted to such an environment. Many have a bluish, waxy coating on the surface of their leaves (glaucous) which affords them protection from salt spray and sun damage, as well as salt glands which excrete excess salt. Among the more common halophytes are the marigold-like yellow sea aster (*Pallenis maritima*) and the pale yellow-flowered rock samphire (*Crithmum maritimum*) which both frequent rocky outcrops, boulders and cliffs throughout the region.

Common halophytes include the yellow sea aster (*Pallenis maritima*) (left) and the rock samphire (*Crithmum maritimum*) (right).

Sand dune plant communities are dominated by stands of the familiar marram grass (*Ammophila arenaria*) throughout the western Mediterranean.

Sand dune plant communities are dominated by stands of the familiar marram grass (*Ammophila arenaria*) which stabilises sand. Other common species in this habitat include the attractive, summer-flowering sea daffodil (*Pancratium maritimum*), the fleshy sea spurge (*Euphorbia paralias*), sea bindweed (*Calystegia soldanella*) and sea rocket (*Cakile maritima*). Beyond the foredunes lie the more nutrient-rich, damp fixed dunes which are dominated by sedges (family Cyperaceae) and various perennials — often local rarities — and which eventually grade into garrigue and maquis.

Saltmarshes and estuaries host only the most salt-tolerant plant species, which are often submerged completely each day by seawater. Saltmarshes in the western Mediterranean are mainly dominated by species of the family Amaranthaceae, such as *Sarcocornia perennis*, *Atriplex halimus* and *Salicornia ramosissima*, which are all rather succulent, bluish or reddish, and salt-secreting, typical of all halophytes. Some saltmarsh systems in the far west, such as those in the Algarve, are home to the striking yellow parasite *Cistanche phelypaea* which stands proudly among swathes of its host plants (often shrubby Amaranthaceae).

Salicornia ramosissima is common to subdominant in saltmarsh communities across the region.

Much of the western Mediterranean coastline is severely threatened by tourist-driven development and its associated habitat destruction, as well as by displacement by alien exotics planted for ornament, such as the highly invasive hottentot fig (*Carpobrotus edulis*) which smothers and outcompetes native plant communities throughout the region.

Delicate sand dune plant communities are threatened by the encroachment of the highly invasive hottentot fig (*Carpobrotus edulis*) across the region.

PINE FORESTS

Pine forests are a common feature of the western Mediterranean. The umbrella pine (*Pinus pinea*) has been cultivated since antiquity for its edible pine nuts, and its umbrella-shaped canopy is iconic of the region. It is frequent on hill slopes as well as in towns and gardens, or can sometimes be seen growing in large stands, such as those of the Maremma Park in western Tuscany. The Aleppo pine (*P. halepensis*) is another very common pine in the region, particularly on higher ground inland, and is extensively cultivated for timber and pulpwood. Finally the maritime pine (*P. pinaster*) is very common along the coast throughout the region, typically on rocky slopes flanking sheltered Mediterranean coves, or in plantations, as a source of timber and turpentine. Pine forests are typically less floristically rich than shrub-dominated habitats, but their understoreys host a unique flora of their own, which is often rich in acid-loving perennials as well as orchid species such as the violet limodore (*Limodorum abortivum*), tongue orchids (*Serapias* spp.), bee orchids (*Ophrys* spp.) and helleborines (*Epipactis* spp.).

Umbrella pine (*Pinus pinea*) plantations (left) are a common feature at lower elevations across the region, whilst cork oaks (*Quercus suber*) (right) are common on higher ground in the southwest, and are an important refuge for wild flowers.

OAK FORESTS

At higher altitudes, the mean annual temperature falls whilst rainfall increases. This cooler, damper climate gives rise to sub-montane oak (*Quercus*) dominated plant communities that are markedly different to those of the maquis and garrigue at lower altitudes. These oak forests form refuges for plants adapted to cooler conditions in the otherwise hot, dry Mediterranean climate, and share many of the species that are common the temperate deciduous forests of Northern Europe. The most common and widespread species of oak at low to moderate elevations is the large, evergreen holm oak (*Q. ilex*), which forms pure and mixed stands throughout the western Mediterranean. Native cork oak (*Q. suber*) forests are also common in southern Portugal and Spain and in northwest Africa, and are cultivated throughout the western Mediterranean as a renewable source of cork, predominantly for bottle stops. The cork oak's thick, insulating bark makes it particularly resistant to fire, and can also be harvested repeatedly without harming the tree. In addition to their significant economic importance as a source of cork, cork oak forests support a rich biodiversity which is relatively well-protected by low-intensity farming methods. Common species in the understory of holm oak and cork oak forests include shrubs such as the strawberry tree (*Arbutus unedo*), heathers (*Erica* spp.), broom (*Cytisus scoparius*) and butcher's broom (*Ruscus aculeatus*), as well as shade-loving perennials including foxgloves (*Digitalis purpurea*), peonies (*Paeonia* spp.), wild arums (*Arum italicum*) and a myriad of orchids, for example helleborines (*Epipactis* spp.). Extensive stands of oak forest which once covered much of the western Mediterranean have been felled for agriculture, and replanted with *Eucalyptus* and various species of pine and fir (*Pinus* and *Abies* spp.) which are fast-growing and realise faster profits.

DISTURBED HABITATS

Disturbed and degraded habitats such as roadsides, car parks, disused land, demolition sites and gardens are very rich in native annuals and short-lived perennials, which are adapted to growing in transient environments. These plants are typically fast-growing and drought-tolerant, and in the western Mediterranean tend to flower in early spring, and set seed and die before the summer, due to acute water stress. Annuals in the daisy family (Asteraceae) are particularly common, for example fleabane (*Conyza canadensis*), various thistle species (*Carduus, Cirsium,* and *Onopordum* spp.), sow thistles (*Sonchus* spp.), and in damper areas, the yellow-flowered *Dittrichia viscosa*. Other plants of interest on disturbed ground include the squirting cucumber (*Ecballium elaterium*), the fruits of which violently propel their seeds into the air in a stream of liquid, and the parasitic broomrapes (*Orobanche* spp.), which lack chlorophyll and rob their nutrients from the roots of other plants.

A high population density, international trade links and land clearing for agriculture and tourism in the western Mediterranean have all contributed to an increase in the emergence of disturbed habitats, and to invasion by non-native plant species. Most of these alien exotics are from other Mediterranean climates around the world and are therefore quite at home in

Glebionis coronaria, a member of the daisy family (Asteraceae), is a very common feature of disturbed habitats across the western Mediterranean.

the western Mediterranean; indeed, it can be hard to tell them apart from the native flora that they displace. Prominent alien species include the century plant (*Agave americana*), originally from Mexico and now a common escape from towns and gardens, which produces enormous succulent rosettes of leaves, followed by tree-like inflorescences which persist long after they have bloomed. Roadsides in the western Mediterranean are commonly lined with yellow-flowered acacia trees (*Acacia* spp.), native to Australia, and with extensive carpets of the South African Bermuda buttercup (*Oxalis pes-caprae*), which smothers native annuals and is a pest, for example in the Algarve region of southern Portugal. The Hottentot fig (*Carpobrotus edulis*) frequents disturbed coastal sites across the region, and whilst attractive, is a deadly threat to the delicate native ecosystems in which it takes a hold.

FRESHWATER AQUATIC HABITATS

Freshwater aquatic habitats such as ponds, rivers, ditches and seasonally flooded grasslands are dominated by macrophytes — plants adapted to submersion or water-logged conditions. Such habitats are relatively rare in the dry, arid climate of the western Mediterranean, and have a markedly different flora to the maquis and garrigue of drier areas. Like the oak forests, freshwater aquatic habitats form refuges for plants adapted to damper conditions in the otherwise hot, dry Mediterranean climate, and therefore share many of the species that occur in the temperate wetlands of Northern Europe. Common examples include marginal plants such as the yellow flag iris (*Iris pseudacorus*), purple loosestrife (*Lythrum salicaria*), flowering rush (*Butomus umbellatus*), broadleaf arrowhead (*Sagittaria latifolia*), sedges (family Cyperaceae) and bullrushes (*Typha* spp.), as well as submerged plants such as the fringed water-lily (*Nymphoides peltata*) and the rootless and carnivorous bladderworts (*Utricularia* spp.) which trap minute organisms with their tiny, bladder-like, underwater traps. Seasonally flooded plains

Water-logged habitats such as this seasonal lake provide an important refuge for specialised aquatic species such as the insectivorous bladderwort (*Utricularia vulgaris*) (inset).

are frequented by shrubs such as tamarisk (*Tamarix africana*), geophytes including the attractive, mauve-flowered *Scilla peruviana*, and annuals such as the delicate coral necklace (*Illecebrum verticillatum*) which scrambles over damp, sandy banks. Due to the isolation of Mediterranean freshwater aquatic habitats, many of the larger wetland systems are also home to a range of rarities and endemics.

BARE AND ARID HABITATS

Parts of the extreme south of the Mediterranean in North Africa border arid and semi-desert zones. These areas are subject to extreme water stress and are typically dry and bare. The plants that grow in such habitats are physiologically adapted to drought (xerophytes) or are ephemeral, short-lived annuals that complete their life cycle quickly during transient wet periods (much like the plants described in *Disturbed habitats*, p.37). Drought-tolerant species in arid habitats include the succulent shrub *Traganum moquinii* which is a common feature of the succulent thicket flora of arid coastal areas in northwest Morocco, as well as the spiny, virtually leafless shrub *Launaea arborescens*, which frequents bare, rocky ground. Other plants respond rapidly to spring rainfall, such as the resurrection plant *Pallenis hierochuntica* and short-lived annuals such as *Kickxia sagittata* and *Lotus glinoides*. These xerophytes are more typical of Saharan and African Atlantic littoral plant communities than the Mediterranean flora in its strictest sense.

Plants that grow in the driest parts of the region which border arid and desert habitats are adapted to extreme drought, for example ephemeral annuals such as *Kickxia sagittata* (left), tough, spiny shrubs such as *Launaea arborescens* (top right) and plants which respond rapidly to seasonal rains such as *Pallenis hierochuntica* (bottom right).

Species descriptions

GYMNOSPERMS

Seed plants without true flowers, in which ovules are 'naked' rather than enclosed within an ovary.

PINACEAE | PINE FAMILY

Evergreen or deciduous trees with needle-like leaves borne spirally and vegetative buds with brown scales. Male cones with copious pollen; female cones borne on the same tree and more substantial, with woody scales, taking years to mature.

Cedrus CEDAR

Evergreen trees with needle-like leaves, 3–5-angled in cross section, borne singly on leading shoots and in dense clusters on side shoots.

Cedrus atlantica ATLAS CEDAR A large evergreen tree to 39 m with horizontal branches (when mature); bark grey and smooth or with shallow fissures. Leaves dark blue-green and spirally arranged or in dense clusters, needle-like, 15–20 mm long. Female cones erect and barrel-shaped, 50–80 mm long, ripening pale purplish-brown with numerous overlapping scales. Mountain slopes. DZ, MA (planted elsewhere).

1. *Abies pinsapo* var. *marocana*
   PHOTOGRAPH: SERGE D. MULLER
2. *Pinus pinea*
3. *Pinus halepensis* (inset: needles)
4. *Pinus halepensis* female cones

Abies FIR

Evergreen trees with stalkless, needle-like leaves borne singly (not clustered) on long shoots.

Abies alba SILVER FIR A large tree to 48 m with greyish, scaly bark and hairy twigs. Leaves needle-like, 15–30 mm long, flexible and dark green above, yellowish below and slightly notched at the tip, *arranged in parallel rows*; *buds not resinous*. Female cones erect, 10–15(20) cm long, cylindrical-ovoid with distinct bracts between the scales. Mountain woods. ES, FR, IT. *Abies pinsapo* is similar, with rather silvery, stiffly-pointed, blue-green needles to 15 mm long arranged *concentrically* (not in flattened parallel rows), with *resinous buds* and without visible cone bracts. Female cones cylindrical, 90 mm–18 cm long. Local in mountains. ES. ***Abies pinsapo* var. *marocana*** (Syn. *A. marocana*) MOROCCAN FIR is sometimes treated as a variant of *A. pinsapo* and differs in having *less strongly blue* needles, and *longer female cones 11–20 cm long*. Rif Mountains only. MA.

Larix LARCH

Deciduous trees with flattened, soft, green leaves borne singly on leading shoots and in dense clusters on side shoots.

Larix × eurolepis LARCH A deciduous tree of hybrid origin (*L. decidua × L. kaempferi*), to 46 m with greyish-brown scaly bark. Leaves needle-like, and borne in clusters of up to 40, green above with 2 paler bands below, to 13 mm long. Female cones woody, erect, *small*, 20–35(45) mm long and with 3-angled, exerted scales; male cones small and yellow. Planted, on higher ground. ES, FR, IT.

Pinus PINE

Evergreen trees with stiff, often long, needle-like leaves borne in clusters of 2, 3 or 5 on very small shoots.

A. Cones large, >(60)70 mm long

P. pinea

P. halepensis

P. halepensis
(female cone)

Pinus pinea UMBRELLA PINE A large, distinctive tree to 30(40) m, *parasol-shaped* when mature. Bark grey-brown with peeling red patches. Leaves green, 70 mm–15(20) cm long, 1.2–1.7 mm wide, with minute, forward-pointing teeth along the margins. Female cones 70 mm–14 cm long, broadly ovoid-spherical, shiny, red-brown; seeds *scarcely winged* or unwinged; scales more or less all the same size. Common on hills and in coastal habitats, often in large stands, frequently planted for the edible seeds. Throughout (possibly not native in North Africa). *Pinus radiata* is similar with clusters of 2(3) leaves and female cones in which with basal scales clearly more prominent than those at the apex. Native to the USA but planted in Spain and probably elsewhere. ES.

Pinus halepensis ALEPPO PINE A medium-sized tree to 20 m with an *irregular, rounded crown*, and distinctly twisted branches; branches and twigs greyish, bark becoming reddish and fissured; buds not sticky. Leaves needle-like, borne in pairs, 60 mm–12(15) cm long and 0.7–1 mm wide, straight or twisted. Female cones egg-shaped, 60 mm–12 cm long, borne on *recurved* stalks; seeds *winged*. Common on dry slopes and cliffs, especially in the centre of the region; isolated populations occur in Portugal and Tunisia (also planted in these countries). Throughout.

P. pinaster (female cone)   *P. pinaster*

Pinus pinaster MARITIME PINE A tall, pyramidal tree to 35 m, often flat-topped and stunted in coastal habitats. Bark blackish, becoming fissured, reddish above. Leaves dark green, *in pairs, long, 15–20(25) cm*. Female cone ovoid, borne in clusters of 2–8, light brown and shiny with spiny scales, 80 mm–22 cm long. Common on maritime sands and widely cultivated for timber, soil conservation and the production of turpentine. DZ, ES, FR, IT, MA, PT. Subsp. *hamiltonii* forms extensive stands in fir forests in the Rif and middle Atlas. MA (planted elsewhere).

B. Cones small, <70(90) mm long

P. sylvestris (female cone)   *P. sylvestris*

Pinus sylvestris SCOTS PINE A tall tree to 36 m. Bark dark grey-brown below, *pale orange-red and flaking* above; *crown flat-topped and lopsided when mature*. Leaves *small and twisted*, 20–80 mm long, 3 mm wide, distinctly stalked, and in pairs with a persistent grey basal sheath; *blue-grey*. Female cone more or less globose, acute, deflexed and yellowish-brown, 25–75 mm long. Mountain slopes throughout, and widely planted elsewhere for timber; absent from Corsica. *Pinus uncinata* is similar, with leaves in clusters of *2–3* and trunk *not* reddish above (grey-brown) with thick (not thin) twigs and *dark green* (not bluish) needle-like leaves. A mountain species (not strictly Mediterranean) that extends into the Pyrenees. ES.

Pinus nigra CORSICAN PINE A variable (with subspecies described), tall, robust species to 42 m with a *straight trunk and open, pyramidal crown*; bark dark grey above and deeply fissured; twigs dark brown with slightly sticky buds. Leaves in pairs, dark green, 80 mm–18 cm long and 2 mm wide, scarcely twisted and not bluish. Female cones solitary or clustered, long, egg-shaped, to 30–90 mm, yellowish-brown, shiny and more or less *unstalked*. Rocky mountain slopes; widely planted elsewhere. FR (incl. Corsica), IT (incl. Sicily).

P. nigra (female cone)

1. *Pinus pinaster* (inset: top, female cone; bottom, male cones)
2. *Pinus sylvestris* (inset: bark)
3. *Pinus nigra* (inset: female cone)

CUPRESSACEAE | JUNIPER FAMILY

Evergreen trees or shrubs with resin, leaves opposite or in whorls, scale-like or needle-like, usually in groups of 3; vegetative buds without bud-scales. Female cones with woody or succulent scales.

Juniperus JUNIPER

Evergreen shrubs with twigs spreading in 3 dimensions; leaves needle-like or opposite and scale-like when mature, borne on spreading (not flat) branches. Typically dioecious; female cones berry-like, with *fused, succulent* scales. Seeds not winged.

Juniperus communis COMMON JUNIPER A dense, greyish shrub to 26 m. Leaves needle-like, 4–20 mm long, borne in *groups of 3* with a *single* white band above. Female cones egg-shaped to rounded, 5–10 mm, green and later bluish-black when ripe (cones take years to ripen); male cones solitary. Throughout (regional subspecies exist).

Juniperus oxycedrus PRICKLY JUNIPER Similar to *J. communis*: a greyish dioecious shrub to 15 m with gradually narrowing, sharply spine-tipped needle-like leaves 8–25 mm long and 1 mm wide, with *2 pale bands* above, in whorls of 3. Female cones 8–15 mm across, ripening red-brown. Dunes and maquis. Throughout. *Juniperus navicularis* is similar but has shorter needles just 4–12 mm long which are *abruptly* spine-tipped and female cones 7–10 mm across. Hill forests. ES, PT.

Juniperus phoenicea PHOENICEAN JUNIPER A tall *shrub to 8 m*, with grey-brown bark peeling in narrow strips. Leaves 5–14 mm long, (just 0.7–1 mm on older branches) and *scale-like with membranous margins,* and adpressed to the stem. Female cones 8–14 mm, spherical, red-brown when mature. Throughout. *Juniperus thurifera* is a similar dioecious *tree to 20 m* but has scale-like leaves *without membranous margins*. Local on mountain slopes. DZ, ES, FR (incl. Corsica), MA. *Juniperus sabina* also has scale-like leaves without membranous margins but is a *prostrate shrub 30 cm–1 m*. Rare, in mountains. ES.

Cupressus CYPRESS

Evergreen trees with twigs spreading in 3 dimensions and opposite, scale-like leaves when mature. Female cones woody, separating into scales; seeds winged.

Cupressus sempervirens CYPRESS A distinctively narrowly-columnar tree to 30 m. *Leaves dark green,* scale-like, just 0.5–1 mm long and closely overlapping. *Female cones spherical, 25–40 mm across* and yellowish. An Aegean native widely planted in towns and gardens. Throughout. *Cupressus macrocarpa* MONTEREY CYPRESS has dark green leaves 1–2 mm long and female cones 20–35 mm long. Native to California, USA, widely planted. *Cupressus arizonica* has *pale blue-grey* scale-like leaves. Native to Arizona, USA, widely planted. *Cupressus dupreziana* var. *atlantica* (Syn. *C. atlantica*) MOROCCAN CYPRESS is similar to *C. sempervirens* but has much *bluer foliage*, with a resinous spot on each leaf, and *small, spherical cones to 25 mm long*. An endemic of the High Atlas Mountains south of Marrakech. MA.

Tetraclinis

Evergreen trees with scale-like leaves. Fruit a woody cone separating into scales; seeds winged.

Tetraclinis articulata ARAR TREE A small, juniper-like, slow-growing tree to 15 m with a solitary, or few trunk(s). Leaves scale-like, 1–8 mm long and 1.5 mm wide; arranged in opposite pairs, free and triangular at the tips. Female cones brown, 10–15 mm long, with 4 thick scales in opposite pairs, splitting; seeds markedly winged. Mainly in mountains (rarer at sea level) in North Africa; local elsewhere (Malta and southeast Spain). DZ, ES, MA, MT, TN.

1. *Juniperus oxycedrus*

2. *Juniperus oxycedrus*
   female cones

3. *Juniperus phoenicea*

4. *Juniperus phoenicea*
   female cones

5. *Cupressus sempervirens*
   (inset: female cones)

6. *Cupressus dupreziana* var.
   atlantica

ARAUCARIACEAE | MONKEY-PUZZLE FAMILY

Very large evergreen trees with columnar trunks and needle-like or flattened leaves borne spirally; vegetative buds without bud-scales. Male cones very large, catkin-like and drooping. Seeds not winged.

Araucaria

Evergreen trees native to Australasia and South America; leaves broad and with many veins or awl-shaped and incurved.

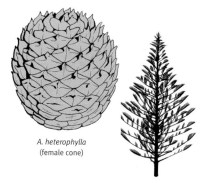

A. heterophylla
(female cone)

Araucaria heterophylla NORFOLK ISLAND PINE A tall, erect tree to 60 m, with horizontal spreading branches arranged somewhat symmetrically in whorls around the trunk; leaves awl-shaped and incurved. Female cones squat and globose, with woody, spirally-arranged scales and wingless seeds. Native to Norfolk Island in the Pacific; widely planted.

A. heterophylla (habit)

EPHEDRACEAE | JOINT PINE FAMILY

Small shrubs with rush- or broom-like stems, small opposite or whorled leaves, often reduced to scales. Dioecious; male 'cones' in small axillary clusters; female 'cones' with several pairs of bracts. 'Fruit' (pseudo-fruit) berry-like with 1–2 seeds.

Ephedra

Broom-like shrubs with branched stems and inconspicuous, scale-like leaves. Male cones with several 2–3-celled stamens.

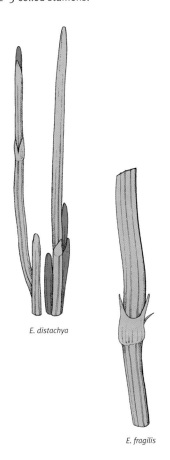

E. distachya

E. fragilis

Ephedra distachya JOINT PINE *A low, scrambling shrub to 1 m* (depending on exposure) with erect stems from creeping rhizomes, forming a thicket. Stems rigid and broom-like, leafless except for ash-coloured scale-like leaves to 2 mm at the joints, *thin, 0.7–1 mm across, greenish-yellow, and not easily broken at the nodes.* Male and female cones separate; greenish yellow and small, the male in clusters of 4–8 pairs, the female in solitary pairs. Fruit 5–7 mm, berry-like and reddish, the seed protruding. Local on dry rocky slopes, sea cliffs, and river banks. ES, FR, IT. *Ephedra major* (Syn. *E. nebrodensis*) is similar, often short (<1 m) and with *very slender* stems 0.4–9.7 mm across, and with dark brown (not whitish) leaves. Fruits yellowish to reddish. Widespread but absent from smaller islands. DZ, ES, FR (incl. Corsica), IT (incl. Sardinia and Sicily), MA, TN.

Ephedra fragilis JOINT PINE Similar to *E. distachya* but often much *taller* (depending on exposure), climbing or prostrate *3–4(5) m* with disordered, *dull dark green, flexible but fragile stems 1.5–2.2 mm across that break easily at the nodes.* Leaves scale-like to 2 mm long, male and female cones separate; the male in clusters of 4–8 pairs, the female in solitary pairs. Fruit 7–9 mm, berry-like and red, the *seed completely enclosed.* Local in dry, sandy and rocky habitats; throughout; common in the centre-west, the Balearic Islands and Sicily.

1. *Araucaria heterophylla*
2. *Ephedra distachya*
3. *Ephedra fragilis*

BASAL ANGIOSPERMS

An ancient flowering plant lineage.

ARISTOLOCHIACEAE | BIRTHWORT FAMILY

Perennial herbs and woody climbers with creeping rhizomes. Leaves simple, alternate, untoothed and cordate. Flowers very distinctive, regular or zygomorphic; stamens 6–12 in 1–2 whorls; ovary inferior; styles 6. Fruit a capsule.

Asarum

Rhizomatous herbs with aerial stems <10 cm long. Flowers regular and bell-shaped with 3 lobes; stamens 12.

Asarum europaeum ASARABACCA A low, evergreen, rhizomatous perennial with creeping, hairy stems and dark green, shiny, kidney-shaped leaves with prominent veins; leaves seemingly opposite. Flowers bell-shaped and dull purplish-red, 12–15 mm long, ending in 3 pointed lobes. Rare, in woodland scrub and thickets. FR, IT.

Aristolochia BIRTHWORT

Herbs with stems >10 cm. Flowers zygomorphic and tubular, swollen at the base (pitcher-like); stamens 6. Root morphology also important but difficult to observe and not included here.

A. Leaves clasping (scarcely stalked); flowers solitary.

A. rotunda

Aristolochia rotunda ROUND-LEAVED BIRTHWORT A scrambling or ascending perennial herb to 70 cm, with oval-heart shaped, virtually stalkless *leaves that clasp the stem*. Flowers 25–50 mm long, more or less straight-tubed, limb that droops when mature. Fruit 20–25 mm. Dry, rocky and scrubby places. Throughout, except Morocco. Populations in Corsica and Sardinia with smaller leaves, fruits and seeds are considered by some to be a distinct species, *A. insularis*.

B. Leaf stalks markedly exceeded by the flower stalks; flowers solitary.

A. pistolochia

Aristolochia pistolochia BIRTHWORT A scrambling or ascending, *hairy* perennial with a stock of numerous tubers. Leaves oval-triangular, short-stalked, and with *minute teeth* on the margin and lower surface. Flowers 20–55 mm, *almost straight-tubed*, brownish with a large, dark brown limb that droops when mature, wider than the tube. Fruit 10–35 mm. Dry, rocky and scrubby places; local. ES, FR, MA, PT.

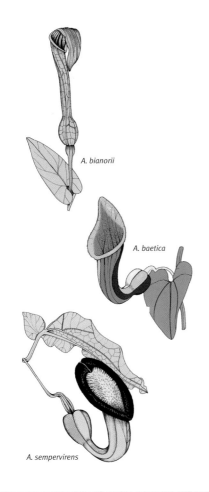

A. bianorii

A. baetica

A. sempervirens

Aristolochia bianorii A *short,* often *prostrate* perennial to 50 cm with small leaves just 16–35 mm long, heart-shaped, *elongated when mature* and very *short-stalked.* Flowers 10–30 mm long, dull yellowish-brown with dark stripes and a dark red-brown limb. Fruit 5–10 mm. Rocky habitats. ES (Balearic Islands).

Aristolochia baetica A *climbing* herb to 60 cm with *glaucous,* heart-shaped leaves with a deep cleft at the base. Flowers *rather large,* 20–70 mm long, *maroon, brown or blackish-purple,* strongly curved and densely hairy within. Fruit large, 20–60 mm long and hanging; green, later yellow then blackish and splitting to shed seeds. A range of habitats including damp and shady places near water as well as open, coastal garrigue. Locally common in the far west. DZ, ES, MA, PT.

Aristolochia sempervirens A climbing perennial herb with stalked, heart-shaped leathery leaves that are hairless, and large flowers, to 50 mm, dull purplish-brown with a distinct yellow throat. Woodland, thickets and damp, shady habitats. Strongly southern in distribution, extending locally into Italy. DZ, IT, MA, TN. Populations in southeast Sicily are considered by some to be a distinct species, *A. altissima.*

1. *Asarum europaeum*
2. *Aristolochia pistolochia*
3. *Aristolochia baetica* (inset: cross-section)
4. *Aristolochia sempervirens* in fruit

C. Leaf stalks equalling, or slightly exceeded by the flower stalks; flowers solitary.

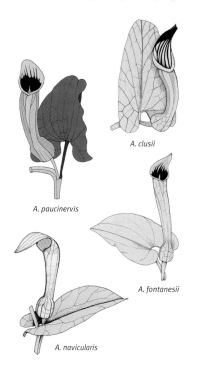

A. clusii

A. paucinervis

A. fontanesii

A. navicularis

Aristolochia paucinervis (Syn. *A. longa*) A virtually hairless scrambling or ascending perennial to 45 cm. Leaves oval-triangular, often with somewhat wavy edges, 31–74 mm long with *stalks equalling the flower stalks* (5–20 mm). Flowers sub-erect, 28–64 mm long, pale green-yellow or brownish *with contrasting dark purple stripes*. Fruit 10–26 mm. Garrigue and sea cliffs. Probably throughout, except for many islands. Specimens described as *A. longa* from hills in southern Portugal have larger leaves and a trailing habit. *Aristolochia clusii* is similar but with leaves *hairy* on both faces, pale yellow, dark-striped flowers with tube exceeded by the limb, IT (south + Sicily), MA. *Aristolochia fontanesii* has *large flowers* 50–81 mm, exceeding the leaves; flowers green-yellow, flushed reddish-purple, with 7–9 veins. An endemic of limestone, clay, fields and coastal rocks. DZ.

Aristolochia navicularis A trailing to sub-erect perennial with leaves hairless on the lower surface. *Flowers pale yellow-green without prominent darker stripes*; limb sub-equal to the tube. IT (Sardinia), (probably TN).

D. Leaf stalks usually longer than flower stalks; flowers solitary.

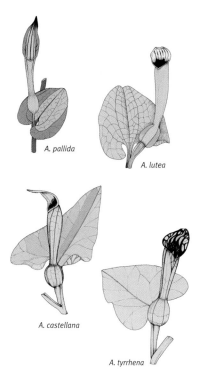

A. pallida

A. lutea

A. castellana

A. tyrrhena

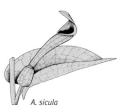

A. sicula

Aristolochia pallida An ascending species, similar to the *A. paucinervis* group, with broad, kidney-shaped leaves 23–72 mm long with stalks shorter than the flower stalks. Flowers 24–57 mm long and greenish-yellow with brownish stripes, the limb darker, short and slightly curved. Local and absent from much of the far west. DZ, FR, IT, MA, TN. *Aristolochia lutea* is similar with leaves 25–62 mm long and pale yellowish flowers 27–69 mm with a purple throat; *limb equal or shorter than tube*. IT (incl. Sicily). *Aristolochia castellana* is very similar to *A. lutea* but smaller, with leaves 10–55 mm long and flowers 20–49 mm. A rare endemic. ES (centre). *Aristolochia tyrrhena* has hairless leaves, glaucous below. Flowers with small limb. Limestone habitats; an island endemic. IT (Sardinia). *Aristolochia sicula* has large leaves relative to the flower size, hairy above, hairless below, long-stalked. Flowers small, greenish-brown. An island endemic. IT (Sicily).

E. Flowers borne in *clusters*.

Aristolochia clematitis BIRTHWORT A vigorous, more or less hairless *erect or ascending* (not climbing or scrambling) perennial to 50 cm (1 m) with numerous unbranched stems. Leaves bluntly-heart-shaped and stalked; light green. Flowers to 20–35 mm long and *yellow* (the only species in the region with flower of this colour); tube straight. A range of habitats including waste places near habitation and damp thickets. Scattered throughout, except the extreme west, though uncommon. DZ, ES, FR, IT, MA, TN.

MAGNOLIACEAE | MAGNOLIA FAMILY

Trees with flowers arranged in rings, with stamens and pistils in spirals on a conical receptacle, an arrangement seen in fossil flowers.

Magnolia

Large trees with broad, leathery leaves. Flowers large and radially symmetrical with 9–13 tepals stamens and carpels numerous. Fruit an aggregate of follicles.

Magnolia grandiflora MAGNOLIA An exotic tree up to 20 m tall (often less) with simple, oval, leathery dark green leaves to 20 cm long and large, white, scented flowers to 30 cm across. Native to the southeast USA; planted in towns and gardens. Throughout.

1. *Aristolochia clematitis*
2. *Aristolochia clematitis* fruit
3. *Aristolochia paucinervis*
4. *Magnolia grandiflora*

LAURACEAE | LAUREL FAMILY

Aromatic, evergreen shrubs and trees with simple, alternate leaves. Flowers small and cosexual, regular, with 8–12 stamens; style 1. Fruit a 1-seeded berry.

Laurus BAY LAUREL

Aromatic trees and shrubs, mainly tropical, with leaves with shining oil glands.

Laurus nobilis BAY LAUREL A monoecious shrub or small tree to 10(18) m tall with blackish bark and hairless shoots. Leaves 50 mm–10 cm long, alternate, oblong-lanceolate, untoothed with slightly wavy edges and dotted with glands. Flowers to 10 mm across and rather inconspicuous: greenish yellow, borne in clusters in the leaf axils, with 4 petals. Fruit a black, ovoid berry. Rocky habitats and ravines, also frequently planted. Throughout.

Persea AVOCADO

Trees and shrubs with large, fleshy, 1-seeded fruits.

Persea americana AVOCADO A tree to 15 m tall with large, oval-elliptic leaves, leathery and pointed at the apex. Flowers small, white, with 5 petals. Fruit an avocado pear. Native to Mexico and Central America; frequently planted in gardens in the region. Throughout.

1. *Laurus nobilis*
2. *Persea americana*
3. *Persea americana* flowers
4. *Arum italicum* fruits
5. *Arum italicum* (inset: flowers)
6. *Arum maculatum* (inset: flowers)

MONOCOTS

Plants typically with a single seed leaf, leaves with parallel (not net) veins, and flower parts in multiples of 3. Mostly herbaceous and with underground storage organs such as bulbs and corms.

ARACEAE | ARUM FAMILY

Perennial, usually hairless, tuberous herbs with leaves all basal and stalked. Individual flowers tiny, borne in a compact spike (spadix), enfolded in a large, often leafy bract (spathe). Fruit a berry.

Arum

Tuberous perennials with arrow-shaped, untoothed leaves. Flowers unisexual, borne at the base of a spadix which is enveloped by the spathe. Fruit a red berry, produced generally after the leaves have withered.

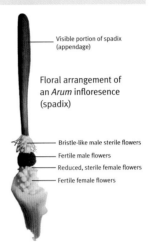

Visible portion of spadix (appendage)

Floral arrangement of an *Arum* infloresence (spadix)

Bristle-like male sterile flowers
Fertile male flowers
Reduced, sterile female flowers
Fertile female flowers

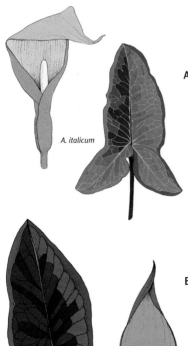

A. italicum

A. maculatum

A. Spring-flowering; spadix yellow, spathe (usually) without purple markings.

Arum italicum ITALIAN ARUM A spring-flowering perennial with a large tuber and long stalked leaves appearing in autumn; leaves large, 15–35 cm long, and arrow-shaped with pointed lobes, semi-erect on long stalks; mid to dark green with variable paler marbling and venation. Visible portion of the *spadix pale yellow*; spathe large to 40 cm, later forward-drooping, pale yellow-green. Berries ripening red, borne in a densely clustered spike. Damp hedgerows and ditches; the most common *Arum* species in the region. Throughout.

B. Spring-flowering; spadix purple to black (sometimes pale), spathe often marked with at least some purple.

Arum maculatum WILD ARUM A spring-flowering perennial with a large tuber and long stalked leaves appearing in spring; leaves similar to those of *A. italicum* but 70 mm–20 cm long, less erect, more wrinkled, and with less divergent lower lobes, sometimes spotted. Inflorescence foetid, the visible portion of the *spadix chocolate-brown* (sometimes grey or yellow) and $<1/2$ the length of the spathe limb; spathe, if flushed purple, darker towards the edges (not centre). Berries ripening red, borne in a densely clustered spike. Cool, damp

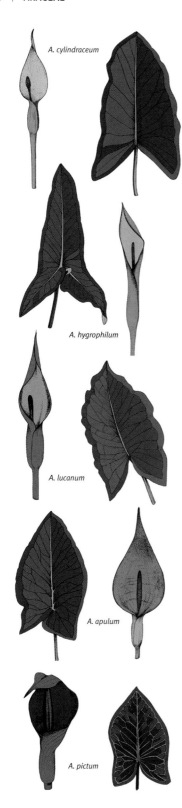

A. cylindraceum

A. hygrophilum

A. lucanum

A. apulum

A. pictum

woods. Absent from much of far south and west. DZ, ES, FR, IT, MA, TN. *Arum cylindraceum* (Syn. *A. alpinum*) is similar to the previous species, but has a *discoid tuber* and long-stalked inflorescence with a *very short and abrubt*, club-shaped spadix with a *long, slender stipe*; inflorescence not foetid. Rocky slopes and woods. Distribution improperly known due to confusion with the previous species; absent from islands. ES, FR, IT, PT. *Arum hygrophilum* is similar to *A. maculatum* but with a discoid tuber, leaves with basal lobes $1/3$ of the length of the blade, and long-stalked inflorescence with a white-green spathe with a purple border and *purple within the spathe tube*; spathe tube *long and attenuated*; inflorescence not foetid. Rare; streamsides and ditches. MA. *Arum lucanum* is similar, but with a long, slender, dark purple spadix which is about $3/4$ the length of the spathe limb (50 mm in length in total). Rare; maquis and mountain slopes. IT (south).

Arum orientale A spring-flowering perennial with a discoid tuber from which leaves appear in the autumn; leaves arrow-shaped and deep green, often with purple flushed petioles (but without dark spots). Inflorescence with a white-green spathe *evenly suffused with purple,* especially along the margin, but white within the tube; visible portion of spadix *dark* purple-black (similar variants in the *A. maculatum* group occur) and *long and narrow* (not club-shaped) $>1/2$ the length of the spathe limb; male flowers in 2 whorls; inflorescence smelling of horse dung. An eastern species. IT (but plants here may refer to *A. apulum*). **Arum apulum** is similar, also with discoid tuber and broadly triangular leaves but with the spathe pale green at the apex and margins and variably *purple-suffused in the centre*; male flowers in 3–4 whorls. Low scrub; rare. IT (southeast).

C. Autumn-flowering; spathe purple-brown throughout.

Arum pictum An *autumn*-flowering perennial with vertical, globose tubers from which leaves appear in the autumn, *just before flowering*; leaves thick and leathery, arrow to heart-shaped and dark green with ligher veins. Inflorescence with a *deep blackish-purple spathe* to 25 cm long and slightly shorter purple spadix. Rocky scrub and pine woods. Larger central islands. Subsp. sagittifolium is possibly distinct, and has arrow-shaped leaves. ES (Balearic Islands). Subsp. pictum has heart-shaped leaves. FR (Corsica), IT (Sardinia and Italy).

Arisarum

Small, tuberous perennials with heart-shaped leaves. Inflorescences like those of *Arum* but smaller and the spathe fused over most of its length, appearing tube-like. Fruit green, several-seeded.

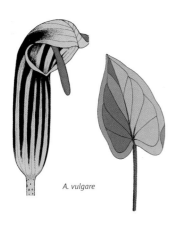

A. vulgare

Arisarum vulgare FRIAR'S COWL An autumn-flowering, short perennial with all basal leaves 20 mm–17 cm long, broadly heart-shaped with rounded basal lobes. Flower stalk more or less equal to the leaf stalks (55 mm–35 cm) and sometimes spotted. Spadix greenish or brown, and slightly protruding from the *hooded spathe* which is white, striped green to chocolate brown; tubular-cylindrical below and forward-drooping above. Berries greenish to blackish. Widespread and common in a wide range of habitats. Throughout. *Arisarum simorrhinum* is very similar but has a swollen (ventricose), *purple-striped spathe*, and spadix not protruding and swollen at the tip. Local in the southwest only. DZ, ES, MA.

1. *Arum apulum* (inset: flowers)
2. *Arum pictum*
3. *Arum pictum* flowers
4. *Arisarum vulgare* (inset: fruits)

Arisarum proboscideum MOUSE PLANT A spring-flowering, short perennial with all basal, heart-shaped leaves (26)60 mm–16 cm long. Flower stalks short, so inflorescences are borne close to the ground. Spathe dark brown and hooded, 20–40 mm long, with the tip extended into a *long appendage 50 mm–15 cm long*; spadix concealed and whitish in colour. Local in shaded and wooded habitats, often amongst rocks and boulders. ES, IT.

Ambrosina

Tiny tuberous aroid with oval leaves and spadix enclosed.

Ambrosina bassii AMBROSINIA A very small, tuberous, autumn to winter-flowering perennial rather like *Arisarum*. Leaves to just 70 mm long, green and unmarked. Spathe greenish yellow and red-spotted, *small*, to 20 mm long and inflated below with a sharp kink and upturned, cylindrical tip; spadix lacks a protruding part and completey enclosed within the base of the spathe. Fruit a many-seeded berry. Scrub and garrigue. FR (Corsica), IT (incl. Sardinia and Sicily).

Biarum

Small tuberous perennials with narrow *unlobed* (not arrow-shaped) leaves and inflorescences virtually stalkless above ground. Fruits greenish or whitish or purplish. Closely related and difficult to distinguish in the field: arrangement of the minute flowers is a key diagnostic.

B. tenuifolium

(spadix)

B. carratracense

(spadix)

(spadix)

B. dispar

A. Spadix with an uppermost ring of sterile male flowers above the fertile male flowers.

Biarum tenuifolium A small, autumn-flowering perennial to 20 cm with leaves appearing after flowering. Leaves 11–29 cm (from the base), oblong lanceolate (linear-lanceolate late in season). Inflorescence sprouting directly from the ground and stalkless; spadix long and slender, brown, and often *long-exceeding* the spathe, with *sterile, thread-like flowers above and below* the male flowers; spathe brownish or purplish. Fruit a *whitish* berry. Garrigue, rocky slopes and stony pastures. Subsp. *tenuifolium* has *unbranched* male sterile flowers. Strongly eastern. IT (south + Sicily). Subsp. *arundanum* (Syn. *B. arundanum*) also widely treated as a true species, has *2-branched* sterile male flowers arranged in 8 regular whorls. Far southwest of region on limestone clay. ES (southwest), MA, PT.

B. Spadix with absent or few uppermost sterile flowers above the fertile male flowers.

Biarum carratracense A small tuberous perennial similar in form to *B. arundanum*, differing in having a slightly inflated, tubular lower portion to the spathe, the spathe tube margins free for $^1/_2$ their length; *spreading* (rather than erect) sterile female flowers *scattered* over the lower portion of the spadix, but *with few or no sterile male flowers* above fertile male flowers. *Fruits purplish.* ES (southwest). *Biarum dispar* (Syn. *B. bovei*) is similar but has a rounded rather than narrowly cylindrical tubular

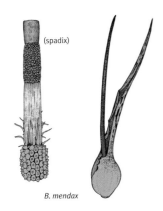

(spadix)

B. mendax

portion of the spathe, and sterile female flowers grouped at the base of the spadix (above fertile female flowers); sterile male flowers very few or absent. Strongly southern. DZ, ES, IT (Sardinia), MA, TN.

Biarum mendax A tuberous perennial similar to the previous species but with *oval* rather than linear-lanceolate leaves and spathe *enclosed at the base* (*margins fused throughout*) *and greatly enlarged and distended*; the *spathe large*, 85 mm–23 cm long, greenish outside and purplish within, *remaining erect* in flower; spadix erect (not drooping) with male sterile flowers absent, sterile female flowers present above the fertile female flowers and running along the length of the spadix at their bases (decurrent). Southwest Iberian Peninsula. ES (southwest); possibly PT.

Zantedeschia ARUM LILY

Rhizomatous herbaceous perennials with long-stalked, broadly oval leaves with cordate bases. Spadix without an appendix. Fruit a yellow berry, several-seeded. Native to southern Africa.

Zantedeschia aethiopica ARUM LILY A hairless perennial with rather fleshy, arrowhead-shaped leaves 10–45 x 10–25 cm, borne on long stalks to 75 cm. Flower stalks somewhat exceeding the leaves; *spadix yellow*, to 15 cm long, and about ¹/₂ the length of the spathe; individual flowers indistinct; *spathe pure white*. Berries yellow, often not developing. Planted and naturalised in ditches. Throughout.

Dracunculus DRAGON ARUM

Large, robust tuberous perennials with deeply palmately lobed leaves. Inflorescences with a long, prominent spadix (to 40 cm). Fruit a several-seeded berry.

Dracunculus vulgaris DRAGON ARUM A distinctive, robust, erect perennial to 1 m with a solitary, trunk-like stem formed from the leaf bases that is greyish-yellow, spotted dull purple. Leaves green with white streaks, *deeply divided into lanceolate, more or less equal segments to 20 cm long*. Spathe maroon-purple and hairless within, large, 20–40 cm long, tapering to the tip, soon drooping; spadix blackish and shiny, held straight; inflorescence foul-smelling. Fruit a red berry, borne numerously in a large, compact, oblong head. Rocky roadsides, ravines and garrigue at low altitude. Local and southeastern in distribution (predominantly eastern Mediterranean); sometimes planted. DZ, FR (Corsica), IT (incl. Sardinia).

Zantedeschia aethiopica

Helicodiceros

Tuberous perennials similar to *Dracunculus* (previous included in the genus) but with a markedly hairy inflorescence.

Helicodiceros muscivorus DEAD HORSE ARUM A robust perennial to 34(50) cm, rather bushy below. Leaves grey-green and deeply *dissected into unequal lobes*. Inflorescence short-stalked (20 mm–16 cm); spathe 11–40 cm long, yellowish-maroon within and *very hairy*, prostrate and often draping on surrounding vegetation when mature; spadix protruding and lying on spathe surface (not erect), *covered in long bristles*; inflorescence foul-smelling. Rocky slopes and sea cliffs. ES (Balearic Islands), FR (Corsica), IT (Sardinia).

ALISMATACEAE

Annual or perennial, hairless, aquatic herbs, rooted in mud, with erect or floating leaves; leaves stalked, with linear to rounded blades. Flowers cosexual and regular, arranged in panicles or racemes; sepals and petals 3; stamens 6-numerous; style absent, ovary superior. Fruit a head of achenes or few-seeded follicles.

Baldellia LESSER WATER-PLANTAIN

Aquatic herbs with all basal leaves in a rosette. Flowers pale mauve or white, long-stalked, borne in umbels; stamens 6; carpels spiralled.

1. *Helicodiceros muscivorus* habit
2. *Helicodiceros muscivorus* flowers
3. *Helicodiceros muscivorus*
4. *Baldellia ranunculoides*
5. *Sagittaria sagittifolia*

B. ranunculoides

Baldellia ranunculoides LESSER WATER-PLANTAIN An erect or spreading perennial to 20 cm with mostly basal, linear-lanceolate leaves, *tapered at both ends* and long-stalked. Flowers pale purplish-white, 10–16 mm wide, borne solitary or in few-flowered clusters (up to 50, usually fewer; occasionally with trailing stems and solitary flowers). Fruit an aggregate of 35–55 *spindle-shaped* (fusiform) achenes. Local in wet ditches and seasonally flooded plains. West of region. ES, MA, PT. *Baldellia repens* is very similar, also having leaves attenuated into the petioles, but fruit an aggregate of *fewer* (<28), *ovoid achenes*. West of region. ES, MA, PT. *Baldellia alpestris* is similar but at least some of the leaves *abruptly contracted* into the petiole, and the inflorescences with fewer flowers (1–7). Northwest Iberian Peninsula, possibly extending into the western Mediterranean. ES, PT.

Damasonium

Aquatic herbs with all basal, usually submerged or floating leaves. Flowers with white petals each with a basal yellow blotch; stamens 6; carpels 6–10 in 1 whorl. Fruits with carpels *long-beaked and spreading in a star*.

Damasonium alisma THRUMWORT A floating or submerged aquatic annual with erect stems to 30(60) cm. Leaves all basal and long-stalked, oval with a heart-shaped base and floating, 30–60(80) mm long; leafless flowering stems carried up to 50 cm above water, bearing 1–3 simple whorls of white, stalked flowers to 6 mm across. Fruit of 6 spreading (star-like) follicles 5–14 mm long; *each containing just 2 seeds 2.5 mm long*. Still water. Rare in the region. ES, FR, IT. *Damasonium bourgaei* is very similar but with seeds <1.5 mm. Southern Iberian Peninsula, all major islands, southeast France and North Africa. DZ, ES, FR, IT, MA, TN. *Damasonium polyspermum* is similar to the previous species but has fruits with *follicles containing 5–16 seeds*. DZ, ES, FR, IT (Sicily), MA, TN.

Luronium FLOATING WATER PLANTAIN

Aquatic perennials with stems submerged or floating and rooting at the nodes. Flowers solitary or 2–5, long-stalked, borne in the leaf axils; stamens 6; carpels in a flattish mass or irregular whorl.

Luronium natans FLOATING WATER PLANTAIN A delicate aquatic perennial with submerged stems to 50 cm, rooting at the nodes, and linear underwater leaves as well as floating leaves with slender stalks. Flowers 1–2 per inflorescence, white with a yellow central spot, 12–18 mm across, borne on slender stalks 30–70 mm long. Stagnant or slow-moving water. Rare in the region. ES, FR.

Sagittaria ARROWHEAD

Aquatic monoecious perennials with leaves all basal. Flowers *unisexual*, borne in whorls on leafless stems; male above, female below; stamens 7-numerous; carpels spiralled in a rounded head. Leaf shape variable and sometimes misleading.

Sagittaria sagittifolia ARROWHEAD A monoecious, rather large-flowered perennial with leaves all basal, of 3 forms: submerged and linear; floating and elliptic, and aerial and arrow-shaped. Flowers 20–30 mm across, *white with a dark purple centre and purple anthers; bracts free at the base*. Stagnant or slow-moving water. A north European species extending locally into the Mediterranean. ES, FR, IT. *Sagittaria latifolia* BROADLEAF ARROWHEAD is similar but with flowers usually lacking a puple blotch and anthers yellow. ES, FR, IT. *Sagittaria montevidensis* (Syn. *S. calycina*) is similar to the 2 previous species but has bracts fused at the base. Native to Central America; naturalised. ES (possibly elsewhere).

Alisma

Aquatic perennials with leaves all basal, linear or narrowly-elliptic to oval. Flowers many, each with 3 petals, white or pale mauve with a yellow basal blotch; stamens 6; carpels numerous in 1 whorl, 1-seeded and curved inwards.

Alisma plantago-aquatica WATER-PLANTAIN An erect, hairless, aquatic perennial with leaves and stems carried above the water, 20 cm–1.2 m high. Leaves all basal, long-stalked and *oval-elliptic* with *rounded lobes at the base*, the blade 36 mm–29 cm long. Numerous whorls of pale flowers form a branched, pyramidal inflorescence; flowers 7–12 mm across, long-stalked; anthers 2 x as long as broad. Seeds without protuberences. Muddy and damp habitats or shallow water. Widespread, including larger islands. Throughout. *Alisma lanceolatum* is similar but with leaf-blades attenuated (graduating into their petioles), the blade 50 mm–30 cm long. Fruits widest in the middle, with straight, erect styles. Similar distribution to the previous though more southern. *Alisma gramineum* is distinguished by its *linear to linear-lanceolate leaves*, the blade just 35–55 mm long. Fruits widest in the upper $^1/_2$ and with recurved styles. Scattered and widespread though rare. Throughout.

BUTOMACEAE | FLOWERING RUSH FAMILY

Perennial, aquatic, hairless herbs rooted in mud, with linear, all basal leaves. Flowers borne in *Allium*-like umbels with flowers in parts of 3; stamens 9; style 1, carpels 6. Fruit a group of 6 splitting follicles.

Butomus

Hairless aquatic perennials with basal leaves. Distinctive for having all similar sepals and petals and 6, virtually free follicles.

Butomus umbellatus FLOWERING RUSH A distinctive, erect, aquatic perennial to 1.5 m. Leaves all basal, linear, pointed and 3-angled, almost as tall as the flowering stems. Flowers pale pink-white, with similar petals and sepals 10–15 mm long, borne in umbels with brownish bracts on tall, rounded stems. Local, in still or sluggish water. Throughout.

CYMODOCEACEAE

A small, mainly tropical family of marine perennials with narrow, flat leaves with 3-numerous veins and serrated edges. Flowers inconspicuous, unisexual, with petals absent, borne solitary or in pairs, or in a cyme; male flowers with 2 anthers; styles 1–3. Fruit stony and compressed.

Cymodocea

Marine, grass-like perennials distinguished by their *solitary, unisexual* flowers.

C. nodosa

Cymodocea nodosa A grass-like perennial with leaves 2–7 together on short shoots; stems with annual scars at the nodes from previously shed leaves. Leaves dark green, to 40 cm long and *narrow*, to just 4 mm wide, 7–9 veined and spiny-toothed towards the tips. Flowers solitary, without a perianth; male flower with a filament-like stalk, *stamens 2*; female flower stalkless with *2 styles* and *2 thread-like stigma lobes*. Fruit laterally compressed. Local on submerged marine sandy substrates. Throughout.

JUNCAGINACEAE | ARROWGRASS FAMILY

A small family of hairless, aquatic herbs with linear, sheathing basal leaves. Flowers small and green, borne in erect spikes or racemes; tepals 6; stamens 6; carpels either 3 or 6, joined to a superior ovary. Fruit 3 or 6 1-seeded units splitting when ripe.

Triglochin

Aquatic, hairless herbs, superficially *Juncus*- or *Plantago*-like but with 3 or 6 1-seeded fruit segments.

Triglochin barrelieri (Syn. *T. bulbosa* subsp. *barrelieri*) A low to short perennial to 35 cm with a bulbous rootstock with fibres 0.8 mm wide. Developed leaves 4–7(18), linear, tapered to a point and 70 mm–35 cm long. *Flowers borne in spring*, virtually stalkless, tiny and greenish, to 2.5 mm, borne in *slender spikes* 25 mm–25 cm long. Fruit elliptic and spreading, 6–10 mm long. Local in salt marshes and on river banks. Probably throughout. *Triglochin laxiflora* (Syn. *T. bulbosa* subsp. *laxiflora*) is similar but has just 1–3(5) developed leaves and fibres 0.3 mm wide. *Flowers borne in autumn*. Similar habitats. Scattered throughout.

Triglochin palustris MARSH ARROWGRASS A hairless, rush-like perennial to 50(75) cm with a basal tuft of slender, fleshy leaves deeply furrowed on the upper surface near the base. Flowers small, to 3 mm and short-stalked; tepals rounded and purplish. Fruit 6–8 mm, club-shaped and opening from the base into 3 sections to form a distinct arrow shape. ES, FR, IT, MA, PT.

POSIDONIACEAE

Submerged marine perennials with flowers in branched, few-flowered spikes borne in the axils of long bracts; flowers actinomorphic; male flowers on the upper spikes and cosexual flowers below; stamens 3; stigma disk-like. Fruit fleshy and about the size of an olive.

Posidonia

Grass-like marine herbs distinguished by their inflorescence of male and *cosexual* flowers borne in *cymes*.

Posidonia oceanica SEAGRASS A grass-like perennial with stout rhizomes *densely covered in numerous brown fibres formed from old leaf bases*. Leaves strap-shaped, 40–70 cm long, dark green with a blunt, rounded tip and 13–17 longitudinal veins. Flowers greenish (rarely produced), 2–3 in cymes, cosexual below, male above. Fruit fleshy and ovoid, 20–32 mm. A marine, often sub-dominant species. Widespread and common; an enormous clonal colony exists south of Ibiza. Throughout.

1. *Alisma plantago-aquatica*
2. *Butomus umbellatus*
3. *Posidonia oceanica*

ZOSTERACEAE | EELGRASS FAMILY

A small family of marine perennial grass-like herbs. Flowers unisexual or cosexual, borne in 2 rows in a congested inflorescence enclosed in sheath-like bracts; perianth absent; stamen 1; style 1 with 2 stigmas. Fruit a stalkless drupe.

Zostera

Grass-like marine herbs distinguished by their complex, *congested* inflorescence of *unisexual* flowers enclosed in a leafy sheath.

Z. marina

Zostera marina COMMON EELGRASS A submerged marine perennial with rhizomes to 2 mm across with 1–2 bundles of roots at each node, and leaves 50 cm–1.2 m long and *broad*, (2)4–20 mm wide, with 5–11 veins, and broad, rounded, *bristle-tipped points*; *basal sheaths entire, not split*. Flowers greenish, borne in terminal, much-branched inflorescences, enclosed in the base of the sheath; stigma 2 x the length of the style; female flowers with *1 style*; male flowers with *1 stamen*. Locally abundant in suitable habitats; fine sandy or silt substrates submerged to 10 m. Rather local but widespread. Probably throughout. *Zostera noltii* is similar but with a *smaller rhizome*, under 1.5 mm across, *with up to 4 bundles of roots* at each node and *narrower leaves,* just 0.5–1 mm wide with 3 veins, to 22 cm long; *leaf sheaths open down 1 side*. Widespread in estuaries. Throughout.

1

2

3

4

5

DIOSCOREACEAE

A primarily tropical family including the yam. Leaves simple and alternate. Plants dioecious; flowers small and greenish borne in spikes or racemes in the leaf-axils; stamens 6 (vestigial in female flowers); style 1, stigmas 3. Fruit a several-seeded berry.

Dioscorea

A twining herb unusual for having dioecious inconspicuous flowers and red berries (*Bryonia* shares these characteristics but has hairy, palmately lobed leaves and tendrils).

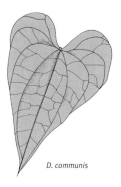

D. communis

Dioscorea communis (Syn. *Tamus communis*) BLACK BRYONY A tall, twining perennial climber to 5 m, dying back annually to a tuber; superficially similar to *Smilax* but *spineless*. Leaves glossy dark green, heart-shaped and long-stalked, 50 mm–15 cm x 40 mm–11 cm. Flowers greenish-yellow, 3–6 mm across, borne in loose racemes; male flowers with 6 stamens, female flowers with 6 minute lobes and a conspicuous ovary. Fruit a rounded red berry 10–13 mm across. Locally common in damp woods and thickets. Throughout.

SMILACACEAE

Climbing lianas with heart-shaped leaves, hooked spines, and paired tendrils at the leaf bases. Flowers unisexual; tepals and stamens 6; style absent, stigmas 3. Fruit a berry with 1–several seeds.

Smilax

Woody vines with spines and tendrils. Flowers greenish with 6 tepals. Berries black when ripe, with 1–3 seeds.

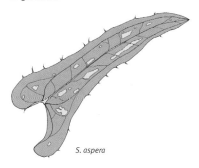

S. aspera

Smilax aspera COMMON SMILAX A variable, creeping, scrambling or climbing shrub 6–10 m with angled, smooth or prickly stems. Leaves dark shiny green and leathery, triangular to heart-shaped, often with prickles on the margins, and with a pair of tendrils at the base of the leaf stalk; blade 42–68 mm (10.3 cm). Flowers unisexual, green-white or yellowish or pinkish, and scented, to 5 mm, borne in branched clusters. Berry red, ageing black. Common on garrigue and coastal scrub, including the larger islands. Throughout. *S. aspera* subsp. *balearica* is treated as distinct by some authors, and differs mainly in its compact, shrub-like habit. Rocky slopes and sea cliffs. ES (Balearic Islands).

1. *Dioscorea communis*
2. *Smilax aspera* habit
3. *Smilax aspera* flowers
4. *Smilax aspera* fruits
5. *Smilax aspera* subsp. *balearica*

LILIACEAE | LILY FAMILY

Bulbous, tuberous or rhizomatous, mostly hairless perennials with leaves with parallel veins. Flowers usually with 6-parted perianth, often with star-like tepals not fused; stamens 6, style 1 or 3. Fruit a capsule or berry. Traditionally a much larger family, now divided into smaller familes following a revision based on DNA analysis.

Gagea STAR-OF-BETHLEHEM

Small perennials with 1–2 bulbs and few basal leaves and 0–few stem leaves. Flowers 1–5, white or yellow and bell-shaped or star-like with tepals not fused. Fruit a capsule. Difficult to distinguish in the field.

A. Plants with a single basal leaf; stems angular and 4-sided. Flowers yellow.

Gagea pratensis A short, slender perennial 40 mm–20 cm high with 2–3 bulbs and a solitary, broadly linear, flat basal leaf 80 mm–30 cm long and 2–4(5) mm wide and a single opposite pair of lanceolate stem leaves with hairy margins. Flowers with tepals 9–16(18) mm long, borne on hairless stems, 2–6, yellow, tinged greenish towards the base, star-shaped with narrowly lanceolate tepals; stigmas 3, yellow. Grassy habitats. Widespread but absent from most islands. ES, FR, IT. *Gagea lutea* is very similar but the basal leaf is wider, 4–11(13) mm, with a hooded tip, and flowers with a *single, green stigma;* bracts 0–1(2); tepals 9–14(18) mm long. ES, FR, IT, PT. *Gagea reverchonii* is similar, but with a channelled (not flat) leaf 1.5–5(7) mm wide, and the inflorescence with 1–4 bracts and flowers with smaller tepals 7–10(12) mm long. Mountains. ES (south).

B. Plants with 2 (rarely 1 or 3) *hollow* basal leaves; stems more or less cylindrical. Flowers yellow.

Gagea liotardii (Syn. *G. fistulosa*) A short, slender spring-to summer-flowering perennial 50 mm–14(20) cm tall with paired bulbs and 2(1–3) basal, lanceolate leaves to 4 mm wide that are hollow (pithy) and hairless, hollow stems. Flowers 3–5, yellow with a greenish exterior, with relatively broad tepals 10–14 mm long. Grassy habitats in mountains. Absent from most islands. ES, FR, IT.

C. Plants with 2 (rarely 1 or 3) basal cylindrical to semi-cylindrical leaves that are *not hollow*. Flowers yellow.

Gagea bohemica EARLY STAR-OF-BETHLEHEM A *very small* bulbous perennial with stems <40 mm high. Basal leaves 2, very narrow, 1.5–4.5 mm and thread-like, often curled and prostrate, stem leaves shorter, wider (2.5–12 mm) and alternate. Flowers 1(–4), yellow, with a greenish exterior, with *blunt tepals* 10–14(15) mm long, borne on hairy or hairless stalks. Dry grassy and rocky habitats or in woods. Rather rare. FR (incl. Corsica), IT (incl. Sardinia and Sicily). *Gagea mauritanica* is similar but with *opposite to sub-opposite stem leaves.* DZ, ES (incl. Balearic Islands), FR, IT (incl. Sicily).

Gagea foliosa A low, spring to summer-flowering bulbous perennial with paired bulbs and 2 (1–3) basal leaves; *basal and stem leaves similar:* flat, narrowly linear to linear-lanceolate, 40 mm–27 cm long and 0.5–2.5(3) mm wide. Flowers 1–4, yellow with a greenish exterior, tepals 6–13(15) mm long. ES, FR (incl. Corsica), IT (incl. Sardinia and Sicily). *Gagea cossoniana* (Syn. *G. algeriensis, G. wilczekii*) is similar but lacking thickened, ascending roots. Basal leaves unequal, triangular in section, 10–25(30) cm long, the largest of which is thicker than it is wide and strongly keeled; plant greyish. ES, MA, PT.

Gagea lacaitae A low, spring-flowering bulbous perennial with paired bulbs and 2 long, hairless basal leaves 10–20 cm long and very narrow, just 2–4.5(6) mm wide, exceeding the inflorescence. Inflorescence to 18 cm tall with just 1–2 yellow, green or purple-tinged flowers with oval-elliptic tepals 9–15 mm long, hairy on the exterior; anthers yellow; flower stalks hairless to slightly hairy. Mountain habitats. Absent from many islands. ES, FR, IT.

Gagea villosa (Syn. *G. arvensis*) A short, greyish, slightly hairy bulbous perennial with 2 basal narrowly linear, channelled leaves with a rounded keel, 12–25 cm x 1–2.5(3) mm, and a single opposite pair of stem leaves (rarely 3) just below the flowers. Flowers *numerous, 3–21 together*, yellow with a greenish exterior, star-shaped with narrow, pointed tepals 10–14(17) mm long, the outer ones often deflexed. Absent from most islands. ES, FR, IT.

Gagea lojaconoi (Syn. *G. longifolia*) A low, bulbous, slender perennial with narrow, hollow, reed-like basal leaves and a large, rounded bulb; stem leaves alternate and rather distant. Flowers nodding in bud, borne on hairless or nearly hairless flower stalks, *numerous* (up to 10–15); tepals broadly oval and rather blunt. IT (incl. Sardinia and Sicily). *Gagea sicula* is a similar low bulbous, slender perennial with flat, *not hollow or rush-like* leaves and a pear-shaped bulb. Flowers *few* (<7) with narrow, slightly pointed tepals. IT.

D. Plants with 2-several basal leaves. Flowers white.

Gagea trinervia A low, spring-flowering bulbous perennial typically with 2–4 hairless, linear basal leaves and alternate, linear-lanceolate stem leaves. Flowers *white, borne solitary*, with anthers to 2 mm long. Shady, rocky habitats. IT (Sicily), TN.

Tulipa TULIP

Bulbous perennials with solitary stems, and few, rush-like leaves. Flowers with tepals all alike; stamens 6. Fruit a 3-parted capsule containing fairly large, flat seeds. A genus better-represented in the eastern Mediterranean.

T. sylvestris

Tulipa sylvestris WILD TULIP A short to medium, hairless, bulbous perennial to 50 cm tall. Leaves 2–4, strap-shaped and channelled, 8–37 cm long and 10–18 mm wide. Flowers 1–2, nodding in bud; yellow tinged with orange and or green, the tepals becoming recurved with age; stamens hairy at the base. Fruit a capsule, often not produced. Garrigue, rocky slopes, scrub and roadsides. ES, FR, IT, MA, PT. Subsp. *sylvestris* has external tepals 33–53 x 8–16 mm, yellowish or greenish. Fruits rarely produced. Throughout (rare or absent in much of Iberian Peninsula). Subsp. *australis* has external tepals 22–35 x 4–10 mm, tinged crimson on the outside. Fruits sometimes produced. Throughout the range of the species. *Tulipa clusiana* grows to 30 cm with leaves 20–25 cm long and

1. *Tulipa sylvestris*
2. *Tulipa clusiana*

10–17 mm wide. Flowers whitish-pink with darker pink stripes externally and dark purple anthers. Native to central Asia from Iraq eastwards, very locally naturalised in the Mediterranean. ES, FR, PT, TN.

Fritillaria FRITILLARY

Small, bulbous perennials with solitary, unbranched stems. Leaves alternate, and mostly on the stems. Flowers tubular or bell-shaped, nodding; tepals 6, all petal-like; stamens 6; style 3-lobed. Fruit an erect, 3-parted capsule containing many flattened seeds.

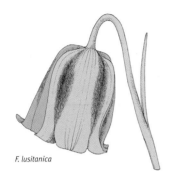

F. lusitanica

Fritillaria lusitanica A short perennial to 30 cm with erect stems and all leaves alternate. Leaves 5–9 and linear, 30 mm–26 cm long. Flowers 1–3; broadly bell-shaped, and large relative to the plant, the tepals 15–52 mm long, green and brownish-purple, often slightly chequered. Fruit a capsule 18–25 mm long, not winged. Garrigue, maquis and dunes; common in southwest Iberian Peninsula. ES, PT.

1. *Fritillaria lusitanica*
2. *Colchicum autumnale*
3. *Colchicum cupanii* PHOTOGRAPH: GIANNIANTONIO DOMINA
4. *Colchicum cupanii*

COLCHICACEAE

A family of about 200 species of herbaceous perennials with rhizomes or corms, previously included in the Liliaceae. Leaves few, alternate and sheathing below. Flowers 1–few, conspicuous; tepals and stamens 6; styles 3. Fruit a capsule.

Colchicum AUTUMN CROCUS

Cormous perennials with basal leaves that appear with, or after the flowers. Flowers crocus-like with 6 spreading tepals; stamens 6; styles 3. Fruit a capsule, borne centrally in the leaf tuft. A genus more diverse in the eastern Mediterranean.

A. Autumn-flowering; tepals fused.

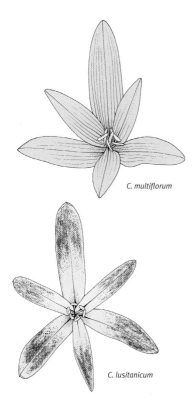

C. multiflorum

C. lusitanicum

Colchicum autumnale MEADOW SAFFRON A small perennial with a dark brown corm 25–40 mm with a long neck. Leaves appear in the spring in groups of 3–4(5), *broadly lanceolate*, 30–48 mm wide. Flowers 1–3 per bulb appearing in the autumn, *without leaves*; flowers solitary or several, 10–26 cm tall, borne on rather long, slender stems; flowers pink with orange, hook-like stigmas and stamens attached at 2 levels; pollen yellow; tepals 30–50 mm long. Damp meadows and other grassy habitats. Local, in grassland and woods. ES, FR, IT. *Colchicum multiflorum* (Syn. *C. neapolitanum*) is similar, but with *outer leaves 20–40(45) mm wide*, and 15–35 cm long. Mainly mountains in northern Iberian Peninsula. ES, PT.

Colchicum lusitanicum A small, autumn-flowering perennial, blooming before the appearance of the leaves. Leaves 3–6, *linear-lanceolate* and dark, shiny green, 25–58 mm wide, and 14–40 cm long. *Flowers numerous, 1–6(13) per bulb*, pink and slightly chequered, 10–20(30) cm long, *anthers blackish-purple to pink* (rarely yellow), styles whitish and curled at the tips; tepals 40–75(85) mm long. Local on dry hillsides; inconspicuous except when in flower. ES (incl. Balearic Islands), IT (central), PT. *Colchicum longifolium* (Syn. *C. neapolitanum*) is similar, but with 2–4 *narrow, linear-lanceolate leaves just 12–20(25) mm wide*. Throughout, including most islands; absent from parts of Italy and northeast Africa.

Colchicum bivonae A low perennial, flowering in autumn, before the leaves appear. Leaves in clusters of 5–9, linear-lanceolate or lanceolate, hairless and sub-erect, 15–25 cm long. Flowers similar to previous *Colchicum* species but more cup-shaped, the tepals 50–60 mm long, borne in clusters; flowers pink to lilac and strongly chequered, the tepals not twisted; anthers brownish-purple. Rocky slopes and maquis. IT (incl. Sardinia)

Colchicum cupanii MEDITERRANEAN MEADOW SAFFRON A low, autumn-flowering perennial, normally with 2 leaves linear to linear-lanceolate, sometimes with hairy margins at the base; *leaves present at flowering-time* but initially short, later to 15 cm long. Flowers pale pink to lilac with markedly *narrow-elliptic tepals* 10–25 mm long; anthers purplish-black and styles yellow. Rocky habitats and maquis from southeast France eastwards. FR, IT.

B. Spring-flowering; tepals fused

Colchicum triphyllum A low, *spring-flowering* perennial with leaves *borne in groups of 3* (rarely 2 or 4), *present at flowering time*; leaves rather short, 20 mm–10(15) cm long initially, later expanding. Flowers pale pink-lilac, not chequered; tepals to 30 mm long, with purple-black or greenish anthers and straight styles. Open, sunny, stony slopes. ES (south).

C. Tepals free (not fused) and anthers attached at the base (not the middle). Widely described under the genus *Merendera* in other floras.

C. filifolium

Colchicum filifolium (Syn. *Merendera filifolia*) A ground-level, *autumn-flowering* perennial with a corm with an extended 'neck'. Leaves 5–12 per flower, to 10 cm long, appearing at or just after flowering time, and *narrow, to just 3 mm wide*. Flowers pink-mauve and crocus-like with oblong to narrowly elliptic tepals, borne solitary (rarely paired) with anthers 5–8 mm long. Capsules oblong, 8–15(20) mm long. Local on dry, stony and bare ground. ES (incl. Balearic Islands), FR, PT. *Colchicum androcymbioides* (Syn. *Merendera androcymbioides*) is a similar, *spring-flowering* species with leaves always developed at flowering time, with pale pink to white flowers with conspicuous yellow anthers 2.5–5.5 mm long. An endemic of Andalucía. ES.

Androcymbium

Herbaceous, cormous perennials. Inflorescences congested with enlarged floral bracts; stamens 6; carpels 3. Species in the western Mediterranean are similar, and capsule dehiscence is a diagnostic character (to be observed in mature fruits only). Sometimes described under the genus *Colchicum*.

Androcymbium europeaum ANDROCYMBIUM A low, winter-flowering, cormous perennial to 12 cm with rosettes of flat, broadly linear to broadly lanceolate leaves 20 mm–16 cm long which form a collar around the flowers. Flowers borne 1–6; *short-stalked and congested*; white to pink with narrow darker stripes, with or without darker speckle-like markings; tepals 22–35 mm long, free to the base (not fused) and long-pointed, with yellowish glands at the base. Fruit a 3-valved capsule, *not splitting when mature*. Bare ground in south Spain and the Atlantic coast of North Africa. ES, MA. *Androcymbium graminea* is very similar but with much *narrower*, strap-like leaves that gradually taper to the tips. Fruiting capsules *splitting when mature*. MA. *Androcymbium wyssianum* is desert-dwelling species, possibly just a form of *A. graminea*; flowering after seasonal rains. DZ, TN.

ORCHIDACEAE | ORCHID FAMILY

One of the largest families of flowering plants, with about 26,000 species worldwide. Perennials with flowers borne in spikes or racemes; flowers each with a bract, usually conspicuous, zygomorphic; perianth with 6 tepals in 2 whorls of 3: the outer 3 tepals ('sepals') all similar, the inner 3 ('petals') with the central the largest and distinct, known as the lip or labellum; ovary inferior; anthers and stigma together form a central column. Fruit a capsule.

Epipactis

Perennials with numerous stem leaves. Flowers borne in racemes, often twisted so that flowers face 1 way; inner perianth whorl of 2 similar upper segments with a 2-parted lip consisting of an inner hypochile and an outer, triangular epichile; ovary not twisted. Most species occur at higher altitudes. A very complex group in which species limits are not yet resolved; the classification here is of the most widespread taxa, supported by existing data, but likely to be revised.

A. Hypochile concave but not distinctly cupped.

E. palustris

Epipactis palustris MARSH HELLEBORINE An erect perennial to 45(60) cm with leaves oblong-lanceolate, pointed. Flowers borne in loose, few-flowered racemes (to 14); lower bract *as long as flower;* the outer tepals oval-lanceolate and brownish or purplish green, the inner upper 2 segments shorter and whitish with purple markings; lip with a *heart-shaped epichile with frilly margins* and a yellow spot at the base; flowers open widely. Damp habitats; rare in the region, particularly in the south. ES, FR, PT, IT.

B. Hypochile concave and markedly cupped; leaves narrow.

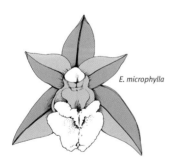

E. microphylla

Epipactis microphylla SMALL-LEAVED HELLEBORINE A slender perennial 20–40(60) cm high with *few (<12), short leaves those on the stems 30–60 mm (12 cm), the largest leaf shorter than its internode.* Inflorescence lax; outer surface of perianth covered in short hairs; flowers nodding and scented; sepals and petals oval, incurved and green-white within, reddish outside; lip 5–7.5 mm long with a shallow cup-shaped hypochile pinched sharply at the front; epichile heart-shaped with crinkled bosses at the base. Rare in the region, in shady, deciduous forests and scrub. ES (incl. Balearic Islands), FR, IT.

C. Hypochile cupped; leaves oval to oval-lanceolate.

E. atrorubens

Epipactis atrorubens DARK-RED HELLEBORINE A slender perennial to 30(60) cm with up to 11 oval leaves usually >50 mm and forming 2 rows. Inflorescence long, 1-sided and thickly covered in short hairs at the top; flowers *purplish to brownish-red with contrasting bright yellow anther,* nodding and vanilla-scented with pointed tepals; epichile frilly and with wrinkled bosses at the base. Rare,

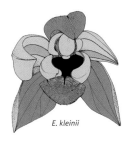

E. kleinii

in high altitude scrub and woodland. ES, FR, IT. *Epipactis kleinii* is more slender with more elongated leaves in 2 rows. Flowers *very small* (lip 4.5–5 mm long) and *scarcely scented* with sepals greenish (like petals) and *3.6–6.5 mm long only* (not to 11 mm long as in previous species); epichile pink in centre. Very rare, in the Pyrenees. ES, FR.

D. Hypochile cupped and light green to brown.

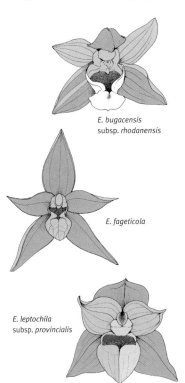

E. bugacensis
subsp. *rhodanensis*

E. fageticola

E. leptochila
subsp. *provincialis*

E. muelleri

E. placentina

Epipactis bugacensis subsp. *rhodanensis* (Syn. *E. rhodanensis*) A greenish perennial with 1–4 flowering stems *densely green-hairy* towards the top. Leaves in 2 opposite-alternate rows, small (30 mm–11 cm long), often shorter than their internodes. Flowers greenish, exceeded by their bracts (30–50 mm), open and horizontal; hypochile greenish-white or pinkish; epichile heart-shaped and whitish or pink with only *slightly* frilly margins; flower stalks *purple-tinted*. Rare on wooded river banks. ES, FR, IT (northwest). *Epipactis fageticola* is similar but lacking pink or purple pigmentation on the flowers and flower stalks (*all white-green and green*). Rare. ES, FR, IT (Aosta valley), PT. *Epipactis leptochila* subsp. *provincialis* (Syn. *E. provincialis*) is a mostly greenish species, notable for its bracts which are *yellowish* at the base. Flowers early. FR (Provence).

Epipactis muelleri A slender, yellowish-green perennial 20–50(70) cm high with 1–2 stems with leaves in more or less opposite rows that are channelled and arching with undulate margins; uppermost leaves bract-like. Inflorescence dense with 10–40 rather bell-shaped flowers that do not open fully; hypochile cup-shaped; epichile whitish and *broadly heart-shaped* (broader than long, 4–5 x 3–4 mm), sometimes turned backwards, with 2 attenuated bosses at the base; anther yellowish. *Flowering early*, in sunny grassland or scrub at high altitude. Rare. ES, FR, IT, PT. *Epipactis placentina* is similar but always with solitary flowering stems carrying flowers that are *intensely tinted with magenta*, particularly the hypochile and epichile. Rare and local. FR (Isere), IT (north, centre and Calabria + Sicily).

1. *Epipactis lusitanica*
2. *Epipactis tremolsii*

E. Hypochile cupped and olive green, brown or blackish, epichile without or with scarcely frilly margins.

E. helleborine

Epipactis helleborine BROAD-LEAVED HELLEBORINE A variable (with many subspecies), erect perennial forming clumps of 1–3 stems. Stem leaves larger than basal leaves, *oval-elliptic*, strongly veined and spirally arranged; *longer than their internodes*. Raceme many-flowered (up to ~100) and rather 1-sided. Locally widespread in a range of habitats, mostly in woodland scrub and hill forests. ES, FR, IT, PT, TN. Subsp. *helleborine* is 20–70(90) cm high with leaves 40 mm–17 cm long. Throughout the range of the species. Subsp. *schubertiorum* (Syn. *E. schubertiorum*) is *much taller, to 1 m* and with *much larger leaves* which are arched and spreading, to 12 cm x 40 mm. IT (Serra San Bruno in Calabria). *Epipactis meridionalis* is similar, but with short stems (often only 12 cm) and *short leaves* to 60 mm, with few flowers borne in a 1-sided inflorescence; bosses at base of epichile *bright crimson*. Deciduous woodland only. IT (incl. Sicily).

Epipactis lusitanica (Syn. *E. tremolsii* subsp. *lusitanica*) Similar to *E. helleborine* but often slender, 20–60 cm, with *small leaves*, the largest just 30 mm long, with wavy margins; flowers few, 5–25(30); *bracts narrow, equalling or scarcely exceeding the lowermost flowers;* flowers fully open, *hypochile pink*. Rare, only in the Algarve and southwest Andalucía. ES, PT. **Epipactis tremolsii** (Syn. *E. helleborine* subsp. *tremolsii*) is also widely treated as a form of *E. helleborine* but genetically more closely related to *E. lusitanica*, has erect-spreading stems, often clumped, and robust and widely oval to round leaves 60 mm–11 cm. Flowers 15–40(60); *bracts broad, exceeding the lowermost flowers;* *hypohile greenish*. DZ, ES, FR, IT (north), MA, PT, TN.

1

2

Cephalanthera HELLEBORINE

Differs from *Epipactis* in that flowers have a less clear demarcation of the hypochile and epichile, and perianth folded forward with tepals forming a bell shape.

A. Flowers white with yellow or orange markings.

C. longifolia

Cephalanthera longifolia SWORD-LEAVED HELLEBORINE An erect, hairless perennial to 60 cm with stems with whitish scales below, and long, *narrow, linear leaves* around the stem, with the tips somewhat drooping. Flowers *pure white* and *open,* with orange markings inside the lip, borne in dense spikes of 2–15(20); tepals pointed; bracts shorter than their ovaries. Local in damp, rocky woods. A temperate to sub-Mediterranean species. Throughout.

Cephalanthera damasonium WHITE HELLEBORINE An erect, hairless perennial with angled stems to 60 cm. Leaves oblong-lanceolate at the tip, to 10 cm. Flowers 3–11(16), each with a leafy bract below; perianth *creamy white*, to 20 mm long, *partially open* and erect, with blunt segments; hypochile with an orange blotch; epichile with orange ridges; bracts longer than their ovaries. Widespread across the European Mediterranean in mountain woodlands, but rarely abundant. ES, FR, IT.

B. Flowers purplish-pink.

Cephalanthera rubra RED HELLEBORINE An erect perennial to 60 cm with short, straight, lanceolate leaves; stems flexuous and with glandular hairs above. Inflorescence lax with 3–8(15) *bright pink flowers* which open widely; epichile intensely coloured. Shaded woods and scrub, usually on limestone. Widespread, rather common in southern France. DZ, ES, FR, IT, MA, TN.

1. *Cephalanthera rubra*
2. *Neottia ovata*
3. *Limodorum abortivum*

Limodorum LIMODORE

L. abortivum

Distinctive perennials that lack green pigment and true leaves, occuring in pine forests.

Limodorum abortivum VIOLET LIMODORE A distinctive, *purplish* mycoheterotroph (plant lacking chlorophyll and parasitising a fungus) 10–47 cm (1.1 m) tall, *without any green leaves*. Flower spike lax with 10–45(65) flowers; lateral segments long, 20–37 mm, violet or whitish; lip yellowish or white with violet veins; bracts exceeding the ovaries; *spur 10–20 mm long*. Local, sometimes common, in pine forests or near pines in mixed evergreen and deciduous woods. Throughout. *Limodorum trabutianum* is similar, but with a *short spur just 0.5–3(4) mm*. Strongly western. ES (incl. Balearic Islands), FR, PT.

Neottia

Perennials either with green, functional leaves or brown, scale-like leaves devoid of chlorophyll. Flowers with lip greatly exceeding the other 5 tepals and divided into 2 terminal lobes; pollinia stalkless.

A. Brownish perennials devoid of chlorophyll (mycoheterotrophic) and with highly reduced, scale-like leaves.

Neottia nidus-avis BIRD'S NEST ORCHID *A brown plant devoid of chlorophyll* that superficially resembles *Orobanche*. Stems 10–52 cm with sheath-like scales; inflorescence with most or all flowers in the upper half; dense and many-flowered (15–70). *Flowers brown-yellow* and short-stalked; sepals and petals similar, 5–7 mm; lip 9–11 mm long, divided into 2 basal, divergent lobes. Fairly common in deciduous or mixed evergreen and deciduous mountain woods. Probably throughout cooler areas.

B. Leaves normally 2 in an opposite pair on the stem; flowers yellow-green or dull reddish; lip deeply divided into 2 apical lobes.

Neottia ovata (Syn. *Listera ovata*) TWAYBLADE A greenish, hairy-stemmed perennial 20–60(75) cm normally with 2 oval leaves low on the stem. Inflorescence lax but many-flowered (up to ~100); *flowers yellowish-green* with equal tepals that are curved forwards, the sepals 2 x the width of the petals; lip to 10 mm long with 2 blunt, divergent lobes. Absent from hot, dry areas and possibly North Africa. FR, ES, IT.

Orchis

Tuberous perennials with 2 ovoid tubers and several leaves in basal rosettes, those on the stem often sheath-like. Flowers often borne in short, dense spikes; upper 5 tepals incurved, the 2 lateral sepals incurved or erect to spreading; lip with 2 lateral lobes and 1 terminal lobe, the latter often larger and 2–3-lobed. Spur absent or short to long.

A. Flowers greenish with upper tepals incurved to form a 'helmet'; lip with 3 lobes, the terminal lobe with 2 sub-lobes; spur absent. Still widely described within the genus *Aceras*.

O. anthropophorum

Orchis anthropophorum (Syn. *Aceras anthropophorum*) MAN ORCHID A short, slender orchid to 12–38(50) cm tall with 4–9 oval-shaped, blunt, shiny green leaves forming a basal rosette, and sheathing the stem. Inflorescence many-flowered, cylindrical and slender; flowers green-yellow streaked with dull red; sepals and petals forming a loose hood above the lip; lip 11–12(15) mm long and pendent, with 2 slender 'arms' and 2 shorter, spreading 'legs' and with 2 swellings near the base; no spur. Locally frequent on maquis and in light woodland scrub or grassland. Throughout.

B. **Flowers white, pink or purple with tepals forming a hood; lip characteristically human-shaped (anthropomorphic) with narrow lateral lobes and a 2-lobed terminal lobe, often toothed.**

Orchis simia MONKEY ORCHID A short to medium perennial to 30(45) cm with 2–5 unspotted basal leaves and 1 or 2 stem leaves. Inflorescence dense and many-flowered, the flowers opening from the top downwards; bracts to 4 mm long; tepals curved to form a hood with upturned points; white to lilac outside; lip deeply 3-lobed with a narrowly rectangular central area spotted with red, all lobes ending in spindly, bright magenta points (to just 1 mm wide), the secondary lobes of the *middle lobe as narrow as the lateral lobes*. Widespread and fairly frequent in the European Mediterranean; distribution in North Africa less well known. DZ, ES, FR, IT.

Orchis militaris MILITARY ORCHID A perennial to 45(60) cm with 3–5 unspotted basal leaves and 1–2 sheathing the stem. Inflorescence dense above, laxer below, with up to 40 flowers; tepals curved to form a hood that is lilac-red (sometimes pale) outside and strongly veined within. Lip pink with a pale centre with groups of dark hairs; deeply 3-lobed, the laterals curved inwards and the middle lobe divided into 'legs'; *secondary lobes of central lobe much broader than the laterals*. Widespread but absent from many islands. DZ, ES, FR, IT.

O. italica

Orchis italica ITALIAN MAN ORCHID An erect, robust perennial 18–43(50) cm high with 5–10 leaves in a rosette, and 2–4 sheathing the stem but not reaching the flowers. Leaves oblong-lanceolate, wavy-edged, and often flecked with brown. Flowers whitish pink with darker veins, borne in dense, many-flowered inflorescences; tepals forming a loose hood; lip 12–21(25) mm long, tipped and spotted with purple, with *slender, pointed* 'arms and legs' with a short 'tail' in the middle; spur down-curved and the length of the ovary; bracts tiny, 1-veined and much exceeded by the ovaries. Locally frequent to abundant on base-rich garrigue, scrub and grassland. Throughout (except France, Corsica and Sardinia).

Orchis purpurea LADY ORCHID An erect, robust perennial to 50 cm (1 m) with up to 6 unspotted, shiny basal leaves and 1–2 sheathing the stem. Inflorescence rather tall with many flowers densely crowded in the upper part only. Bracts to 3 mm long. Tepals curved to form a *strongly brownish-purple spotted green 'helmet'*; lip paler with numerous tufts of brownish hairs; lip 3-lobed and almost entire with an apical notch forming a 'skirt'. Open woods and thickets. DZ, ES, FR, IT.

1. *Orchis anthropophorum*   3. *Orchis italica*
2. *Orchis militaris*   4. *Orchis purpurea*
   PHOTOGRAPH: NIELS FAURHOLDT   PHOTOGRAPH: NIELS FAURHOLDT

C Flowers white, pink or purple with lateral sepals spreading to erect; petals folded over the column; lip 3-lobed, the laterals broad and the middle lobe longer and divided into 2 further lobes.

Orchis mascula EARLY PURPLE ORCHID A variable, slender perennial with broadly lanceolate, shiny-green leaves with darker spots. Bracts lanceolate and shorter than the ovary. Inflorescence dense, with up to 20 sweet-smelling flowers; tepals mauve or reddish, the middle sepal curving with the petals to form a hood, the laterals reflexed; *lip 8–16 mm long, 3-lobed and convex from the centre, the centre pale and spotted*, the middle lobe notched, sometimes with white spots towards the base; lateral lobes scarcely reflexed; spur *at least as long as the ovary* (as long or 2 x length of lip), pointing upwards and thickened at the apex. Open woods. Throughout. The following are often treated as subspecies of *O. mascula*: *Orchis langei* has a cylindrical spur as long as its lip, and lip *abruptly inflected downwards* with a kinked appearance. ES, FR, MA. *Orchis ichnusae* has a centrally folded lip, *strong pleasant fragrance* and less elongated inflorescence; spur scarcely longer than lip (or equal), 9–14 mm long. IT (Sardinia). *Orchis ovalis* has stem and leaf bases with *numerous small dark streaks and tepals with long, wavy tips;* lip 3-lobed with a very wavy (to scalloped) margin. Mountains. Distribution poorly known. FR (Corsica), IT.

Orchis olbiensis SOUTHERN EARLY PURPLE ORCHID A medium-sized perennial 16–43(50) cm tall. Leaves oblong-lanceolate, 50 mm–18 cm long, glossy green with numerous longitudinal dark blotches, sometimes unspotted. Flowers borne in loose, cylindrical spikes; tepals red purple, the lip shallowly 3-lobed with the lateral lobes deflexed, and with a paler central area spotted with red; spur robust, curved outwards, and *exceeding the ovary*. Related to the North European *O. mascula*, which has denser inflorescences, and spur equalling the ovary. Maquis and woodland. DZ, ES (incl. Balearic Islands), FR, MA, PT, TN. *Orchis laeta* is similar but with yellow flowers. (Not to be confused with other yellow-flowered species in the group: *Orchis provincialis* and *O. pauciflora*). DZ, TN.

Orchis spitzelii A medium-sized, slender perennial 21–41 cm high, the upper of the stem brownish-red, stems slightly flexuous. Basal leaves 2–5(7), spreading to erect and *unmarked*, those on the stem sheathing. Flowers with dorsal sepals and petals slightly curved to form a hood, lateral sepals erect or curving inwardly; sepals dark crimson or pink outside and tinted *olive green within*; *lip very convex, appearing folded*: pink, spotted red *without* a white basal area; 3-lobed, the middle lobe indented and with a slightly wavy margin; spur conical and *pointing downwards*, 6–10 mm long (*just shorter than ovary, and more or less equalling lip*). High altitudes. DZ, ES, FR, MA, TN. *Orchis cazorlensis*, also treated as a subspecies of *O. spitzelii* has leaves *sometimes spotted brown* and the flower lip *scarcely convex* and spreading; spur *short* (5.2–8.8 mm), the length of the lip. Mountains of southern Spain (possibly also Balearic Islands and Morocco). ES. *Orchis patens* is also similar to *O. spitzelii* but has sepals *greenish in the centre*, a lip deeply 3-lobed and folded, white with dark spots in the centre, and spur *not* distinctly pointing downwards (100° with lip) rather cone-shaped and thick at the base. DZ, IT, TN. *Orchis × ligustica* is a hybrid of *O. patens* and *O. provincialis* with *dark* crimson-pink (not pale to mid pink-crimson) flowers, a horizontal to ascending, *purple* spur and *densely spotted* leaves. Chestnut woods. IT (Liguria).

D. Flowers cream, yellowish or white.

Orchis provincialis PROVENCE ORCHID A slender plant 14–39 cm with 4–9 densely spotted leaves below, and up to 3 sheathing the stem. Bracts narrow with 1–3 veins, equalling or exceeding the ovaries. Inflorescence lax and cylindrical with *light yellow flowers*; dorsal sepals erect, laterals spreading, petals curved inwards to form a partial hood; lip 3-lobed, the middle lobe bent downwards in the centre and red-spotted; spur curved upwards and blunt at the tip. Maquis and deciduous or evergreen woodland; uncommon. Throughout, except Portugal. *Orchis pallens* is similar but with unspotted leaves and an unmarked lip. Distribution patchy. DZ, ES, FR, IT. *Orchis pauciflora* has unspotted, channelled leaves and a lax inflorescence with only up to 7 (not 20) flowers; lip much darker in the centre than the edges (to sulphur yellow) with small, dark specks and scalloped to toothed along the margin. DZ (rare), IT (incl. Sardinia).

Neotinea

Perennials with 2 ovoid tubers and leaves 2–4 at the base; those on the stem reduced. Flowers with upper 5 tepals incurved; lip with 2 large lateral lobes and a larger terminal lobe; spur short and rounded at the tip.

Neotinea conica (Syn. *Orchis conica*) A short perennial 55 mm–26 cm high with dense, cylindrical inflorescences with a "whiskered" appearance. Flowers with pinkish-white tepals curving to form a hood, the sepals green below and terminating in long, fine points, and a 3-lobed, *flat* or slightly concave lip; flowers spotted with darker markings. Rather uncommon, sometimes in large numbers on limestone scrub. DZ, ES, FR, PT. *Neotinea tridentata* (Syn. *Orchis tridentata*) is similar but with more intensely coloured flowers with pink to lilac sepals that are *not greenish* at the base. DZ, FR, IT.

Neotinea tridentata (Syn. *Orchis tridentata*) A short perennial 55 mm–26 cm high with a dense, cylindrical inflorescence. Flowers with pinkish-white tepals, greenish at the base and with darker spots, curving to form a hood, the sepals green below and terminating in long, fine points, and a 3-lobed, *flat* or slightly concave lip. DZ, ES (incl. Balearic Islands), FR, PT. Subsp. *conica* (Syn. *N. conica, Orchis conica*) has paler flowers with sepals greenish at the base in short, dense inflorescences with a "whiskered" appearance. Much of the range of the species.

Neotinea lactea (Syn. *Orchis lactea*) MILKY ORCHID A small perennial to 20 cm with up to 8 foliage leaves and a further 1–3 sheathing the stem. Inflorescence very dense and ovoid-elongated with a whiskered appearance. Bracts as long as ovaries. Tepals dull whitish-pink with dark veining and green centre, curved to form a hood; *lip convex,* 3-lobed with a further-divided middle lobe, pale with darker spots; spur cylindrical and curving downwards. DZ, FR, IT (records from Iberian Peninsula probably refer to *N. tridentata* subsp. *conica*).

1. *Orchis olbiensis*
2. *Orchis provincialis*
PHOTOGRAPH: NIELS FAURHOLDT

Neotinea ustulata (Syn. *Orchis ustulata*) BURNT ORCHID A small perennial to 15(30) cm with unspotted leaves which increase in size up the stem. Inflorescence many-flowered, opening from below and dense and *dark-coloured at the top*; tepals curved to form a blackish-red hood (unopened buds form the 'burnt' tip to the flower spike); lip to 8 mm long and white with red spots, deeply 3-lobed with spreading, linear laterals and a rectangular central lobe. Widespread in Europe but rare in the Mediterranean. DZ, ES, FR, IT.

Neotinea maculata (Syn. *Orchis intacta*) DENSE-FLOWERED ORCHID A small, pale perennial 10–25(40) cm tall with normally densely spotted leaves (sometimes unspotted) forming a rosette. Inflorescence *small and dense; flowers very small* (appearing dwarfed by bracts) and scented; dull pinkish-white with purplish markings; *sepals forming* a *pale hood*; lip 3-lobed, 3–5 mm long only, the middle lobe rectangular and the laterals pointed. Maquis and pine and deciduous woods at higher altitude; rather rare. Throughout.

Anacamptis

Perennials with 2 rounded tubers and several leaves, decreasing in size up the stem. Upper tepals incurved; lip 3-lobed, the middle lobe largest; spur long and slender to thick.

A. **Tepals forming a tight hood; lip 3-lobed, the middle lobe entire and longer than the laterals; spur thick and sequestering nectar.**

Anacamptis coriophora (Syn. *Orchis coriophora*) BUG ORCHID A rather small perennial 14–37(60) cm tall. Leaves 4–11, lanceolate and folded. Inflorescence ovoid and dense with many flowers. Flowers pinkish-brown and scented; tepals converging to form a beaked hood; lip spotted; spur downward-pointing and the length of the ovary; spur usually shorter than lip. Mountains and damp habitats, absent from the Balearic Islands and distribution poorly known in North Africa. ES, FR, IT, PT.

1

2

B. Plants occuring in damp habitats, leaves narrow and unmarked, not in a basal rosette but arranged on the stem; bracts leaf-like.

Anacamptis laxiflora (Syn. *Orchis laxiflora*) LOOSE-FLOWERED ORCHID A rather tall perennial to 50(80) cm with up to 8 channelled, unspotted leaves. Inflorescence *lax*; stems red and flowers rather uniformly purplish-red with outwardly-spreading sepals; petals incurved to form a loose hood; lip 3-lobed and *strongly convex*, the central lobe shorter than the down-folded laterals (or absent), usually forming a tooth between them; centre of the lip white and *unspotted*. Spur to the length of the ovary and *as long or longer than the lip* and thickened at the end. Local in the European part of the region, in wet grassland, marshes and near open water. ES, FR, IT, PT. *Anacamptis palustris* (Syn. *Orchis palustris*) is similar but with differing lip characteristics: central lobe longer than the laterals, the whole lip flatter (laterals scarcely down-turned); central area of lip light purple-red *with small darker spots*; spur tapering to the apex, rather thick and horizontal. Similar (damp) habitats to the previous species, but later flowering (early summer). Rare. DZ, ES (incl. Balearic Islands). Robust forms (sometimes referred to as *Orchis robusta*) to 90 cm with *large leaves* to 30 cm, and larger flowers with conical spurs *shorter than the lip* occur in Mallorca, Algeria and Morocco.

C. Sepals oval with green veins, the upper tepals forming a hood; lip broad and weakly 3-lobed, the laterals rounded.

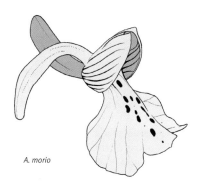

A. morio

Anacamptis morio (Syn. *Orchis morio*) GREEN-WINGED ORCHID A low to medium, often *robust* perennial 12–37(50) cm tall. Leaves mostly in a basal rosette, with several sheathing the stem and crowded, unspotted. Flowers 4–16, usually pink (sometimes white), with tepals forming a hood, the lateral sepals strongly *veined and suffused in green*; lip bluntly 3-lobed and almost folded in 2, with a central paler patch with red spots; spur *thick*, equalling the ovary and horizontal or pointing upwards. Throughout. Forms with *lax inflorescences with fewer flowers* (7–10) with lips with a large *white, unmarked to scarcely marked* centre in Iberian Peninsula, France and North Africa are sometimes referred to as *A. champagneuxii* (Syn. *Orchis champagneuxii*). *Anacamptis longicornu* (Syn. *Orchis longicornu*) is often treated as a subspecies of *A. morio*, and has inflorescences with 5–14 flowers with a *long, more or less straight spur* 11–13(17) mm; lateral lobes darker than hood and coloured differently. Rare and local. DZ, ES (Balearic Islands), FR (Corsica), IT (Calabria), TN.

D. Lip entire and fan-shaped to diamond-shaped.

Anacamptis papilionacea (Syn. *Orchis papilionacea*) PINK BUTTERFLY ORCHID A variable, short to medium perennial with stems 18–38(55) cm high with 4–9 leaves in a basal rosette, and up to 5 sheathing the stem; stems reddish above. Inflorescence robust with few, (6–22) large flowers; bracts reddish and exceeding the ovary; tepals pink-red, pointing forward and not forming a hood; lip not divided, *with a crinkly margin* with streaks of darker coloration, often upturned at the sides; spur 8.7–13.5 mm long and *slender* (1.4–2.5 mm). Widespread and locally common. Throughout. *Anacamptis collina* (Syn. *Orchis collina*) is very similar but with a large, *expanded, sack-like spur* (4.6–7.8 x 3–4.7 mm). Similar distribution but more rare and local, possibly absent from France and the islands; also on Malta.

1. *Neotinea ustulata*
   PHOTOGRAPH: NIELS FAURHOLDT

2. *Neotinea maculata*

E. Lip with 2 ridges at the base.

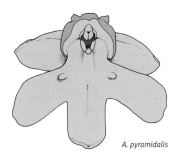

A. pyramidalis

Anacamptis pyramidalis PYRAMIDAL ORCHID An erect, robust perennial to 60 cm. Leaves lanceolate, grey-green and unspotted. Flowers pink-purple (rarely white), borne in distinctly *cone-shaped or dome-shaped dense spikes*; tepals all pink, the lip broad and deeply 3-lobed, the spur long and slender (12–14 x 1 mm). Readily distinguished by the shape of the inflorescence. Widespread and common in sunny, grassy habitats. Throughout.

Himantoglossum

Tall and robust perennials (generally the largest in the region), with 2 ovoid tubers. Flowers with a very long, narrow lip with 2 lateral lobes; spur short.

Himantoglossum robertianum (Syn. *Barlia robertiana*) GIANT ORCHID A very robust perennial with stout stems to 80 cm. Foliage leaves 5–10, matt green, to 30 cm long, sometimes with faint markings. Inflorescence dense and many-flowered, the lower bracts prominent and exceeding the flowers; lateral sepals form a hood with the petals; lip distinctly 3-lobed, the laterals with wavy edges, the middle with 2 parallel ridges; flowers greenish to purplish brown; spur cone-shaped, to 6 mm long. Frequent on scrub and open woodland on base-rich substrates. DZ, ES, FR, IT, MA, TN.

Himantoglossum hircinum LIZARD ORCHID A very robust and *tall* perennial to 70(90) cm with numerous foliage leaves decreasing in size up the stem. Inflorescence dense, smelling of goats, with up to ~100 flowers with very long lips, giving a straggling appearance; tepals form a hood; lip 3-lobed with a whitish, spotted central area, the lateral lobes linear-pointed, the middle lobe *markedly long and twisted, to 60 mm in length*; spur cone-shaped. Rare in the far west and in North Africa, common in the centre of the region. Probably throughout. *Himantoglossum adriaticum* is similar but with laxer inflorescences with fewer flowers with little scent; lateral sepals to 11 mm long (9 mm in previous), the middle lobe with a *pronounced incision at the tip* (to 18 mm, only to 7 mm deep in previous); spur shorter, to 3 mm. High altitude grassland. IT.

Gennaria

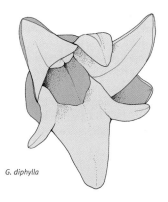

G. diphylla

Perennials with rhizomes, stem with 2 alternate leaves. Flowers greenish.

Gennaria diphylla TWO-LEAVED GENNARIA A short, sometimes clumped perennial 16–36 cm high with 2 oval to heart-shaped leaves at the base, 45 mm–12 cm long. Inflorescence dense, 1-sided with 10–47(85) flowers; tepals all of similar length, 3 mm long, with backwardly curved tips; greenish yellow; lip 3-lobed, the middle lobe the largest; spur grooved. Garrigue, pine woods and laurel forests at a range of altitudes, often in shade. Absent from many areas including Balearic Islands and centre and east of region. DZ, ES, FR (Corsica - rare), IT (Sardinia - rare), MA, PT, TN.

1. *Anacamptis morio*
2. *Anacamptis coriophora*
   PHOTOGRAPH: NIELS FAURHOLDT
3. *Anacamptis pyramidalis*
4. *Himantoglossum hircinum*
5. *Himantoglossum robertianum*
   PHOTOGRAPH: NIELS FAURHOLDT
6. *Himantoglossum robertianum*
   PHOTOGRAPH: NIELS FAURHOLDT
7. *Gennaria diphylla*

Spiranthes LADY'S TRESSES

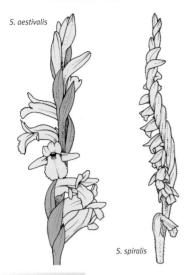

S. aestivalis

S. spiralis

Perennials with tuberous roots and several leaves (all basal or some stem leaves). Flowers arranged spirally in tight spikes, usually white; lip virtually unlobed and close to the other tepals forming an almost tubular perianth; pollinia unstalked.

Spiranthes aestivalis SUMMER LADY'S TRESSES A small perennial 10(40)cm tall with stems glandular above, with 3–6 basal, upright, linear-lanceolate leaves, flowering in early summer. *Inflorescence stems leafy*, with up to 20 flowers in a spiral; flowers pure white with tepals forming a tube which opens slightly at the end; lip tongue-shaped and yellowish at the base. DZ, ES, FR, IT, MA, PT. *Spiranthes spiralis* is similar but with leaves in a tight rosette, often withered when in flower in early autumn, and the smaller flowers borne closely set in a *tight spiral* in a *leafless* inflorescence. Dry, shady, grassy and wooded habitats. Throughout.

Platanthera

Tuberous perennials with paired or few leaves at the stem base. Flowers white or greenish with spreading lateral lobes; lip entire; spur long and slender.

Platanthera chlorantha GREATER BUTTERFLY ORCHID A tuberous perennial with tall stems to 60(80) cm with 2, rarely 3 basal leaves and up to 4 smaller leaves above. Inflorescence lax and many-flowered; flowers white (often greenish or yellowish) and scented; central sepal broad and heart-shaped, forming a helmet with the 2 petals; lip tongue-shaped, to 18 mm long; *anther lobes divergent downwards*; spur short and thick, 19–28 mm long. Open woods and meadows on limestone. DZ, ES, FR, IT. *Platanthera bifolia* is similar but more slender with smaller, sweetly scented flowers in which the *anthers run parallel and close together*; spur 15–20 mm. Open woods and meadows on acid substrates. DZ, ES, FR, IT, TN. *Platanthera algeriensis* is similar to both species but has *entirely green flowers* and occurs only in bogs and wet grassland. Apparently widespread but very rare. DZ, ES, FR (incl. Corsica), IT (incl. Sardinia), MA, TN.

Gymnadenia

Perennials with several tapering tubers and several leaves decreasing in size up the stem. Flowers fragrant, with 3-lobed lip and horizontal marginal petals; spur long and slender.

Gymnadenia conopsea FRAGRANT ORCHID An erect perennial 16–81 cm tall with 4–8 glossy, linear, unspotted leaves at the base to 19 cm long, and few stem leaves. Flowers numerous (17–82), borne in *dense, cylindrical spikes*; flowers pink to lilac (rarely white) with lateral segments spread out and the remaining 3 forming a hood; lip with 3 blunt, equal lobes, *spur long, slender and downward-pointing*, 11.4–15(18) mm, almost 2 x the length of the ovary. Damp habitats and woodland fringes. ES, FR, IT.

Dactylorhiza

Perennials with tubers lobed or clustered, leaves several, often spotted. Flowers with all tepals separate, equal or with inner lobes smaller; lip shallowly 3-lobed with a spur; leaf-like bracts *as long as or longer than flowers*. Highly variable and hybridising; a difficult group to distinguish in the field (only a small number of this complex genus described here).

Dactylorhiza insularis Barton's orchid A tuberous perennial with stems 21–33(43) cm high with 6–7(9) foliage leaves ascending the stem (not all in a rosette). Inflorescence cylindrical and lax with 8–14(18) flowers; bracts as long, or shorter than the flowers (19–24 mm); *flowers lemon yellow* with lateral sepals held more or less horizontally, 8–9 mm long, and the central sepals shorter and curving over the petals; lip to 9 mm long, almost flat and equally 3-lobed, with a pair of reddish blotches at the base; *spur cylindrical and straight, held horizontally* (or slightly down-curved), 7–9 mm. Pinewoods and chestnut woods. Widespread across European part of the region. ES, FR (incl. Corsica), IT (incl. Sardinia), PT.

Dactylorhiza romana Roman orchid A tuberous perennial with stems 15–35 cm and 3–7 foliage leaves in a basal rosette, and 1–3 along the stem. Inflorescence dense, ovoid-cylindrical in shape, the bracts rather longer than the flowers; flowers variable in colour, including *cream, yellow or magenta*; lateral sepals erect and turned outwards, to 10 mm long, the central sepal slightly shorter and curving over the lateral petals which are wider; lip rather flat and 3-lobed, the middle lobe raised; *spur cylindrical and bent to point upwards, exceeding the ovary* (to 25 mm). Higher altitude maquis and open woodland to 2,000 m on base rich substrates. A primarily eastern species. IT (Sardinia and Sicily). *Dactylorhiza sulphurea* (Syn. *D. markusii*) is similar but *always with yellow flowers* with *lip spreading and broader than long* (5–7 x 7–8(10) mm), exceeded by the even *shorter* (9–10 mm) spur held at an angle, *not erect*. Rare but widespread. DZ, ES, IT (incl. Sardinia and Sicily), MA, PT.

Dactylorhiza sambucina elder-flowered orchid A tuberous perennial with hollow stems 17–26(39) cm high. Leaves 4–7, along the stem or in a loose rosette. Inflorescence dense, oval and many-flowered; flowers variable shades of yellow or magenta, scented of elderflower; lateral sepals upright, the central sepal inclined over the petals; lip to 8–9 mm long and 8–12(14) mm wide: elliptic, folded and scarcely 3-lobed. Spur cylindrical-tapering, 12–15(17) mm long and *curved downwards, parallel with ovary*. Mountains, open woods and meadows. DZ, ES, FR, IT.

Dactylorhiza elata robust marsh orchid *A very tall and robust perennial often >1 m,* sometimes clumped. Foliage leaves unspotted to lightly spotted, 5–6(8), narrow to oval-lanceolate and unspotted, pointing upwards. Inflorescence long, with bracts exceeding the flowers; flowers bright pink with darker markings; lateral sepals 9–10(13) mm long, turned upwards, twisted and with incurved margins, sometimes spotted; lip weakly 3-lobed with a small central lobe and the lateral lobes often turned down, with a series of concentric markings; spur robustly cylindrical and curved downwards, parallel with ovary. Wet meadows and hillsides. Throughout except Italy and most islands. *Dactylorhiza saccifera* is possibly a subspecies of *D. elata*, with distinctly *spotted leaves* and flowers with robust cylindrical to almost conical spurs that are up to 4 mm thick and held horizontal to slightly downward-pointing. Mainly eastern. FR (Corsica), IT (Tuscany). **Dactylorhiza durandii** (Syn. *D. maurusia*) is a rare North African endemic with unspotted leaves, pale flowers and narrow spurs 2–3 mm thick that point downwards, equalling the ovaries. The Rif, in marshes. MA.

1. *Gymnadenia conopsea*
2. *Dactylorhiza durandii*
   PHOTOGRAPH: SERGE D. MULLER
3. *Dactylorhiza durandii*
   PHOTOGRAPH: SERGE D. MULLER

Ophrys BEE ORCHID

Distinctive perennials with 2 ovoid tubers and with flowers that dupe male insects into attempting to mate with them to bring about pollination (pseudocopulation). Sepals large and often greenish or pink; petals 2, smaller and hairy; lip large and variously patterned, usually clothed in fur-like hairs. Numerous species described but DNA data suggest the genus has been excessively split.

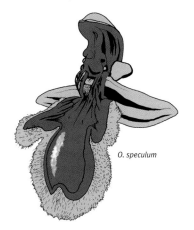

O. speculum

A. Column rounded; stigmatic cavity about as wide as the anther.

Ophrys speculum MIRROR ORCHID A short perennial 7 mm–50(65) cm. Basal leaves oblong and blunt-tipped, stem leaves pointed. Flowers borne in spikes of 1–15(18); sepals green or yellowish, usually striped brown; petals dark purple, hairy, and $1/3$ the length of the sepals; lip to 10–16 mm long, *broad*, 7–14 mm, 3-lobed, the middle lobe with a *blue, shiny mirror* framed with yellow, and fringed with brown or blackish hairs. Widespread and locally common, including most islands (rare in France). Throughout. Subsp. *speculum* has a *broad* lip *without distinctly recurved margins* and reddish-black hairs. Throughout the range of the species. Subsp. *lusitanica* has a lip with recurved margins (therefore narrow), the mirror with a deep yellow margin, and fringed with *yellowish-brown (not blackish) hairs*. Garrigue and maquis; local in the Algarve. PT.

B. Column rounded; stigmatic cavity about twice as wide as the anther.

Ophrys insectifera FLY ORCHID A slender perennial 10–60 cm, with few (2–15), sparse flowers. Sepals pale green, petals dark brown or yellowish-green; lip brown (sometimes with yellow margin) with recurved margins, 3-lobed, the *middle lobe much longer than the side lobes* and with a deep cleft; mirror distinct and dull blue-grey; column rounded. Grassland and open woodland. Absent from most of the region and only at higher altitude. ES, FR, IT. Subsp. *insectifera* is the common (but variable) form, distinguished by its brown lip that is longer than wide and brown petals. Throughout the range of the species. ES, FR, IT. Subsp. *aymoninii* is a French endemic with a lip as long as wide with a *broad yellow margin* (superficially like *O. lutea*). FR.

O. insectifera

O. bombyliflora

Ophrys bombyliflora BUMBLEBEE ORCHID A short, loosely clump-forming perennial 50 mm–35 cm. Basal leaves oval-lanceolate, forming a flat rosette, stem leaves erect and clasping the stem. Flowers few, borne in short spikes of 1–6; sepals green; petals green with a purplish base, triangular, and $1/3$ the length of the sepals; *lip small*, 6–10 mm long, 3-lobed with the lateral lobes deflexed, brown with a central bluish, shield-shaped mirror. Locally common in the west of the region, rarer further east; occurs on larger islands. Throughout.

1. *Ophrys speculum* subsp. *speculum*
2. *Ophrys speculum* subsp. *lusitanica*
3. *Ophrys bombyliflora*
4. *Ophrys speculum* subsp. *speculum*
5. *Ophrys lutea* subsp. *lutea*
6. *Ophrys lutea* subsp. *lutea*

O. lutea

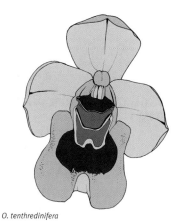

O. tenthredinifera

Ophrys lutea YELLOW OPHRYS A short perennial 70 mm–50 cm. Basal leaves oblong and pointed, stem leaves smaller and narrower. Flowers borne in spikes of 3–12; sepals green, petals greenish or yellowish and ¹/₂ the length of the sepals; lip 9–18 x 10–15 mm, 3-lobed, with a broad, flat, *yellow margin* surrounding a red-brown area with a blue-grey mirror. Common. Throughout. Subsp. *lutea* has a lip to 19 mm wide with a yellow margin 3–6 mm wide. The common form, found throughout the range of the speces. Subsp. *galilaea* has a *smaller lip* to 12 mm wide with a *narrow* yellow margin to 2 mm across. Very local in the region. ES (incl. Balearic Islands), IT (incl. Sicily), PT.

Ophrys tenthredinifera SAWFLY ORCHID A short to medium perennial 10–60 cm. Basal leaves oval-lanceolate and blunt or pointed, stem leaves similar. Flowers borne in short spikes of 3–8(11); sepals *purplish-pink* (rarely green or white), the central erect; petals similar in colour and ¹/₃ the length of the sepals; *lip broad and square, scarcely lobed or unlobed*, 9–18 mm long, brownish-purple with a broad yellowish margin, and a small, brown-spotted, 2-parted mirror. Locally common but strongly southern. Throughout.

O. fusca

Ophrys fusca SOMBRE BEE ORCHID A short to medium perennial 80 mm–44 cm. Basal leaves oblong-lanceolate and blunt-pointed, stem leaves smaller and narrower. Flowers borne in spikes of 3–6(9); sepals green or pinkish; petals green and $^1/_2$ the length of the sepals; lip 8–20 mm long, horizontal to down-curved, 3-lobed and purplish or yellowish brown, often yellow-edged, and with a bluish or greyish W-shaped mirror. Common throughout. Subsp. *iricolor* is distinguished by the lip which is wine-red underneath and has a bluish, shining mirror. Local and predominantly eastern. ES (Balearic Islands), IT (Sardinia), PT. Subsp. *fusca* has a lip *pale green underneath*. Throughout the range of the species. Subsp. *pallida* is easily distinguished by its *white sepals*. Rare. IT (Sicily, possibly Sardinia), TN.

Ophrys omegaifera A variable, slender to robust perennial to 50 cm. Flowers with pale green-yellow and oval-elliptic sepals; petals yellowish, greenish or brownish and almost flat at the margins; lip grey-brown or blackish, straight to abruptly curved, 3-lobed and velvety to hairy; *mirror fish-tail shaped* and violet to grey (sometimes marbled), delineated by a distinct white to *blue omega-shaped band*; column rounded. Widespread across the south, including larger islands. DZ, ES, MA, PT, TN. Subsp. *hayekii* has a straight lip, sepals broadest at the middle and unmarked mirror. Very rare. IT (Sicily), TN. Subsp. *dyris* has a inclined-spreading lip and reddish-brown mirror. Rare. ES (incl. Balearic Islands), PT.

Ophrys atlantica A slender perennial 15–30 cm, with 3–6(9) flowers borne in a lax spike. Sepals green or yellowish and narrowly oval; petals brownish green, linear and wavy-margined and recurved; lip purplish-black, narrow at the origin, *broadly saddle-shaped* towards the apex and 3-lobed with incurved margins; column rounded. A predominantly North African species extending into Andalucía. DZ, ES (south), MA, (possibly TN).

C. Column extended into an S-shaped tip; pollinia with drooping stalks.

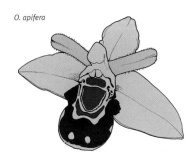
O. apifera

Ophrys apifera BEE ORCHID A very variable, short to medium perennial 15–50 cm. Basal leaves oval-lanceolate and blunt or pointed, stem leaves similar but smaller. Flowers borne in short spikes of 5–10(15); sepals bright pink with a green mid vein (rarely green or white); petals green or purplish and <$^1/_3$ the length of the sepals; lip 3-lobed, the *central lobe curved backwards,* with a shield-shaped brown or violet mirror with a yellowish margin; stamen elongated into a snout-like appendage. Common and widespread. Throughout.

1. *Ophrys tenthredinifera*
2. *Ophrys fusca*
3. *Ophrys atlantica*
   PHOTOGRAPH: NIELS FAURHOLDT
4. *Ophrys apifera*

D. Column acute; pollinia with firm (not drooping) stalks.

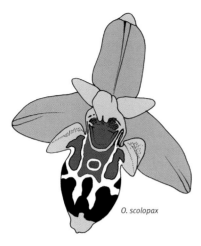

O. scolopax

Ophrys scolopax woodcock orchid A short to medium perennial 10–50 cm. Basal leaves lanceolate and pointed, stem leaves narrower and more pointed. Flowers borne in rather long spikes of 3–15; *sepals pink* (rarely green or white), the central erect; petals similar in colour and ¹/₂ the length of the sepals; lip 8–14 mm, oval, 3-lobed, and brownish purple and velvety with a large brownish-blue, H-shaped or spotted mirror with a narrow yellow margin. Fairly common in the far west of the region, rare in Italy. Throughout. Subsp. *scolopax* has triangular sepals, and the middle lobe of the lip *not markedly narrowed* above the base. Locally common but absent from many areas including most of mainland Italy and larger islands. DZ, ES, FR, IT, PT, TN. Subsp. *apiformis* (Syn. *O. sphegifera*) has more or less *linear sepals*, and the *lip narrowed* (strongly recurved) above the base. Predominantly western. ES (incl. Balearic Islands), IT (Sardinia and Sicily), TN. Subsp. *conradiae* has *long petals* (longer than wide) and mirror covering the basal ¹/₂ of the lip. Fr (Corsica), IT (Sardinia).

Ophrys fuciflora A variable, tall and many-flowered perennial to 60 cm. Sepals purplish, white or green, the dorsal petal boat-shaped or flat; petals bright pink (less often purplish or green) and triangular with recurved margins; *lip broad and square*, yellowish-brown and entire to weakly 3-lobed with a distinct H-shaped, but often complicated mirror that is dull blue or reddish brown with a *conspicuous cream border*; column acute in shape. Absent from much of west, south and many islands. FR (incl. Corsica), IT (incl. Sardinia and Sicily). Subsp. *fuciflora* is the widespread and variable form with sepals to 16 mm long, and leaves remaining fresh when in flower. Subsp. *elatior* is similar but flowering later with withered leaves. Rare, mainly northern. FR, IT. Subsp. *apulica* has a lip with sigmoidly recurved sides. IT (Mount Gargano and Sicily - rare). Subsp. *chestermanii* has white to violet sepals and a trapezoid-shaped, blackish lip with uncomplicated markings. IT (Sardinia). Subsp. *lacaitae* is a very distinctive form with green sepals and the *lip yellow almost throughout*. IT (incl. Sicily), MT. Subsp. *oxyrhynchus* has an almost straight lip that is reddish, and has a mirror on >¹/₃ of its surface. IT (incl. Sicily). Subsp. *biancae* has a trapezoid-shaped lip that has a broad yellow margin and reddish centre, and white-green sepals. IT (Sicily: Etna and near Palermo).

Ophrys argolica A variably compact to slender perennial to 50 cm with large flowers. Sepals green, pink, white or violet and rather flat; petals green or pink and elliptic with flat margins; lip reddish or yellowish brown, *straight and flat*, entire to weakly 3-lobed with a mirror consisting of *horseshoe-shaped figure of 2 central spots* which are grey-blue and not distinctly shiny, with or without a paler border; column acute to obtuse. A mainly eastern species. IT. Subsp. *crabronifera* has a broad stigmatic cavity, to 6 mm wide at the base. Coastal. IT. Subsp. *biscutella* has a narrow stigmatic cavity, to <4 mm wide at the base, and reflexed sepals. IT (south). *Ophrys bertolonii* is similar to *O. argolica* but shorter, to 35 cm, with normally bright pink sepals, pink petals, and a broad blackish-brown lip that is *bent forwards* at the base; mirror nearly square and shining blue-grey. Column acute. Flood plains of the river Po southwards. IT (incl. Sicily).

Ophrys sphegodes A widespread and very variable perennial (with many forms described). Slender, 10–70 cm tall with 3–10 flowers in a lax spike. Sepals green or yellowish, sometimes violet-flushed but *not pink*; petals also greenish or yellowish; lip brownish or blackish, often with a yellowish or reddish paler margin, rather *round, straight and flat* and entire or very weakly 3-lobed; mirror usualy H-shaped and variably complicated and bordered. Column normally acute. Very widespread and common in the European Mediterranean; rare in Portugal; present on the larger islands. ES, FR, IT, PT. Subsp. *sphegodes* is the common form, but highly variable: lip *to 20 mm long* (large) and with lateral sepals all of 1 colour (which varies). Mainly western. ES, FR, IT. Subsp. *litigiosa* has a smaller lip (to 10 mm) that is shorter than the dorsal sepal, and sometimes with a yellow margin. Northern Spain eastwards. ES, FR, IT. Subsp. *atrata* has a lip with a border of distinctly longer hairs. Common on the larger islands and in southern Italy. ES, FR, IT. Subsp. *passionis* has a lip with an H-shaped mirror that has 2 additional exensions from the base. ES (northwest), FR (southwest), IT (incl. Sardinia (rare) and Sicily).

1. *Ophrys scolopax*
2. *Ophrys fuciflora* PHOTOGRAPH: NIELS FAURHOLDT
3. *Ophrys fuciflora* PHOTOGRAPH: NIELS FAURHOLDT
4. *Ophrys argolica* PHOTOGRAPH: NIELS FAURHOLDT
5. *Ophrys argolica* PHOTOGRAPH: NIELS FAURHOLDT
6. *Ophrys sphegodes* PHOTOGRAPH: NIELS FAURHOLDT

Ophrys lunulata A slender perennial with large flowers. Sepals pale pink to white, and petals similar; lip reddish-brown and deeply 3-lobed at the middle; mirror distinctly crescent-shaped, shiny blue-grey or reddish and central. Column acute. Frequent in the southeast of Sicily, unlikely to occur elsewhere (though erroneous records exist). IT (Sicily).

Serapias TONGUE ORCHID

Tuberous perennials with 2(–5) ovoid tubers, similar to *Ophrys* but with an elongated lip that is downward-pointing and tongue-like with 2 short, upturned lateral lobes. Potentially hybridising with other genera including *Ophrys*, the extent of which is not fully known.

A. Base of flower lip with an entire or deeply channelled blackish protuberance.

Serapias lingua TONGUE ORCHID A short, tuberous, clump-forming perennial to 25 cm, stems sometimes spotted. Leaves narrowly lanceolate and grey-green. Flowers borne in lax spikes of up to 9; sepals and petals purple; lip to 32 mm long, *maroon-coloured with a single, coffee-bean like blackish protuberance at the base; epichile oval-lanceolate and to 12 mm wide;* bracts shorter than the sepal hood. Common in a range of habitats including garrigue, woodland and wet meadows. Throughout (north to Dordogne, France). Subsp. *duriaei* (Syn. *S. strictiflora*) has a *much narrower,* maroon-coloured lip with straight margins; petals acuminate with an oval base 2–4 mm wide; *epichile triangular, to 4.5 mm wide.* Similar habitats to the previous species. Locally abundant in the far west of the region; absent from the islands. DZ, ES, (south), MA, PT. Slender forms with pale flowers with a *pale yellowish to orange-red epichile* in Iberian Peninsula are sometimes referred to as *S. elsae.* Populations in southern France with *gradually* attenuated, drop-shaped petals (not drop-shaped then abruptly attenuated) have been described as *S. gregaria* but are now widely accepted to be *S. lingua* subsp. *duriaei.*

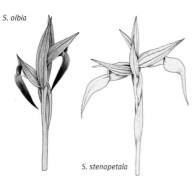

S. olbia

S. stenopetala

Serapias olbia is very similar to *S. lingua* in most respects; a tuberous, often clumped perennial to 30 cm tall. Flowers with a lip to 31 mm long, pointed and *bent strongly* backwards, *deep blackish-red/purple* with dense purple hairs over the surface, and a deeply grooved blackish protuberance at the base. Possibly of hybrid origin. Very rare. FR.

Serapias stenopetala is an Algerian endemic of cork oak forests with very pale yellow or pinkish flowers, with a *lemon-yellow* single protuberance at the base of the lip. DZ.

B. Base of flower lip with 2 distant protuberances; petals acuminate-drop shaped.

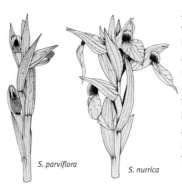

S. parviflora

S. nurrica

Serapias parviflora A perennial similar to *S. lingua* but with *paler* greenish-maroon flowers; petals drop-shaped, lip narrow with 2 (not 1) brownish-red (not blackish) ridges at the base, and *smaller,* just as long as the hood (15–22 mm long), *not distinctly longer* than hood. Limestone areas, damp grassland and marshes. Similar distribution to *S. lingua* but rather rare. Throughout. *Serapias nurrica* is similar to *S. parviflora* but the sharply pointed lip has a *distinct pale and incurved margin.* Rare. ES (Menorca, Balearic Islands) FR (Corsica), IT (incl. Sardinia and Sicily), TN (islands only).

1. *Serapias lingua*	3. *Serapias cordigera*
2. *Serapias lingua* subsp. *duriaei*	4. *Serapias negleta*
PHOTOGRAPH: SERGE D. MULLER	PHOTOGRAPH: NIELS FAURHOLDT

S. cordigera

C. Base of flower lip with 2 distant protuberances; petal base orbicular in shape.

Serapias cordigera HEART-FLOWERED ORCHID A short to medium perennial 12–40 cm. Leaves narrowly lanceolate, channelled, and sharply pointed; the lowermost leaves sheath-like, purplish and spotted at the base. Flowers borne in short spikes of 2–12; sepals and petals reddish; lip to 30–45 mm long, reddish, the prominent central lobe *broad and heart-shaped, and 2 x as large as the sepals*, hairy at the mouth, and with 2 dark ridges at the base. Damp grassland and marshes. Throughout. Subsp. *cossyrensis* (Syn. *S. cossyrensis*) has broad leaves, short inflorescence and flat, very broad, hairy, maroon lip, blooms early (March). Volcanic garrigue, scrub and woodland on the island of Pantelleria only. IT (Sicily). *Serapias orientalis* var. *siciliensis* is distinguished by its (variably coloured) shorter, broader and extremely *hairy lip*. (A rare variant of an eastern Mediterranean species). IT (Sicily).

Serapias vomeracea LONG-LIPPED SERAPIAS A rather *tall and slender* perennial 15–60 cm. Leaves linear, channelled, grey-green, and sharply pointed; the lowermost leaves sheath-like, and green (not purple). Flowers borne in short spikes of 3–10 with equal or longer bracts; sepals and petals purplish; lip 21–45 mm long - much exceeding the sepals; pale yellowish-red to maroon, with 2 parallel, similar-coloured ridges at the base; the prominent middle lobe triangular; *bracts longer than the hood*. Damp grassland and marshes. DZ, ES, FR, IT, MA, (PT, where populations not typical, requiring further investigation). *Serapias perez-chiscanoi* is similar but has a shorter inflorescence (20–35 cm), and smaller, *white-green* flowers. South-central Iberian Peninsula (Extremadura region). ES, PT.

S. vomeracea   S. perez-chiscanoi

Serapias neglecta SCARCE SERAPIAS A medium perennial to 30 cm with unspotted leaves; basal leaves scale-like. Inflorescence short and dense with disproportionately large flowers; bracts shorter than hood; lip pointed and heart-shaped, *pale* in colour: *salmon-pink or yellowish* (not dark purple or red). Open pine woods garrigue, olive groves and damp meadows (usually on acid soils) along the coast from Provence to Elba. FR (incl. Corsica), IT (incl. Sardinia). Subsp. *apulica* (Syn. *S. apulica*) has a notably robust habit and heart-shaped but long-pointed lip that is orange-red with a tuft of pink or whitish hairs in the centre. IT (south).

S. neglecta

IRIDACEAE | IRIS FAMILY

Bulbous, tuberous or rhizomatous perennials, usually with linear leaves, all basal or alternate. Flowers with perianth of 6 tepals, the outer 3 often different from the inner 3, enclosed in 1–2 often papery spathes when in bud; stamens 3; style often with 3 branches, ovary inferior. Fruit a 3-parted capsule.

Iris IRIS

Perennials with rhizomes or bulbs with flat, vertical leaves (or 4-angled). Flowers with the outer 3 tepals (falls) horizontal to down-turned, and the inner 3 (standards) erect; styles rather petal-like and arching over the falls, each with a stamen beneath. Some classifications recognise several genera which are here all described under *Iris*, consistent with most floras.

A. Beardless irises with rhizomes.

*Iris unguicularis*ALGERIAN IRIS A low, tufted, rhizomatous, winter-flowering perennial with tough, dark green, strap-shaped, pointed leaves to 15 mm wide. Flowers solitary, with a rather open appearance, to 70 mm across, blue-lilac with white markings and dark purple venation, and a central orange band on the falls; falls oblong-elliptic, to 80 mm long, the standards similar in length but narrower; perianth tube long (10–20 cm) with the ovary at ground level; flowers scented. Dry, exposed and sunny habitats in North Africa. DZ, TN.

Iris foetidissima (Syn. *Chamaeiris foetidissima*) ROAST BEEF PLANT A medium, tufted, rhizomatous, spring-flowering perennial with shiny green, pointed, sword-shaped leaves 10–25 mm wide which smell faintly unpleasant when crushed (vaguely reminiscent of roast beef). Flowers 50–70 mm across, dull pale purple to liver-coloured (rarely grey-yellow) with darker veins on the falls, borne on a branched, flattened stem with several flowers opening in succession. Fruits large and distended, splitting to reveal rows of bright orange seeds which persist for months. Woods, thickets and other shady habitats. Fairly common and widespread more or less throughout.

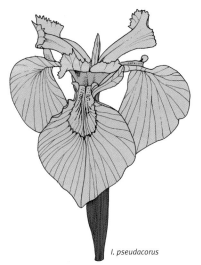

I. pseudacorus

Iris pseudacorus (Syn. *Limniris pseudacorus*) YELLOW FLAG A tall and erect rhizomatous perennial herb to 1.5 m. Leaves sword-shaped, 10–30 mm wide, green or greyish. Flowers large, 70 mm–10 cm across and *bright yellow* with faint red-purple veins, borne 1–3, each with a papery spathe below; unbearded, with standards markedly smaller than the falls. Capsule large and splitting into 3 segments when ripe, exposing large brown seeds. Damp habitats, river banks and near seasonally running water. Common in suitable habitats throughout.

Iris spuria BLUE IRIS A tall and erect, tufted, rhizomatous perennial herb to 90 cm with a thick brown rootstock covered in old leaf bases. Leaves grey-green, rather tough, 6–20 mm wide. Flowers 2–4, rather *small* and grey-blue or lilac with darker veins and a yellow strip in the centre of the falls; falls to 50 mm long. A variable species with intra-specific taxa described. Damp, often saline habitats. Widespread from south France westwards, absent from many islands. DZ, ES, FR (incl. Corscia), MA, TN.

B. Beardless irises with smooth bulbs.

Iris xiphium (Syn. *Xiphion vulgare*) SPANISH IRIS A medium bulbous perennial 30 cm–2 m high with narrow, channelled leaves 20–70 cm long and 1–12 mm wide. Flowers large and solitary, *violet-blue* with a yellow ring on the centre of the oval falls (rarely all yellow-flowered variants, which have been described as *I. lusitanica* and *I. juncea*); standards erect and somewhat smaller than the falls. Frequent in rocky, grassy areas and scrub. Throughout, except the Balearic Islands and Sicily. *Iris tingitana* (Syn. *Xiphion tingitana*) has *silvery-green* leaves that are long and arching. Flowers pale to deep blue with yellow bands on the falls and *sharply pointed* standards. North Africa only. DZ, MA. *Iris filifolia* (Syn. *Xiphion filifolium*) has very narrow leaves just 1–2(5) mm wide and *dark reddish-purple (not bluish) flowers* with a bright orange area on each of the falls. Dry rocky habitats in south Spain and North Africa. DZ, ES, MA.

I. xiphium

1. *Iris foetidissima* (inset: fruits)
2. *Iris xiphium*
3. *Iris pseudacorus*
4. *Iris albicans*
5. *Moraea sisyrinchium*

Iris serotina (Syn. *Xiphion serotinum*) An *autumn-flowering* perennial 40–80 cm tall with leaves 20–60 cm long and 1–7(13) mm wide. Flowers purple, with internal tepals just 6–10 mm (up to 25 mm in other species in group) which are linear-oblong (rather than oblong-lanceolate). Rare. ES (south), MA (north).

Iris planifolia (Syn. *Juno planifolia*) A low to short, bulbous perennial with stems to just 30 mm with longer leaves to 35 cm, 30 mm wide; leaves all basal, shiny green, produced in a fan-like arrangement. Flowers bluish-violet with darker veins, and *small, horizontal standards,* to 12–30 mm long, and toothed. Local on limestone rocks and exposed mountain slopes. Iberian Peninsula eastwards (exluding France and most islands). DZ, ES, IT (Sardinia), MA, PT, TN.

C. Beardless irises with corms or spreading tubers (including *Iris* and *Moraea*).

Iris tuberosa (Syn. *Hermodactylis tuberosus*) SNAKE'S-HEAD, WIDOW IRIS A spring-flowering, perennial 20–40 cm high with spreading, finger-like tubers (unlike other species described here). Leaves 1.3–4 mm wide and rush-like, 4-angled (square in cross-section) and *long,* exceeding the flowering stems which are up to 40 cm tall. Flowers solitary with a broad, leafy sheath to 20 cm long; sweetly scented, yellow-green, usually with velvety brown-purple falls 40–50 mm long, the standards about $^1/_2$ as long, and narrow. Grassy and rocky habitats, garrigue and roadsides. Southeast France eastwards. FR, IT.

M. sisyrinchium

Moraea sisyrinchium (Syn. *Gynandriris sisyrinchium*) BARBARY NUT A cormous perennial with a fibrous tunic; short, to 10–30(40) cm high with slender stems. Leaves 1 or 2, sheathing at the base, the free portion longer than the flowering stems, linear and channelled, to 40 cm long. Flowers borne in groups of 1–4, violet-blue, the falls with white and yellow markings, and the standards slightly shorter. Common on pathways, roadsides and other dry, open habitats near the coast; often patch-forming. Flowers open in the afternoon sun. Throughout.

D. Bearded irises with rhizomes.

Iris × *germanica* BEARDED IRIS A robust, perennial herb to 90 cm with branched stems and a thick rhizome visibly spreading above ground. Leaves broad and sword-shaped, grey-green, (20)30–60 mm wide; shorter than the flowering stems. Spathes beneath the flowers green at the base but papery at the tips, often tinged purple. *Flowers large,* to 10 cm across, borne in groups; violet-blue (standards often paler) and fragrant, the falls with a *conspicuous yellow beard* along the centre. A range of rocky, dry and grassy habitats. Cultivated and widely naturalised. DZ, ES, FR, IT, MA, TN. White-flowered, cultivated forms in France and Italy are sometimes described as *Iris florentina.*

Iris pallida A robust perennial to 1 m similar to *I. germanica,* with a thick rhizome, and leaves distinctly *blue-grey,* thick and finely-ribbed. Spathes beneath the flowers silvery and *papery throughout.* Flowers uniformly pale lilac-blue. Gardens and maquis in north Italy, possibly naturalised elsewhere. IT. Subsp. *cengialtii* is generally smaller with green leaves that are strongly ribbed, brown and papery spathes and *deep violet* flowers. IT (northeast).

Iris albicans A short, erect rhizomatous perennial herb 40–75 cm tall with rhizomes thick and spreading partially above ground, similar in most respects to *I. germanica.* Leaves broad and sword-shaped, 13–50 mm wide. Flowers borne in clusters along the stems, each with a spathe that is green at flowering time, or papery at the tip; *flowers large and pure white,* the falls with yellow beards. Native to the Middle East; locally commonly naturalised and patch-forming on roadsides in the region, probably throughout.

Iris lutescens A short, erect rhizomatous perennial with unbranched stems 30 mm–40 cm. Leaves sword-shaped and *straight* or scarcely curved, grey-green 4–18(25) mm wide, not withering in the winter. Spathes green, *keeled* and membranous at the tip. Flowers yellow, blue and brown bicoloured, to whitish, with a *white to yellow beard* on the centre of the falls; perianth tube rather long, 22–43 mm. Grassy and rocky habitats. Iberian Peninsula eastwards, excluding most islands. ES, FR, IT. Forms with *violet-purple flowers with* a *white to violet beard* in Iberian Peninsula are sometimes referred to as *I. subbiflora*. ES, PT. *Iris pseudopumila* is similar to *I. lutescens* with leaves slightly curved, to 10 mm wide and *long, rounded spathes* to 12 cm which *lack a keel* and tightly sheath the perianth tube which is *long, to 70* mm; large flowers to 80 mm across are borne singly; flower colour variable (not a diagnostic). Rocky scrub. IT (southeast + Sicily), MT (incl. Gozo).

Freesia

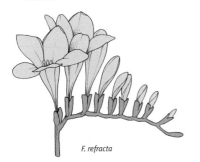

Cormous perennials with *Iris*-like leaves. Flowers slightly zygomorphic with 6 tepals, borne in 1-sided racemes, scented; style slender with 3 deeply split branches.

Freesia refracta A perennial to 30(50) cm with an underground corm and leafy, branched stems. Leaves 5–6, linear-lanceolate, to 30 cm long and 15 mm wide. Flowers yellow (rarely white), tubular, and 2-lipped, fragrant. Native to South Africa, and cultivated as an ornamental. ES, FR, IT, PT.

F. refracta

Crocus

Small perennials with corms. Leaves basal, enclosed beneath by sheaths, with a central whitish channel. Flowers cup-shaped, with 1–2 spathes; tepals 6, all similar; stamens 3; style solitary with 3 (or more) branches, each with variably divided stigmas.

A. Spring-flowering.

Crocus nevadensis A low, cormous perennial with leaves partially developed when in flower; leaves 3–5, each with a pair of lateral grooves below, to 10 cm long. Flowers *white-violet*, the tepals with a darker keel and white-yellow stigma; flowers appear in *spring*. Stony mountain slopes. Rare. ES (south and east), MA.

Crocus cambessedesii A low, cormous, *autumn, winter or spring-flowering* perennial with 3–5 leaves just 0.5–1 mm wide, developed or not at flowering time. Flowers *very small*, the tepals 14–18 mm long; flowers usually very pale lilac (or white, cream or dark lilac), the outer tepals streaked with dull blackish-purple, all tepals blunt and elliptic; stamens white and styles orange-red. Rocky habitats (growing directly from crevices), scrub and pine woodlands. ES (Balearic Islands).

Crocus corsicus A low, cormous, winter to spring-flowering perennial with corm tunics netted in the upper part, similar in most respects to *C. cambessedesii* but with 1 spathe (rarely 2) at the base of the perianth tube, that is undivided, membranous and often brown-spotted. Flowers larger than the previous species, wide-opening when mature, borne singly or in groups of 2–3; tepals to *35 mm long*; flowers pale lilac, the outer tepals streaked with purple; stigmas 3-parted and red and stamens orange - both contrasting the pale flowers. FR (Corsica).

Crocus etruscus A late winter-flowering cormous perennial to 80 mm with a large bulb and *fibrous tunic*. Flowers *pale purple* with distinct longitudinal veins externally, and contrasting orange stigmas; *bract ascending*. IT (central-southern Tuscany).

Crocus minimus A low, cormous, winter to spring-flowering perennial with leaves present at flowering time, similar to *C. corsicus* and *C. cambessedesii* but with *very unequal spathes* spotted with green or brown. Flowers *small*, the tepals to just 27 mm long, pale purple-lilac, the outer tepals streaked with brown-purple; anthers and stigmas yellow (not bright red). Scrub. FR (Corsica), IT (Sardinia).

Crocus versicolor A low, cormous, spring-flowering perennial in which fibres on the corm are parallel only (not netted). Leaves grey-green and partially developed at flowering time, 1.5–3 mm wide. Flowers pale lilac or white, the 3 outer tepals striped with purple, to 35 mm long; throat white or pale yellow. Meadows, woods and stony pastures. FR (south), IT (northwest).

C. serotinus

B. Autumn-flowering with leaves present (at least partially) at flowering time.

Crocus serotinus A low, cormous, autumn-flowering perennial with leaves partially developed when in flower or appearing immediately after; leaves 4–7, narrow, 1.5–3.5 mm wide. Flowers 1–2, pale lilac-pink with a pale yellow throat, orange stigma and yellow to orange stamens and style, scented. Coastal sands and pine woods. ES, PT. *Crocus clusii* (Syn. *C. serotinus* subsp. *clusii*) has a bulb tunic with netted fibres (not parallel), but is otherwise similar. ES, PT.

Crocus sativus SAFFRON CROCUS A low, cormous, autumn-flowering perennial with long, filiform leaves present at flowering time. Flowers large, tepals to 50 mm long, bright lilac-purple to whitish, with dark purple veining which is darker at the base. Style branches divided into 3 and *very long*, to 32 mm, often collapsing against the inner flower. Native to Turkey but cultivated. ES, FR. *Crocus thomasii* is similar, with large, pale white-lilac flowers which lack prominent veining, and have a *pale yellow throat*; stigma 3-parted and red; anthers yellow. Primarily Adriatic. IT.

Crocus longiflorus A low, cormous, autumn-flowering perennial with leaves usually partially developed at flowering time; leaves 4–6, white-striped and 1–3 mm wide. Flowers lilac to purple with darker veining externally and a prominent yellow throat; fragrant; tepals blunt, oval and broadest above the middle, to 43 mm long; stamens yellow; style 3-branched and orange-red. Grassland and woods. IT (incl. Sicily), MT.

C. Autumn-flowering with leaves absent at flowering time.

Crocus nudiflorus A low, cormous, autumn-flowering perennial with leaves *absent at flowering time* (appearing in winter-spring); leaves borne in groups of 3–4 and 2–4 mm wide. Flowers purple to lilac, indistinctly veined on the outside; corolla tube very long, 10–22 cm; stamens with white filaments and yellow anthers, stigmas orange and much-branched. Damp meadows, often on higher ground. ES, FR.

Romulea

Low, cormous perennials with basal, linear leaves that are often 4-grooved. Flowers normally enclosed in bud by 2 spathes; flowers *Crocus*-like but borne on a green stem; tepals 6, contracting into a very short perianth tube; stamens 3, style solitary and 3-branched.

A. Flowers white, pink or violet and (usually) distinctly yellow-orange in the throat.

Romulea bulbocodium A low, cormous spring-flowering perennial. Leaves 3–7, deep green and curved or erect-patent. flowers crocus-like, borne on scapes 30 mm long that often have >1 flower; white to lilac, *always with a yellow throat*, and striped purple; tepals elliptic and pointed, to 30 mm long; stamens about $^1/_2$ the length of the tepals, stigmas variable but usually *over-topping the anthers. Spathe papery almost throughout and weakly green-veined*. The most common species in the region, in a range of habitats from mountain rocks to coastal sands. Throughout. *Romulea clusiana* has flowers with larger tepals to 46 mm long, and anthers to 11 mm long (not <8 mm); flowers transversally coloured yellow-orange in the lower $^1/_2$ and more or less white in the upper $^1/_2$, and flushed violet towards the tips (those of the previous species more uniform). ES.

R. ramiflora

Romulea ramiflora A low, cormous spring-flowering perennial with 4–6 recurved to erect leaves. Flowers pale to deep blue-violet, often with darker veins and greenish on the outside; throat normally yellowish; tepals oblong and blunt, 14–21 mm long. *Spathes almost entirely green with a narrow papery margin*. Coastal sandy habitats; common. Throughout. *Romulea melitensis* is a similar, low, cormous *late spring-flowering* (April) perennial with erect to spreading, long narrow leaves. Spathes green with a *narrow or absent papery margin*. Damp habitats. IT (Sicily), MT (incl. Gozo).

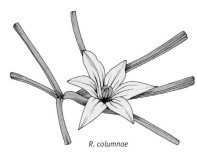

R. columnae

Romulea columnae SAND CROCUS A low, cormous perennial with 3–8 erect to spreading, recurved leaves 0.6–1 mm wide. Flowers solitary or in groups of 2–3 and *small*, to 12 mm across, the corolla tube to 5 mm long and tepals 10–19 mm long; tepals narrowly lanceolate and long-pointed; flowers mauve to white, usually with a whitish-yellow to yellow throat, sometimes greenish externally; anthers to just 2.5 mm. Damp coastal turf and sand; widespread across the region. Throughout. Subsp. *rollii* (Syn. *R. rollii*) has tepals violet with darker veins within and *white-green with dark blackish-purple streaks and veining* in the lower $^1/_2$ externally, the outer tepals with a central greenish band; throat white-yellow; pollen yellow. Outer spathe green with a *wide papery border*, bracteole papery. Coastal rocks, cliffs and sandy habitats. DZ, FR, IT (incl. Sardinia and Sicily), MA, MT (incl. Gozo), TN.

B. Flowers white, pink or violet with little or no bright yellow-orange in the throat.

Romulea bocchierii A short to medium, cormous perennial to 40 cm (including the leaves). Flowering scapes always with a *single flower* and with rush-like stem leaves *erect to sub-erect*. Flower bracts papery with 2 green veins. Flowers *white inside, with a dull greenish-yellow (not bright yellow-orange) throat* (the same colour as the anthers). IT (Sardinia).

Romulea requienii A low, cormous perennial with leaves *lying flat on the ground*; to 1.25 mm wide. Spathes papery with a single green vein, sometimes speckled with violet or brown. Flowers violet with darker veins, the throat *white-violet (not yellow)*; tepals to 21 mm long and rounded at the tip; pollen yellow. Coastal sands and seasonally flooded coastal habitats. FR (Corsica), IT (south + Sardinia). *Romulea ligustica* is a similar, with erect to patent leaves and flowers violet with a *white throat*, the perianth tube short, 5–7 mm long and stigmas not exceeding the anthers; anthers with *white (not yellow) pollen*. Grassy and stony coastal habitats. FR (Corsica), IT (incl. Sardinia). *Romulea × limbarae* is an island endemic possibly of hybrid origin between *R. ligustica* x *R. requienii*, and is intermediate in morphology. IT (Sardinia: Mount Limbara). *Romulea corsica* is an island endemic possibly of hybrid origin of *R. revelieri* x *R. requienii*, and is intermediate in morphology. Flowers violet-purple. FR (Corsica: east coast).

C. Flowers deep reddish- or violet-purple with no yellow in the throat.

Romulea revelieri A low, cormous spring-flowering perennial. Leaves 3–6, spreading to recurved. Flowers with bluntly pointed to rounded lobes; *deep purple* with no yellow in the throat (throat whitish or pale purple); corolla tube to 5 mm long and tepals to 11 mm long; *stigmas equalling or exceeding the anthers*; spathe to 14 mm long and green with a papery margin; bacteole papery and brown-spotted, green in the centre. Damp and flood-prone sandy habitats near the coast. FR (Corsica), IT (incl. Sardinia).

1. *Romulea bulbocodium*
2. *Gladiolus illyricus* (inset: stamens)
3. *Gladiolus italicus* (inset: stamens)
4. *Watsonia meriana*

Romulea linaresii A low, cormous spring-flowering perennial. Leaves 3–6, spreading to recurved. Flowers deep violet-purple without yellow in the throat; tepals elliptic and broadest above the middle, bluntly pointed or rounded, to 19 mm long; *anthers exceeding the stigmas*; spathe green below, papery towards the tip. Sandy coastal habitats. IT (Sicily).

D. Flowers bright pink with a yellowish throat with dark purple veins.

Romulea rosea A short to medium, cormous perennial with erect, rush-like leaves and *many flowers* (often >3); flowers *bright pink* with lanceolate, pointed tepals; throat yellow with dark purple veins; anthers yellow and exceeded by the white stigma. An alien exotic from South Africa naturalised in sandy habitats. IT (Sardinia).

Gladiolus

Cormous perennials with fans of flat, sword-shaped and ribbed leaves. Flowers borne in long, rigidly erect stems, each flower with a green bract; tepals 6, unequal, fused into a short tube at the base; stamens 3; style slender with 3 short branches. A confused genus with conflicting keys in regional floras, possibly due to high levels of hybridisation (particularly Iberian Peninsula); reliable identification in the field may not be possible.

A. Anthers longer than their filaments, or aborted.

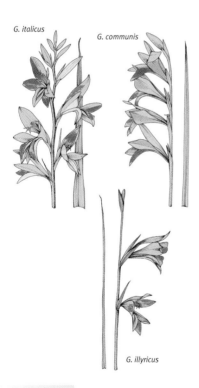

G. italicus

G. communis

G. illyricus

Gladiolus italicus A tall, cormous perennial to 80 cm, with *broad, sword-shaped leaves* 40–80 cm long and 17 mm wide, arising from a rounded corm with netted fibres. Flowers 6–15, borne in a 2-sided inflorescence, each flower with an equally long leafy bract; flowers 40–50 mm long, pink-red, the segments very unequal: the upper longer and broader than the laterals; anthers *longer* than the filaments. Seeds winged. Common on garrigue, maquis, roadsides and cultivated land. Throughout.

B. Anthers equalling, or shorter than, their filaments.

Gladiolus communis A perennial 50 cm–1 m, similar to *G. italicus* in form with leaves 30–70 cm long and 5–22 mm wide. Flowers bright pink, borne in a lax 10–20-flowered weakly 2-sided inflorescence. Anthers spear-shaped, and the *same length or shorter than their filaments*. Seeds winged. Stony fields, woods, maquis and cultivated places. ES, FR, IT, PT. *Gladiolus illyricus* has traditionally been differentiated from *G. communis* in being shorter, 25–50 cm high with leaves 10–40 cm long and 4–10 mm wide and with fewer (3–10) flowers. Recorded throughout, but its distinction from the previous species remains unclear.

Watsonia

Herbaceous perennials with corms and Iris-like leaves. Flowers borne in erect spikes, slightly zygomorphic; style slender with 3 split branches. Native to South Africa.

Watsonia meriana WATSONIA A clumped cormous perennial 50 cm–2 m high with broad, pointed, rush-like pale green leaves 25 cm–1 m long and 15–50 mm wide. Flowers orange-red with a long tubular throat, borne in erect racemes. Ditches and wet grassland. Naturalised in Iberian Peninsula. ES, PT.

XANTHORRHOEACEAE

A family of many genera formerly included in the Liliaceae in its broader, traditional description, characterised by dense tufts of long narrow leaves and stout, woody spikes of flowers; perianth of 6 tepals, often conspicuous; stamens 6; style 1 with minute or 3-lobed stigma. Fruit a capsule.

Aloe

Succulent shrubs with robust, spiny-margined leaves forming a rosette. Flowers in racemes; corolla tubular, often orange or red. Fruit a capsule. Native to Africa and Madagascar but widely cultivated in the region.

A. perfoliata

Aloe arborescens TREE ALOE A *large, much-branched shrub* to 2(3) m. Leaves succulent and narrowly triangular-lanceolate, 50–60 mm long, crowded in large rosettes and deflexed-spreading; greyish without pale markings, and toothed along the margin. Inflorescence unbranched and dense; flowers scarlet, 35–45 mm long. Native to southern Africa, frequently cultivated, naturalised but rarely far from habitation. ES, FR, PT. **Aloe perfoliata** (Syn. *A. mitriformis*) is similar but ground-hugging with weakly spiralled leaves and erect racemes of red-orange tubular flowers. Planted in the region; naturalised on the coasts of the southern Iberian Peninsula, possibly elsewhere. ES, PT.

Aloe vera A stoloniferous perennial succulent with numerous basal leaf rosettes with grey-bluish or reddish-tinged leaves to 40–50 cm long. Inflorescences to 1 m, with *yellow flowers* 25–30 mm long with protruding stamens. Planted and locally naturalised. Throughout (mainly southern). *Aloe brevifolia* is similar to *A. vera* but with oval-lanceolate leaves with raised protuberances on the underside, and unbranched inflorescences with orange-red flowers. Naturalised. FR (south).

1. *Aloe vera*
2. *Aloe perfoliata*
3. *Aloe arborescens*

Aloe aristata A non-stoloniferous perennial with rather small leaves to 10 cm long; leaves erect or incurved and dark green-purple, *covered in white, raised protuberances*, and a soft, white terminal spine. Flowers orange-red, 30–40 mm long. Naturalised. FR (south).

Aloe succotrina A large succulent shrub to 2 m, branched or not with triangular-lanceolate leaves to 50 cm long; leaves *grey-green* (sometimes with some white spots), the margins with dark-purplish teeth. Flowers pink-red, to 40 mm long, borne in tall, unbranched racemes. Naturalised on coasts. ES, FR.

Simethis KERRY LILY

Rhizomatous perennials with leaves all basal. Flowers borne in panicles; *filaments woolly* and inserted into a pit in the anther.

Simethis mattiazzii (Syn. *S. planifolia*) KERRY LILLY A hairless rhizomatous perennial to 40 cm with flopping, leek-like, linear, bluish leaves to 50 cm x 7.5 mm. Flowers white, purplish outside, borne in lax, few-flowered panicles; *stamens markedly woolly*. Seeds black and shiny. Garrigue and maquis, often on acid soils; local but widespread. Throughout.

Asphodelus ASPHODEL

Robust, hairless, herbaceous and tuberous perennial (rarely annual) plants with leafless stems and linear leaves in basal tufts. Tepals free, all similar; flowers borne on simple or branched stems; filaments hairless. Many species very similar; fruit shape is an important diagnostic.

A. Leaves flat. Flower-stalks *thickened in fruit* (1–1.3 mm wide).

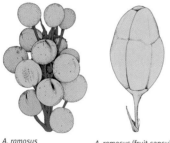

A. ramosus   A. ramosus (fruit capsule)

Asphodelus ramosus A tall, robust herbaceous perennial 50 cm–1.6(2) m tall with numerous fleshy roots. Leaves flat, strap-shaped and grey-green. Flowers white and star-like, with tepals 11–20 mm long, with brownish mid-veins, borne in tall, erect, much-branched inflorescences in which the *lateral branches are almost as long as the terminal. Capsules ovoid, and large*, 6–12 x 4–9 mm. Common in the far west, in stony pastures and goat-grazed maquis; more local elsewhere. ES, FR, IT, MA, PT.

B. Leaves flat and keeled. Flower-stalks *not* thickened in fruit (<1 mm wide).

A. aestivus

Asphodelus aestivus A tall, robust, spring-flowering herbaceous perennial 70 cm–1.8(2) m tall with numerous fleshy roots. Leaves flat, strap-shaped, keeled beneath and grey-green. Flowers white and star-like, with tepals 14–19 mm long, borne in tall, erect, much-branched inflorescences; *filaments much-exceeding the tepals*; the lateral branches almost as long as the central raceme; bracts whitish. Capsules *small and round-egg-shaped, scarcely narrowed at the base*, 5–6 x 6–7 mm. Common and widespread in a range of habitats including maquis, rocky slopes and waste places. Throughout. *Asphodelus serotinus* is similar with rounded *fruits surrounded at the base by the top of the fruit stalk*. ES, PT.

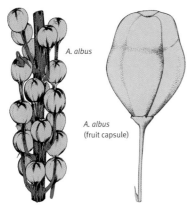

A. albus

A. albus
(fruit capsule)

Asphodelus albus WHITE ASPHODEL A medium to tall perennial 60 cm–1.5 m tall, similar to the previous species. Leaves linear and grey-green, with a central keel, 60 cm–1.1 m x 25 mm. Inflorescence a *narrow, spike-like raceme*, rarely branched at the base, and usually *unbranched* above or with few, short lateral branches; flowers white, to 30 mm across, the tepals with greenish mid-veins. Capsules 7–13 x 6–11 mm, borne densely along the spike. Locally common on rocky ground and grazed maquis. Spain eastwards but absent from most islands. ES (incl. Balearic Islands), FR, IT (incl. Sardinia). *Asphodelus macrocarpus* is similar but has *large, spherical, orange to brown capsules 10–18 x 12–18 mm* borne on erect, narrow stalks with dark brown bracts. ES, FR, IT, MA, PT.

Asphodelus cerasiferus A tall, hairless perennial with smooth stems 80 cm–2 m tall. Leaves flat and keeled, to 90 cm long and 8–25 mm wide. Flowers white, borne in simple or branched racemes. Fruits *very large, 10–20 x 10–20 mm*, rounded and grooved, brown-yellow or orange when ripe. DZ, ES, FR, IT, MA.

1. *Simethis mattiazzii*
2. *Asphodelus aestivus*
3. *Asphodelus ramosus*
   (inset: flowers)
4. *Asphodelus ramosus*
   (fruits)

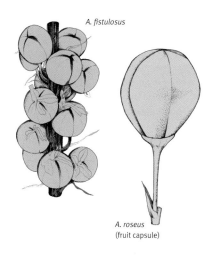

A. fistulosus

A. roseus
(fruit capsule)

C. Leaves hollow.

Asphodelus fistulosus HOLLOW-LEAVED ASPHODEL A small, tufted *annual* or short-lived perennial 30–60 cm (1.5 m) with numerous fleshy roots. *Leaves linear and slender, 1–35 mm wide, hollow, and cylindrical.* Flowers white, borne in a lax raceme of 10–15 on a long, smooth stalk with membranous, whitish bracts; tepals 8.5–12.5 mm, with brownish mid-veins. Capsule 5–6 x 4–6 mm. Common in grazed pastures, stony ground and waste places across the region. Throughout. *Asphodelus roseus* is similar but has *pale pink flowers*. ES (south), MA. *Asphodelus tenuifolius* is similar but with fibrous roots, narrower leaves to 2.5 mm and smaller flowers with tepals 5.5–7.5(8) mm. Capsule 3–4 mm long. Arid habitats, sometimes abundant; common in North Africa (rare in Europe). DZ, IT (south + Sicily), MA, TN.

Asphodeline

Similar to *Asphodelus* but rhizomatous (not tuberous) and flowers borne on *leafy stems*.

Asphodeline lutea YELLOW ASPHODEL A medium to tall herbaceous perennial to 1 m with fleshy rhizomatous roots. Leaves linear and blue-green, 2–3 mm wide, untoothed, and *ascending the flowering stems* (not all basal). *Flowers yellow,* to 40 mm across, borne in a dense raceme; tepals with green mid-veins; stamens 6, of 2 different lengths, the longer with curved filaments. Capsule to 12 mm across. Rocky slopes, garrigue and mountain slopes. IT (incl. Sicily).

1

2

3

4

AMARYLLIDACEAE

Bulbous or rhizomatous (or cormous) perennials with leafless stems, similar to Liliaceae. Flowers solitary or in umbels, enclosed in papery bracts before flowering; tepals 6, usually all petal-like; stamens 6; style 1, ovary 3-parted. Fruit a 3-parted capsule, often succulent.

Sternbergia WINTER DAFFODIL

Bulbous perennials, not smelling of garlic. Flowers with funnel-shaped, crocus-like perianth with 6 yellow, all similar tepals.

Sternbergia lutea WINTER DAFFODIL An autumn-flowering bulbous perennial with strap-like leaves 70 mm–10 cm x 4–15 mm which appear before or during flowering. *Flowers crocus-like, bright yellow* and held erect on stems 25 mm–10(20) cm; flowers short-tubed with 6 oval and pointed tepals; stamens 6, the filaments greatly exceeding their anthers. Rocky maquis. Locally common, but distribution scattered; absent from France, the far west, and much of northern Africa. DZ, ES, IT, TN. *Sternbergia colchiciflora* is similar but with much *narrower leaves* (2–5 mm) that appear *after flowering*. Flower stems just 10–20 mm long, largely concealed by their bulb tunics; flowers bright yellow, funnel-shaped with linear tepals, broadest above the middle. Stony and rocky pastures at high altitude. Spain eastwards, absent from most islands. ES, FR, IT, MA, TN.

Acis

Bulbous perennials with *solid flower stalks and narrow leaves*. Flowers borne in umbels of 1–5 flowers; corolla bell-shaped with 6 tepals, pink or white; anthers greatly exceed their filaments. Recently separated from *Leucojum*.

A. Flowering in late summer to autumn.

A. autumnalis

Acis autumnalis (Syn. *Leucojum autumnale*) AUTUMN SNOWFLAKE A *late summer flowering*, low to short, delicate bulbous perennial 11–30 cm high with thread-like *leaves appearing after flowering*. Flowers borne on scapes of 1–2(4); white, tinged with pink at the base (rarely all pink), bell-shaped, 8–12 mm long, the outer tepals 3-toothed at the tips; spathes with 2 valves. Rocky slopes and hillsides; widespread across the region, absent from many islands. ES, IT (incl. Sardinia and Sicily), MA, PT. *Acis rosea* (Syn. *Leucojum roseum*) is similar but usually with solitary flowers (rarely 2 or 3), *pink* with pointed tepals and 1-valved spathes. FR (Corsica), IT (Sardinia).

---

1. *Asphodelus fistulosus*
2. *Asphodelus macrocarpus*
3. *Asphodelus tenuifolius*
4. *Asphodeline lutea*

B. Flowering in spring to early summer.

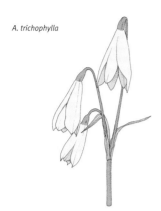

A. trichophylla

Acis trichophylla (Syn. *Leucojum trichophyllum*) THREE-LEAVED SNOWFLAKE A small, slender, winter to spring-flowering bulbous perennial 14–38 cm high. Leaves usually 3, linear and long and narrow to 18 cm long and 1 mm wide, present at flowering time. Flowers white, slightly pink-tinged, 13–19(20) mm long and bell-shaped, borne in clusters of 2–4 on slender stalks 23–46 mm long; outer tepals pointed; style thread-like and slightly longer than the stamens; spathe with 2 valves. Widespread in sandy habitats, particularly on dunes. ES, MA, PT. *Acis longifolia* (Syn. *Leucojum longifolium*) is similar, but with all-white flowers, *small*, to just 11 mm long with blunt (not pointed) tepals, and style slightly *shorter* than the stamens. Rocky slopes. FR (Corsica).

Acis nicaeensis (Syn. *Leucojum nicaeensis*) FRENCH SNOWFLAKE A small, slender, winter to spring-flowering bulbous perennial. Leaves linear, and well-developed at flowering time. Flowers white and widely bell-shaped with pointed outer sepals, to 1 mm long, usually soliatary (rarely 2–3) with styles slightly longer than the stamens; spathe with 2 valves; ovary disk-like and 6-parted. Maritime Alps. FR (southeast). *Acis valentina* (Syn. *Leucojum valentinum*) is similar but with very narrow, *thread-like* leaves 1.5–2 mm wide which appear *after flowering* in the autumn; flowers borne in spring by which time leaves have withered; flowers with a promiment 6-lobed disk at the base of the perianth. Rocky pastures in Valencia. ES (east).

Leucojum

Bulbous perennials with *hollow stalks and broad leaves*. Flowers borne in umbels of 1–5 flowers; corolla bell-shaped with 6 tepals, pink or white; anthers greatly exceed their filaments.

Leucojum aestivum SUMMER SNOWFLAKE A *robust, clump-forming*, bulbous, spring-summer perennial 31–61 cm high with strap-shaped leaves 50 cm x 5–15 mm. *Flowering stems flattened, twisted and 2-winged,* with up to 8 flowers 11–13(15) mm long; corolla bell-shaped and white, each tepal with a green spot at the tip. River banks, ditches and other damp or wet habitats. FR, IT.

1. *Acis trichophylla*
2. *Acis trichophylla*

1

2

Narcissus DAFFODIL

Bulbous, usually early-blooming perennials with basal leaves and hollow, leafless scapes. Flowers with 6 tepals; stamens 6, surrounded by a cup- or trumpet-like corona. Capsule 3-parted. A difficult genus, often with high variability and hybridisation, in which species limits are much disputed.

A. Flowers with white or green tepals, flowering in autumn.

N. serotinus

Narcissus serotinus A delicate, bluish-green, bulbous perennial with stems 11–29 cm with 1 flower, and very *narrow*, cylindrical, rush-like leaves 0.7–1.3 mm wide appearing in spring, and *absent when in flower* in the autumn; flowers fragrant, to 30 mm across with white, spreading tepals and a *very small*, orange corona to just 2 mm long; spathes membranous. Garrigue, pastures and hillsides. Widespread but absent from France. DZ, ES, IT, MA, PT, TN.

Narcissus elegans A delicate, bulbous perennial similar to *N. serotinus* but with, *broader, channelled leaves 1.4–5.2 mm wide* appearing before or *with the flowers*; flowers borne in *clusters of 2–9*; fragrant, to 38 mm across with white, spreading tepals and a small, orange corona 0.6–2 mm long; spathes membranous. Dry, rocky habitats. ES (Balearic Islands), IT (incl. Sicily).

Narcissus viridiflorus GREEN-FLOWERED NARCISSUS A short, delicate, bulbous perennial 17–47 cm high with rush-like, hollow leaves 1.4–5 mm wide, often borne singly, appearing after the flowers. Flowers distinctive for being dull *olive green*, to 30 mm across, slightly foetid, borne in clusters of 2–5; tepals narrow-oblong, pointed and reflexed; corona shallow and 6-lobed. Damp and sandy habitats, often coastal. ES (southwest), MA.

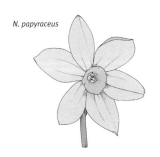

N. papyraceus

Narcissus papyraceus PAPERWHITE NARCISSUS A hairless, bulbous perennial with broad, strap-like leaves, blue-green and slightly channelled, 6.9–9(13) mm wide. *Flowers pure white*, borne in umbels of 4–11(20); flowers to 40 mm and strongly fragrant. Dry rocky places and fallow ground. Throughout except for the Balearic Islands, Corsica and Sardinia.

B. Flowers with cream, white or green tepals, flowering in spring.

N. tazetta

Narcissus tazetta BUNCH-FLOWERED DAFFODIL A medium, hairless, bulbous perennial with stems 23–60 cm and blue-green flat, broad, strap-like leaves 6.2–12 mm wide, equalling the stems, present when in flower. Flowers borne in umbels of 3-many (15); fragrant, white (sometimes cream to yellow), with a prominent yellow corona 3–6 mm long that is 2 x as wide as high, on a 2-edged, flattened scape. Spathe at the base of the inflorescence membranous. Meadows and pastures. Throughout, except for mainland Italy and Sicily.

Narcissus poeticus PHEASANT'S-EYE DAFFODIL A medium, hairless, bulbous perennial with stems to 60 cm and with blue-green flat, linear leaves 5–9(13) mm wide, equalling the stems, present when in flower. Flowers rather large and showy, *normally solitary*, nodding and scented; corolla white with a greenish tube; tube to 30 mm long, tepals wide-spreading, to 22–38 mm long; corona to 2.3–5.2 mm long, yellow with a *bright red, frilly rim* and 3 stamens protruding. A high-altitude species, locally abundant in deciduous woods and mountain pastures. Absent from most islands. ES, FR, IT.

Narcissus cantabricus A similar species to the more common, yellow-flowered *N. bulbocodium*: A delicate, low to short bulbous perennial, often clumped. Leaves linear, 0.8–1(1.2) mm wide, often borne singly per bulb. Flowers white, borne singly and horizontally at the end of the scape with linear-lanceolate, reflexed tepals and a *large and prominent corona* that is cone-shaped, 8–17 mm long and flared and toothed at the rim. Garrigue and grassland. ES (incl. Balearic Islands), MA.

Narcissus triandrus ANGEL'S TEARS A spring-flowering, low to short bulbous perennial to 30 cm with dark green, flattish to semi-cylindrical leaves to 3 mm wide, keeled beneath. Flowers *distinctly downwardly nodding*, in groups of 2–8(15), white to pale cream; corona yellow and cup-shaped; tepals *strongly reflexed* and therefore parallel with the ovary. Scrub, screes and hillsides in the southern Iberian Peninsula and Pyrenees. ES, PT.

1. *Narcissus bulbocodium*
2. *Narcissus bulbocodium*
3. *Narcissus gaditanus*

C. Flowers with yellow tepals, the corona much larger than the tepals.

Narcissus bulbocodium HOOP PETTICOAT NARCISSUS A short, tufted perennial 80 mm–35 cm tall. Leaves 2–4, more or less upright, semi-cylindrical, 1.6–2.3 mm wide and dark green. Flowers held horizontally and with a *large, broadly cone-shaped corona*, 8.4–24 mm long and bright yellow; tepals erect-spreading, and pointed; stamens inserted (not exceeding the corona); flowers borne on stalks slightly exceeding the leaves. Common in rocky places, coastal scrub and other dry open habitats (also in mountain pastures). ES, PT.

D. Flowers with yellow tepals, the corona of similar length to tepals.

Narcissus pseudonarcissus WILD DAFFODIL A medium, clump-forming perennial 14–76(95) cm high with many flattened, strap-shaped light green leaves 6–9(17) mm wide. Flowers with spreading pale yellow tepals and an orange-yellow corona with a frilly edge, 22–30 mm long, as long as the tepals. Meadows and hillsides and cultivated. ES, FR, IT.

E. Flowers with yellow tepals, the corona shorter than the tepals.

N. assoanus

Narcissus gaditanus A small bulbous perennial 90 mm–32 cm with numerous leaves greatly exceeding the flowering scapes. Flowers bright yellow, usually borne in pairs (4–10) and *very small* to 20 mm across with a *curved tube* and cup-like corona 3–4 mm long and just 6–10 mm wide, and *large, backward-pointing tepals* 5–6 mm long; flower stalks *shorter* than their spathes when mature. Open slopes, screes and coastal cliffs. ES, PT. *Narcissus assoanus* (Syn. *N. willkommii*) is very similar but with larger flowers to 24 mm across with tepals restricted at the bases and *flower stalks exceeding their spathes* when mature. Local in damp grassland. ES, FR, PT. *Narcissus cuatrecasasii* is similar to *N. gaditanus* but with *leaves more or less flat* (not semi-cylindrical), to 2.6–3.7 mm wide, with *2 keels* running along the length. Flowers yellow with spreading tepals, each ending in a point. Rocky habitats. ES (centre and south).

Narcissus jonquilla JONQUIL A delicate, often clumped, bulbous perennial with linear leaves 2–4 mm wide which are deep green, semi-cylindrical and slightly channelled. Flowers borne in clusters of 2–5, golden yellow and strongly sweetly scented, to 32 mm across, the corolla tube to 30 mm long, the corona to 4 mm long. Damp meadows. Native in the far west, naturalised in France and Italy. ES, FR, MA, PT.

F. Flowers with yellow tepals, the corona reduced to scales.

Narcissus cavanillesii (Syn. *N. humilis*) TAPEINANTHUS A short, slender, bulbous perennial 80 mm–18 cm with rush-like leaves 0.7–1.1 mm wide, often borne singly. Flowers yellow, to 22 mm across, solitary or paired with narrow and pointed tepals; corona very shallow, formed by *6 scale-like appendages*; stamens long-protruding. Open woods and grassland. ES.

Pancratium SEA DAFFODIL

Bulbous perennials with flowers borne on long-stemmed, terminal umbels; corolla funnel-shaped with 6 narrow tepals and a corona with 12 teeth.

P. maritimum

P. foetidum

P. illyricum

Pancratium maritimum SEA DAFFODIL A distinctive, summer-flowering short to medium, clump-forming, bulbous perennial. Leaves fleshy, grey-green, hairless and long 16–37(75) cm. Flowers large and white, to 15 cm long, the perianth tube *<2 x as long as the segments;* fragrant, borne in umbels of 4–9(19); tepals all similar; stamens 6, borne on the rim of the cone-like corona. Fruit a 3-parted capsule. Characteristic of the Mediterranean dune flora. Throughout. *Pancratium trianthum* is similar but the perianth *larger*: 16–22 cm long, the *tube 2–3 x as long as the tepals*. North African coasts. DZ, MA, TN. *Pancratium foetidum is similar to P. maritimum, with very strongly fragrant flowers,* borne rather few, on solitary stems or in small clusters before the emergence of the leaves. Coastal rocks, woodlands and Scrub. DZ, MA, MT, TN.

Pancratium Illyricum ILLYRIAN SEA-LILY A medium, clumped, summer-flowering bulbous perennial with stems to 45 cm long, and numerous basal, strap-shaped, blue-green leaves 15–30 mm wide. Flowers white, fragrant, 60–80 mm long, each with a straight, greenish corolla tube about a $^1/_4$ the length of the flowers; corona short, deeply divided into 6 lobes, each with 2 teeth (thus appearing 12-toothed), widening abruptly at the mouth. Rocky habitats, often coastal. FR (Corsica), IT (Capri and Sardinia).

Allium

Distinctive bulbous perennials smelling of onion or garlic when crushed. Flowers borne in terminal, often spherical umbels; tepals 6 all similar; stamens 6. Fruit a 3-parted capsule. Some species produce bulbils in the flower-heads.

A. Innermost *stamens with 3-parted filaments* (filaments with teeth).

A. ampeloprasum

Allium ampeloprasum WILD LEEK *A large and robust, leek-like* bulbous perennial with a membranous bulb, with stout flowering stems to 1(2) m tall. Leaves 4–10, pale blue-green, V-shaped in cross-section (at least below), rough along the margins and with a central keel beneath; leaves often withered when in flower. Flower-heads large and spherical with numerous lilac flowers with yellow anthers, borne on long stalks; innermost stamens with 3-parted filaments, the *lateral teeth greatly exceeding the central (fertile) part*; spathe usually soon-falling. Common, particularly near the coast. Throughout. *Allium baeticum* is similar but with a fibrous (not membranous) bulb tunic that extends to the base of the stem, and innermost stamens with 3-parted filaments, the lateral teeth only just exceeding the central (fertile) part. DZ, ES, MA, PT, TN. *Allium atroviolaceum* is similar to *A. ampeloprasum* but less robust and with *dark purple flowers* borne in smaller umbels 30–60 mm across. Dry fields. A primarily eastern Mediterranean species. IT.

Allium commutatum A leek-like bulbous perennial 40 cm–1.2 m high, similar to the previous species group. *Leaves many:* 6–12, *not rough* along the margins. Flowers whitish-pink or dull purple, with greenish or purplish keels along the tepals, to 3.3–4.2 mm long; *margin of inner tepals finely-toothed.* Coastal sands and cliffs. DZ (islands only), ES (Balearic Islands), FR (Corsica), IT (incl. Sardinia), TN (islands only).

Allium sphaerocephalon ROUND-HEADED LEEK A medium to tall bulbous perennial 20–70(90)cm high with flower scapes round in cross-section. Leaves 2–4(5) semi-cylindrical with a groove along the surface, sheathing the stem below, 1–2(3) mm wide; spathes 2-valved, and shorter than the umbels. Flowers dark purple-red (rarely pale), each tepal with a greenish keel, borne in *dense, spherical heads* to 40 mm across; stamens greatly protruding; lateral teeth of 3-parted filaments only just *equalling* the central (fertile) part. Common in a range of dry habitats more or less throughout (rare in far west). ES, FR, IT, MA, PT, TN. *Allium guttatum* is similar but with often, sparser heads of paler *greenish or whitish* flowers, in which the lateral teeth of the 3-parted filaments *much exceed* the central (fertile) parts. Throughout, except the Balearic Islands, France and Corsica.

B. Innermost stamens entire (without teeth) and leaves flat.

1. *Pancratium maritimum*
2. *Allium ampeloprasum*
3. *Allium commutatum*
4. *Allium sphearocephalon*
5. *Allium chamaemoly*

Allium chamaemoly A *very small*, bulbous perennial just 10–85 mm tall. Leaves 4–8, *spreading flat on the ground*, sheathing only the very base of the flower scape, linear with hairy margins; spathes 1-valved and 2–4 lobed. Flowers white and star-like, 10–16 mm across, borne in umbels of 7–18; tepals with greenish mid-veins. Flower stalks *curving in fruit.* Dry, coastal habitats. ES (incl. Balearic Islands), IT (incl. Sicily), MA, PT.

A. trifoliatum

Allium subvillosum A small bulbous perennial 11–45(65) cm tall. Leaves 2–5, 11–60 cm long, straight, narrowly linear with *hairy margins,* almost all basal. Stem cylindrical. Flowers white and cup-shaped, borne in hemispherical, many-flowered umbels; *perianth completely white* with yellow *anthers slightly protruding*; spathe 1-valved and persistent. Flower stalks *upright in fruit.* Common on maritime sands in the far west. ES, MA, PT. *Allium subhirsutum* is very similar, but with a smooth (not pitted) bulb tunic, and stamens *not protruding.* Maritime sands. One of the most common white-flowered species of *Allium* in the region (rare in far west). Throughout. *Allium trifoliatum* is similar to the previous 2 species with umbels 25–40 mm across with flower stalks to 20 mm long; *flowers pink or pink-keeled*, with narrow, pointed stamens ¹/₂ as long as the petals; spathe equal to the flower stalk. Fields and hillsides from France eastwards. FR, IT, MT.

C. Innermost stamens entire (without teeth) and leaves flat or folded; flowers *yellow.*

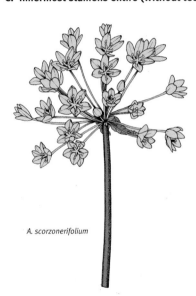

A. scorzonerifolium

Allium scorzonerifolium A medium bulbous perennial 14–36(58) cm tall. Leaves linear and sub-adpressed to the stem, 3–4(14) mm wide; stem angular. Spathe papery. Flowers *bright yellow*, rather long-stalked, borne in lax umbels; *ovary crested.* Rare. ES (south). *Allium moly* is similar, but with *wider* elliptic (not linear) leaves 0.8–2.8(41) mm wide, not adpressed to the stems, and stems cylindrical, not angular. *Ovary not crested.* ES, FR.

D. Innermost stamens entire (without teeth) and leaves flat or folded; flowers *white to pink.*

Allium triquetrum THREE-CORNERED LEEK A short to medium bulbous perennial 17–59 cm. Scapes triangular in cross-section. Leaves 2–3(5), linear and tapered with a distinct keel, 6–9(16) mm wide. Flowers *white*, the tepals with greenish mid-veins, bell-shaped, 10–18 mm long, nodding, borne in *1-sided umbels* with only 5–15 flowers. Damp woodland, thickets and riverbanks, also cultivated in gardens, more or less throughout the European Mediterranean. ES, FR, IT, PT. *Allium pendulinum* is similar, also with triangular stalks which are shorter, to 25 cm, with leaves 3–8 mm wide, and 2–9 *smaller* flowers (just 3–5 mm long), *initially erect*, later all nodding; tepals each with 3 green veins; umbels not 1-sided. FR (Corsica), IT (incl. Sardinia and Sicily).

1. *Allium subvillosum*
2. *Allium roseum*
3. *Allium massaessylum*
4. *Allium schoenoprasum*
5. *Allium vineale*
6. *Allium vineale*

A. neapolitanum

Allium neapolitanum NAPLES GARLIC A short to medium bulbous perennial to 50 cm tall. Leaves few (2–3), linear, 11–24(36) mm wide, and sheathing the stem at the base and keeled along the back. Flower stems triangular in cross-section and smooth, carrying white flowers in loose clusters to 70 mm across; *spathe 1-valved* and shorter than the flower stalks. Locally common in grassy habitats and pastures. Widespread. Throughout. **Allium massaessylum** is distinguished by its narrower, flat leaves 3.7–5(6) mm wide, more or less cylindrical stems, *2-valved spathe* that exceeds the flower stalks, and white flowers, the *tepals white with a purple keel*. ES, MA, PT. *Allium trifoliatum* is similar to *A. neapolitanum* but with leaves and sheaths *hairy*, at least along the margins, cylindrical stems and white flowers with pink mid veins along the tepals, or flushed pink throughout. Rocky habitats. IT (incl. Sardinia and Sicily).

Allium nigrum A medium to tall bulbous perennial, difficult to distinguish from *A. neapolitanum*. Leaves 2–5, all basal, *broadly linear*, to 24–33(70) mm wide, and tapered to a point, much exceeded by the scapes which are rounded in cross-section. *Spathes 2–3(4)-valved* with pointed tips, free to the base. Flowers 30–90, white or very pale lilac with green mid-veins, and *blackish ovaries*, borne in dense, spherical umbels; anthers yellow. Various dry, grassy habitats. Throughout, though rare or absent in parts of the far west. *Allium cyrillii* is similar but with tepals narrow (1–1.5 mm not 1.5–3 mm wide) with incurved tips. IT (south).

A. roseum

Allium roseum ROSY GARLIC A medium, bulbous, hairless perennial 40–65(85) cm with 2–4, linear, keeled leaves, 4–12 mm wide, similar to *A. neapolitanum* in most respects but with *cylindrical* (not angled) stems and pale *pink*, bell-shaped flowers to 12 mm long, borne in loose umbels that are sometimes mixed with bulbils; stamens not protruding; *spathe with 2–4 valves, fused at the base.* A range of dry habitats; common. Throughout. *Allium lineare* (Syn. *A. confertum*) is similar to *A. roseum*, also with pale pink flowers, in umbels to 30 mm across on short stems (<15 cm), and *small flowers 5–7 mm long*. FR, IT (incl. Sardinia and Sicily).

E. Innermost stamens entire (without teeth) and leaves *cylindrical, subcylindrical or filiform.*

A. paniculatum

A. vineale

Allium paniculatum (Syn. *A. pallens*) A small, bulbous perennial 26–73 cm tall. Leaves 2–4, 40 mm–20 cm long, semi-cylindrical and *very narrow, 0.9–1.8 mm wide*, sheathing the stem at the base. Flowers borne in compact, hemispherical umbels to 35 mm across; perianth *bell-shaped to cylindrical and open at anthesis*; to 6 mm, white or pale pink, often on drooping stems; stamens included within the perianth or with the yellow anthers slightly protruding; spathe persistent, *2-valved*, and greatly exceeding the flower stalks. Dry, open habitats. ES, FR, MA, PT.

Allium vineale CROW GARLIC A medium, bulbous perennial 22 cm–1 m with rounded (not angular) scapes. Leaves 2–4(6), cylindrical, channelled and sheathing the base of the scape. Spathe 1-valved and beaked, papery, *not longer than the flowers* and soon-falling; flowers reddish or greenish and bell-shaped, 3–4 mm long, often few: *most or all of the flowers replaced by bulbils, often sprouting green shoots before falling*; stamens of flowers *protruding*. Fairly common in dry grassy habitats. Throughout.

Allium amethystinum A rather robust, bulbous perennial to 1.2 m. Leaves 3–7, narrowly-cylindrical and sheathing the lower part of the stem, often *withered* by flowering time. Flower stalks reddish above, cylindrical in cross section, with purple, tubular-bell-shaped flowers, to 4.5 mm long, borne in *large, regular* spherical to drop-shaped umbels to 50 mm across; stamens protruding. Rocky and disturbed habitats. IT (incl. Sicily).

Allium schoenoprasum CHIVES A *tufted* bulbous perennial 12–46 cm with many leaves and flowering stems. *Leaves cylindrical and hollow,* grey-green, 10–29 cm long and 1.5–2.6(4) mm wide. Flower stalks cylindrical, with dense, small umbels to 40 mm across, with 7–50 *pink-purple flowers*, and no bulbils; anthers *yellow and not protruding; spathe 2-valved*, shorter than the umbel. Generally mountain-dwelling in rocky habitats, or cultivated as a culinary herb. Throughout.

ASPARAGACEAE

A large and variable family comprising many genera formerly included in an even larger broadly defined family Liliaceae. Shrubs or large perennials. Flowers cosexual (rarely dioecious e.g. *Asparagus* and *Ruscus*); perianth with 6 tepals; stamens 6; styles 1–3. Fruit a capsule or berry with 1-numerous seeds.

Asparagus ASPARAGUS

Shrubby, hairless, rhizomatous perennials with tough stems (becoming woody) and with *cladodes* (rather than true leaves). Flowers small and inconspicuous, bell-shaped with 6 tepals. Fruit a small berry.

Asparagus officinalis WILD ASPARAGUS A tall, hairless, *herbaceous* perennial with erect to spreading, smooth (not spiny) stems, to 1.5 (2) m, with or without clusters of small green cladodes in the axils, giving a *feathery appearance*. Flowers very small, to 6 mm long, borne in the leaf axils; unisexual, greenish-yellow and bell-shaped. Fruit a globose, red berry 10 mm across. Various habitats; common. Throughout.

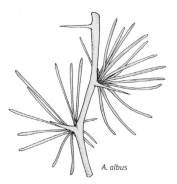

A. albus

Asparagus albus A tall, autumn-flowering perennial 90 cm–1 m, with woody, whitish, flexible stems, smooth or grooved; spines 3–12(16) mm; *cladodes not leaf-like, long, 5–25 mm* soon-falling to leave bare, spiny stems, borne in clusters of 8–30. Flowers borne in late summer, small, 3 mm, and sweetly scented, borne in clusters of 5–15 on stalks 3–7(10) mm long; *perianth white* with 6 spreading tepals. Fruit a berry turning red through black. Strongly southern. DZ, ES (south + Balearic Islands), FR (Corsica), IT (south + Sardinia and Sicily), PT, TN. *Asparagus pastorianus* has larger spines 5–15(20) mm long and cladodes 12–40(70) mm in clusters of 8–30. Maquis in North Africa. MA.

1. *Asparagus officinalis*
2. *Asparagus officinalis*
3. *Asparagus albus* (flowers)
4. *Asparagus albus*
5. *Asparagus pastorianus*

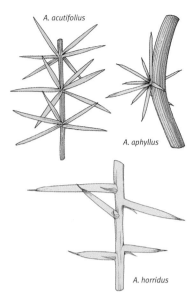

A. acutifolius

A. aphyllus

A. horridus

Asparagus acutifolius A *climbing* subshrub with much-branched, woody stems to 1.5(3) m. *Cladodes more or less equal*, borne in *numerous clusters, 10–35(70)*; *small, just 2–8 mm long*, and *spine-tipped*, to 0.5 mm thick. Flowers borne in clusters of *2–4*, 4 mm long, *pale yellowish-green*, sweetly scented and mixed among the cladodes; flower stalks surrounded by bracts at the base. Berry black when ripe. Throughout.

Asparagus aphyllus is similar to *A. acutifolius* but *shorter, and non-climbing,* to 60 cm (1 m) tall with hairless green stems. *Cladodes markedly unequal,* in clusters of *3–10(15)*, to 20 mm long and 0.6 mm thick. Flowers in groups of *3–8(10)*. Berries black when ripe; flower stalks surrounded by bracts at the base. Rocky scrub and garrigue. Throughout. **Asparagus horridus** (Syn. *A. stipularis*) is similar to *A. acutifolius* but with the flower stalk bases with just *1–2 (not surrounded by) bracts*, or bracts absent; stems with *solitary cladodes (or 2–3) per fascicle*. ES, MA, PT.

1. *Asparagus acutifolius* (flowers)
2. *Asparagus acutifolius* (fruits)
3. *Asparagus horridus* (fruits)
4. *Cordyline australis*
5. *Asparagus horridus* (flowers)

Cordyline

Trees and shrubs with dense, strap- or sword-shaped leaves and numerous white or cream flowers.

Cordyline australis A small tree up to 10 m (usually less) with a stout grey trunk which is simple or branched. Leaves sword-shaped, erect to 1 m long. Flowers white, scented and borne in dense panicles to 1 m long. Native to New Zealand, widely planted in towns and gardens. Throughout.

Ornithogalum STAR-OF-BETHLEHEM

Bulbous, spring-flowering perennials with basal leaves in a rosette. Flowers normally white and star-shaped, borne in racemes; each flower with 1 bract; tepals free and petal-like; stamens 6. Fruit a 3-valved capsule with many seeds.

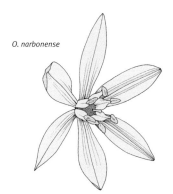

O. narbonense

A. Inflorescence a raceme; leaves without central white mid-veins.

Ornithogalum narbonense A medium, spring-flowering perennial 25–57 cm tall. Leaves 4–6, linear with a sheathing base, to 16 mm wide, and persistent until after flowering. Flowers 20–30 mm wide, white and scentless, borne erect on stalks the same length in a *tall, many-flowered, slender raceme*, with tepals wide and spreading, with a prominent green strip along the back; anthers yellow; ovary to 5 mm with a flattened top, *distinctly exceeded by the narrow style*. Common in rocky pastures; one of the most widespread species in the region. Throughout.

Ornithogalum pyrenaicum BATH ASPARAGUS A medium, spring-flowering perennial to 80 cm (1 m) with slender green stems somewhat asparagus-like in bud. Leaves 5–8, linear and slightly channelled, without a white mid-vein, often withered by flowering-time. Flowers many (>20), *yellow-green*, 18–23 mm across, star-shaped, borne in *long, slender, spike-like racemes,* faintly scented; anthers yellow. Various habitats including woods and scrub; widespread in the European Mediterranean. ES, FR, IT, MA, PT.

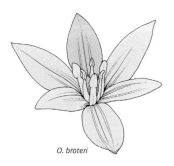

O. broteri

Ornithogalum broteri is easily distinguished by its *single narrow leaf* 4–9(12) mm wide, and few (3–7), white flowers without a green stripe on the back of the tepals, with a yellowish (not blackish) ovary. Local in dry, rocky places. ES, MA, PT. *Ornithogalum reverchonii* A medium bulbous perennial, with bulbs 50–60 mm across, similar to *O. broteri* but with *4–6 leaves*; leaves 50–80 mm long and 15–25 mm wide. Flowers 35–40 mm across, borne in racemes to 60 mm–20 cm long; bracts single and 3–5 x the length of the flower stalks; tepals to 20 mm long, elliptic-lanceolate. ES, MA, PT.

B. Inflorescence a raceme; leaves with white mid-veins.

Ornithogalum nutans DROOPING STAR-OF-BETHLEHEM A perennial to 60 cm with 4–6, strap-like leaves to 15 mm wide. Flowers white, broadly bell-shaped, to 45 mm and nodding, borne in *1-sided racemes*; tepals somewhat recurved, with an external grey-green stripe; bracts longer than their flower stalks. Open woodland, scrub and pastures. Native to the eastern Mediterranean but cultivated and naturalised. ES, FR, IT, PT.

C. Inflorescence *corymb-shaped*; leaves with white mid-veins.

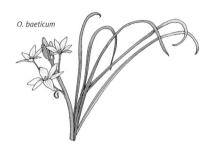

O. baeticum

Ornithogalum baeticum (Syn. *O. umbellatum, O. orthophyllum*) A short, bulbous perennial 30 mm–11(19) cm with leaves to 4–10 mm wide, *with central white stripes 1–3 mm wide; young leaves glaucous*. Inflorescence corymb-like with 4–18 flowers; flowers 30–40 mm wide, white, the tepals with green external stripes. Capsule oblong, *furrowed and hexagonal* in cross-section; *flower-stalks erect in fruit, slightly curved at the apex*; seeds not smooth. DZ, ES, MA, PT. (References to *O. exscapum* and *O. collinum* in Iberian Peninsula probably refer to *O. baeticum*). *Ornithogalum exscapum* has tepals to just 15 mm long, flowers borne on short scapes, and *flower-stalks clearly deflexed in fruit*. Strongly eastern. IT.

Ornithogalum collinum A low to short, spring-flowering perennial to 20 cm. Leaves 4–15, linear, *channelled* with a central white stripe. Flowers white, 20–32 mm, borne in broad, corymbs *at ground level* on very short scapes (<40 mm); tepals white with a bright green stripe on the external surface; anthers yellow or reddish. Garrigue and rocky slopes. FR, IT.

D. Inflorescence corymb-shaped with star-like flowers; leaves with white mid-veins; *fruiting stalks ascending*.

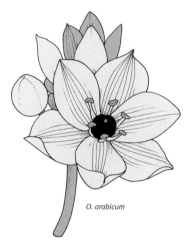

O. arabicum

Ornithogalum arabicum A stout, bulbous perennial 40–60 cm tall with numerous offsets from the main bulb. Leaves 5–8, flat or channelled, plain green and sheathing at the base, 20–45 mm wide. Flowers white or cream, 40–50 mm wide with a *conspicuous violet-black ovary* in the centre, borne in a lax raceme with up to 25 flowers; flowers with broad tepals *without a central green stripe*. Capsule cylindrical. Local in dry, rocky places. Throughout. *Ornithogalum concinnum* is similar but with *2–3 leaves* 10 mm wide. *Flowers very densely arranged at the top of long, slender stalks*. Portugal and west-central Spain only. ES, PT.

1. *Ornithogalum narbonense*   3. *Drimia maritima*
2. *Ornithogalum broteri*   4. *Scilla peruviana*

Drimia SEA SQUILL

Similar to *Ornithogalum* but with very large bulbs and numerous flowers borne in a long, terminal spike, flowering in late summer-autumn.

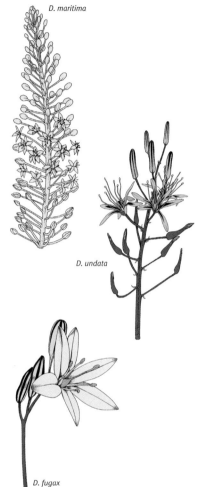

D. maritima

D. undata

D. fugax

A. Leaves >3 mm wide and more or less flat (not semi-cylindrical).

Drimia maritima (Syn. *Urginea maritima*) SEA SQUILL A stout perennial 60 cm–1.2(1.5) m tall with a very large bulb to 6–15(18) cm across, sitting close to the soil surface. Leaves all basal, large and broadly lanceolate, slightly wavy-margined and shiny green, prominent in spring, but disappearing before the flowers appear in late summer; *leaves at least 20 mm wide*. Flowers white and star-like, to 16 mm across, borne in a long, stout, spike-like inflorescence arising erect from the ground. Common in a range of dry habitats including coastal scrub and garrigue; a conspicuous and characteristic plant in the Mediterranean region in late summer. Throughout.

Drimia undata (Syn. *Urginea undulata*) UNDULATE SEA SQUILL A short to medium perennial 25–40(55) cm tall, with a bulb to 17–35(45) mm across. Leaves linear-lanceolate borne in basal tufts, 15 mm wide, wavy-margined, and withering before flowering. Flowers dull pink, greyish or greenish purple, to 26 mm across, borne in lax racemes of up to 30 flowers; tepals with a red mid-vein; styles exceeding the stamens. Rocky habitats. ES, FR (incl. Corsica), IT (incl. Sardinia), MA.

B. Leaves <3 mm wide and semi-cylindrical.

Drimia fugax (Syn. *Urginea fugax*) RED SEA SQUILL A perennial similar to *D. undata*, with a bulb 18–35(50) mm across, but shorter stems 12–32(42) cm tall with thread-like, semi-cylindrical leaves to *to just 1–2 mm wide*. Flowers pale pink, the tepals with red mid-veins. Dry hillsides, similar distribution to the previous species. ES, FR, IT, MA.

Scilla SCILLA

Bulbous perennials with basal leaves and leafless stalks. Flowers star-shaped and usually blue, borne in racemes or solitary, with 0 or *1 bract at the base of each flower stalk*; tepals 6, separate and spreading. Fruit a 3-parted capsule.

A. Basal bracts >4 mm, linear and persistent; leaves appear before flowering.

Scilla peruviana A robust medium, spring-flowering perennial 6–35(40) cm high. Leaves numerous, long and flat, 10–30(46) mm wide and 15–45(60) cm long. Inflorescence large, showy, hemi-spherical, with 30–100 flowers; flowers blue and star-like with yellow anthers (populations with yellowish-brown flowers occur in North Africa); bracts papery and long. Locally common, often in large numbers in damp, grassy habitats at low altitude. Throughout, except for most islands.

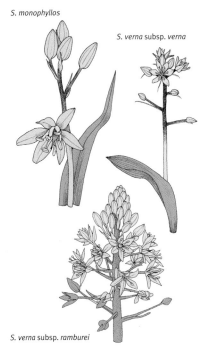

S. monophyllos

S. verna subsp. *verna*

S. verna subsp. *ramburei*

Scilla monophyllos A small, bulbous perennial with stems 60 mm–25(30) cm and a *solitary leaf* 9–15(30) mm wide, linear-lanceolate and sheathing at the stem. Inflorescence a 4–8(12)-flowered raceme becoming lax above; bracts to 7 mm long, exceeded by the flower stalks, at least below; flowers blue and star-like. Local on sandy ground and cliff-tops. ES, MA, PT. *Scilla verna* is similar but with 1-few, linear leaves 2–18 mm wide, and lower *bracts exceeding the flower stalks*. Seasonally wet places and lake margins in Iberian Peninsula and Morocco. ES, MA, PT. Subsp. *verna* (Syn. *S. odorata*) has short inflorescences to just 55 mm, and leaves shorter than the stem, and to just 50 mm wide. ES, PT. Subsp. *ramburei* (Syn. *S. ramburei*) has much larger inflorescences to 40 mm–20(33) cm and leaves to 18 mm wide. ES, MA, PT.

Scilla lilio-hyacinthus PYRENEAN SQUILL A bulbous perennial 15–40 cm high with bulbs consisting of a series of yellowish scales (like those of *Lilium*). Leaves 6–18(22), shiny, flat and rather broad to 20–30(40) mm wide with an abrupt tip. Flowers 5–15(22), blue and star-like, to 15 mm across. Bracts to 10–19(21) mm long and papery, white and tapering. Woods and meadows at high altitude in the Pyrenees. ES, FR.

B. Basal bracts to 3 mm, triangular to oval; leaves appear before or during flowering.

Scilla hyacinthoides A *tall, spring-flowering* bulbous perennial 30–55(60) cm with leaves to 30–50(60) cm long and 8–30(70) mm wide with shortly-hairy margins. *Flowers borne in long racemes of 100 or more blue-violet flowers*. Of eastern origin but naturalised. ES, MA, PT. *Scilla numidica* is similar in form, but *autumn-flowering*, with stems to 14–30(45) cm with *purple-lilac flowers*. ES (Balearic Islands), FR (Corsica), IT (Sardinia), MA.

Prospero

Bulbous perennials previously grouped with *Scilla*, differing in having strap-shaped to filiform leaves, and scapes of flowers appearing without, or adjacent to, the leaves in autumn.

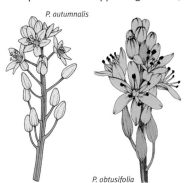

P. autumnalis

P. obtusifolia

Prospero autumnalis (Syn. *Scilla autumnalis*) AUTUMN SQUILL A low to short *autumn-flowering* perennial with stems to 14–30(45) cm, and 5–10(14) hairless, narrowly linear, *channelled* leaves only 1–2(4) mm wide, absent when in flower. Flowers 10–25(66), usually *pink-purple* (sometimes lilac or blue) and star-like, borne in a spike-like raceme; bracts absent. Locally common in a range of dry, rocky and grassy habitats; widespread in the region. Throughout. The following forms are considered by some authors to be distinct: *Prospero corsicum* has just 2–6 blue-violet flowers. FR (Corsica), IT (Sardinia). *Prospero hierae* has 4–12 whitish flowers. IT (Sicilian islets).

Prospero obtusifolia (Syn. *Scilla obtusifolia*) A perennial similar to *P. autumnalis* with just 2–4(5) recurved, *broad* leaves 5–15(25) mm wide that are *not channelled* and have a minutely *hairy margin*. Similar habitats, but local. ES (incl. Balearic Islands), FR (Corisca), IT (Sardinia and Sicily).

Hyacinthoides BLUEBELLS

Bulbous perennials similar to *Scilla* but with *2 papery bracts at the base of each flower.*

A. Spring-flowering; Mediterranean-Atlantic in distribution (except *H. flahaultiana*)

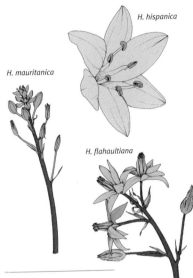

H. hispanica

H. mauritanica

H. flahaultiana

1. *Scilla monophyllos*
2. *Hyacinthoides hispanica*
3. *Hyacinthoides mauritanica*
4. *Dipcadi serotinum*

Hyacinthoides hispanica SPANISH BLUEBELL A bulbous perennial to 35(40) cm with 2–4(8), linear-lanceolate leaves that are normally of length equal to, or shorter than, the flower stalks, 12–36(50) cm long. Perianth bell-shaped and pale blue with spreading lobes; flowers spreading, borne in *rather long racemes*, not scented; anthers bright blue; stamens inserted in the *middle* of the perianth. Frequent in damp, shady woods in Portugal to west and southern Spain. ES, PT.

Hyacinthoides mauritanica A small, bulbous perennial 8–30(36) cm high with spherical bulbs with 3–5 spreading to recurved leaves, *long-tapering*, 11–30(41) cm. Inflorescence with flowers *clustered at the top* (not long and narrow), the flowers *distinctly pointing upwards*, on stalks which elongate in fruit. Small plants with undulate leaves from the Cape St. Vincent area of southwest Portugal have been described as *H. vincentina* but research shows that they do not differ significantly from *H. mauritanica*. MA, PT. *Hyacinthoides flahaultiana* is similar and also has long-pointed leaves, but has *narrowly pyramidal* racemes in which the flower stalks are less erect. Spiny scrub in the Anti-Atlas Mountains. MA.

1

2

3

4

H. reverchonii

H. italica

Hyacinthoides reverchonii A small, bulbous perennial 11–17(38) cm high with pear-shaped bulbs with 3–4(5) long, but rather abruptly pointed leaves 12–15(35) cm with *dark red bases*. Flowers similar to the previous species, star-like and pale blue, but *rather large* - the tepals exceeding 7 mm, with a slight purple-violet tinge; anthers blue. Mountain woodlands of the Sierra Cazorla and adjacent Sierra Segura. ES (southeast).

Hyacinthoides italica A bulbous perennial with narrow leaves to 8 mm wide that have broad (not long-tapering) leaf-tips, and flowers with long tepals (up to 7 mm when fully open) and *stamens inserted at the base* of the perianth (not the middle) with yellow anthers in some populations in the region. Woods and shady, rocky habitats. FR (southeast), IT (northwest).

B. Autumn or winter-flowering; North African mountainous regions.

H. lingulata

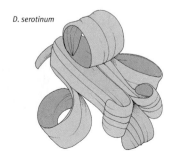

H. ciliolata

Hyacinthoides cedretorum A bulbous perennial with revolute tepals, to which the stamens are attached at their bases (not fused to the tepals along much of their length), and the tepals forming a bell-like shape; anthers and pollen typically violet-blue (populations in the High Atlas mountains have creamy-yellow pollen). Cedar woods in mountain ranges from Morocco to north Algeria. DZ, MA.

Hyacinthoides aristidis A bulbous perennial with *few* (2 or sometimes 3), *broadly linear-lanceolate, abruptly pointed leaves* with a distinct keel along the centre of the undersurface. Flowers very pale blue with contrasting darker blue anthers, opening from the apex of the raceme downwards. DZ, TN. *Hyacinthoides lingulata* An autumn- to winter-flowering bulbous perennial similar to *H. aristidis* but with *many* (>3) shorter and flatter leaves. DZ, MA, TN. *Hyacinthoides ciliolata* is similar to the previous species with a more easterly distribution with flowers opening *from the base* of the raceme upwards. TN (northeast).

Dipcadi

Bulbous perennials with tubular (not star-shaped) flowers, and numerous black, flattened, disk-like seeds.

D. serotinum

Dipcadi serotinum A distinctive, spring-flowering hairless bulbous perennial 20–40(50) cm high. Leaves all basal and linear-lanceolate, 15–35(60) cm long. Inflorescence an erect, lax raceme of 6–20(23) flowers facing more or less 1 way; *perianth brownish orange*; bracts exceeding their flower stalks. Easily distinguishable by its flower colour. Locally common in rocky and sandy habitats in the far west; rare in France and Italy and absent from many islands. ES (incl. Balearic Islands), FR, IT, MA, PT. var. *fulvum* (Syn. *D. fulvum*) flowers in the autumn, and is sometimes treated as a distinct species. ES, MA.

Bellevalia

Bulbous perennials with lax spikes of dull-coloured flowers; corolla with 6 deeply divided teeth; bell-shaped and tubular but not constricted into a throat. Fruit a 3-parted capsule.

A. Flowers bright blue or violet in bud, <10 mm long.

Bellevalia dubia A small, bulbous perennial 13–35(40) cm tall with 4–5(6) leaves. Leaves held erect or recurved, equalling or slightly exceeding the flower stalks, linear and tapered towards the tip. Inflorescences cylindrical with 15–30(40) rather long-stalked (4–8 mm) flowers; perianth to 5–8 mm, bell-shaped, and *dark blue* when open, the lobes with a whitish margin. Various grassy and scrubby habitats. IT (incl. Sicily). *Bellevalia hackelii* (Syn. *B. dubia* subsp. *hackelii*) is widely treated as a subspecies of the former and differs chiefly in having horizontal to recurved flower stalks (not erect-spreading), and more intensely coloured, less greenish flowers. Common in dry, open and grassy habitats in the Algarve. PT.

B. Flowers dull purple in bud, ≥10 mm long.

Bellevalia ciliata A small, bulbous perennial with linear-lanceolate leaves with *long-hairy margins*. Flowers blue-violet in bud, with greenish teeth, turning dull brown as they open, borne in broad, *conical-pyramidal flower-heads* (not cylindrical) of up to 50 flowers; flowers to 11 mm long, borne on *long stalks*, 30–40 mm long. Cultivated land and field margins in southern Italy, also naturalised in France. FR, IT. *Bellevalia trifoliata* is similar but has *cylindrical* flower-heads of *long-tubular* flowers (to 16 mm) which are violet-purple in bud, and pale yellow-brown at the apex when fully open, borne on *short stalks*, under 10 mm long. Predominantly eastern Mediterranean. FR (south), IT (Tuscany).

C. Flowers dull white in bud.

1. *Bellevalia hackelii*
2. *Muscari neglectum*
3. *Leopoldia comosa*

Bellevalia romana A small to medium bulbous perennial with stems to 35 cm, exceeded by the leaves; leaves 3–6, linear-lanceolate, 5–15 mm wide and with smooth margins. Flowers 20–30, borne in lax, cylindrical racemes on spreading to erect stalks to 20 mm long; perianth *whitish*, becoming brown upon opening; 7–10 mm long. Damp meadows and cultivated land. France eastwards (not Sardinia). FR, IT.

D. Flowers dull purple in bud, just 5–6 mm long.

Bellevalia webbiana is similar to *B. romana*, and also has smooth leaf margins and flowers <10 mm long (typically 5–6 mm in length), but the perianth deep, dull *purple*, becoming brown with age. *Quercus* woods, fields and olive groves. IT (Tuscany and Emilia-Romagna).

Leopoldia TASSEL HYACINTH

Bulbous perennials with dense to lax racemes. Fertile flowers constricted at the throat, often brownish or greenish, and apical sterile flowers blue, violet or pink, often highly reduced. Capsule 3-lobed, splitting and often compressed.

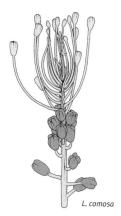

L. comosa

Leopoldia comosa (Syn. *Muscari comosum*) TASSEL HYACINTH A short to medium bulbous perennial 20–60(90) cm tall. Leaves 2–3(5), linear, recurved, grooved and tapered gradually to the tip, to 40 cm long and 6–25 mm wide. Flowers borne in a lax raceme with *small, blue, sterile flowers held erect on long, slender blue stalks* to 15 mm long, *forming a tuft at the apex*; fertile flowers brownish-green, *shortly tubular-bell-shaped* with 6 outwardly curved short, cream-coloured or whitish teeth. Common in open rocky and grassy habitats; the most widespread species. Throughout. *Leopoldia gussonei* (Syn. *Muscari gussonei*) is similar to *L. comosa* but with yellow fertile flowers borne on short stalks <5 mm long; the uppermost sterile flowers stalkless. Coastal dunes. IT (Sicily).

Muscari GRAPE HYACINTH

Bulbous perennials similar to *Leopoldia* with dense racemes (laxer in fruit). Fertile flowers usually blue, purple or blackish, constricted at the throat or not, and sterile flowers smaller and fewer (0–10). Capsule 3-lobed and splitting.

A. Spring-flowering.

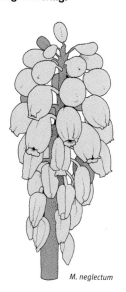

M. neglectum

Muscari neglectum COMMON GRAPE HYACINTH A short, bulbous perennial 12–25 cm high. Leaves 3–9, linear, channelled and all basal, 2–3.5(4) mm wide. Inflorescence a short, dense raceme of touching ovoid blue flowers borne on spreading or recurved stalks; the uppermost flowers sterile and paler. Common in open rocky and grassy habitats; a widespread and variable species. Throughout. *Muscari commutatum* has *violet-black flowers* with *no white teeth*. Grassy and rocky habitats. IT.

B. Autumn-flowering.

Muscari parviflorum A short, bulbous perennial 10–25 cm high with 2–4 linear, thread-like leaves just 1.5–3 mm wide, exceeded by the scape. Flowers sky-blue and bell-shaped, to 5 mm long, borne in sparse racemes; sterile flowers few or absent, tiny if present. Coastal garrigue or grassy habitats from Spain eastwards, excluding France, Corsica and Sardinia. ES, IT.

Agave

Large, imposing succulents with fibrous, very robust leaves in rosettes. Flowers cosexual and regular; stamens 6; ovary superior or inferior and 3-parted. Fruit a splitting capsule; seeds numerous. Native to the Americas.

Agave americana CENTURY PLANT A very large and imposing succulent 6–8 m when in flower; plants take at least 10 years to flower, after which they die, but perennate by offsets. Leaves very large and succulent, to 2 m long and 15–22(30) cm wide, with a stout *black spine at the tip 30–50 mm long*; blue-green or variegated with yellow stripes. Flowers borne on a *tree-like inflorescence* that persists for years, greenish-yellow, to 90 mm long, borne on at least $^1/_2$ the length of the inflorescence scape. Fruit an oblong capsule. Common, particularly near developed areas; cliffs and roadsides. Throughout. *Agave atrovirens* is similar but has broader leaves to 40 cm with a *longer terminal spine, to 15 cm* (always >40 mm) and flowers on $^1/_4$ or less of the length of the inflorescence scape. Naturalised in southern Portugal and possibly elsewhere. PT. *Agave fourcroydes* is similar to both the above species but *with a distinct basal trunk 50 cm–1(2) m* (not stemless). Planted and naturalised.

A. attenuata

ES (southeast). *Agave sisalana* has very straight, rigid leaves that are much narrower, 5–11 cm wide which when mature are spineless along the margins; terminal spine 15–25 mm long. Flowers blue-green, and *mixed with leafy bulbils* which detach and form new plants. Naturalised in Spain and possibly elsewhere. ES.

Agave attenuata is similar to the above species but smaller, with less succulent leaves with a *pale blue-white waxy bloom,* and *flowers borne in an extremely dense, fox-tail like inflorescence, drooping to the ground when mature*. Planted and naturalised in coastal areas in southern Portugal and possibly elsewhere. PT.

1. *Agave sisalana*
2. *Agave sisalana*
3. *Agave americana*
4. *Agave americana*
5. *Agave fourcroydes*

Yucca YUCCA

Perennial succulents with sword-shaped leaves in rosettes. Inflorescence an erect panicle of creamy-white flowers; perianth bell-shaped with lobes longer than their tube. Fruit a non-splitting capsule. Garden escapes, native to North America. Naturalised plants in the region are generally infertile.

Yucca gloriosa YUCCA A tough, erect, succulent shrub 1–2(3) m high with a thick, woody trunk, branching with age and a terminal leafy rosette. Leaves *blue-grey* green with reddish margins, rigid and tough, tapering gradually into a spine-tip; *margins entire* or inconspicuously toothed. Flowers cream-white, bell-shaped and pendent, 45–60 mm long, borne in stout, terminal panicles; tepals 6, all similar. Fruits dry, not fleshy. Widely planted in towns, parks and gardens. Naturalised, especially on dunes. ES, IT. *Yucca aloifolia* is similar but has dark green (not glaucous) leaves with toothed margins. Fruits fleshy. Naturalised in Spain, possibly elsewhere. ES. *Yucca filamentosa* is similar to the previous species but has leaves with *whitish margins from which fibrous filaments are shed*. FR, IT.

Anthericum

A genus of about 300 rhizomatous perennials with long, narrow leaves and simple or branched stems carrying starry white flowers with 6 tepals. Mainly in tropical and southern Africa and Madagascar.

Anthericum liliago St. BERNARD'S LILY A semi erect perennial to 60 cm with a short root stock of fleshy roots. Leaves more or less equalling the scapes, linear and narrow, 5–7(8) mm wide. Flowers white and rather with tepals 16–23 mm long (larger than other species) which markedly *exceed the stamens*, borne in *loose, erect racemes* of up to 10(30); tepals 3-veined and the style curved. Capsules ovoid, 7–9 mm wide. Stony and scrubby ground, woodland and mountain slopes. Throughout except for many islands. *Anthericum baeticum* is similar but smaller with very slender leaves 1–2(4) mm wide which are shorter than the scapes, and with an unbranched inflorescence with 1–10 flowers with tepals 10–15 mm long, more or less equalling the stamens (not exceeding by 3 mm). Capsules narrowly ovoid, 5–6 mm wide. ES (south-centre), MA.

Anthericum ramosum A semi-erect perennial similar to *A. liliago*, with a basal tuft of linear leaves to 5 mm wide which are exceeded by the flowering scapes. Flowers borne on *branched panicles* (rarely simple racemes) and *smaller, with tepals 9–13(14) mm long* which are reflexed when fully open; *style straight*. Dry scrub and sunny slopes. ES, FR.

Aphyllanthes

A small genus with only 1 species present in the region. Perennials with blue, terminal flowers and rush-like leaves. Highly distinctive, no similar species occur in the region.

Aphyllanthes monspeliensis A distinctive spring- to summer-flowering, tufted perennial 10–50 cm high with numerous blue-green, linear, ribbed, rush-like, flopping stems just 0.7–1 mm thick (true leaves reduced to basal reddish-brown, membranous sheaths). *Flowers borne in heads of 1–3, terminal, blue* (rarely white) with darker mid veins, surrounded by bristle-tipped bracts, 10–15 mm long. Fruit a capsule with 3 seeds. Open woods and maquis. Absent from the far west and from some islands. ES (incl. Balearic Islands), FR, IT (incl. Sardinia).

1. *Yucca gloriosa*
2. *Ruscus aculeatus*
3. *Ruscus aculeatus* (in fruit)

Ruscus BUTCHER'S BROOM

Woody evergreen shrubs with stems flattened into leaf-like blades (cladodes) on which the flowers and fruits are borne directly; flowers dioecious; tepals free. Fruit a berry.

R. aculeatus

Ruscus aculeatus BUTCHER'S BROOM A sclerophyllous, shrubby perennial 20–80 cm (1 m) high with dark green, ribbed, *intricately branched stems* (especially above), spiny and bushy with cladodes 40–65 mm long arranged in 2 rows, *ending in a spiny tip*. Flowers male or female on separate plants, inconspicuous, greenish-white, borne on the middle of the upper suface of the cladodes. Fruit a large red berry 10–15 mm across. Common and widespread in woods at higher altitude. Throughout. *Ruscus hypophyllum* is similar but rather shorter, 10–80 cm, and less sclerophyllous, with simple stems (or with few branches) with large, flexible cladodes 50 mm–10 cm long *without spine-tips*; flowers borne on the upper or lower surface of the cladodes; bracts *papery*. Local. ES, FR, IT (incl. Sicily), MA. *Ruscus hypoglossum* A short to medium perennial to 40 cm, similar to the previous species with generally unbranched stems and cladodes without spine tips bearing flowers on the upper surface; bracts *green and leathery* (not papery). Often cultivated, rarely naturalised. IT.

ARECACEAE | PALM FAMILY

Trees with large, pinnately or palmately divided leaves. Flowers unisexual or cosexual, small and greenish, usually borne in spikes or panicles with papery sheaths; petals 3; stamens 6. Fruit a berry or drupe. Species besides *Chamaerops humilis* in the region are planted (rarely naturalising); only the most commonly planted species are described.

Chamaerops

Trunk short; leaves fan-shaped and palmately cut into stiff segments. Native in the region.

Chamaerops humilis DWARF FAN PALM A bushy, clumped, sometimes stemless dwarf palm to 4(7) m; trunk covered in grey fibres. Leaves green or grey-green and fan-like, tough and divided >^{2}/$_3$ their length into 20 pointed, narrowly lanceolate segments; stalk long and with a spine-toothed margin. Flowers yellow, borne in dense panicles 12–25 cm long, with sheathing bracts. Fruit oblong, 7–16 mm, yellow or brown when ripe. Common and widespread on garrigue and in rocky ravines or pastures; the only native European palm. Throughout.

1. *Chamaerops humilis* (inset: fruits)
2. *Washingtonia filifera*
3. *Phoenix canariensis*
4. *Phoenix dactylifera*
5. *Phoenix hybrid* (probably
   P. canariensis x *dactylifera*)

Trachycarpus

Trunk long and slender; leaves with long bare petioles terminating in a rounded fan of segments. Native to Asia.

Trachycarpus fortunei CHUSAN PALM A tall, slender palm to 15 m with a solitary trunk densely covered in brown, matted leaf sheaths from previous years' growth. Leaves large and bright green, fan-shaped and divided into numerous narrow pointed segments fused in the centre of the blade; stalk with spines along the edge. Flowers green-white, borne in stout, lateral panicles to 60 cm long, the male and female flowers on separate panicles of the same tree. Fruit grape-like, to 14 mm long. Native to Southeast Asia. Planted throughout.

Washingtonia

W. robusta

Trunk long and slender; leaves with long bare petioles terminating in a rounded fan of segments. Native to North America.

Washingtonia robusta Rather similar to *Trachycarpus* but much more robust with a very stout, smooth trunk with diagonal furrows to 25 m tall and 40–70 cm wide, the upper part *clothed with the long-persistent withered leaves*. Leaves large and fan-like. Native to Mexico. Planted throughout. *Washingtonia filifera* is similar to the previous species but shorter, to 15 m, with a trunk to 1 m across and leaves with white threads hanging from the leaf intersections. Native to California. Planted throughout.

Phoenix

Trunk long and stout; leaves pinnately cut into numerous leaflets. DNA-analysis shows species are freely hybridising and accurate identification of true species in the field may not be possible.

P. canariensis

Phoenix canariensis CANARY PALM A tall large palm to 20 m tall with a solitary trunk 50 cm–1.2 m across and crown with fronds 1.8–6 m long with pinnately divided leaflets 21–89 cm long. Fruits 15–23 mm long, orange ripening purple-brown. Native to the Canary Islands but widely planted as an ornamental in towns and gardens in the region. Throughout. *Phoenix dactylifera* is superficially similar but taller, to 30 m with a *narrower trunk 20–50 cm across* and smaller crown. Fruit (edible dates) larger, 25–75 mm long, orange-brown and sticky when ripe. Widely cultivated. DZ, ES, MA, PT, TN. *Phoenix canariensis* × *dactylifera* is a hybrid of the above species, intermediate in characterstics, and frequently planted in the region.

Jubaea

Trunk very long and smooth (not covered in fibrous leaf bases).

Jubaea chilensis CHILEAN WINE PALM A palm similar to *Phoenix canariensis* but with shorter leaves and a smooth, grey trunk that is rather swollen in appearance; leaves leave diamond-shaped markings when they fall. Native to Chile but widely planted. Throughout.

JUNCACEAE

Erect, grass-like annuals or perennials with white, pith-filled stems. Differing from grasses and sedges in having regular flowers with 6 similar tepals and (3)6 stamens; style 0 or 1; stigmas 3. Fruit a capsule with 3-numerous seeds. Relative anther to filament length important, and should be measured in late flower or in fruit. Only the more widespread and common species in the region are described.

Juncus

Annuals to perennials with 1–2-faced leaves; hairless. Flowers with 6 tepals (in contrast to other grass-like families); stamens 6. Capsule with numerous seeds.

A. Leaves (normally) all basal; flower clusters *lateral*.

J. effusus

J. capitatus

Juncus effusus SOFT RUSH A perennial with almost smooth stems (30–50 obscure ribs) and acute but *soft-tipped* leaves, and inflorescences of pale brown flowers with the lowermost bract *long and with a narrow sheath,* and anthers 0.4–0.7 mm, *equalling or scarcely longer* than the 3 filaments; tepals more or less equal. Seasonally wet habitats and pool margins. Throughout. *Juncus inflexus* is a similar, densely tufted perennial with *glaucous, leafless, strongly ribbed stems* (12–18 clear ribs) with an interrupted central pith, and 1-sided, brownish inflorescences, and with purple-black basal sheaths; inflorescence with many ascending, unequal, lax branches; anthers 0.8–0.9 mm, equal to 1.5 x the filament length. Pastures and damp habitats. Throughout. *Juncus conglomeratus* is similar to the previous species, differing in that the stems are dull green (not glaucous) with 10–30(35) ridges and with a continuous internal pith; bract adjacent to inflorescence flat and opened out. ES, FR, IT, MA, PT.

B. Leaves (normally) all basal; flower clusters *terminal*.

Juncus capitatus DWARF RUSH A dwarf, tufted, firmly rooted rhizomatous annual just 17–50 mm (18 cm) tall with all basal bristle-like leaves 0.2–0.6 mm wide and several stiff, unbranched, leafless (except for leafy bracts at the top) stems with compact heads of flowers, overtopped by 1 or 2 bracts. Outer perianths segments greenish, later reddish with curved segments, exceeding the inner segments. Damp, sandy habitats. Throughout.

1. *Juncus acutus*
2. *Juncus acutus*

C. Leaves all basal or stems with 1 or more leaves; *maritime habitats.*

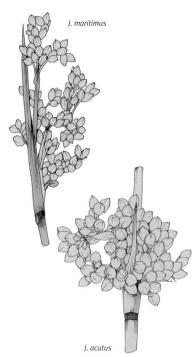

J. maritimus

J. acutus

Juncus maritimus SEA RUSH A stiffly erect perennial with stems 30 cm–1.5 m, and 2 mm wide, and a short, creeping rhizome. Leaves sharp, almost all basal, generally shorter than the stems, and *sharply pointed.* Inflorescence an interrupted panicle with 2–3 flower-heads. Flowers straw-yellow and 6-parted with unequal tepals; anthers 2 x as long as the filaments. Bracts sharply-pointed, the largest longer than the inflorescence. *Capsules triangular-ovoid and pointed, 2.5–3.5 mm long, not longer, or slightly longer than the tepals.* Salt-marshes, dunes and coastal grassland. Throughout. *Juncus rigidus* is similar but with the capsule 3.5–5 mm and clearly *exceeding* the tepals. Coastal sands and marshes. Rare and local. DZ, IT (incl. Sardinia and Sicily), MA, TN.

Juncus acutus SHARP RUSH Similar to *J. maritimus* but taller, 70 cm–1.8 m, with more *densely tufted stems* leaves sharply pointed and longer than the stems, flowers reddish-brown, borne in dense, rounded inflorescences exceeded by their bracts; anthers up to 5 x longer than the filaments; tepals more or less equal. *Capsule 3.2–6 mm long, much longer than the tepals; inner tepals with membranous margins extended into lobes.* Coastal sands and salt-marshes. Throughout.

1

2

D. Stems with 1 or more leaves; (often) *inland habitats.*

Juncus articulatus JOINTED RUSH A variable, rhizomatous, medium to tall perennial with leafy stems, 12–95 cm tall. Leaves linear and *jointed (with clear transverse markings)*. Inflorescence terminal, with erect to spreading branches and without exceeding bracts. Flowers brown to blackish; anthers 0.3–1 mm, more or less equalling their filaments. Capsule 2.5–3.5 mm long. Dune slacks and other damp, sandy habitats. Throughout. *Juncus striatus* is similar, and also has jointed leaves that are distinctly *grooved,* and anthers 1–1.8 mm, 2–3 x as long as the filaments, and a *beaked* capsule. Damp habitats. Widespread but scattered. Throughout. *Juncus bulbosus* is similar to *J. articulatus* but *not rhizamatous*: a tufted, short to medium perennial 30 mm–35 cm with prostrate, spreading (or even floating) stems which are *swollen at the base.* Leaves indistinctly jointed, rather grooved. Inflorescence a sparse cluster with erect to spreading branches carrying green to brown flowers; outer tepals acute. ES, FR, PT. *Juncus subnodulosus* is similar to *J. bulbosus* but a rhizomatous species with *erect stems without markedly swollen bases*; outer tepals obtuse. ES, FR, IT, MA, PT.

Juncus bufonius A small perennial with usually *solitary flowers* (1–3) each with 3 stamens, with anthers *shorter* or as long as the filaments and unequal tepals; inner tepals longer than the capsule; capsule 3–5 mm. Damp places. Throughout. *Juncus hybridus* is very similar to *J. bufonius* but with fascicles of 3–6(10) flowers, each with 6 stamens (not 3), with anthers often to only $^1/_2$ the length of the filaments, and inner tepals equalling or scarcely exceeding the capsules. Damp habitats, including saline areas. Throughout.

CYPERACEAE

A large family of hairless, herbaceous sedges with rhizomatous root systems; leaves grass-like but stems solid, and often triangular. Flowers wind-pollinated, borne on spikelets, highly reduced, the perianth often in the form of bristles; monoecious or sometimes dioecious; stamens 1–3; style 0 or short; stigmas 2–3. Fruit a small, 2–3-angled nut. Only a small subset is described here.

Carex SEDGE

Rhizomatous, spreading or tufted perennials with stems triangular in section. Inflorescence of 1-flowered spikelets grouped in spikes; lowest bract leaf-like or glume-like; flowers *unisexual,* either mixed in 1 spike or dioecious (often the upper spikes male and the lower female); perianth bristles 0; stamens and stigmas 2–3.

1. *Cyperus capitatus*
2. *Cyperus capitatus*

C. vulpina

C. divisa

Carex vulpina GREATER FOX SEDGE A stout, tufted perennial to 1 m with *sharply 3-angled to almost winged* stems; leaf sheaths wrinkled. Ligule surrounded by an excess of papery material, and rather truncated; inflorescence a wedge-shaped, red-brown flower cluster to 80 mm long, often interpted below, with few, short bracts; flowers male above, female below with long-pointed glumes with green mid-veins and brown margins. Damp rocks and woods, streamsides, possibly throughout the European region, rare in the west but probably often confused with the next species. ES, FR, IT, PT. *Carex cuprina* (Syn. *C. otrubae*) is difficult distinguish in the field (distinguishing characters are microscopic), differing mainly its unwrinkled leaf sheaths and acute ligules. Distribution unclear due to confusion with the previous species. *Carex paniculata* is a similar, large and densely tufted sedge, forming clumps to 1 m across, with 3-angled, greenish to blackish-brown stems to 1.5 m high. Spikes all similar in appearance, often close together, with both male and female flowers, borne stalkless and overlapping. ES, MA, PT.

Carex divisa A *creeping*, hairless perennial with *slender, wiry* stems 15–50(70) cm high and flattened leaves; stems often clustered. Flowers male above, female below, in scarcely interupted spikes; lowest bract bristle- or leaf-like and just exceeding the inflorescence. Damp grassy and sandy places. Throughout.

Cyperus

Tufted annuals to perennials with stems triangular in section and grass-like leaves. Inflorescence an umbel or umbel-like raceme with many-flowered grass-like spikelets clustered in dense heads; lowest 2–10 bracts leaf-like and exceeding the inflorescence; flowers cosexual without a perianth; stamens 1–3; stigmas (2)3.

C. capitatus

C. eragostris

A. Inflorescence a *very dense capitulum*.

Cyperus capitatus A small, tough, hairless, glaucous rhizomatous perennial with few, wiry leaves; stems solitary, 30(40) cm tall. Leaves blue-green, becoming yellow with age, with inrolled margins. Bracts leaf-like, erect and exceeding the inflorescence; later brown and withered. *Flowers borne in dense, brown, terminal spikelets;* stamens 3. Common, sometimes abundant, on coastal dunes. Throughout.

B. Inflorescence rather compact to lax.

Cyperus eragrostis A shortly rhizomatous perennial with erect, often solitary stems to 60(80) cm. Flowers borne in greenish-yellow to brownish spikelets 8–13 mm long, rather compact; stamen 1. Native to tropical America, naturalised in the region, sometimes abundantly. Throughout.

C. longus

C. laevigatus subsp. *distachyos*

Cyperus longus A rather robust perennial 37–78 cm (1 m) tall with thick, far-spreading rhizomes. Leaves to 2–5 mm wide, shorter than or equalling the stems. Inflorescence diffuse: a simple or compound umbel with 6–10(12) rays *of brownish or reddish spikelets* 12–30 cm long; stamens 3; bracts 2–6, the outer exceeding the inflorescence. A weed of damp places and pool margins. Throughout.

Cyperus esculentus EDIBLE CYPERUS A perennial to 60 cm (1 m) with underground tubers and leaves to 2–10 mm wide. Inflorescence an umbel of 3–8 straw-coloured spikelets 10–12 cm long forming lax clusters on the branches; glumes densely overlapping; stamens and stigmas 3; bracts 2–3 x longer than the inflorescence. Cultivated in Mediterranean Europe. ES, FR, IT, PT.

Cyperus laevigatus subsp. *distachyos* (Syn. *C. distachyos*) A low to medium, tufted perennial to 50 cm with mat-forming rhizomes; stems erect and bunched or solitary, and rounded or triangular in section. Leaves few, and reduced, 30–60 mm long. Inflorescence a fascicle of 2–4(9) stemless spikelets with greenish-brown flowers; stamens 3; *bracts 2, erect, and forming an apparent elongation of the stem,* and exceeding the inflorescence. Local in marshes, stream margins, and other wet habitats. Throughout.

1

Eleocharis

Perennials with rounded to ridged stems and 0 leaf blades. Inflorescence a terminal spikelet, the lowest bract glume-like; flowers cosexual; perianth of 0–6 bristles; stamens 3; stigmas 2 or 3.

E. palustris

Eleocharis palustris COMMON SPIKE-RUSH A hairless, aquatic perennial with creeping rhizomes with solitary (1st year) and later numerous, *tufted*, leafless stems to 60(75) cm, bearing cylindrical spikelets 5–30 mm long; *stigmas 2*. Stems reddish at the base with *leafless sheaths*; sheaths pale brown. Stems with approximately equal air canals in cross-section. Marshes and seasonally flooded habitats. Throughout. *Eleocharis multicaulis* is very similar to the previous species, differing chiefly in having *3 stigmas*. Nuts 3-angled. Throughout.

Schoenus

S. nigricans

Tufted perennials with leaves crescent-shaped in cross-section. Inflorescence with several spikelets in dense terminal heads of 1–4-flowered flattened spikelets; lowest bract leaf-like or glume-like; flowers cosexual with 0–6 perianth bristles; stamens and stigmas 3.

Schoenus nigricans BLACK BOG-RUSH A densely tufted perennial to 75 cm with all leaves basal, and terminal inflorescences. Leaves shorter than or roughly equalling the stems, dark grey-green, hard and wiry. Inflorescence with 5–10 rather flattened spikelets, with the lowest bract usually clearly exceeding it; stamens and stigmas 3. Nuts 3-sided and creamy-white. Locally common on maritime sands and peat. Throughout.

Isolepis CLUB-RUSH

Typically slender annuals. Inflorescence a head of 1–several spikelets with numerous florets; bracts if present exceeding the inflorescence, leafy and falling early; glumes spirally arranged; stamens 1–3, style 2–3 parted.

I. setacea

I. cernua

Isolepis setacea BRISTLE CLUB-RUSH A small, slender, tufted, sedge-like herb wth narrow stems <0.5 mm wide and 10–15(30) cm long. Leaves few, and shorter than the stems. Inforesence *much exceeded by the bract* which is 5–20(35) mm long, spikelets 1–4, 2–4 mm long; empty bracts (glumes) purple-brown with green midribs and translucent margins. Damp, marshy habitats. Throughout. *Isolepis cernua* is similar but with *bracts shorter or scarcely exceeding* the inforesences which are borne on long, often drooping stems; glumes greenish with a red-brown spot each side of the midrib. Damp coastal, sandy habitats. Native to East Africa; naturalised. ES, PT.

---

1. *Schoenus nigricans*

POACEAE | GRASS FAMILY

Annual or perennial, tufted, often rhizomatous or creeping plants. Leaves alternate, linear and sheathing the stem, generally with a membranous ligule at the base of limb. Inflorescences very variable, often a spike or panicle; flowers not brightly coloured, wind-pollinated and with (1)3(6) stamens, and a pistil with normally 2 styles, enclosed within 2 bracts; the whole called a floret; florets arranged into spikelets with 2 empty bracts (glumes) at the base. A large and important family; it is well beyond the scope of this book to describe all species in the region; only a very small subsection is included here.

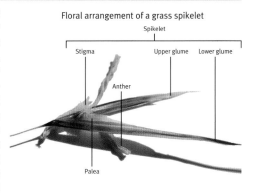

Floral arrangement of a grass spikelet

Spikelet

Stigma   Upper glume   Lower glume

Anther

Palea

A. Florets all unisexual and dissimilar (the male differing from the female).

Zea MAIZE

Leaves 50 mm–12 cm wide. Male flowers in slender terminal panicles, female spikelets in simple racemes (cobs) among the lower leaves.

Zea mays MAIZE A very robust annual grass to 5 m with stout stems and many broad leaves 50 mm–12 cm wide. Male flowers borne in a terminal inflorescence of spikelets to 20 cm long; female spikelets borne below on swollen, lateral branches and enclosed in overlapping leafy bracts, with *long, projecting styles* (to 25 cm). Fruit very distinctive: hard, shiny, yellow, white or purple, arranged around a very swollen axis ('corn on the cob'). Native of South America but widely cultivated in suitable areas of the European Mediterranean. ES, FR, IT, PT.

B. Florets all similar (cosexual or unisexual), spikelets virtually stemless and borne in a solitary, terminal, spike-like cluster.

Lolium RYE GRASS

L. perenne

Annuals to perennials. Inflorescence simple, unbranched and spike-like with stalkless spikelets alternately arranged edgeways onto a jointed axis, flattened; glumes solitary in lateral spikelets (2 in terminal); lemma 5–9-veined and awned or not.

Lolium perenne RYE GRASS A hairless, tufted, wiry perennial grass to 50(90) cm. Stems smooth and slender, bent below. Leaves narrow, up to 3 mm wide, and *folded* until mature. Ligule to 1 mm, abruptly pointed. Inflorescence *simple*, to 15 cm long with compressed, oval spikelets to 15 mm; lower lemmas 3.5–9 mm and almost always *awnless*. Cultivated as a fodder crop. Throughout. *Lolium multiflorum* is similar but an annual with *awned lemmas* (awns to 15 mm); glume shorter than the spikelet. Throughout.

Catapodium

C. rigidum

Annual grasses with a simple or little-branched, stiff, 1-sided, spike-like inflorescence; spikelets rather compressed with many (3–14) florets in 2 ranks; glumes 2, nearly equal and papery; lemma blunt and leathery, 5-veined and awnless.

Catapodium rigidum (Syn. *Desmazeria rigida*) FERN-GRASS A small, stiff, hairless, *tufted*, *bluish* annual with erect stems to 15(60) cm with several to numerous erect or spreading stems. Leaves often purplish, fine-pointed, to 2 mm wide and flat or with inwardly rolled margins. Inflorescence a more or less 1-sided panicle to 80 mm long, often branched below, with sparse, tiny spikelets to 7 mm long, each with 5–10 minute florets; glumes 1.3–2 mm and pointed; lemmas longer, 2–2.6(3) mm and blunt. Coastal sands, dry, bare habitats and walls and rocks. Throughout.

Aegilops

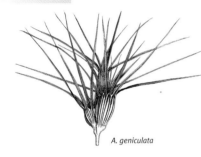

A. geniculata

Inflorescence with stalkess spikelets arranged broadside along the axis in a distinctive compact ovoid or cylindrical cluster; glumes large and tough and strongly veined with 2–4 awns at the apex; lemmas also awned.

Aegilops geniculata A low, tufted annual with erect stems to 40 cm. Leaves with a flat blade 2–3 mm wide, finely hairy on the upper surface; ligule very short. Inflorescence congested, with 1–2 vestigial spikelets at the base of the fertile ones; fertile spikelets often just 2–4, with virtually equal glumes; lemma with a long bristle 15–30 mm long. Common on bare and fallow ground and in olive groves. Throughout.

Hordeum

H. murinum

Annuals, sometimes perennials. Inflorescence spike-like, dense and long-awned with spikelets in clusters of 3 arising from each joint in the axis; glumes narrow, long-awned and 1–3-veined; lemmas 5-veined with long awns. Other, very similar species occur.

Hordeum murinum WALL BARLEY An annual to 60 cm with tufted, erect, smooth stems. Leaves linear, to 4 mm wide and hairy on both surfaces with shiny sheaths; ligule membranous and small, to 1 mm long. Inflorescence a more or less bilaterally symmetrical, bristly, dense, spike-like panicle, lemmas with awns 10–50 mm long; glumes with awns 10–30 mm long. A common grass in a range of habitats. Throughout.

Secale RYE

Annuals. Inflorescence with spikelets with 2 (3–4) cosexual florets; glumes very narrow, linear-lanceolate, 1-veined and acute or short-awned; lemma long-awned and keeled with stiff hairs.

Secale cereale RYE A rather glaucous, erect, hairless annual to 1–2 m with soft, rough leaves to 8 mm wide, and a long, slender, nodding, dense and spike-like inflorescence 50 mm–15 cm long with numerous long-awned, stalkless spikelets; glumes keeled, linear and narrowed into long, awn-like tips; lemma with ciliate keels and awns 20–50 mm. Cultivated and naturalised in the European Mediterranean. ES, FR, IT, PT.

Elytrigia

E. juncea

Perennials. Inflorescence with 1 spikelet per node, with several to many florets, flattened broadside onto the rachis; glumes 3–11-veined, rarely awned; lemmas 5-veined, unawned or short (rarely long)-awned.

Elytrigia juncea (Syn. *Elymus farctus*) A tough, clump-forming, rhizomatous perennial grass to 60(80) cm, often co-occurring with *Ammophila arenaria*. Leaves inrolled at the margins, 2–6 mm wide, and minutely hairy above, hairless beneath; *ligules short*. Inflorescence a slender spike with dense, erect, virtually stalkless, laterally compressed and hairless spikelets; *lemma unawned*. Common on coastal dunes across the region. Throughout.

1. *Aegilops geniculata*
2. *Hordeum murinum*
3. *Elytrigia juncea*
4. *Cynodon dactylon*

Parapholis HARD-GRASS

Annuals with very slender, whip-like inflorescences with alternate spikelets arranged broadside and set into hollows of the axis, each with a single floret; glumes equal and 3–5-veined; lemma finely 3-veined.

P. incurva

Parapholis incurva CURVED SEA HARD-GRASS A distinctive, short, tufted annual with *spreading, curved* stems 10–20 cm long. Leaves flat or inrolled, linear and pointed, to 2 mm wide, rough above and along the margins with reddish sheaths. Inflorescence 10–80 mm (15 cm), often not exserted from its sheath, slender, rigid, cylindrical, *strongly curved and jointed* with spikelets adpressed to the stem; spikelets to 7 mm long, a little longer than the joints of the axis; glumes equal, closing the cavities of the axis. Saline coastal habitats, particularly salt-marshes and cliffs. Throughout.

C. **Florets all similar (cosexual or unisexual), spikelets virtually stemless and borne in several, often terminal, spike-like clusters.**

Cynodon BERMUDA GRASS

Perennials. Inflorescence a compound *umbel* of 3–6 slender branches with stalkess spikelets arranged in 2 rows; glumes 1-veined; lemma 3-veined.

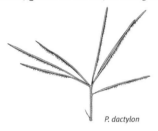

P. dactylon

Cynodon dactylon BERMUDA GRASS A spreading, low to short perennial with creeping stems to 30 cm. Leaves linear and flat, hairless or hairy along the margins. Inflorescence distinctive and star-like: *3–6 spikes 20–50 mm long, outwardly spreading from a single central axis*; spikelets 2–3 mm long, stalkless. Native to tropical Africa but a commonly naturalised weed in a range of disturbed habitats; common. Throughout.

Hyparrhenia

Inflorescence branched with paired, slender, spike-like clusters arising from leaf-like bracts.

H. hirta

Hyparrhenia hirta A tufted, medium to tall perennial grass to 1.2 m with smooth stems. Leaves linear, to 3 mm wide and more or less hairless. Inflorescence a panicle of *paired racemes*, each enclosed in leaf-like bracts; lemma to 4.5 mm long with a stout bristle-tip to 20 mm long, twisted and hairy below. Fairly frequent in dry, rocky habitats. Throughout.

D. **Florets all similar (cosexual or unisexual), spikelets stalked or short-stalked and forming a compound (narrow or broadly-branched) inflorescence.**

Cynosurus

Annuals to perennials. Inflorescence a compact spike-like panicle of fertile spikelets with (1)2–5 cosexual florets and sterile spikelets with sharp-pointed lemmas in a herring-bone arrangement.

C. echinatus

Cynosurus echinatus ROUGH DOG'S-TAIL A short to tall hairless annual with erect or spreading stems to 75 cm (1 m) and flat leaves 2–10 mm wide. Inflorescence a dense, plume-like, 1-sided, oblong panicle 10–40(80) mm long of *shiny green or purplish spikelets*; the outer spikelet of each pair comb-like with several pairs of spreading, long-awned, sterile lemmae; inner spikelet fertile and wedge-shaped. Dry rocky and grassy scrub. Throughout.

Tragus

Annuals. Inflorescence a spike or spike-like with 2–5 spikelets on very short branches at each node, each with 1 cosexual floret; glumes unequal, the upper longer and 5–7-veined, each vein with hooked bristles; lemma 3-veined.

Tragus racemosus STALKED BUR GRASS A creeping, branched, spreading annual, rooting at the nodes with erect stems to 40 cm and short, flat leaves to 3 mm wide with spines along the margins. Inflorescences spike-like, long, cylindrical, purple; spikelets 3–5 per node, the upper glumes *with 7 rows of fine-crooked bristles on the backs*. Dry sandy areas, dunes, waste places and olive groves; rarer in the far west. Throughout.

Lamarckia GOLDEN DOG'S TAIL

Annuals. Inflorescence with spikelets of 2 kinds: the upper with 1 fertile floret and 1 rudimentary floret, the lower with several pairs of overlapping, blunt, sterile lemmas in 2 ranks.

L. aurea

Lamarckia aurea GOLDEN DOG'S TAIL A more or less hairless, low annual grass with tufted, erect stems to 20(30) cm. Leaves linear, 2–6 mm wide with hairy margins; ligule membranous, 5–10 mm long, pointed or blunt. Inflorescence 30–90 mm long, dense, rather 'fluffy' and 1-sided with the outer spikelets sterile, greenish and later golden. Common on fallow and cultivated ground and roadsides; widespread. Throughout.

Dactylis COCK'S-FOOT

Perennials. Inflorescence a more or less 1-sided panicle or compound with spikelets crowded into dense clusters at the ends of the side branches; spikelets flattened, short-stalked with 2–5 florets; glumes keeled and 3-veined; lemma keeled and 5-veined, very shortly awned or awnless.

D. glomerata

Dactylis glomerata COCK'S-FOOT A perennial, bluish clump-forming grass with erect or spreading stems to 1.4 m. Leaves rough, with ligules 2–10 mm long. Inflorescence an erect, rather unequal and 1-sided tufted panicle of *laterally compressed spikelets borne in dense clusters on lateral branches*, often with prominent yellow stamens. Common in grassy habitats. Throughout.

Lagurus HARE'S TAIL

Annuals. Inflorescence very compact, compound and spike-like, densely silkily hairy; spikelets with single florets, falling as a unit when ripe; glumes bilobed, 1-veined and awned, longer than the obscurely 5-veined lemma.

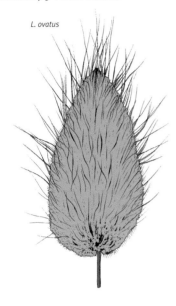

L. ovatus

Lagurus ovatus HARE'S TAIL A low to medium, softly hairy, grey-green annual to 60 cm. Leaves linear-lanceolate and flat; ligule hairy and membranous, to 3 mm long. Inflorescence distinctive; *egg-shaped, dense, 'fluffy' soft and white*, 5–20 mm long; lemma semi-transparent and with awns 8–20 mm long. Common in sandy coastal environments and rocky slopes inland; also cultivated as an ornamental. Throughout.

1. *Tragus racemosus*
2. *Tragus racemosus*
3. *Lagurus ovatus*
4. *Lamarckia aurea*

Gastridium NIT-GRASS

Annual grasses with compact, spike-like inflorescences with ascending to adpressed branches; spikelets laterally compressed, solitary with a single cosexual floret; glumes unequal and exceeding the florets, 1-veined and papery at the tips; lemma membranous, 5-veined and awned or not (awns shorter than the lemmas).

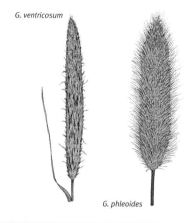

G. ventricosum

G. phleoides

Gastridium ventricosum NIT-GRASS A small, annual grass to 50(90) cm tall with flat, hairless leaves, often withered when in flower; ligules to 3 mm long, and pointed. Flowers borne in green, *strongly erect* (at least at first), more or less bilaterally symmetrical and laterally compressed panicles 5 mm–10(16) cm long; *spikelets short, 3–5 mm long with a single floret*. Scattered in the west, possibly elsewhere. ES, MA, PT. Forms with lax panicles and florets to 3 mm long are traditionally described as *G. laxum*. *Gastridium phleoides* is native to Asia but naturalised as a casual weed in the region, and distinguished by its very dense panicles with *longer spikelets 5–8 mm long*; lemma *densely pubescent*, with an awn 4–7(8) mm. Throughout, though absent from many islands.

1. *Setaria viridis*
2. *Macrochloa tenacissima*
3. *Macrochloa gigantea*
4. *Stipa gigantea*

Setaria BRISTLE GRASS

Annuals or perennials. Inflorescence a cylindrical and spike-like panicle with numerous, densely clustered, stalkess spikelets with 2 florets; glumes unequal; lemma 5-veined, awnless.

Setaria viridis GREEN BRISTLE GRASS A loosely tufted annual to 50 cm (1 m) with wide, flat, *hairless leaves*; ligule a ring of hairs. *Inflorescence dense, very bristly, cylindrical and erect;* bright green up to 12(17) cm long; spikelets 2–2.5(2.7) mm long. Common on sandy, fallow and waste ground. Throughout. *Setaria faberi* CHINESE FOXTAIL is very similar, with sparsely hairy leaves and larger spikelets, 2.7–3 mm long. A widespread weed. Throughout.

Stipa (incl. *Macrochloa*)

Annuals or perennials with a membranous ligule fringed with hairs; lemma with forward-pointing bristles and a long terminal, persistent awn.

S. capensis

(i) *Stipa* group: lemmae entire (not split).

Stipa capensis An *annual or biennial* to 60 cm, with leaves to 15 cm long with revolute margins; glaucous and hairy or hairless. Inflorescence a *dense, slender panicle*; enclosed at the base by a subtending leaf, to 15 cm long and 10 mm wide; spikelets solitary, the fertile spikelets stalked; glumes persistent and *all more or less similar, 15–20 mm long* and exceeding the florets and 3-veined; awns 70 mm–10 cm. Common to abundant, often coastal. Throughout. *Stipa parviflora* is similar but a *perennial* with *dissimilar glumes*: those below to 15 mm and those above to 7.5 mm. Usually coastal. ES, FR, MA.

Stipa pennata A densely tufted glaucous perennial grass to 60 cm, distinctive for having narrow, feathery inflorescences with very long, trailing awns which have many conspicuous silvery-white hairs; spikelets yellow-green; glumes to 20 mm with a hairless slender awn 2–3 x as long; lemma to 25 mm with a *very long, feathery and twisted awn to 35 cm long*. Local; rocky and stony habitats. Throughout, though absent from many islands.

(ii) *Macrochloa* group (also widey described under *Stipa*): lemmae markedly split.

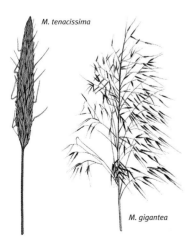

M. tenacissima

M. gigantea

Macrochloa tenacissima (Syn. *Stipa tenacissima*) ESPARTO GRASS A *very large*, tough, tufted perennial grass to 1.5 m tall, initially forming dense clumps becoming open with age. Sheaths glabrous or hairy; ligules to 0.8 mm and hairy, with a feathered tip. Leaves 30 cm–1.2 m, linear, often with downturned margins, about 1 mm across and sharply pointed. Inflorescence a dense panicle 25–35 cm with 3–4 branches per node; branches adpressed, and densely hairy in the axils densely; stalks shorter than their spikelets; glumes lanceolate, membranous, glabrous; lemmae pubescent, and 2-lobed at the tip. Local but forming extensive colonies in open, often sandy habitats. ES, MA, PT. *Macrochloa gigantea* (Syn. *Stipa gigantea*) is very similar to *M. tenacissima* but with *very lax* panicles of flowers and a lanceolate (not split) ligule. Similar habitats, often on dune slacks; mostly in the west (recorded in Italy). ES, IT, MA, PT.

Ammophila MARRAM

A. arenaria

Rhizomatous perennials. Inflorescence spike-like, dense, cylindrical with large spikelets with single florets; glumes papery and keeled, 1–3-veined; lemma also papery, 5–7-veined, lanceolate and hooded at the tip.

Ammophila arenaria MARRAM GRASS A familiar, tough, clump-forming rhizomatous perennial grass with smooth stems to 1.2 m. Leaves to 5 mm wide but appearing narrower due to the inrolled margins; *ligules narrow and pointed*, 10–30 mm long. Inflorescence a slender, spike-like panicle with dense, erect spikelets to 16 mm long; lemma with a very short, stiff bristle-tip with a *ring of hairs at the base*. Very common to dominant on coastal dunes. Throughout.

E. Florets all similar (cosexual or unisexual), spikelets long-stalked and forming a compound, branched, open or spreading inflorescence; spikelets with *1 fertile floret*, and often with additional sterile florets.

Sorghum

Annuals or perennials. Inflorescence large and much-branched with paired, shiny spikelets with 1 stalkless fertile floret and 1 or more stalked, sterile or male florets; glumes all compressed and 3-pointed at the tip.

S. halepense

Sorghum halepense JOHNSON GRASS A *large*, erect, deeply rhizomatous perennial to 1.5 m with stems *silkily hairy at the nodes*. Leaves hairless with rough margins, <20 mm wide. Inflorescence a large terminal, rather lax, pyramidal panicle to 30 cm long; spikelets shiny, to 5.5 mm long; lemma notched at the apex and with an awn to 12 mm long; sterile floret violet, hairy and lanceolate. Locally common in damp, disturbed habitats. Throughout. *Sorghum bicolor* is larger (to 2 m), with broader leaves (>20 mm) and lacks creeping stems; *inflorescence dense, oval* to 50 cm long. Cultivated for livestock in the Mediterranean, sometimes a casual weed (native to Asia and Africa). Throughout.

1. *Ammophila arenaria*
2. *Sorghum halepense*
3. *Corynephorus canescens*

Oryza

Annuals. Inflorescence with many spikelets, each with a single floret; glumes much shorter than the lemma, 1-veined; lemma compressed and keeled, strongly 5-veined; stamens 6.

Oryza sativa RICE A hairless, aquatic annual with leafy stems to 1.3 m. Leaves flat and smooth, to 10 mm wide. Inflorescence large, lax and erect or curved with numerous long lateral branches bearing numerous spikelets; glumes very small and equal, to 2 mm long; lemma 7–9 mm long, hairy above and short-pointed. Native to the tropics but widely cultivated in suitable areas (deltas). ES, IT, PT (probably elsewhere).

Piptatherum (including *Oryzopsis*)

Shortly rhizomatous perennial grasses with transparent ligules. Spikelets short and dorsally compressed; lemma with a long, straight, often falling terminal awn. A much confused genus; *Oryzopsis miliacea* now established to be genetically distinct from *Piptatherum*.

P. coerulescens

O. miliacea

Piptatherum coerulescens (Syn. *Oryzopsis coerulescens*) A tall, erect, perennial grass to 70 cm (1 m) with hairless stems. Leaves 15–31 cm long and 1–12 mm wide, rough on both surfaces, with *long ligules* to 11 mm long, transparent. Flowers borne in panicles 30 mm–15 cm long with spikelets 5–14 mm; glumes sub-equal, exceeding the florets, 3–9-veined; lemma 2.6–6.5 mm long, *leathery*, with awns 1–15 mm long, falling. Throughout.

Oryzopsis miliacea (Syn. *Piptatherum miliaceum*) A tall, erect, perennial grass to 1.5 m with hairless stems, and leaves rough above. Flowers borne in light green to straw-coloured panicles to 40 cm long; 1-sided, drooping and lax, with several (to 20) branches at each node along the stalk; glumes 3–4 mm, lemma 2–2.5 mm, hairless, stiff, membranous (*not* leathery) with awns 3–5 mm. Dry open habitats among shrubs and other vegetation. Throughout.

F. Florets all similar (cosexual or unisexual), spikelets long-stalked and forming a compound, branched, open or spreading inflorescence; spikelets with *2 or more fertile florets*, and often glumes (empty bracts) *equalling or longer than the florets*.

Corynephorus

Densely tufted perennial grasses with compact panicles; spikelets with 2 florets; glumes almost equal, 1-veined; lemmas 5-veined with a bent awn club-shaped at the apex.

Corynephorus canescens GREY HAIR-GRASS A low, *tufted, bluish-grey perennial* grass to 35 cm with dense fasicles of rigid bristle-like leaves, of more or less similar length. Spikelet purple and white with 2 florets, compressed, in a compound panicle, exceeded by the 1-veined, membranous glumes; anthers orange or purple, much longer than broad. Locally common on coastal dunes. ES, FR, IT, MA, PT. *Corynephorus divaricatus* is an *annual* with hairless leaves, mainly all on stem, lacking dense fasicles of rigid basal leaves. Inflorescence a lax panicle with spikelets to 4 mm long; glumes 3–4 mm long; lemma to 1.9 mm long, elliptic and more or less entire. Coastal sands and plains; far west only. ES, MA, PT. *Corynephorus fasciculatus* is very similar to the previous species but with smaller spikelets 2.8–3.5 mm long, and minute anthers to 0.6 mm long (not 1.3 mm long). Similar habitats and distribution. ES, MA, PT.

Arundo

A. donax

Large perennials. Inflorescence large, compound, feathery with numerous spikelets, each 1–7-flowered; glumes lanceolate, papery and keeled, 3-veined; lemma papery with dense silvery hairs.

Arundo donax GIANT REED The largest grass in the region: an *extremely robust*, rhizomatous perennial with *bamboo-like stems* to >6 m. Leaves grey-green, to 60 mm wide. Inflorescence a large panicle to 60 cm long; silky and silvery, with spikelets to 20 mm long, each with usually *3 florets*; lemma notched at the apex with a short bristle-tip; glumes papery, keeled and 3-veined, usually (not always) longer than florets. Common in damp places, sometimes planted in long banks along roadsides. Throughout.

G. Florets all similar (cosexual or unisexual), spikelets long-stalked and forming a compound, branched, open or spreading inflorescence; spikelets with *2 or more fertile florets*, and often glumes (empty bracts) *much shorter than the florets*.

Bromus BROME

Annuals. Inflorescence with flattened, long-stalked spikelets with many overlapping florets; glumes unequal and often awned, 3–7(9)-veined; lemma 7–9(11)-veined, minutely split at the apex. Many species occur in the region, all variable; dwarf specimens in dry areas are not reliable for identification.

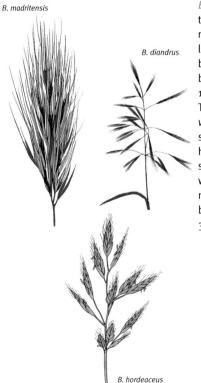

B. madritensis

B. diandrus

B. hordeaceus

Bromus madritensis COMPACT BROME A short to medium tufted annual with erect stems. Leaves tapered, to 5 mm wide and *softly white-hairy*. Inflorescence a rather lax, erect, wedge-shaped panicle with short, bunched branches; spikelets to 60 mm long; lemma with a long bristle-tip to 16 mm long; hairy or hairless; glumes 1–3-veined. Common on fallow and cultivated ground. Throughout. *Bromus diandrus* is similar but with a long, *very lax*, *nodding* inflorescence (not wedge-shaped) with spreading spikelets. Throughout. *Bromus hordeaceus* has an erect, rather *short and dense inflorescence* with spikelets on *short stalks*, exceeded by their spikelets which have a slightly inflated appearance; lemma 6.5–11 mm. ES, FR, IT, PT. *Bromus squarrosus* is distinguished by its *reflexed awn*, at right angles to the lemma; glumes 3–9-veined. Dry ground and wasteland. Throughout.

1. *Arundo donax*
2. *Briza minor*
3. *Briza maxima*
4. *Avena barbata*

Briza

Annuals or perennials. Inflorescence distinctive: spikelets flattened and inflated, ovoid- to heart-shaped and awnless, pendulous with 4–30 overlapping all cosexual florets.

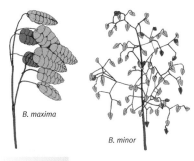

B. maxima

B. minor

Briza maxima LARGE QUAKING GRASS A hairless, low annual grass with often solitary, erect stems. Leaves flat and linear, to 4 mm wide. Inflorescence a lax panicle, with up to 15 large, *drooping papery spikelets* appearing inflated, on slender stalks, green then purplish ripening pale brown, 8–25 mm long. Common on fallow ground, roadsides and maquis; widespread. Throughout. *Briza minor* is similar but with *numerous (>20), smaller spikelets 2.5–5 mm long*. Open habitats. Throughout.

Avena OAT

Annuals. Inflorescence a compound, diffuse panicle, branched with large, long-stalked and drooping spikelets with 2–3 florets; glumes papery and exceeding the florets, 7–11-veined; lemma leathery and 7–9-veined with a stout, bent, long awn. A variable and difficult genus.

A. barbata

Avena barbata BEARDED OAT A medium to tall, erect grass with solitary or grouped stems to 1 m. Leaves linear, to 15 mm wide and hairless or slightly hairy on the margin; ligule membranous, to 5 mm long. Inflorescence a 1-sided, very lax panicle with spikelets drooping on slender stalks; *lowest lemma 12–18 mm with 2 short bristles at the tip, 3–5 mm long*. Fallow and cultivated ground and roadsides. Throughout. *Avena sterilis* is similar but the *lowest lemma 16–25 mm long with 2 short points <1.5 mm, strongly bent below* forming a distinct dog-leg. Cultivated and fallow land. Throughout.

1 2 3 4

Phragmites

P. australis

Spreading perennials. Inflorescence a large, feathery and compound panicle with numerous slender spikelets with numerous 2–6(10) florets with hairy stalks; glumes unequal, 3–5-veined; lemma hairless and 1–3-veined, awnless.

Phragmites australis (Syn. *P. communis*) A bed-forming large, reed-like grass, rather similar to *A. donax* with tall, rather slender stems to 3.5 m which do not overwinter. Leaves up to 50 cm long and 50 mm broad, grey-green and tapered to the tip; sheaths smooth and hairless, surrounding the leaf nodes. Flowers borne in drooping, 1-sided, more or less cylindrical, bunched greenish or purplish panicles with spikelets 8–16 mm long, each *with up to 10 florets*. Aquatic habitats such as lake margins. Throughout.

H. Large and imposing, woody perennials. Ligule with a single row of hairs; lemma with ridges.

Cortaderia PAMPAS GRASS

Densely tufted perennials. Inflorescence a large, plume-like, spreading panicle; spikelets laterally compressed, with 2–7 florets; glumes slightly unequal and 1-veined; lemma silky-hairy, 3–5-veined, acuminate and long-awned.

Cortaderia selloana PAMPAS GRASS A very tall, dense perennial to 3 m tall and >1 m across. Leaves long and slender, to 2 m long and 1 cm wide, with sharp serrated edges. Inflorescence plume-like and overtopping the leaves; panicle open; ovoid and dense; 25 cm–1 m long; spikelets solitary, lanceolate and stalked, laterally compressed; 12–16 mm long, each comprising 3–7 fertile florets with diminished florets at the apex; glumes all similar, exceeded by their spikelets. Native to South America and widely planted in towns and gardens, probably throughout, and widely naturalised, particularly in eastern Spain. Throughout.

I. Ligule with a single row of hairs or hairs absent; lemma without ridges.

Pennisetum

P. villosum

Perennials. Inflorescence a dense, narrow panicle with fascicles of spikelets interspersed with bristles.

Pennisetum villosum A large, rather exotic-looking, sparsely clump-forming perennial grass to 1 m tall. Leaves to 70 cm long and rigid. Inflorescence terminal and rather like those of *Lagurus ovatus* but larger; 'fluffy' and white and broadly cylindrical, borne in larger numbers terminally, drooping on slender stems; spikelets to 14 mm, interspersed with bristles. Native to eastern Africa; a casual weed in the western Mediterranean, naturalised in dry habitats. DZ, ES, MA, TN.

Panicum

Annuals. Inflorescence a compound, diffuse and much-branched panicle with slender branches; spikelets flattened, with 2 florets, the upper fertile and the lower male or sterile; glumes unequal; lower lemma 5–11-veined, awnless.

Panicum repens A perennial grass with creeping underground stems and stiff erect stems to 80 cm. Leaves in 2 ranks, glaucous and stiff, to 6 mm wide, the uppermost more or less equalling the inflorescence. Inflorescence erect and narrow with slender ascending branches with numerous whitish branches of spikelets, each to 2 mm long, without awns; glumes unequal, the upper exceeding the lower. Damp, sandy habitats, particularly on the coast. Throughout.

Panicum miliaceum A robust, annual grass, similar to *Zea* or *Sorghum* before flowering: leaves to 20 mm wide and *sheaths with long hairs*. Inflorescence rather *dense and flopping,* to 20 cm long, with numerous slender branches of *plump* (bead-like), often purplish spikelets each 4.5–5.5 mm long; glumes unequal, the upper exceeding the lower. Widely cultivated as a cereal and forage crop. Throughout.

---

1. *Phragmites australis*
2. *Cortaderia selloana*

TYPHACEAE | BULLRUSH FAMILY

Herbaceous, aquatic (typically marginal) perennials with stout rhizomes and rush-like, linear leaves sheathing at the base. Flowers unisexual, borne in distinctively dense, cylindrical spikes, the male flowers above and the female below; perianth of bristles or scales (difficult to distinguish from stamens); stamens 1–8; styles and stigmas 1. Fruit a 1-seeded spongy drupe or capsule.

Typha BULLRUSH

Unmistakable aquatic perennials with flowers borne in cylindrical spikes.

Typha latifolia BULLRUSH, GREAT REEDMACE A robust, herbaceous, marginal perennial to 3 m with stout, creeping rhizomes and rush-like, erect, thick and flat, linear leaves 8–20 mm broad, sheathing at the base. Flowers unisexual, borne in dense, cylindrical spikes: male flowers borne above in a narrow, yellow spike, later falling; female flowers below, forming a broad, cylindrical spike which turns brown when mature, 18–30 mm wide; *female flowers lack secondary bracts*. Fruits 1-seeded. Widespread and locally common in suitable habitats (pond and river margins, swamps and ditches). Throughout. *Typha angustifolia* is similar but with narrower leaves 3–6(10) mm wide, sheaths with few or no mucilaginous glands and female flowers *with abruptly pointed secondary bracts, and separated from the male flowers by a naked section of the stem 30–80 mm (12 cm) long*. Similar habitats. Throughout. *Typha domingensis* is similar to *T. angustifolia* with leaf sheaths internally covered with *numerous brown, mucilaginous glands*, and rather slender, honey-brown cylindrical spikes of female flowers with *tapering secondary bracts*. Similar habitats. Throughout.

Sparganium BUR-REED

Aquatic, herbaceous perennials rooted in mud with simple or branched stems and erect or floating leaves. Flowers unisexual, crowded into globular heads, the female below and subtended by leafy bracts; perianth with 1–6 membranous scales. Fruit dry, not splitting.

Sparganium erectum BRANCHED BUR-REED An erect, hairless perennial to 1.5 m, usually standing in shallow water. Leaves all erect (sometimes floating), triangular in section, *keeled* and linear. Inflorescence branched, with the globular yellow male heads borne above the female on each branch; flower clusters unstalked; tepals thick with dark tips. Water-logged and aquatic grassy habitats. Throughout.

MUSACEAE | BANANA FAMILY

Herbaceous (though tree-like) perennials with leaves with persistent overlapping basal sheaths forming a 'trunk'. Monoecious with flowers borne in clusters (effectively a raceme of cymes); stamens 5–6; carpels 3. Fruit a berry.

Musa BANANA

Herbaceous, robust perennials with trunk-like stems and 6–20(–numerous) leaves in a canopy. Flowers borne 12–20 per cluster, unisexual. Fruit the familiar banana.

Musa acumunata (Syn. *M. cavendishii*) BANANA A herbaceous (tree-like), tropical-looking perennial 2–9 m high. Leaves very large and broad, to 2 m long and 50 cm across with a prominent midrib, sometimes splitting at the margins; pinnately veined. Flowers unisexual, borne in large pendulous clusters. Fruit an elongated berry (a banana), 30 mm–40 cm long. Widely cultivated. DZ, ES, FR, IT, MA, TN.

CERATOPHYLLACEAE | HORN-WORT FAMILY

A family with elusive evolutionary origins, with a single genus of few species characterised by a submerged aquatic habit and minute unisexual flowers with a superior ovary; stamens 10–25; style 1. Fruit an achene.

Ceratophyllum

Submerged, aquatic, rootless perennial herbs with *whorled, forked leaves*.

Ceratophyllum demersum HORN-WORT A submerged aquatic perennial with slender, flexible stems to 1 m with whorls of brittle, dark green leaves *forked once or twice*. Flowers minute, arising from the axils though rarely formed; unisexual, green and stalkless. Fruit a 1-seeded nut 4–5 mm with a pair of spreading basal spines and a solitary terminal spine. Common in slow moving or still water, often in man-made aquatic habitats. Throughout. *Ceratophyllum submersum* SPINELESS HORN-WORT is very similar but with leaves *forked 3–4 times* and fruits that lack a basal spine and have a short or nonexistent apical spine. Distribution improperly known due to confusion with *C. demersum* but probably throughout.

1. *Typha latifolia*
2. *Typha latifolia*
3. *Typha angustifolia*
4. *Sparganium erectum*
5. *Musa acumunata*
6. *Ceratophyllum demersum*

EUDICOTS

Plants typically with 2 cotyledons (seed leaves), netted veins radiating from a central main vein (rarely parallel veins), flower parts in multiples of 4 or 5 (rarely 7), and unlike most monocots, often undergoing secondary growth, forming trees and shrubs.

RANUNCULACEAE | BUTTERCUP FAMILY

Herbaceous annuals and perennials or woody climbers with alternate, simple or compound leaves. Flowers typically regular (sometimes zygomorphic), with 5 sepals and petals; stamens numerous. Fruit an aggregate of achenes or follicles (or a berry or capsule).

Helleborus

Herbaceous perennials with leaves spirally arranged or all basal, with long, toothed leaflets. Flowers borne in *winter or spring* in branched clusters; sepals 5, petals 5–12 in the form of small nectaries; stamens numerous; carpels 2–5. Fruit an aggregate of follicles.

Helleborus foetidus STINKING HELLEBORE An erect, hairless perennial with persistent, leafy stems to 80 cm with palmate leaves divided into 7–11 leaflets, aromatic when crushed. Flowers 10–30 mm across, cup- or bowl-shaped with yellow-green sepals, often tinged purplish. Follicles 2–4. Woods and scrub; rare in hot, dry areas. ES (incl. Balearic Islands), FR (incl. Corsica), IT (incl. Sardinia and Sicily), PT. *Helleborus viridis* has annual stems to 40 cm, bare below, with *larger flowers 30–50 mm across*. Follicles 3(4), *fused* at the base. Woods and pastures. ES, FR, IT. *Helleborus lividus* has *trifoliate* stem leaves. Follicles 5–7. Wooded slopes; an island endemic of Mallorca. ES (Balearic Islands).

Anemone ANEMONE

Perennial herbs with palmate or palmately lobed basal leaves and a whorl of leafy bracts beneath the solitary (or few) flowers. Perianth with 1 whorl of 5–20 petal-like sepals; stamens numerous. Fruit an aggregate of many-seeded carpels, spirally arranged.

A. Stem leaves *short-stalked* and similar to basal leaves. Carpels *shortly hairy*.

Anemone nemorosa WOOD ANEMONE A delicate perennial 50 mm–30 cm tall with creeping rhizomes and palmately lobed basal leaves with oval, toothed or lobed segments, and similar stem leaves borne on an erect stem bearing solitary, white, drooping flowers; palmately lobed leaf-like bracts groups of 3 below the flowers. Sepals 6–7(9) flushed pink or purple; anthers yellow. Fruits drooping. Deciduous woods in cooler areas. ES, FR, IT. *Anemone apennina* is similar but with *blue* (rarely white) flowers with 10–15(18) sepals, and *erect fruits*. FR (Corsica), IT.

B. Stem leaves *short-stalked* and similar to basal leaves. Carpels *densely woolly*.

Anemone sylvestris SNOWDROP WINDFLOWER A short to medium, hairy, non-rhizomatous perennial to 50 cm with deeply palmately lobed basal leaves which are similar to the stem leaves. *Flowers white,* to 70 mm across, usually solitary; petals 5, broadly oval and silky below; anthers yellow. Fruit head erect and globular with woolly hairy carpels. Deciduous woods and grassy slopes in cooler parts of the European Mediterranean northwards. ES, FR, IT.

---

1. *Helleborus foetidus*
2. *Anemone palmata*
3. *Anemone pavonina* (red-flowered form)
4. *Anemone pavonina* (purple-flowered form)

C. Stem leaves *stalkess* and dissimilar to basal leaves. Carpels *densely woolly*.

Anemone coronaria CROWN ANEMONE A short to medium, hairy, tuberous perennial with basal leaves 2–3-times cut into narrow segments; stem leaves deeply cut into numerous narrow segments and stemless. Flowers pink, red, blue or purple, often with a paler centre, bowl-shaped, to 75 mm with elliptic, overlapping petals; anthers bluish. Common on cultivated land. Throughout. *Anemone hortensis* is similar to *A. coronaria* but has less divided basal leaves, narrow, lanceolate bract-like and entire to 3-lobed stem leaves, and narrow-elliptic petals. Double-flowered forms are often cultivated. Throughout. *Anemone pavonina* may be a form of *A. hortensis* and has bract-like stem leaves which are entire to 3-lobed, and flowers with *broader, oval petals* and blue-black anthers. FR, IT.

Anemone palmata YELLOW ANEMONE A low, hairy perennial with a tuberous rhizome. Basal leaves more or less circular with 3–5 shallow, toothed lobes. Upper leaves united at the base. Flowers bright yellow, to 35 mm across with up to 15 petals. Locally common in rocky habitats and maquis. Frequent in the Algarve; absent from mainland Italy. ES, FR, IT (Sardinia and Sicily), PT.

Ranunculus BUTTERCUPS

Terrestrial or aquatic herbs with entire or lobed leaves. Flowers borne solitary or in cymes; sepals and petals normally 5, the petals often shiny and white or yellow; stamens numerous (or 5–10); carpels numerous. Fruit a head of 1-seeded achenes. Many species occur; not all are described.

A. Leaves as long as broad and entire to palmately divided, flowers yellow and carpels *not smooth* (spiny or warty).

Ranunculus muricatus ROUGH-FRUITED BUTTERCUP A short, usually hairless annual with a stout, much-branched stem to 40 cm with *kidney-shaped*, lower leaves with 3–7-shallow-tooth-lobes; upper leaves wedge-shaped with up to 5 lobes or occasionally entire. Flowers yellow, 6–16 mm across. Fruit rounded, and made up of *large, strongly keeled, flattened achenes with spines to 1 mm on the faces* (not the grooves) and tapered into an abruptly curved beak 2–3 mm. One of the most common species across the region in ditches and in other damp, grassy places. Throughout.

Ranunculus parviflorus SMALL-FLOWERED BUTTERCUP A short, spreading, hairy annual to 40 cm with lower leaves with 3–5 toothed lobes, and the upper leaves simple or more often with deeply cut oblong lobes. Flowers distinctive: pale yellow and *minute,* just 3–6 mm across with reflexed sepals and 1–5 *narrowly oval petals, or petals absent.* Carpels to 3 mm with hooked spines on the sides; style short and hooked. Scattered throughout the west and centre of the region in dry, grassy habitats. ES, FR, IT, MA, PT. *Ranunculus chius* is similar but has 3-lobed leaves and flower-stalked *distinctly swollen when in fruit.* Ditches and wooded habitats. FR (Corsica), IT (Sardinia and Italy).

1. *Ranunculus muricatus*
2. *Ranunculus bullatus*
   PHOTOGRAPH: GIANNIANTONIO DOMINA
3. *Ranunculus flammula*
4. *Ranunculus lingua*

R. bullatus

B. Leaves as long as broad and *entire to shallowly palmately divided*, flowers yellow and *carpels smooth*.

Ranunculus bullatus A low, hairy perennial with leaves all basal, stalked, oval to rounded and toothed but not lobed. Flowers borne solitary on leafless stalks, yellow, to 26 mm across with 5–12 petals, and *sweetly scented*. Local on rocky slopes. Throughout.

C. Leaves as long as broad and *deeply palmately lobed*, flowers yellow; carpels smooth, wrinkled or spiny.

Ranunculus arvensis CORN CROWFOOT A short to medium, often hairless annual to 60 cm with spatula-shaped, simple or more commonly toothed to dissected leaves with narrowly lanceolate to linear lobes. Flowers bright pale green to lemon-yellow, 4–12 mm across, borne in branched clusters. Fruiting heads with achenes 6–8 mm with *prominent, long, rigid spines >1 mm long*. Common on cultivated and disturbed land. Throughout. *Ranunculus marginatus* is similar but sparsely hairy with more deeply cut leaf lobes, deep, golden yellow flowers and achenes with *blunt, wrinkled tubercles on the faces and a short beak to 1 mm*. Native to Asia but widely naturalised. FR (Corsica), IT (incl. Sardinia).

Ranunculus sceleratus CELERY-LEAVED CROWFOOT A medium, rather stout and fleshy, more or less hairless annual with *shiny green leaves* that are deeply divided, those above into narrow segments, on much-branched, hollow, grooved stems. Flowers numerous, borne in branched clusters, yellow, 5–10 mm across, with reflexed sepals. Fruit a cylindrical head of numerous hairless achenes, each to 1 mm. Wet, marshy habitats. Throughout.

Ranunculus bulbosus BULBOUS BUTTERCUP A medium, hairy *perennial* to 40 cm with a swollen underground stem base; *hairs spreading below but adpressed to the stem in the upper parts*. Basal leaves stalked and 3-lobed. Flowers borne on furrowed stalks, 15–30 mm across and *bright yellow*. Vegetative parts often wither after flowering. Throughout. *Ranunculus sardous* is similar to *R. bulbosus* but a hairy *annual* with a scarcely swollen underground stem base. Leaves 3-lobed and shiny. Flowers 12–25 mm across and *pale yellow*; sepals have dark markings along the margins (best seen in bud). Achenes smooth except for a row of small tubercles surrounded by a green border. Coastal grasslands, damp places, meadows and other grassy habitats. Throughout.

Ranunculus velutinus A rather tall, silkily downy hairy perennial, especially below. Leaves geranium-like, *broadly oval* with 3 wedge-shaped toothed to lobed divisions; stem leaves similar but smaller. Flowers yellow, to 25 mm across, borne on slender stalks; sepals reflexed. Fruiting heads globular and hairless. FR (incl. Corsica), IT. *Ranunculus monspeliacus* is a similar white to silvery hairy to woolly perennial with variable leaves that are generally oval with 3 shallow to deeply lobed segments. Flowers yellow, to 26 mm across with reflexed sepals. Fruit oblong and slightly hairy. Rocky and grassy habitats from Spain eastwards (absent from Sardinia and Balearic Islands). ES, FR, IT.

Ranunculus paludosus JERSEY BUTTERCUP An annual to 40(60) cm. Leaves stalked and pale green, those at the base with saddle-shaped lobes, the upper leaves 1–2, small with numerous narrow linear lobes. *Flowers pale lemon yellow,* 20–30 mm across. Achenes 2.5–3 mm long, mostly smooth with a straight or scarcely hooked beak 1 mm long. An arable weed on disturbed ground. Throughout.

D. Leaves entire and markedly *longer than broad*, flowers yellow.

Ranunculus flammula LESSER SPEARWORT A rather fleshy, hairless (or nearly so), hollow-stemmed perennial; stems to 50 cm and rooting at the base. *Leaves narrow and entire*; broader and more heart-shaped at the base. Flowers rather few, borne on slightly furrowed and hairy stalks, small, shiny and yellow, 7–20(25) mm across; sepals hairless and spreading. Carpels numerous, hairless and pitted,

1–2(2.3) mm long. Locally common in damp and water-logged habitats. Throughout. *Ranunculus lingua* is similar but more robust, to 1.2 m, erect and with much larger flowers, 20–50 mm across borne on unfurrowed stalks. Similar habitats. Throughout.

Ranunculus bupleuroides A perennial 25–60 cm tall with entire, broad, stalked blue-green, oblong-lanceolate leaves which are often inwardly curved, with *distinct paler venation*. Flowers yellow, 18–27(32) mm across with 5–6 petals, each with a basal nectary. Local in cool, damp habitats. PT.

E. *Flowers pink or white* and plant *aquatic* (species similar and difficult to distinguish).

Ranunculus aquatilis COMMON WATER-CROWFOOT An annual or perennial aquatic herb with floating and submerged leaves, the latter sometimes absent; floating leaves deeply divided into *3–5(7) straight-sided segments with teeth at the tips. Flowers up to 18 mm across; petals adjacent when mature.* Fruit stalk up to 50 mm long (*shorter* than stalk of opposite leaf). Locally common in suitable habitats; occurs in still or slow-moving water. Throughout. *Ranunculus peltatus* POND WATER-CROWFOOT is similar but with divergent leaf segments with *rounded tips*, and *larger flowers* to 30 mm across, and a fruit stalk to 15 cm long (*longer* than stalk of opposite leaf). Locally common in aquatic habitats, but varies in abundance with seasonal rainfall. Throughout. *Ranunculus trichophyllus* is similar but with petals *not touching when mature* (widely spaced). Still or slow-moving water. Throughout.

Ficaria

Tuberous perennial herbs with heart-shaped leaves. Flowers yellow, with 7–12(13) petals; stamens and carpels numerous. Fruit a head of achenes.

Ficaria verna (Syn. *Ranunculus ficaria*) LESSER CELANDINE A variable (with several subspecies described), hairless perennial to 25 cm with long-stalked, triangular *heart-shaped*, wavy-margined to shallowly-lobed, fleshy, dark green leaves. Flowers 10–30 mm across, shiny and yellow, turning white on ageing with 7–12(13) narrow petals and *3 sepals*. Very common in damp, grassy places. Throughout.

Nigella LOVE-IN-A-MIST

Annual herbs with solitary flowers and rather feathery, pinnately divided leaves with narrow segments. Flowers 5-parted, usually bluish. Carpels fused along their inner margins and many-seeded.

Nigella damascena LOVE-IN-A-MIST An erect hairless annual to 50 cm (usually less) with alternate, finely 2–3-pinnately divided leaves with narrow segments, the *uppermost in a feathery whorl just below the flowers*. Flowers solitary, and pale blue, 15–30(35) mm across with 5 petals and a central cluster of stamens and carpels; carpels joined over most of their length. Common on cultivated, sandy, disturbed and waste ground. Throughout.

Nigella gallica (Syn. *N. arvensis*) A short, hairless annual 25–40 cm with leaves alternate, and finely divided into linear segments. Flowers 30–35(40) mm across, bright blue, with reddish anthers. Carpels fused for most of their length, and densely minutely glandular; erect-spreading in fruit. Frequent in a range of habitats. ES, MA, PT. *Nigella papillosa* (Syn. *N. arvensis*) is similar but with large flowers 35–60(70) mm across and styles spreading in fruit. ES, MA, PT.

Delphinium

Annuals or perennials with palmately, broadly divided leaves and, typically blue, flowers borne in erect racemes, each with 5 petal-like outer segments, the uppermost spurred at the back, and 4 inner petal-like segments, the 2 uppermost with spurred nectaries. Fruit 3–5 follicles.

A. Plant an annual or biennial.

D. staphisagria

Delphinium staphisagria A medium to tall hairy annual or biennial 30 cm–1 m with alternate leaves along the stem, 1-pinnately divided into 5–7 lobes (sometimes 2-pinnate). Flowers deep blue, large to 25 mm across, borne in long racemes. Flowers with *short, blunt, down-turned spurs* 30–50 mm long - shorter than the petals. Fruit with inflated follicles, bearing few seeds. Local on rocky slopes and on garrigue and maquis inland. The most widespread and common species in the region. Throughout. *Delphinium pictum* is a similar, hairy annual or biennial with lanceolate, rather lobed segments and pale blue flowers borne in long racemes with bracts at the base; the spur 6–8(9) mm (about 2/5 the length of the petals). Fruit with inflated follicles bearing *numerous seeds*. Rocky habitats on all the larger islands and mainland Italy. ES, FR, IT. *Delphinium requienii* is similar but covered in *long, silky hairs* and bracts in the middle (not at the base) of the racemes. FR (incl. Corsica).

1. *Nigella damascena*
2. *Nigella gallica*
3. *Delphinium peregrinum* (pale-flowered form)
4. *Delphinium peregrinum*
5. *Delphinium gracile*

Delphinium peregrinum VIOLET LARKSPUR A slender, medium to tall hairy annual or biennial to 1 m; stems with a white bloom of spreading hairs. Lower leaves greyish and palmate with narrow segments, the upper leaves unlobed. Flowers borne in slender racemes, dull blue to violet, or dirty white, *small* to 18 mm across, the spur *upturned and longer than the petals*; the nectariferous petals hairless. Dry scrub. IT (incl. Sicily). *Delphinium halteratum* is very similar but shorter, to 40 cm, with leaves without a white bloom and hairs adpressed to the stem (not spreading) and *brighter blue flowers* in which the spur exceeds the petals x 1.5–2.5, the sepals are broadly oval, and all tepals are finely downy all over. FR, IT (Sardinia and Sicily). *Delphinium gracile* is similar to *D. peregrinum* but with paler flowers with a *long spur*, 17–19 mm long, 2–4 x as long as the tepals; lateral petals are oblong and *heart-shaped* at the base; inflorescence not branched below. Coastal scrub. ES, PT. *Delphinium nanum* is similar to *D. peregrinum* but the lateral petals are *not heart-shaped* at the base, and the inflorescences branched below. Coastal scrub; widespread. Throughout.

B. Plant a perennial; restricted to the high altitude regions of Iberian Peninsula.

Delphinium montanum A bushy perennial with numerous inflorescences of *pale blue flowers with dark brown inner petals*. Seeds have a densely pitted surface. Mountains. ES. *Delphinium fissum* is a lax perennial with leaves *clasping the stem* and inflorescences of dull dark blue flowers 22–26 mm across which are densely hairy and *partially closed*. Seeds with <25 pits on the surface. Rocky woods. ES. *Delphinium bolosii* is a perennial similar to *D. fissum* but with larger (32–35 mm) *fully open flowers*. Mountains of Tarragona. ES. *Delphinium emarginatum* is a small, lax perennial distinguished from *D. montanum* by its leaves which are only *semi-clasping the stem* (around $^1/_2$ the circumference). Restricted to the Sierra Nevada. ES.

Consolida LARKSPUR

Annuals similar to *Delphnium* with palmate leaves with *numerous thread-like segments*. Flowers in terminal racemes; sepals 5, the upper long-spurred, petals 4. Fruit a *solitary* follicle.

C. ajacis

Consolida ajacis LARKSPUR A medium to tall, downy annual to 1 m with a simple or branched stem and deeply dissected lower leaves. Flowers borne in rather lax inflorescences, typically *bright blue* (pink, white and pale blue forms are cultivated), the upper petal with a backwardly projecting spur, 12–18 mm long. Local on disturbed ground, sandy places and field margins. Widely naturalised. Throughout. *Consolida orientalis* is similar but stickily hairy, and with short-stalked *purple-violet* flowers borne in a dense raceme; *spur shorter, 6–10(12) mm*. Seeds red-brown. Cultivated land and waste ground. Throughout.

Consolida regalis FORKING LARKSPUR A slender, medium, rather downy, widely branching annual to 50 cm. Leaves divided into linear lobes. Flowers violet-blue to dark blue, to 28 mm, borne in lax panicles; spur 12–25 mm long. Fruit hairless; seeds black. Arable and disturbed habitats. Throughout, except most islands. *Consolida pubescens* is similar but *distinctly hairy*, with dissected lower bracts. Flowers variably whitish, violet or pale blue; spur 12–15 mm long. Seeds grey to reddish. ES, FR, IT (incl. Sicily). *Consolida mauritanica* is similar *C. regalis* but the upper lobe of the nectary is shorter (0.5–1 mm, not 2–3 mm) and the spur longer, 17–20(22) mm. ES, MA, PT.

Clematis

Woody climbers with opposite, pinnately divided leaves. Flowers borne in branched cymes. Fruit a cluster of achenes, each usually with an elongated, often feathery style.

A. Plant not climbing or scrambling, stems hollow below.

Clematis recta A rather bushy, *erect,* herbaceous perennial to 1.5 m tall with pinnately divided leaves and *erect to spreading (non-climbing) stems* that are hollow, at least below. Flowers white and hairless, except on the margins, similar in appearance to *C . flammula* (see below). Widespread in the European Mediterranean on rocky ground and scrub. ES, FR, IT.

1. *Consolida ajacis*
2. *Consolida regalis*
3. *Clematis recta* (inset: flower)
4. *Clematis vitalba*
5. *Clematis flammula* (inset: flower)

B. Plant climbing or scrambling, stems solid.

Clematis vitalba TRAVELLER'S JOY A woody climber to 30 m with *1-pinnately divided leaves* with pointed leaflets. *Flowers dull greenish white or cream* with segments *hairy on both sides*, fragrant, borne in terminal or lateral loose clusters. Styles feathery in fruit. Scrubby habitats. Rarer in the far west and the south. ES, FR, IT, PT. *Clematis flammula* FRAGRANT CLEMATIS is similar but shorter, to 6 m, and has *2-pinnately divided leaves* with stalked, oval to circular leaflets. *Flowers white*, to 20 mm across, fragrant, with 4 pointed segments *hairy only on the margins and under-surface*. Maquis and scrub. Throughout.

Clematis cirrhosa VIRGIN'S BOWER An evergreen climber to 4 m with shiny green leaves that are variably divided, 3-lobed or entire, toothed or not. Flowers cream, often red-spotted within, nodding and *bell-shaped*, to 20 mm long, and silky hairy outside. Woods and scrub; widespread but local. Throughout.

Clematis viticella A deciduous climber to 4 m with pinnately divided leaves with oval, untoothed segments. *Flowers blue or purple, 30–60 mm across*, opening widely; styles not feathery. Scrub and thickets; cultivated forms are also widely grown. IT. *Clematis campaniflora* is similar in form, though more slender, with *small, pale violet* (or whitish) flowers to 30 mm across with *feathery styles*. ES, PT. *Clematis orientalis* is a climber with *yellow, bell-shaped flowers*. Native to Asia, naturalised. ES.

Aquilegia COLUMBINE

Herbaceous perennials with leaves spirally arranged. Flowers regular with 5 sepals and petals, each with a backward-pointing spur; stamens numerous; carpels 5(10). Fruit a follicle.

Aquilegia vulgaris COLUMBINE A variable, erect, hairy perennial, often branched, to 60 cm (1 m). Leaves stalked and toothed or lobed. Flowers large, to 50 mm long, *blue-purple* (sometimes white or pink) with 5 similar tepals, the *petal-like segments elongated into erect, curving spurs* 15–22 mm long. Follicles 15–20 mm. Grassy, shady habitats and wetlands. ES, FR, IT, MA.

Adonis PHEASANT'S EYE

Erect, hairless annuals with feathery, 2–3-pinnately divided leaves. Flowers yellow or red and buttercup-like; sepals 5, petals 5–8; stamens and carpels numerous. Fruit an aggregate of achenes.

A. Flowers always scarlet.

Adonis annua PHEASANT'S EYE An erect, hairless annual to 40 cm with simple or branched stems. Leaves 3-pinnately divided with numerous narrow, feathery segments; *the lower leaves unstalked*. Flowers 15–25 mm across, borne erect and terminal, with 5–8 scarlet petals, blackish at the base, and *hairless, spreading sepals*. Disturbed, damp and grassy habitats. Throughout. *Adonis flammea* differs in having *hairy sepals closely adpressed to the petals*. ES, FR, IT, MA. *Adonis aestivalis* is similar to both the previous species but has *hairless sepals adpressed to the petals* and *stalked lower leaves*. ES, FR, IT, MA.

B. Flowers yellow (sometimes orange or red).

Adonis microcarpa YELLOW PHEASANT'S EYE A short to medium annual, similar in general appearance to *A. annua* but with *flowers yellow throughout* (sometimes orange or red) to 28 mm across. Fruiting achenes borne in *elongated seed heads* (not rounded to egg-shaped). ES, FR, IT, MA, PT. *Adonis dentata* is similar but *short, usually to just 12 cm*, and with *flowers yellow, blackish in the centre*. DZ, TN.

Ceratocephala

Small annuals with trifoliate leaves with *narrow, forked segments*. Flowers yellow; stamens 5–10. Fruit an aggregate of achenes.

Ceratocephala falcata A low, hairy, rather tufted annual 20 mm–12 cm high with all basal, long-stalked trifoliate leaves divided into narrow, forked segments. Flowers regular, solitary, yellow, (8)10–16 mm across; petals 1.5 x as long as sepals. Fruits borne in a teasle-like heads of long, upward-curving, spine-like achenes. Rare and local. DZ, ES, FR, IT, TN.

1. *Aquilegia vulgaris*
2. *Adonis annua*
3. *Papaver somniferum*
4. *Papaver rhoeas* (flowers)
5. *Papaver rhoeas* (fruit)

PAPAVERACEAE | POPPY FAMILY

Annuals or perennials with milky or watery sap. Leaves shallowly 1–2-pinnately lobed. Flowers solitary or in racemes. Sepals 2(3), petals 4(6), often crumpled when newly opened; stamens numerous; style 1. Fruit a splitting capsule.

Papaver POPPY

Annuals or perennials often with a white latex. Flowers solitary with red, mauve or white petals; stigma a stalkless 4–20, rayed, flat disk.

A. Fruits lacking prominent bristles.

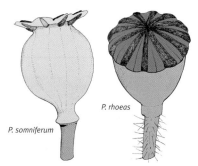

P. rhoeas

P. somniferum

Papaver somniferum OPIUM POPPY A vigorous, erect, *whitish-glaucous* annual to 50 cm (1 m) with pinnately divided, oval leaves. Lower leaves with a short petiole, the upper leaves clasping the stem. Flowers large, petals 25–50 mm long, pale purple with a dark centre, anthers yellow. Capsule hairless. A relic of cultivation and widely naturalised in waste places. Throughout. Subsp. *setigerum* has stems *with sparse long, fine bristles*. The leaves are often more deeply lobed, and *end in a bristle*. Capsule rather narrow. Native on coastal sands. Probably throughout.

Papaver rhoeas COMMON POPPY An erect, bristly annual 60(80) cm tall. Leaves pinnately lobed, to 15 cm long with pointed segments, often 2-pinnately divided. Flowers solitary on long stalks with long, bristly hairs; petals crimson with or without a dark centre, 30–45 mm long; anthers bluish. Capsule more or less *rounded, and hairless,* <20 mm. Common on cultivated and disturbed ground. Throughout.

1. *Papaver dubium*
2. *Papaver hybridum*
3. *Papaver dubium* (fruit)
4. *Glaucium flavum*

1

2

3

4

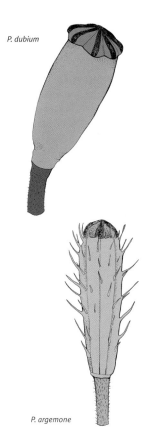

P. dubium

P. argemone

Papaver dubium LONG-HEADED POPPY Similar to *P. rhoeas* but with adpressed hairs on the upper parts of the stem, the leaf segments blunt, not pointed, flowers a paler or more orange-red, usually without a dark centre; petals 15–35 mm long; capsule *oblong*, somewhat widened towards the apex. Common on cultivated and disturbed ground. Throughout. *Papaver guerlekense* (Syn. *P. stipitatum*) is similar but with smaller flowers with petals <32 mm, and the capsule stalked above the petal bases. Native to the eastern Mediterranean, naturalised. ES. *Papaver pinnatifidum* is similar to the above species but has oval-triangular leaf segments, and yellow or brownish rather than violet anthers. ES, FR, IT.

B. Fruits with prominent bristles.

Papaver argemone PRICKLY POPPY A short to medium, bristly annual to 45 cm with adpressed hairs on the stem, and pinnately divided leaves. Flowers scarlet, petals 15–25 mm long, often with a dark centre, and the *petals not overlapping*. Capsule <25 mm, shortly *cylindrical*, ribbed, and *sparsely bristly*. Common on sandy waste ground, often coastal. Throughout. *Papaver hybridum* is similar but with darker crimson-red flowers with petals 10–25 mm long, and an ovoid to *spherical capsule that is covered in pale, stiff bristles*. Similar habitats. Throughout. *Papaver apulum* is similar to *P. hybridum* but slightly hairy (rather than bristly), and with *hairless flower buds*, and bristly but not ribbed capsules. IT.

Roemeria

Annuals. Flowers with 4 petals; stamens numerous; stigmas 2–4. Similar to *Papaver* but with *linear fruits* that split at the base into 2–4 parts.

Roemeria hybrida (Syn. *Chelidonium hybridum*) ROEMERIA A short, slightly hairy annual 20–40(50) cm high with yellow sap. Leaves alternate, 3-pinnately divided into linear segments. Flowers poppy-like and *violet to purple* with a darker centre; petals 15–30 mm long; sepals 10–13 mm; anthers pale blue or cream. *Capsule long and linear,* bristly and 4-parted, 50 mm–10 cm long. DZ, ES, FR, IT, MA, TN.

Glaucium

Annuals to perennials with a watery latex. Flowers solitary with 4 petals; stamens numerous. Fruits *very long and narrow,* splitting into 2 parts.

Glaucium flavum YELLOW HORNED-POPPY A glaucous, branched biennial to perennial to 90 cm with oblong, wavy and pinnately lobed leaves, the upper leaves clasping the stem. The plant has a yellowish latex. Flowers yellow, 60–90 mm across. Fruit narrowly cylindrical and *very long*, 15–30 cm; curved, and *hairless*. On coastal sands and shingle, or disturbed habitats inland. Throughout. *Glaucium corniculatum* RED-HORNED POPPY is similar but with *orange to red flowers* (sometimes yellow) to just 50 mm across. *Capsule hairy*. Throughout.

Hypecoum

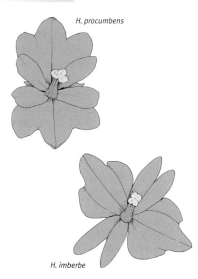

H. procumbens

H. imberbe

Annual herbs. Flowers small with 4 rather unequal petals; stamens 4; stigmas 2. Fruit a capsule, usually curved.

Hypecoum procumbens A delicate, low to short, hairless, glaucous annual with wide-spreading stems to 15 cm. Leaves 2-pinnately lobed, segments lanceolate or linear. *Bracts leaf-like.* Flowers borne in small, branched clusters, flowers to 15 mm across, with 4 yellow petals: 2 large 3-lobed petals, and 2 small, lateral petals. Capsule erect and jointed. Local in maritime habitats. Throughout. *Hypecoum imberbe* is similar but more erect and with *linear bracts.* Flowers orange-yellow with more or less *evenly-lobed large outer petals.* Fruits scarcely jointed. Cultivated land and waste places. Throughout. *Hypecoum pendulum* is similar to *H. procumbens* but with the *outer 2 petals unlobed or scarcely lobed. Capsule pendent, straight and scarcely jointed.* Various disturbed habitats. DZ, ES, FR, IT, MA, TN.

Fumaria

Trailing or scrambling annual herbs with 2–4-pinnately divided leaves and distinctive leaf-opposed racemes of tubular 2-lipped flowers with 2 small sepals, 2 outer petals and 2 narrower inner petals; stamens 2. Fruit a more or less spherical 1-seeded achene. Only the most widespread species are described; species similar and difficult to distinguish.

F. agraria

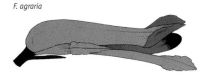

Fumaria agraria An annual to 3 m with broad, flat, oval leaf segments. Flowers 2-lipped, pale pink, only the inner petals with dark purple tips, and the lower petal spreading at the margin, borne in a racemes of 8–15, longer than their stalks; corolla 11–15 mm. Fruits notched at the top. Frequent in Iberian Peninsula. ES (incl. Balearic Islands), PT. *Fumaria barnolae* is similar but with 15–25 darker pink flowers, 12–14 mm with purple upper lip and inner petals. *Fruit not notched, borne on slender, recurved stalks.* Northern Spain and the islands eastwards. ES, FR, IT. *Fumaria gaillardotii* is similar *F. barnolae* but with *fruits borne on robust, upward-pointing stalks.* ES, FR, IT.

F. capreolata

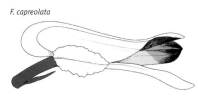

Fumaria capreolata WHITE RAMPING-FUMITORY Rather similar to *F. agraria*; a tall, hairless, blue-green, scrambling annual without tendrils. Leaves wedge-shaped with blunt lobes. *Flowers held sub-erect,* 2-lipped, 10–13(14) mm, the *upper petal compressed with upturned margins not concealing the keel,* creamy white, often tinged with pink, and *tipped with reddish black,* borne in a cylindrical raceme of 14–25(30) flowers; fruit stalk strongly curved. *Racemes shorter than their stalks.* Very common on disturbed ground. Throughout. *Fumaria bicolor* is similar to *F. capreolata* but with short racemes of 8–15 flowers, 11–12.5 mm *gradually turning pink,* tipped with dark purple. Recorded from

1. *Fumaria capreolata*
2. *Fumaria agraria*
3. *Fumaria officinalis*

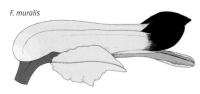
F. muralis

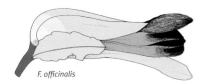
F. officinalis

southern Iberian Peninsula, but probably only present on the islands. ES (Balearic Islands), FR (Corsica), IT (Sardinia). *Fumaria muralis* is similar to both species, with 10–17 pink flowers, 9–11(12) mm, the upper petal bearing broad wings concealing the keel; sepals toothed at the base, and *racemes as long as (or longer) than their stalks*. ES, FR, PT.

Fumaria officinalis COMMON FUMITORY A very delicate hairless, blue-green, scrambling annual with broad, flat, oval-lanceolate leaf segments. Flowers 7–8(9) mm, *numerous*, 10–45, *mauve*, tipped with blackish purple on the wings of the upper petal and apex of the inner petals, borne in a raceme longer than the *short stalk*. Fruits wider than long. Common on disturbed ground. Throughout. *Fumaria parviflora* is similar but with *chanelled leaf segments*. Flowers small, 5–6 mm long and *white, flushed very pale pink* in almost stalkess racemes; *bracts at least equalling fruiting stalks*. Throughout. *Fumaria vaillantii* is similar to *F. parviflora* but with *pink flowers*, 5–6 mm; *bracts shorter than fruiting stalks*. ES, FR, IT, MA, PT.

Platycapnos

Annuals with pinnately divided leaves. Flowers zygomorphic; stamens 2. Like *Fumaria* but with fruits strongly compressed (not spherical); 1-seeded.

Platycapnos spicatus A grey-green annual to 30 cm tall with 2-pinnately lobed leaves. Flowers tiny, borne in dense rounded heads, superficially clover-like; petals creamy to pink, tipped red. Fruit oval. Local on disturbed ground. ES, FR, IT, PT.

1   2   3

PLATANACEAE | PLANE TREE FAMILY

A small family of trees (with only 1 genus), with peeling bark and simple, palmately lobed leaves. Flowers unisexual, regular; stamens 3–4; carpels 5–8. Fruit a hairy achene.

P. x hispanica

Platanus × hispanica (Syn. *P. × hybrida, P. × acerifolia*) PLANE TREE A large, deciduous tree to 35(44) m with bark peeling in flakes, and 5–7 sharply lobed palmate leaves. Flowers borne in dense globular heads, often in pairs; fruit globular and pendent, 20–35 mm across. Not native (of uncertain horticultural origin) but widely planted on roadsides in the area. Throughout. *Platanus orientalis* is similar, with the base of the leaves with numerous smaller lobes, the lobes longer than wide and toothed, and usually 3–6 fruits per cluster. Native to the Himalayas; widely planted throughout.

BUXACEAE | BOX FAMILY

Evergreen trees and shrubs with simple, opposite leaves. Flowers borne in early spring, unisexual, small and in spikes or lateral clusters; sepals and stamens 4; styles 2–3. Fruit a capsule.

Buxus BOX

Shrubs with small, leathery, opposite leaves and unisexual flowers with 4 stamens. *Fruit a 2–3-horned capsule.*

Buxus sempervirens BOX An evergreen shrub to 5 m with downy, 4-angled shoots. Leaves small, oval, 13–25(30) mm long and 7–12(15) mm wide, frequently upward-swept, pale and sometimes hairy below. Male and female flowers separate but on the same plant, in axillary and terminal inflorescences, greenish and rather inconspicuous. Fruit a shiny, green-yellow, ovoid capsule 8–11 mm long with 3 horns. Widespread and fairly common on garrigue, especially in southern France, but absent from most islands. DZ, ES, FR, IT, MA, TN. *Buxus balearica* is similar but has larger leaves 30–45(50) mm long and 15–25(30) mm wide, the whole plant is virtually hairless and the fruiting capsule 12–14 mm. Similar habitats. ES (incl. Balearic Islands), IT (Sardinia), MA.

DROSOPHYLLACEAE

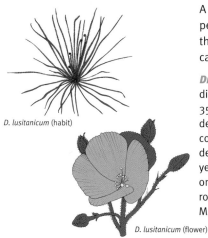

D. lusitanicum (habit)

A small family (with 1 genus) of insectivorous perennials with conspicuous sticky hairs covering the leaves. Flowers yellow with 5 petals; stamens 10; carpels 5. Fruit a capsule.

Drosophyllum lusitanicum PORTUGUESE SUNDEW A distinctive slender, *very sticky-glandular* perennial 20–35(45) cm high with ascending stems and leaves in a dense rosette. Leaves linear, spreading and long-tapering, covered in red-tipped glandular hairs (often covered in dead insects), 10–20 cm long and 2–3 mm wide. Flowers yellow, with 5 petals 20–30 mm long, borne in lax clusters on a scape to 30 cm long. Very rare and local, in disturbed rocky habitats, tracks and dry forests on acid soils. ES, MA, PT.

D. lusitanicum (flower)

TAMARICACEAE | TAMARISK FAMILY

Deciduous shrubs or trees with alternate, small (often scale-like), stalkless leaves. Flowers typically in catkin-like racemes, with (4)5 sepals and petals; stamens 5; styles 3–4. Fruit a capsule; seeds with hairy tufts.

Tamarix

1. *Platanus orientalis*
2. *Buxus sempervirens*
3. *Drosophyllum lusitanicum*
4. *Drosophyllum lusitanicum*

Shrubs or small trees with a distinctive habit and simple, alternate, small, scale-like leaves. Flowers tiny, borne in catkin-like spikes. Fruit a capsule with hairy, wind-borne seeds. Species often very similar and difficult to distinguish in the field.

A. Flowers 5-parted.

Tamarix africana TAMARISK A feathery, more or less hairless tree to 3(6) m with blackish bark. Leaves small, 1.5–3 mm long, pointed and growing close to the stem. Flowers white or pale pink, borne in *almost stalkless racemes* 30–70 mm long and *6–8 mm wide*, borne on the previous year's wood *often before the leaves*. Petals 5, to 2–3.3 mm long and usually *persistent* in fruit. Bracts hairy. Widespread in coastal marshes and near streams. Throughout. *Tamarix mascatensis* is very similar to *T. africana*, but with racemes mainly arising from young stems. DZ, MA, TN. *Tamarix canariensis* is similar to *T. africana* but minutely hairy with reddish brown bark, with small, pale pink *oblong petals* 1.3–1.9 mm that *fall in fruit; racemes just 3–5 mm wide*. Similar habitats; absent from many islands. DZ, ES, FR, IT, MA, PT. *Tamarix gallica* is very similar to *T. canariensis* but entirely hairless with blue-green leaves. Racemes 20–50 mm long and 4–5 mm wide, *petals elliptic, 1.7–2 mm long. Inflorescence stalk hairless*. Absent from many islands. ES, FR, IT.

B. Flowers 4-parted or both 4- and 5-parted.

Tamarix parviflora A shrub or tree to 5 m with brown-purple bark, hairless or minutely hairy. Leaves pointed, 3–5 mm long with membranous margins. Flowers white or pink, typically with *4 petals*, borne in racemes to 30 mm long and up to 5 mm wide on the previous year's wood; sepals finely toothed; bracts tiny; both *bracts and sepals purple at the tips*. Riverbanks and roadsides. Native to the east Mediterranean, naturalised elsewhere. ES, FR (Corsica), IT, TN. *Tamarix dalmatica* is similar, also hairless, and with *thick (not slender) racemes to 12 mm wide*; flowers pale pink and typically 4-parted. Coastal marshes. ES (incl. Balearic Islands), IT (Sicily).

Reaumuria

Typically small, prostrate shrubs in arid habitats. Similar to *Tamarix* but with *fewer, larger* flowers; stamens ~70; styles 3–5. Fruit a 3–5-valved capsule.

Reaumuria vermiculata REAUMURIA A hairless erect or ascending, *small shrub* to just 30 cm. Leaves semi-cylindrical, to 12 mm long, long-pointed, blue-green and crowded around the base of the branches, sparser above; stems with leafy lateral shoots. *Flowers solitary*, white and 5-parted, the calyx concealed by numerous long-pointed bracts; stamens in 5 clusters; styles 5. Coastal rocks and cliffs, predominantly in North Africa. DZ, IT (Sicily), TN.

FRANKENIACEAE | SEA HEATH FAMILY

Dwarf shrubs (sometimes annual herbs) with opposite, entire leaves and no stipules. Flowers usually cosexual with 5 partly fused sepals and 5 free petals; stamens usually 6; style 1, divided. Fruit a small capsule.

Frankenia

Woody-based subshrubs, easily recognisable by their distinctive flower structure and small, *Erica*-like leaves.

F. pulverulenta

A. Plant an *annual* (scarcely woody at the base).

Frankenia pulverulenta ANNUAL SEA HEATH An *annual with numerous prostrate branches* 50 mm–17(30) cm, not particularly woody, often spreading in a circle, with oval leaves 5–6(8) mm long that are hairless above, crispy hairy below, and often reddish. Flowers borne solitary in clusters in the axils of the branches and upper leaves; stalkess, pink, the petals notched. Various dry and saline habiats. Throughout.

B. Plant (usually) a perennial, with flower *clusters scattered along the terminal stems* (not all terminal).

Frankenia laevis SEA HEATH A low, prostrate, mat-forming, woody and often minutely hairy, often reddish *perennial* with branches 80 mm–50(60) cm, with tiny white-encrusted, *linear* leaves 2.5–4.5(10) mm long with inrolled margins. Flowers pale to deep pink borne mostly in rather crowded short *lateral branches* (not restricted to terminal clusters). Dunes and cliff-tops. A primarily Atlantic species scattered across the Mediterranean. ES, IT, MA, PT. Subsp. *composita* (Syn. *F. composita*) is a weak perennial (sometimes annual) slightly woody at the base, with spreading prostrate reddish branches 10–30(50) cm long and oblong leaves 5–8 mm long, hairless above and with rather broad, flat, *spatula-shaped bristles on the underside and on the petioles*. Flowers purplish, borne in lateral or terminal clusters. Sand dunes. ES (incl. Balearic Islands), MA.

F. laevis

1. *Tamarix africana*
2. *Tamarix gallica*
3. *Frankenia pulverulenta*
4. *Reaumuria vermiculata* PHOTOGRAPH: GIANNIANTONIO DOMINA

C. Plant a perennial, with *flowers arranged in dense terminal clusters or spikes*.

Frankenia hirsuta HAIRY SEA HEATH A low, prostrate, mat-forming perennial with branches 10–20(40) cm long, similar in form to *F. laevis, but white-bristly-hairy* at least above, not strongly white-encrusted, but often white powdery above, and with pale to mid pink *flowers borne at the ends of the branches* (not in lateral clusters). Saline habitats. Absent from much of Iberian Peninsula. ES (Balearic Islands), FR, IT, MA.

Frankenia thymifolia THYME-LEAVED SEA HEATH A rather large, spreading, *ascending perennial with arching branches* 15–25(30) cm long and triangular-oval leaves 2–2.5(3) mm long. *Flowers bright pink, borne often numerously in long, arching, rather 1-sided spikes*. Saline habitats. Rare, in the southwest only. ES, MA.

F. boissieri

Frankenia boissieri A bushy (often dense) perennial with erect branches 90 mm–30(40) cm long with *oval-heart-shaped leaves* 4–7(10) mm long, not white-encrusted, *hairless above*, and purplish *flowers borne in dense terminal clusters;* calyx with white hairs. Saline maritime habitats in southern Iberian Peninsula and northwest Africa. ES, MA, PT. *Frankenia corymbosa* is a similar erect or ascending perennial which differs in having *linear, short-hairy leaves* 2.5–5(8) mm long, often partially white-encrusted. Saline habitats. Far south only. ES, MA.

1. *Frankenia laevis*
2. *Frankenia boissieri*
3. *Armeria pungens*

PLUMBAGINACEAE | THRIFT FAMILY

Perennial herbs with basal, untoothed, simple leaves without stipules. Flowers regular, borne in lax or tight clusters, 5-parted with papery calyx lobes that persist in fruit; petals fused in the lower part to form a tube; stamens and styles 5. Fruit a 1-seeded capsule.

Plumbago

Herbaceous perennials and shrubs with clasping leaves. Inflorescence a dense, terminal spike; corolla tubular; stamens 5, free; style 1 with 5 stigmas. Fruit dry, 1-seeded and 5-valved.

Plumbago europaea A tall, erect, much-branched perennial to 1.2 m with alternate, oval to oblong leaves to 10 cm long, clasping and heart-shaped at the base, *glandular along the margins*. Flowers violet or lilac-pink, the lobes darker *20 mm long*, borne in a spike-like inflorescence; lobes oval. Fruit a 5-parted capsule. Coastal sands, disturbed habitats and hedgerows. Throughout. *Plumbago auriculata* is a scrambling shrub to 6 m with leaves *without* glandular margins. Flowers sky-blue, *40 mm long*. Native to South Africa, planted throughout, sometimes naturalised. ES, PT.

Armeria THRIFTS

Tufted perennials with basal leaves. *Flower-heads dense and hemispherical* with papery bracts, borne on long, leafless stalks; stamens and styles 5. Genereally rare and local in the Mediterranean outside of Iberian Peninsula; species are described by region here.

A. Southern Iberian Peninsula (difficult group with many taxa described which are similar, hybridising, and questionably all true species).

A. gaditana

Armeria pungens A compact, *dwarf shrub with a woody stock,* to 50 cm high. Leaves linear, 40 mm–10 cm long, 1–2.5(3.5) mm wide, rather grass-like and tufted. Flowers white or pale pink in heads to 30 mm across borne on stalks 10–35 cm long. Generally the most common species on coastal sands in the southwest Iberian Peninsula. ES, FR (Corsica), IT (Sardinia), PT. *Armeria macrophylla* is similar but with short branches, and longer, *thread-like* leaves 10–23(30) cm long and just 0.5–1.5(2) mm wide. Scrub and sandy or stony places. Southwest Iberian Peninsula. ES, PT. *Armeria gaditana* is less woody, vertical and little-branched, with much *broader*, strap-like leaves 90 mm–25 cm long and 9–25(35) mm wide, and deep pink flowers. Only in estuarine meadows. Southwest Iberian Peninsula. ES, PT. *Armeria velutina* is a *herb covered in dense, soft, silky hairs; leaves >2 mm wide, with 3–5 veins*. Bracts with papery margins. Scrub on dry acid soils. Southwest Iberian Peninsula; rare. ES, PT. *Armeria rouyana* is similar, with *leaves >2 mm wide with a single vein*. Rare, in Alentejo region. PT.

B. Northeast Spain and southern France.

Armeria ruscinonensis A dwarf shrub with a branched stock. Leaves all similar in shape, linear to spatula-shaped and rather leathery. Flowers pale pink or white, borne in heads 13–20 mm across on flexible stalks 80 mm–30 cm long. Coastal cliffs and mountains. ES, FR. *Armeria soleirolii* is similar but with the inner leaves longer and narrower than the outermost. Coastal cliffs and rocks. FR (Corsica).

C. Mainland Italy and Sicily.

Armeria canescens A variable, short to medium perennial. Leaves variable in dimension, the outermost linear and broadest above the middle, the inner longer and linear to linear-lanceolate, flat or with curled edges, and with membranous margins. Flowers pink, borne in heads to 25 mm across on stalks to 70 cm long; bracts pale brown and with papery margins. Rocky habitats, generally on higher ground. IT (incl. Sicily).

D. North Africa.

Armeria ebracteata A tufted perennial with strap-shaped leaves, gradually narrowed at the base. Flowers pink, borne in dense clusters on long-stalked heads; bracts ending in a sharp point, *the outermost longer than the inner*. DZ, MA, TN.

Limonium SEA LAVENDER

Perennial herbs with simple, leathery leaves forming basal rosettes and flowers in *branching cymes*; flowers persisting and papery in fruit; stamens and styles 5. Numerous species occur in the area, including many local endemics. Some of the most common species, as well as rarities of local interest, are described here.

A. Stems winged above; leaves present.

L. sinuatum

L. vulgare

Limonium sinuatum WINGED SEA LAVENDER A short to medium bristly perennial 10–40 cm tall that could perhaps be confused with species in the Boraginaceae. Leaves lobed, 40 mm–15 cm long and 8–30 mm wide, borne in a basal rosette. Stems with 3–4 undulate wings. Flowers cream; calyx 11.5–14 mm, *conspicuous, blue-purple and with a papery margin*. Saline coastal areas. Throughout.

B. Stems rounded, not winged; leaves present.

Limonium vulgare COMMON SEA LAVENDER A short to medium, hairless perennial to 40(60)cm with bluish, spoon-shaped, pinnately veined leaves 20–30 cm long and 15–30 mm wide, borne in a lax rosette. Flowers crowded, reddish to lilac, borne on dense spikes 10–20 mm long with >4 spikelets per cm. Mud flats and marshes. Far southwest only. ES, PT. *Limonium bellidifolium* often has withered leaves when in bloom, with very short dense spikes <10 mm long; stems with non-flowering lateral branches below. Sandy salt-marshes. ES, FR, IT.

Limonium oleifolium (Syn. *L. virgatum*) An extremely variable (and often hybridising), smooth, hairless perennial 10–50 cm tall with linear-spoon-shaped, *prominently 1-veined* leaves 30–85 mm x 3.5–8 mm withered or not when in bloom. Flowers pink, 7–8 mm across borne on spikes 20–80 mm long with 3–5 spikelets per cm, in zig-zagging inflorescences with numerous non-flowering branches. Various coastal habitats including coastal rocks and sands. DZ, ES, FR, IT, MA, TN. *Limonium algusae* is virtually indistinguishable in the field, but considered to be distinct; a very rare endemic. IT (Sicily: islet of Linosa).

---

1. *Limonium sinuatum*   4. *Limonium algusae* PHOTOGRAPH: GIANNIANTONIO DOMINA
2. *Limonium vulgare*   5. *Limonium camposanum*
3. *Limonium echioides*

Limonium camposanum A small, hairless perennial 15–35 cm tall with *basal rosettes of flat, spatula to spoon-shaped, bluish-green leaves*. Inflorescence with many lax, ascending-spreading branches, branched at 30°–45°, carrying numerous violet flowers; spikelets 7–9 per cm; *calyx tube densely hairy*. Rare, restricted to coastal sands on Mallorca. ES (Balearic Islands).

Limonium echioides A slender annual to 45 cm with *small rosettes of flat, bristly, leaves* which age red-purple; leaves broadest above the middle, gradually narrowing at the base, 7–55 x 3–16 mm. Flowers pink-white and short-lived, borne on widely branching, slender, red, 1-sided branches; spikes 20 mm–18 cm with 1–2 spikelets per cm; calyces persistent; calyx lobes with up to 10 curved spines. Fixed dunes and cliff-tops; frequent. Throughout.

C. Leaves usually few or *absent* when in flower; plant with numerous non-flowering branches.

Limonium insigne A hairless, shrubby perennial 15–80 cm tall with numerous tufted, *branched, leafless, succulent stems* among the fertile, flowering branches. Leaves oblong, with petioles, 30–90 mm long and 7–22 mm wide, in a rosette (leaves often absent). Flowers borne on branched inflorescences, cosexual, pink-purple, to 10 mm across, on spikes 10–35 mm long with 3–6 spikelets 8–10 mm per cm. A rare endemic halophyte of sea cliffs and semi deserts in southern Spain. ES.

L. ferulaceum

Limonium ferulaceum A densely branched dwarf shrub 10–65 cm tall with brownish, reddish or green *dense, feathery tufts of flowering and non-flowering branches.* Leaves elliptic, unlobed, and notably *absent when in flower.* Flowers pink, to 6 mm, borne at the ends of the branches; spikes reduced to spikelets 5–7 mm; calyx 4.5–5.1 mm. Rocky and sandy habitats and saline, marshy areas. ES, PT (probably elsewhere). *Limonium diffusum* is similar but often with some leaves present when in flower, and flowers along the branches (not restricted to the tips). *Spikelets short, 3–3.5 mm,* flowers pale violet; *calyx just 3–3.5 mm.* Coastal sands and salt marshes. An Atlantic species, extending to southwest Iberian Peninsula. PT.

Limonium cancellatum A small perennial to 18 cm with dense rosettes of ascending, small, bluish, narrowly spatula-shaped leaves, gradually tapered at the base. Flowers borne on inflorescences with numerous non-flowering branches; *branches intricately zig-zagging.* Flowers pale violet. Coastal rocks and cliffs in southeast Italy only. IT.

D. Vegetative parts with a compact, cushion-like habit.

Limonium lopadusanum A distinctive, *cushion-forming* perennial subshrub with densely clustered hemispherical *rosettes of spatula-shaped leaves* and pale lilac flowers borne on numerous rather sparse, flexuous and fragile inflorescences that generally lack sterile branches. Coastal rocks, an island endemic. IT (Sicily).

Limonium minutum A small virtually hairless perennial with *ball-like, very dense clusters of leaves; leaves minute, to just 10 x 4 mm.* Flowers pale violet, borne on short, zig-zagging branches. Rare, restricted to coastal rocks on Mallorca; frequently hybridising with closely related species on the island. ES (Balearic Islands).

Limonium ovalifolium A perennial with large, spreading, *dense, cushion-like clusters* of leaves, 35 mm–13 cm long; leaves green, often tinged purple and rather concave, with indistinct venation. Flowers lilac, borne on spreading branches; *flowering stems small and few,* 80 mm–35 cm. Coastal rocks and cliffs. A predominantly Atlantic species extending to southern Iberian Peninsula and northwest Africa. ES, MA, PT.

---

1. *Limonium insigne*
2. *Limonium diffusum*
3. *Limonium cancellatum*
4. *Limonium lopadusanum*
   PHOTOGRAPH: GIANNIANTONIO DOMINA
5. *Limonium ovalifolium*
6. *Limonium minutum*

Limoniastrum

L. monopetalum

Similar to *Limonium* but *large, spreading shrubs* with leaves alternate along the stems and with chalky glands. Flowers with 5(6) stamens.

Limoniastrum monopetalum LIMONIASTRUM A fleshy, spreading shrub 50 cm (2 m) with silvery-green, fleshy spoon-shaped leaves 20–60(90) mm long covered in white scales, sheathing the stem at the base. Flowers bright pink and conspicuous, later violet, to 16 mm borne in loosely branched spikes; corolla with 5 spreading, oval petals. Maritime sands and salt marshes. ES, MA, PT.

POLYGONACEAE | DOCK FAMILY

Herbs or small shrubs without latex, alternate leaves and stipules that form a membranous sheath around the stem. Flowers cosexual or unisexual, often small and greenish or reddish; tepals 3–6; stamens (3)6–9; stigmas 2–3; sessile or with styles. Fruit an achene.

Polygonum

Annuals or perennials with tap-roots. Flowers single or few (<6) in the leaf axils, exceeded by the leaves; stamens 8; stigmas 3, virtually sessile. Achene with 3 rounded angles.

P. oxyspermum

Polygonum maritimum SEA KNOTGRASS A prostrate or ascending, branched perennial with stems 60–80 cm with a woody stock. Leaves narrowly elliptic, glaucous and sessile, with down-turned margins, 15–25 mm long and 6–9(16) mm wide. Stipules silvery, reddish at the base with *8–12 conspicuous branched veins*. Flowers white or pink, 5-lobed and 3–4 mm across, solitary or 2–3 in the nodes. Common on maritime sands and shingle. Throughout. *Polygonum equisetiforme* is similar but has long, wiry branches, and stipules above each leaf *shorter than the internodes;* leaves 2–5(10) mm wide. Flowers borne in *lax spikes which are bare above* or with small, leafy bracts only. ES, IT, MA. *Polygonum romanum* is similar to *P. maritimum* but with green (*not glaucous*), linear to linear-lanceolate, spreading leaves *just 1–2(2.5) mm wide*, and stipules with *few, faint veins*. ES (Balearic Islands). *Polygonum oxyspermum* is an *annual*, with stipules much shorter than the internodes with 4–6 unbranched veins, and *broadly elliptic to linear-lanceolate leaves with scarcely inrolled margins*. Coastal sands. ES, FR, IT, PT.

1. *Limoniastrum monopetalum*
2. *Polygonum maritimum*
3. *Rumex bucephalophorus*
4. *Rumex tunetanus* PHOTOGRAPH: SERGE D. MULLER

Persicaria

Annuals or perennials with rhizomes. Flowers borne numerously in terminal or axillary, leafless, spike-like clusters; stamens 8; style 1. Achene 3-angled or winged.

Persicaria maculosa (Syn. *Polygonum persicaria*) REDSHANK An erect or sprawling *hairless annual* to 80 cm with branched stems reddish below and swollen at the leaf nodes. Leaves lanceolate and tapered at the base, *often with a dark central spot. Flowers pink*, borne in dense terminal (or in axillary) leafless, cylindrical spikes. Widespread and common in damp waste areas. Throughout. *Persicaria lapathifolia* (Syn. *Polygonum lapathifolium*) PALE PERSICARIA is similar but slightly hairy with greenish stems, the flower stalks with yellow glands, and *greenish-white flowers*. Similar habitats. Throughout.

Rumex DOCK

Perennials with terminal or axillary racemes, or panicles with whorled flowers; stamens 6; styles 3. Achene with 3 acute angles.

A. Plant herbaceous, with basal leaves *heart-shaped or rounded, ending abruptly or gradually into the stalk* (not distinctly lobed at the base).

Rumex tunetanus TUNISIAN DOCK An erect annual to 80 cm. Leaves with long petioles to 18 cm; leaf blade to 50 cm long, wavy-edged and heart-shaped at the base. Flowers borne on loose, leafless panicles. Fruit valves triangular-heart-shaped with prominent veins and 5–8 pairs of small unequal teeth. Freshwater habitats. A very rare endemic of northern Tunisia, close to extinction, in only 1 site in the Mogods. TN.

Rumex bucephalophorus HORNED DOCK A variable, reddish low to medium, erect annual to 30(50) cm, branched or not. Leaves normally small, 6–35(65) mm long, lanceolate, oval or spoon-shaped, stalked, and greyish green. Flowers borne on variously sized flower stalks, very small, red and in clusters of 2–3 in the leaf axils, forming a long, dense spike. *Valves triangular-oval or narrow, with 3–4 teeth*. Common on waste ground, maritime sands and sea cliffs. Throughout.

4

1. *Rumex intermedius*
2. *Rumex vesicarius*
3. *Rumex induratus*

Rumex rupestris SHORE DOCK An erect pennial to 50(70) cm with zig-zagging stem branches and thick, oblong *blunt, greyish leaves* with rounded to almost heart-shaped leaf bases, the lower leaves stalked. Inflorescence *only leafy at the base,* with long branches spreading at 25°–50°. Fruit valves entire, each with a tubercle at the base. Local on sand dunes. ES (north), FR (south).

B. Plant herbaceous, with basal leaves triangular or arrow-shaped (*generally lobed at the base*).

Rumex acetosella SHEEP'S SORREL A spreading to erect, *slender* perennial to 45(80) cm with *small leaves*; leaves *arrow-shaped with small, forward-directed lower lobes and a large, oval-lanceolate central lobe* 6–60 mm long. Flowers greenish or reddish, borne in simple or branched racemes, unisexual, the male and female flowers borne on separate plants. Fruit valves equalling the achenes. Dry, bare habitats. ES, FR, IT.

Rumex thyrsiflorus (Syn. *R. acetosa*) COMPACT DOCK A little-branched perennial to 1 m (often less) with basal arrow-shaped leaves, the middle lobe long and lanceolate, 20–90 mm (12 cm) long, and the lateral lobes backward-pointing; stem leaves becoming progressively narrower, and eventually very narrowly linear and clasping the stem. Flowers greenish or red, and borne in *very dense-flowered panicles*. Locally common on dunes or bare ground. ES, FR, IT, PT.

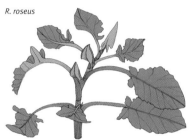

R. roseus

Rumex roseus (Syn. *R. tingitanus*) TANGIER DOCK A robust, rhizomatous perennial to 80(90) cm, branched at the base with greyish, oval-spear-shaped leaves 17–75 mm (10 cm) long that are *wavy-lobed*, often tapered towards the apex. Inflorescence with large, persistent sheaths. Fruit valves exceeding the achenes, heart-shaped at the base, and pink. Locally common on maritime sands and shingle. ES, FR, IT, PT.

Rumex intermedius A perennial with stems to 60 cm, often clumped. *Leaves linear*, 10 cm long and 10 mm wide, with narrow divergent lobes at the base and *undulate margins*; sometimes with leaves of differing shape on the same plant. Fruit valves longer than the nut. Locally common on maritime sands and shingle. ES (incl. Balearic Islands), FR, PT.

Rumex vesicarius BLADDER DOCK An annual with oval-triangular to rounded leaves. Flower stalks each with 2 flowers, one of which is smaller and concealed by *conspicuous*, inflated, rounded bracts 12–18(23) mm long, often *flushed bright pink or crimson* with darker netted veins. Arid and desert fringe habitats in North Africa. DZ, MA, TN.

R. intermedius

C. Plant a *much-branched, woody-based perennial* with basal leaves triangular or arrow-shaped (lobed at the base).

Rumex scutatus A woody-based, much-branched perennial to 40(65) cm with leaves 10–23(45) x 5–26(40) mm, variably triangular-heart-shaped with diverging basal lobes, rather thick and glaucous. Flowers unisexual, reddish, borne in branched clusters. Fruits with oval to rounded valves, heart-shaped at the base, *as long as wide* (4.5–6.5 mm). ES, FR, IT. ***Rumex induratus*** (Syn. *R. scutatus* subsp. *induratus*) is a stouter, greyish, *much-branched shrubby perennial* to 80 cm (1 m) with basal small, arrow-shaped leaves with a lanceolate or oval central lobe, 8–35 mm long; leaves obscured by numerous, dense ascending branches in summer, often *giving the appearance of a hemispherical shrub*. Flowers lax, borne in terminal and axillary clusters. Fruits with *large rounded valves*, heart-shaped at the base, *wider than long* (5.5–10.5 mm), exceeding the nut, and with prominent veins; pinkish. Dry, bare slopes, roadsides and shingle beaches. ES (south), MA, PT (south).

Emex

E. spinosa

Annuals superficially similar to *Rumex*. Flowers unisexual; stamens 4–6; styles 3. Fruit a *spiny nut*.

Emex spinosa EMEX A hairless, short, somewhat fleshy annual with sprawling stems to 50(60) cm. Leaves oval, heart-shaped at the base, 12–14 x 8 cm, and long-stalked (to 25 cm). Male flowers stalked in terminal clusters, female flowers sessile at the base. *Fruit a spiny nut.* Locally common on maritime sands and disturbed ground. Throughout.

Fallopia

A small genus of annuals to herbaceous perennials and woody vines, previously included in a wider treatment of the genus *Polygonum*. Flowers with 8 stamens; styles 3. Achene 3-angled.

Fallopia convolvulus An annual, clockwise twining, climbing or prostrate vine to 1 m with angular stems. Leaves heart or arrow-shaped, pointed and mealy beneath, 30–70 mm long. Flowers greenish or yellowish-white, borne in loose clusters in the leaf axils. Fruit a triangular nut borne on a short stalk 1–3 mm long. A common ruderal weed. Throughout. *Fallopia dumetorum* is similar but has more rounded stems and *fruits borne on stalks 4–8 mm long.* Hedges, scrub and degraded woodland. Throughout.

Fallopia baldschuanica RUSSIAN VINE A woody climber to 10 m. Leaves oval to triangular, periolate, and 25 mm–10 cm long. Flowers borne in lax, spreading to ascending inflorescences to 15 cm long; flowers to 10 mm across, 5-parted and white to pale green or pink; flower stalks 4–10 mm. Fruit a shiny black achene 3–4.5 mm across. Field borders, river banks and disturbed habitats. Native to Asia but now a cosmopolitan weed. DZ, ES, FR, IT, MA, TN.

1. *Emex spinosa*
2. *Fallopia convolvulus*
3. *Fallopia baldschuanica*
4. *Arenaria montana*

CARYOPHYLLACEAE | PINK FAMILY

Herbs with opposite leaves. Flowers cosexual, regular, with 4–5 free or fused sepals and 4–5 petals (absent in some species), typically 8–10 stamens and 2–5 styles. Fruit a many-seeded capsule, sometimes a berry or 1-seeded achene.

Arenaria SANDWORT

Leaves opposite and stipules present. Flowers with 10 stamens and 3 styles. Capsule with 6(–10) teeth. Many mountain-dwelling species occur besides those described here.

A. Sepals exceeding the petals.

Arenaria serpyllifolia THYME-LEAVED SANDWORT A variable, downy, grey-green prostrate or ascending annual with unstalked, oval leaves. Flowers small, to 8 mm across, white, with unnotched petals, *exceeded by their sepals;* sepals 3–4 mm long. Fields and dry habitats. Throughout. *Arenaria leptoclados* is similar (also widely treated as a subspecies) with conical, straight-sided capsules; sepals 2–3 mm long. ES, FR, IT, PT.

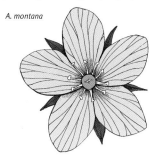

A. montana

B. Sepals shorter than the petals.

Arenaria montana MOUNTAIN SANDWORT A spreading shortly pubescent, grey-green perennial with spreading stems. Leaves opposite, entire, oblong-lanceolate or linear, 1-veined. Flowers with 5 petals, solitary or in 2–11-flowered clusters, flower stalks longer than sepals, petals 2 x as long as sepals, and white, yellow-centered. Capsules ovoid, more or less shorter than sepals. Common on maquis and in woods at various altitudes. ES, FR, MA, PT.

Arenaria balearica BALEARIC SANDWORT A spreading perennial with a slender stock and delicate, prostrate, rather hairy shoots which only ascend at the tips (to 50 mm) and root at the nodes. *Leaves small, oval to rounded, blunt, all stalked, the stalks roughly equalling the blade.* Flowers with 5 petals, solitary and terminal, held on stalks above the foliage; petals 2 x the length of the sepals. Damp, shady rocks on the islands. ES (Balearic Islands), FR (Corsica), IT (Sardinia).

Arenaria bertolonii A low, sparsely mat-forming perennial with slender stems and lanceolate, hairy, pointed leaves. Flowers rather numerous, solitary or in groups of 2–3, white and *large to 30 mm across* with *notched petals*. Rocky habitats. FR (Corsica), IT (incl. Sardinia and Sicily).

Minuartia SANDWORT

Small, slender annuals or perennials. Flowers with 0 or 5 petals, 10 or fewer stamens and and 3(–5) styles. Capsule with 3(–5) teeth.

Minuartia mediterranea MEDITERRANEAN SANDWORT A low, erect or sprawling, hairless annual 20–60 mm (12 cm) high with linear-lanceolate, flat, pointed leaves. *Flowers borne in dense clusters*, with purple-flushed sepals; *flower stalks shorter than the sepals; petals not notched,* $^1/_2$ *the length of the sepals.* Fields and pastures. Probably throughout. *Minuartia hybrida* FINE-LEAVED SANDWORT is a similar, slender annual 30 mm–20 cm without non-flowering shoots, and very *lax inflorescences* and petals absent or $^1/_2$–$^1/_3$ of the length of the sepals. Probably throughout.

Cerastium MOUSE-EAR

Annuals or perennials, often hairy. Flowers with 4–5 sepals and petals (or absent), 3–5(6) styles and 4, 5 or 10 stamens. Fruit a capsule with 2 x as many teeth as styles.

A. Plant not mat-forming; variably hairy or sticky.

C. glomeratum

Cerastium glomeratum STICKY MOUSE-EAR A small to medium, erect or ascending annual to 45 cm, covered in sticky glandular hairs. Leaves oval-elliptic, to 20 mm long, hairy. White flowers borne in *dense clusters*, the *sepals hairy to the apex*, petals more or less equalling the length of the sepals. Very common on bare ground and in dry, grassy places. Throughout. *Cerastium siculum* is now normally regarded as a form of the former species, but traditionally separated on the basis of having *deeply notched petals that are shorter than the sepals*. Flower stalks shorter than the sepals. Rocky hillsides and scrub. ES, FR, IT, PT. *Cerastium ligusticum* LARGE-FLOWERED MOUSE-EAR is similar but has *long-stalked flowers, not borne in dense clusters*. FR (Corsica), IT (incl. Sardinia).

Cerastium fontanum COMMON MOUSE-EAR A very variable, non-sticky, short-lived, tufted or matted *perennial* to 40 cm, *with short, non-flowering basal shoots* with stalkless leaves. Flowers white, borne in loose clusters, petals slightly shorter than or slightly longer than the sepals, and deeply notched. Common in grassy habitats. Throughout.

B. Plant mat-forming and densely grey-woolly.

Cerastium tomentosum SNOW-IN-SUMMER A spreading, *mat-forming, grey-woolly* perennial to 40 cm high with linear-elliptic leaves, densely white-matted. Flowers white, borne in cymes; petals deeply notched. Rocky habitats and widely cultivated as an ornamental. IT (incl. Sicily).

Paronychia

Perennial herbs with small opposite leaves and *conspicuous silvery stipules.* Flowers small, borne in dense clusters *surrounded by silvery bracts*; stamens 5; styles 1–2.

P. argentea

P. echinulata

A. Sepals with papery margins and *bristle-tips.*

Paronychia argentea A branched, prostrate, *mat-forming perennial* with stems 50 mm–50 cm long. Leaves oval-lanceolate, greyish and in opposite pairs, bristle-tipped and almost hairless. Stipules membranous, and shorter than the leaves. Flowers borne in well-defined lateral and terminal clusters 10–15(25) mm across; with prominent *membranous, silvery bracts* (most conspicuous in mature or fruiting heads); sepals with bristle-tips 0.4–0.65 mm. Common in maritime sandy places. Throughout. *Paronychia echinulata* is similar but an *annual*, spreading to erect, and with flower clusters 3–6.5 mm across with less conspicuous bracts (even in fruit) and sepals with bristle-tips 0.6–1 mm. Similar habitats. ES, FR, IT, MA, PT.

B. Sepals with green margins, *without bristle-tips.*

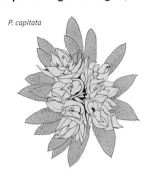
P. capitata

Paronychia kapela A green, spreading, mat-forming perennial similar to *P. argentea* but with very conspicuous, dense, terminal clusters of flowers 8–15 mm across with blunt, silvery bracts; *sepals not bristle-tipped, not recurved and all equal.* Generally in mountains. ES, MA. *Paronychia capitata* is a grey-green, hairy, mat-forming perennial similar to *P. kapela* but with hairier leaves equalling the length of the stipules, and denser flower-heads and *markedly unequal sepals inwardly curved at the tips.* Central and southern. ES, FR, IT, MA, (PT probably).

Illecebrum

I. verticillatum

Similar to *Paronychia* but with white, spongy sepals which persist in fruit. Stamens 5; stigmas 2. Fruit a 1-seeded, 5-valved capsule.

Illecebrum verticillatum CORAL NECKLACE A distinctive, low, hairless annual superficially similar to *Paronychia*. Stems creeping, rooting at the base, reddish, to 20 cm. Leaves 2–5 mm, opposite and with small stipules, Flowers borne in 2 clusters at each node, 4–6-flowered with silvery bracts, looking like white 'pseudo-whorls'. Damp, sandy and gravelly places; rare in the far south and east. ES, FR, IT, MA, PT.

1. *Cerastium glomeratum*
2. *Cerastium tomentosum*
3. *Paronychia argentea*
4. *Illecebrum verticillatum*

Herniaria RUPTURE-WORT

Similar to *Paronychia* but often with tiny leaves and flowers, and *inconspicuous bracts*. Stamens 5; stigmas 2. Fruit an achene. Many virtually identical species have been described in Iberian Peninsula which, in the absence of DNA sequence data, are questionably true species.

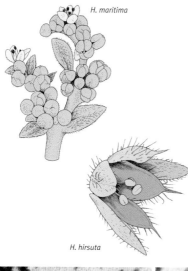

H. maritima

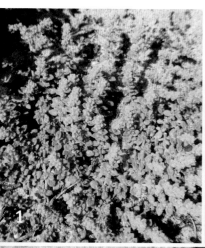

H. hirsuta

Herniaria maritima An inconspicuous, low, prostrate, spreading and more or less mat-forming, *succulent* annual with older stems woody at the base, younger stems hairy on 1 side only. Leaves tiny, to 7 mm, broadly elliptic and with adpressed hairs on both surfaces. Flowers tiny, 1.5–2 mm across, 5-parted, borne in dense clusters of 4–7(18). Rare, on sand dunes in the Algarve only. PT.

Herniaria hirsuta HAIRY RUPTURE-WORT A variable annual (with many forms, sometimes treated as species) with bright green to grey-green leaves 4–8(11) mm long which are clothed in *dense, straight, white, spreading hairs*, and flowers 1.3–1.6 mm with 2–5 stamens, borne in roundish clusters of 7–12 in the leaf axils; *calyx hairy*. Various bare habitats. Throughout. *Herniaria glabra* is similar but with leaves without hairs except for the margins, and *calyx hairless*; flowers 1.3–1.5 mm. Sandy and fallow ground. ES, FR, IT.

Polycarpon ALLSEEDS

Small herbs with forking stems and opposite or whorled leaves with papery stipules. Flowers with *keeled and hooded sepals;* stamens (1)3–5; stigmas 3. Fruit a 3-valved capsule.

P. tetraphyllum

Polycarpon tetraphyllum FOUR-LEAVED ALLSEED A tiny, hairless micro-annual without a woody stock, 20 mm (35 cm) high. Leaves mostly in *whorls of 4*, oval, green or purple. Flowers in branched clusters, white, tiny to 2 mm across with notched petals shorter than the sepals; stamens 3–4. Seeds brownish, with protuberances. Common on dunes and bare, sandy places inland. Throughout. *Polycarpon polycarpoides* is similar but a perennial, often succulent, with a woody base and rounded leaves lax inflorescences; stamens 5. Coastal rocks. ES, FR, IT. *Polycarpon alsinifolium* is very similar to *P. tetraphyllum* (also widely treated as a subspecies) with flowers with 4–5 stamens and mostly whitish, smooth seeds. Dunes. ES (incl. Balearic Islands), PT.

Corrigiola

Annual or perennial herbs with prostrate shoots and alternate, linear to oblong, greyish leaves with stipules. Flowers with 5 stamens; stigmas 3. Fruit an achene.

Corrigiola litoralis STRAPWORT A low, flat, *greyish* annual with spreading or prostrate stems to 50 cm bearing rather fleshy, blunt, entire *narrowly oblong to linear leaves* with papery stipules at their bases. Flowers white, tiny (1–2 mm), and borne in dense, rounded clusters at the end of leafy stems; sepals green with white margins; anthers purplish. Local on damp coastal sands. ES, MA, PT. *Corrigiola telephiifolia* is similar (also widely treated as a subspecies) with *grey-green spatula or spoon-shaped, rather Euphorbia-like leaves*. ES, MA, PT.

Loeflingia

L. hispanica

Low, much-branched annuals with linear, opposite leaves. Flowers stalkess with 5 petals; stamens 3 or 5; style 1. Fruit a 3-valved capsule.

Loeflingia hispanica LOEFLINGIA A low, glandular-hairy, reddish or purplish, and much branched annual with ascending branches to 15(20) cm and linear, long-pointed leaves fused at the base, 3–6(14) mm long. Flowers greenish, tiny to 3 mm across in branched spikes. Petals 5, *shorter than the capsules when mature; stamens 3–5.* Common in dry, sandy habitats. Throughout. *Loeflingia boetica* is similar but always has 5 stamens, and petals exceeding the length of the mature capsules. ES, MA, PT.

1. *Herniaria maritima*
2. *Herniaria hirsuta*
3. *Loeflingia hispanica*
4. *Polycarpon tetraphyllum*

Sagina PEARLWORT

Small, often tufted, moss-like herbs. Flowers minute with 4–5 sepals and petals or petals absent; stamens 4, 5, 8 or 10; styles 4–5. Capsule 4–5-valved.

S. apetala

Sagina apetala ANNUAL PEARLWORT A small, annual herb with very slender, *sub-erect stems to 15 cm, all producing flowers*. Leaves linear, and tapered at the tip. Flowers solitary, small with 4 oval sepals that are often hooded, petals minute and falling early. Common in sandy places. Throughout. *Sagina procumbens* PROCUMBENT PEARLWORT is a similar moss-like, bright green *mat-forming perennial* with a short, non-flowering main stem bearing a central, dense leaf rosette and numerous lateral stems to 20 cm ascending from rooting bases. Sepals and stamens 4(5). Common in sandy and urban waste places. Throughout.

Spergularia SEA-SPURREY

Annuals or perennials with narrow, opposite (sometimes apparently whorled) leaves and leafy tufts at each node; leaves fleshy or not. Petals purple, pink or white; stamens 5–10; stigmas 5. Capsule 5-valved.

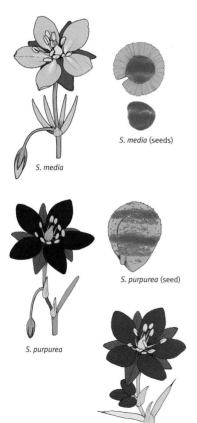

S. media (seeds)

S. media

S. purpurea (seed)

S. purpurea

S. fimbriata

A. Petals normally *exceeding* the sepals; seeds winged or partially winged.

Spergularia media GREATER SEA-SPURREY A low-short almost hairless perennial to 40 cm with fleshy leaves ending in an abrupt point, rounded beneath. Stipules broadly triangular. Flowers 10–12(13) mm across with white or pink *petals that equal or exceed the sepals*; sepals 4–6 mm; stamens 0–10. Capsule 7–9 mm, greatly exceeding the calyx; seeds dark brown, all (or mostly) winged. Common on sandy shores and in salt marshes. Throughout. *Spergularia nicaeensis* is very similar but with a robust and woody base and *short capsules just 3–4.5 mm long*; seeds brown and scarcely winged. ES, FR, IT.

B. Petals normally *exceeding* the sepals; *seeds not winged*.

Spergularia purpurea PURPLE SPURREY A short annual or biennial to 25 cm high with slender stems, often growing in large masses. Leaves not, or slightly fleshy, stipules silvery and lanceolate. Flowers to 9 mm across, with *uniformly mauve petals that distinctly exceed the sepals*; stamens 10. Capsule 2.5–4 mm; seeds brown or blackish and wingless. Common in non-saline sandy, and waste places in the southwest. ES, MA, PT. *Spergularia fimbriata* is similar but more robust, woody below, with *long-awned leaves*, and *longer stipules 6–10 mm*; on coastal sands. ES, MA, PT. *Spergularia diandra* is similar to *S. purpurea* but with *narrowly elliptic* (not oval), lilac petals, bristly wingless seeds, and occurs in saline sandy and waste places. Throughout.

S. marina

S. marina (seeds)

C. Petals *equalling or shorter* than the sepals; seeds winged, not winged or both.

Spergularia marina LESSER SEA-SPURREY An annual to 35 cm with very fleshy leaves, normally a slender (not woody) stock, and short stipules that form a sheath. Inflorescence sparingly branched; flowers 5–8 mm across, petals pink above and white below, *not exceeding the sepals*; sepals 2.5–4 mm; stamens 2–7(10). Capsule 3–6 mm, exceeding the calyx; seeds light brown, unwinged, winged or mixed. Common on sandy shores and in salt marshes. Throughout. *Spergularia tangerina* is similar but *smaller*, with very slender stems just 50 mm (15–20 cm), and a condensed inflorescence, flowers very small, just 3 mm across; stamens 2–3(5). Local, on saline soils. Southwest Iberian Peninsula. ES, PT.

1. *Sagina apetala*
2. *Sagina procumbens*
3. *Spergularia purpurea*
4. *Spergularia media*
5. *Spergularia purpurea*
6. *Spergularia marina*

D. Petals equalling or shorter than the sepals; seeds not winged.

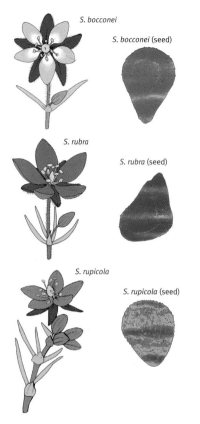

S. bocconei

S. bocconei (seed)

S. rubra

S. rubra (seed)

S. rupicola

S. rupicola (seed)

Spergularia bocconei A slender annual or biennial to 20 cm, *densely hairy on the inflorescence*. Leaves not in dense clusters, stipules triangular. Petals pink with a white base, equalling or shorter than the sepals which are 2–4 mm. Capsule 2–4 mm shorter than the calyx; seeds grey-brown and not winged. Ruderal habitats and fields. Throughout. *Spergularia heldreichii* is similar but with heart-shaped stipules, pink or lilac petals and smooth or rough, wingless *black seeds*. Coastal sands. ES, FR, IT, PT.

Spergularia rubra SAND SPURREY A *sticky-hairy* annual or perennial with spreading stems to 25 cm. Leaves ending in an abrupt point, *not fleshy*, grey-green and strongly whorled. Stipules silvery and lanceolate. *Petals equalling or shorter than the sepals*, uniformly pink; sepals 3–4(5) mm. Capsule 3.5–5 mm; seeds not winged. Sandy (not saline) soils. Local. Throughout.

Spergularia rupicola A rather *robust* perennial to 35 cm with woody stems below, *densely sticky-hairy throughout*; stems often purple. Leaves fleshy, flattened, narrowly linear (1–2 mm wide); stipules oval-triangular and rather silvery. Petals uniformly pink, more or less *equalling the sepals;* stamens 10. Seeds dark brown wingless and rough. Local, in maritime rocky places in the far west only. ES, PT. *Spergularia australis* is similar but with leaves narrowed towards the base (1.2–2 mm wide), *stipules very silvery* and *smooth, wingless, black seeds*. Similar habitats and distribution. ES, PT.

Agrostemma

Annuals with sepals fused into a tubular calyx bearing long, green leaf-like teeth; stamens 10, styles 5. Fruit a capsule with 5 teeth.

Agrostemma githago CORNCOCKLE A short to tall annual to 1 m covered in adpressed, greyish hairs. Leaves narrowly lanceolate and pointed. Flowers borne on long individual stalks, pale reddish-pink with shallowly notched petals shorter than their long-pointed, linear sepals. A rather local but widespread arable weed. DZ, ES, FR, IT, MA, TN.

Saponaria SOAPWORTS

Annuals or perennials with a smooth, tubular calyx, without an epicalyx; styles 2. Fruit a capsule with 4 teeth.

A. Erect or ascending herbs.

Saponaria calabrica CALABRIAN SOAPWORT A *short*, reddish, stiffly spreading-ascending perennial, similar to *Silene* in general appearance. Leaves spoon-shaped to oval. Flowers pink-purple, to 10 mm across, borne in lax clusters; petals not notched; calyx long-tubed, purple and hairy. Garrigue and scrub. IT.

Saponaria officinalis SOAPWORT A *tall and robust*, hairless perennial to 90 cm with oval, strongly veined leaves, the largest >50 mm. Flowers borne in rather compact clusters, 25 mm across, pale pink with unnotched petals; calyx flushed red and inflating in fruit. Damp woods. Throughout the European region. ES, FR, IT, PT.

B. Compact shrubs.

1. *Spergularia bocconei*
2. *Agrostemma githago*
3. *Saponaria officinalis*
4. *Saponaria sicula*
PHOTOGRAPH: GIANNIANTONIO DOMINA

Saponaria sicula SICILIAN SOAPWORT A *compact, mat-forming* subshrub with woody stems below; stems to 15 cm. Leaves opposite, greyish and with rough margins, spoon-shaped below and lanceolate above. Flowers pink with petals *deeply-lobed,* for ¹/₂ their length; calyx long, tubular, hairy and purple. An endemic of Mt. Etna. IT (Sicily).

Silene CAMPION

Herbs with flowers in branched inflorescences. Sepals fused into a tube, often with 5 teeth; petals separate; styles 3 or 5, protruding. Fruit a capsule. Numerous species occur in the region, only a small subset are described here.

A. Perennials with white flowers; *calyx hairy.*

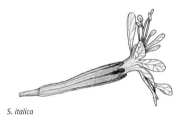

S. italica

Silene italica ITALIAN CATCHFLY An erect, branched perennial to 70 cm with a woody base, sticky above. Leaves oval-elliptic, pointed and sticky-hairy, downy below. *Flowers held erect on branched stems with >1 flower*, with deeply bilobed (often inrolled during the day), *white petals*, yellowish above and greenish or reddish below; calyx 14–21 mm, purplish, cylindrical and *hairy*. Common in dry, grassy places. Throughout.

B. Perennials with white or pink flowers; *calyx slightly hairy to hairless.*

Silene longicilia (Syn. *S. patula*) is similar to *S. italica,* also with more or less erect flowers, but with a *more or less hairless calyx* with dimorphic teeth, and pale reddish-purple petals with darker veins; calyx 11–17 mm. Far west only. PT.

S. latifolia

Silene latifolia WHITE CAMPION A variable, densely *sticky-hairy* medium-tall perennial to 1 m with oval leaves, the lowermost stalked. Flowers unisexual and more numerous on male plants: large and white with deeply notched petals, the calyx red-veined and *not inflated*; calyx 15–22 mm and 10-veined (male) or 20–30 mm and 20-veined (female). Common on disturbed, fallow and cultivated ground, and on cliff-tops. Throughout.

1. *Silene italica*
2. *Silene latifolia*
3. *Silene vulgaris*

1

2

3

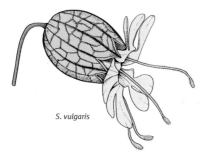

S. vulgaris

Silene vulgaris BLADDER CAMPION A variable perennial to 80 cm with a branching woody stock, and several erect or ascending shoots, all flowering, usually (not always) hairless. Leaves elliptic to oval, the upper stalkless. Flowers white, with *strongly inflated 20-veined calyx* 18–20 mm that persists in fruit. Capsule with erect-spreading teeth. Very common and widespread in a range of habitats. Throughout. *Silene uniflora* (Syn. *S. vulgaris* subsp. *maritima*) has a *low, cushion-like habit* and larger flowers, rarely >4 per stem. Capsule with spreading or curved teeth. Coastal environments. ES, PT. *Silene behen* has similar flowers with inflated, hairless calyx to the previous species, but is an *annual* with broad, bluish leaves. IT (incl. Sicily).

Silene rothmaleri An erect or ascending perennial 14–23 cm high with a woody stock, and lanceolate-spathulate leaves. Inflorescence lax, flowers white, borne in rather *dense, many-flowered panicles*, the *calyx long,* 20–23 mm, and *stems sticky*. A rare endemic restricted to just a few coastal shales in the Algarve. PT.

C. *Annuals* with white (pinkish or yellowish) flowers *opening only at night.*

Silene noctiflora NIGHT-FLOWERING CATCHFLY An erect annual to 50 cm with simple or branched stems which are hairy below and sticky-hairy above. Leaves narrowly oval and unstalked except at the base. Flowers whitish-yellow with deeply notched petals, rolling inwards (exposing yellow) during the day and opening fully at night, and fragrant; calyx 20–30 mm. Common, except in the far south and west. ES, FR, IT.

D. Annuals with *pink flowers* opening in the day.

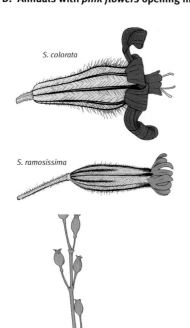

S. colorata

S. ramosissima

S. ramosissima (habit)

Silene colorata A variable, erect or spreading, short annual. Leaves oval-linear, the lower leaves stalked. Flowers with *bright pink* (rarely white) *deeply notched petals*, borne in lax clusters; calyx club-shaped, 10–15 mm long, variably hairy. Very common on waste land and in sandy places. Throughout. *Silene bellidifolia* is similar in form to *S. colorata*; a sparsely bristly annual 25–70 cm with simple or branched stems. Leaves bristly-hairy, lanceolate to spatula-shaped below and oval-lanceolate above. *Flowers more or less sessile* in closely spaced inflorescence; petals pink and deeply notched; *calyx longer, 14–18 mm*. Distribution patchy. DZ, ES, PT, TN.

Silene ramosissima An *extremely sticky* annual with stems 40 mm–40(60) cm and very hairy. Leaves fleshy, narrowly spatula-shaped to linear-lanceolate and pubescent. Inflorescence a long and slender, symmetrical, more or less flat-topped cluster with a single pink flower on each axis, the lower flower stalks long, 20–35 mm; calyx 9–14 mm. Local on maritime sands. Far west only. DZ, ES, MA, PT.

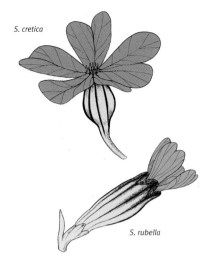

S. cretica

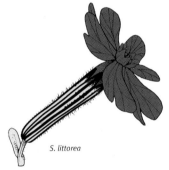

S. rubella

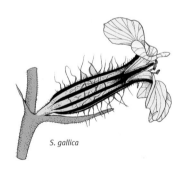

S. littorea

S. gallica

Silene cretica CRETAN CATCHFLY A *slender*, erect annual 20–50 cm high, hairy at the base, *hairless* above, and sticky. Basal leaves oval and broadest above the middle and stalked; upper leaves pointed and unstalked. Flowers bright pink, borne in lax, branched clusters with notched petals; *calyx distinctly ridged and hairless*, 11–16 mm. Cultivated and disturbed ground. ES, MA, PT. *Silene laeta* is a similar small, upright annual with *pale pink flowers,* and calyx 7–10 mm, generally *rough* as well as ridged. ES, FR, PT.

Silene rubella A glandular annual with stems 10–60 cm, lower leaves spatula-shaped and pubescent, often with rolled-in margins; calyx 9–12(13), red-tinged calyx with oval and blunt teeth, *long and narrow-cone shaped below, inflated towards the apex*; petals bright pink. Arable and cultivated land. Widespread but rather local. Throughout.

Silene littorea A clumped, low to short, glandular-hairy, *sticky* annual to 20 cm (or a short-lived micro annual following seasonal rain) with oblong to linear leaves. Flowers with bright pink, notched petals, borne in a few-flowered, raceme-like, leafy inflorescence; calyx 13–18 mm long, straight, or scarcely inflated above the middle when in flower. Common on maritime sands in the west. DZ, ES, MA, PT. *Silene psammitis* is similar but has linear to *linear-lanceolate leaves* (not oval) and longer calyx, 15–19(24) mm. Similar habitats to *S. littorea* but less frequent. ES, MA, PT. *Silene pendula* is similar to *S. littorea* but with an *inflated calyx* widest above the middle, with darker, bristly ridges. IT (incl. Sicily).

E. Annuals typically with pale or *white* (sometimes tinged pink, green or yellow) flowers opening in the day.

Silene gallica SMALL-FLOWERED CATCHFLY A downy, erect annual with simple or branched stems 80 mm–45(60) cm, sticky above. Leaves oval and stalked beneath, narrower and unstalked further up the stem. Flowers small, yellowish white or pinkish, borne in a more or less *1-sided inflorescence with alternating, short-stalked flowers*; calyx small, 6.5–10(12) mm, cylindrical to ovoid with long teeth, $^1/_4$ the length of the tube, *sticky-hairy and 10-veined*. Very common in waste places. Throughout. *Silene sclerocarpa* (*Syn. S. cerastoides*) is similar but ascending and *woolly-hairy* (especially below), the flower stalks short, to 24 mm (not exceeding the calyx); calyx 9–11 mm long and *strongly contracted at both ends and with net veins*. Similar habitats. Far south only. DZ, ES, MA, PT, TN.

1. *Silene uniflora*
2. *Silene rothmaleri*
3. *Silene noctiflora*
4. *Silene colorata*
5. *Silene laeta*
6. *Silene pendula*
7. *Silene littorea*
8. *Silene gallica*
9. *Silene nicaeensis*
10. *Silene sedoides*
11. *Silene baccifera*

S. nutans

Silene nutans Nottingham catchfly A variable, erect perennial 15–80 cm, usually unbranched and sticky above, downy below. Leaves spatula-shaped at the base, narrower and unstalked above. Flowers white, greenish or reddish below with *deeply lobed petals and inward-rolling lobes* in a loose, 1-sided and *drooping inflorescence*; calyx 9–12(14) mm, with teeth to 2 mm. Rarer in the south and in hot, dry areas. Throughout.

Silene nicaeensis A short to medium annual, *densely hairy throughout*, 20–60 cm high. Leaves *narrow* (8 mm wide), often curved. Flowers borne on flower stalks to 15 mm; calyx 10–12 mm, greenish-white with darker striping; petals white or faintly reddish-pink and *deeply notched*, erect, borne up to 8 on a simple or branched inflorescence. Common on dunes in southern Iberian Peninsula. ES, PT. *Silene obtusifolia* is similar but less hairy and with *broader, spatula-shaped leaves* with hairy margins, and flowers borne in *branched raceme-like, flat-topped clusters*; calyx 11–12.5 mm. DZ, ES, MA, PT. *Silene scabriflora* is similar to *S. obtusifolia* but the whole plant is rather woolly-hairy, and the calyx *elongated, and very narrow towards the base;* 12.5–26 mm. ES, MA, PT.

Silene sedoides A *small to minute,* often tufted, *densely hairy,* purplish annual 40 mm–10(15) cm high. Leaves fleshy, spatula-elliptic and <10 mm long. Flowers pinkish or whitish; calyx 5–6.5(7) mm, purple, long and contracted at the base, and hairy. Coastal rocks; local but frequent on the islands. Throughout.

S. inaperta

F. Annuals with flowers with *petals absent or inconspicuous.*

Silene inaperta An erect, slender annual with stems 17–80 cm, erect, usually simple below and branched above, shortly-hairy. Leaves lanceolate below, narrow and linear above; calyx 7.5–13.5 mm with short teeth, *petals absent or within the calyx.* Local in grassy places. DZ, ES, FR, IT, MA, PT.

G. Fruit a *shiny berry* loosely surrounded by a bell-shaped calyx.

Silene baccifera (Syn. *Cucubalus baccifer*) berry catchfly A downy, brittle-stemmed perennial to 1 m, often straggling around surrounding vegetation; not a typical member of its family. Leaves oval, pointed and short-stalked. Flowers rather large and conspicuous, drooping with green-white petals; calyx 8–15 mm. Fruit a globular green, later black berry 6–8 mm. Grassy habitats and woods. Throughout though rarer further south, and absent from hot, dry areas.

Dianthus PINK

Annual to perennial herbs typically with 5 pink petals and calyx base surrounded by an epicalyx of paired bracts; stamens 10; styles 2. Capsule with 4 teeth. Many new species have been described (particularly in the *D. sylvestris* group).

A. Flowers borne in dense heads surrounded by conspicuous, leafy bracts.

Dianthus armeria Deptford pink An erect, hairy *annual or biennial* to 60 cm with dark green oblong basal leaves and linear stem leaves. Flowers open in sunlight, *bright red-pink*, across, subtended by *long leafy green bracts*; *petals rather pointed and toothed*; calyx 15–20 mm. Widespread in grassy habitats. ES, FR, IT, PT.

B. Flowers borne in lax heads, not surrounded by conspicuous, leafy bracts.

Dianthus sylvestris WOOD PINK A very variable, erect, tufted perennial with flower stems to 60 cm. Basal leaves slender and wiry, <1 mm wide, rough margined and often recurved. Flowers generally mid-pink, *scentless and with hairless petals*; epicalyx blunt and leathery; calyx 12–29 mm long. Rocks, cliffs, walls and dry wooded slopes. Many new species traditionally described as *D. sylvestris* have been described. Throughout (in its traditional, wider description). *D. lusitanus* is similar but has grooved, glaucous *cylindrical, fleshy leaves* and a very woody stock. ES, MA, PT.

Petrorhagia

Similar to *Dianthus* with epicalyx of papery scales at the base of a single flower, or with several bracts at the base of a dense flower-head; stamens 10; styles 2. Capsule with 4 teeth.

A. Flowers solitary, not clustered, borne in a lax inflorescence.

Petrorhagia saxifraga (Syn. *Tunica velutina*) TUNIC FLOWER A hairless, *mat-forming* perennial 50 mm–45(50) cm with spreading, then ascending flowering shoots. Leaves linear, pointed and rough-edged. *Flowers solitary*, pale pink to white, borne on *long stalks*; petals notched; *bracts much shorter than the calyx; calyx small, just 3–6(7) mm. Seeds minutely netted, not warted.* Rocks and stony or sandy habitats. ES, FR, IT. *Petrorhagia illyrica* is similar in habit, with linear, 3-veined leaves and white to pink or yellow flowers, often veined or spotted with pink in the centre, borne in spreading to erect panicles; petals oblong and *not notched*. Rocky habitats. IT (incl. Sicily).

B. Flowers 1-few, clustered, borne in dense flower-heads.

Petrorhagia prolifera (Syn. *Kohlrauschia prolifera*) PROLIFEROUS PINK A lax, *hairless annual* 60 mm–50(70) cm. Leaves linear, greyish and rough-edged, fused at the base into *leaf sheaths that are about as long as wide*. Flowers small and pale to mid pink, borne in a dense cluster but opening 1 at a time, *surrounded brown, papery bracts*; bracts equalling the calyx; calyx 10–13 mm. *Seeds minutely netted, not warted*. Maquis. Throughout, except for most islands. ***Petrorhagia nanteuillii*** (Syn. *Tunica nanteuillii*) CHILDING PINK is very similar but with stems hairy along the middle, and leaf sheaths 2 x as long as wide. *Interior* inflorescence bracts obtuse or short bristle-tipped (mucronate); calyx 12–15 mm. *Seeds covered in dense, tiny warts*. Maquis. Common in the far west. ES, MA, PT. *Petrorhagia dubia* (Syn. *Kohlrauschia velutina*) KOHLRAUSCHIA is very similar to the previous species but with calyx smaller, 8–14 mm, *all* inflorescence bracts short bristle-tipped (mucronate) and *seeds covered in sparse, cone-like warts*. Maquis. Throughout.

1. *Dianthus armeria*
2. *Dianthus sylvestris*
3. *Petrorhagia nanteuillii*

AMARANTHACEAE | AMARANTH FAMILY

Herbs and shrubs with alternate, simple leaves. Flowers unisexual, greenish with 2–5 tepals (female) and 3–5 secondary bracts (male and female); stamens 1–5; styles 2–3. Fruit an achene or 1-seeded capsule. Often in maritime or waste habitats. Many genera were traditionally described under the family Chenopodiaceae.

Amaranthus

Annuals of waste land and disturbed habitats. Flowers inconspicuous with a brownish, papery perianth; tepals (2)3–5 (or absent). Fruit 1-seeded. A hand lens is required to examine the tepals.

Amaranthus retroflexus PIGWEED An erect, robust, hairy annual to 1 m with few side branches, green (rarely red). Leaves oval to lance-shaped and stalked. Flowers small and inconspicuous, greenish-white with 5 *linear, tapering tepals with mid veins ending below the tips*; borne in compact green spikes that are leafless towards the top. Fruit capsule splitting when ripe. A North American native widely established on wasteground. Throughout. *Amaranthus hybridus* is similar, with less dense and more branched inflorescences; *tepals 3(5), tapering to a long point*; secondary bracts in female flowers exceeding the tepals. Fruit splitting. Similar habitat and distribution. *Amaranthus viridis* is similar to above species, with diamond-shaped leaves, often with paler markings, and *non-splitting*, densely wrinkled (rough) fruit capsules. Throughout.

Haloxylon

Desert-dwelling shrubs or small trees with thick basal trunks and leaves reduced to scales. Flowers male, or cosexual with 2 short stigmas. Fruit a winged achene.

Haloxylon salicornicum WHITE SAXAUL A rather diffuse to intricately branched shrub to 1 m with virtually leafless, slightly succulent, jointed, woody stems. Leaves minute, short-triangular and scale-like, with membranous margins, woolly within. Flowers borne on short spikes at the end of lateral young shoots, each with 2 oval-concave bracts, woolly at the base; stigmas 2. Fruiting body (perianth) distinctly winged, to 8 mm across. Most frequent in the northern Sahara, extending locally into drier, semi-arid areas of the Mediterranean. DZ, TN.

Achyranthes

Perennial herbs with stems woody at the bases. Leaves opposite. Flowers small and numerous, borne in a lax inflorescence; tepals 5. Fruit an achene.

Achyranthes aspera (Syn. *A. sicula*) A lax, slender perennial to 70 cm with more or less hairy stems. Leaves oval-elliptic and pointed at the tips, 20–60 mm long. Flowers borne in *very slender, long terminal and lateral spikes*, lax below and crowded towards the apex and pointing downwards when fully open; each flower has 3 spine-like bracts and a 5-parted perianth. Fruit a pendulous achene. Usually in bare or scrubby coastal disrubed ground. Spain and North Africa, possibly elsewhere. ES, MA.

Atriplex ORACHE

Herbs and shrubs, often greyish, with flat leaves and inconspicuous flowers. Flowers unisexual, the male flowers with 5 tepals, female flowers with no tepals but 2 secondary bracts, enlarged in fruit.

A. halimus

A. Shrubby perennials in maritime habitats.

Atriplex halimus SHRUBBY ORACHE An erect to ascending, woody shrub to 2.5 m. Leaves small, to 30 mm long, alternate, *silvery-white*, narrowly oval, slightly diamond-shaped, or angled leaves. Flowers yellowish, borne in leafless branched terminal spikes; *secondary bracts 1.5–3 mm, fused only at the base.* Common on sandy shores, cliffs and estuaries. Throughout. *Atriplex glauca* is a similar whitish, shrubby perennial to 50 cm with *prostrate to ascending stems.* Leaves stalkless, 10 x 7 mm. Secondary bracts *oval-diamond-shaped* and with protuberances on the surface. Saline habitats. ES, MA, PT.

Atriplex portulacoides (Syn. *Halimione portulacoides*) SEA PURSLANE A low shrub, similar in form to *A. halimus,* with spreading and often rooting branches, more or less mat-forming. Leaves upward-pointing, silvery, the *lowermost opposite*, narrowly oval and untoothed. Flowers small and green or reddish, borne in more or less leafless panicles; *secondary bracts 2.5–5 mm, fused to just >1/2 their length*. Common to abundant in salt-marshes. Throughout.

---

1. *Amaranthus retroflexus*   4. *Achyranthes aspera*
2. *Amaranthus hybridus*   5. *Atriplex halimus* (inset: flowers)
3. *Amaranthus viridis*   6. *Atriplex glauca*

B. Leafy annuals with green or mealy leaves.

Atriplex prostrata (Syn. *A. hastata*) SPEAR-LEAVED ORACHE A variable, tall, hairless and branched annual to 2 m, mealy when young. Leaves green-grey with spreading triangular basal lobes, the *largest lower leaves with straight-edged lower lobes at right angles to the leaf stalk*. Flowers reddish, borne in dense clusters in leafy panicles; secondary bracts small, entire and triangular, 2–6 mm. Common on sandy shores and disturbed ground. Throughout.

Atriplex patula COMMON ORACHE A very variable, erect or prostrate annual to 1 m with ridged, often reddish stems. Leaves mealy, usually diamond-shaped or arrow-shaped and coarsely toothed below, with *forward-pointing lobes,* the upper leaves linear and slightly toothed. Flowers greenish; *secondary bracts 3–7(20) mm, diamond-shaped and scarcely toothed or entire*. Arable areas and coastal habitats. Throughout.

C. Leafy annuals with silvery-white leaves.

Atriplex rosea A *silvery-white* erect or ascending, much-branched annual to 1.5 m with oval to diamond-shaped leaves with wavy-toothed margins, and mainly *lateral leafy flower spikes* all along the stem; secondary bracts 2–3(4) mm, pinkish or whitish, diamond-shaped and toothed. Cultivated ground and waste places. Possibly absent from the far south and east, otherwise more or less throughout.

1. *Atriplex portulacoides*
2. *Atriplex prostrata*
3. *Chenopodium album*
4. *Chenopodium murale*

Chenopodium GOOSEFOOT

Herbs or small shrubs with alternate, often *mealy* leaves and inconspicuous cosexual or female flowers; secondary bracts absent, tepals 4–5; *stamens 5*. Leaves and seeds are important diagnostics. A complex genus; species described here are the most widespread and easily identifiable. The genera *Oxybasis, Dysphania* and *Blitum* were previously grouped within this genus and are very similar but now established to be genetically distinct.

A. Leaves *heart-shaped* at the base.

Chenopodium hybridum MAPLE-LEAVED GOOSEFOOT A tall and erect hairless annual to 1 m with leaves scarcely mealy and *large, to 15 cm long, oval to triangular with few, long, acute lobes and slightly heart-shaped at the base*. Tepals rounded. Seeds with deep oval pits on the surface. Throughout.

B. Leaves not heart-shaped at the base; *perianth not mealy.*

Chenopodium polyspermum MANY-SEEDED GOOSEFOOT A spreading to erect annual to 60 cm (1 m), not especially mealy (except sometimes the leaf undersides) with *markedly angled (square) and usually reddish stems*. Leaves *untoothed* and entire, sometimes with lower lobes. Tepals rounded. Seeds black. Waste habitats. Throughout. *Chenopodium urbicum* UPRIGHT GOOSEFOOT is similar, hairless to slightly mealy, with stems *ridged* (not square) and (variably) *coarsely toothed* triangular leaves. Tepals rounded. Seeds *black, with shallow grooves*. Throughout.

C. Leaves almost *entire*, not heart-shaped at the base; *perianth mealy.*

Chenopodium vulvaria STINKING GOOSEFOOT An erect annual to 40 cm that is *not mealy* (except for perianth) and *unpleasant-smelling* when crushed. Leaves small (<25 mm), diamond-shaped with basal lobes but more or less entire. Tepals rounded. Seed *with faint furrows and irregular thickenings*. Saline waste places. Throughout.

D. *Leaves toothed or lobed*, not heart-shaped at the base; perianth mealy.

Chenopodium album FAT HEN A tall, erect, *deep green* but *grey-mealy* plant to 1.5 m. Stems stiff and often marked with red. Leaves oval to diamond-shaped and markedly longer than broad, (to 80 mm) and *entire to coarsely and bluntly toothed*. Inflorescence usually leafless at the very top; tepals slightly keeled. Seeds smooth or faintly ridged. Very common on disturbed ground. Throughout. *Chenopodium murale* is similar and also mealy, occasionally reddish with leaves with course, *irregular, forward-pointing teeth* (nettle-shaped). Flowers mealy, with blunt keels on the back of the *toothed tepals*; inflorescence leafy to the top. Seeds minutely pitted. Sandy, disturbed and maritime waste habitats. Throughout.

Oxybasis

Virtually hairless annuals morphologically similar to *Chenopodium* but identified to be a genetically distinct group. *Stamens 1–3*.

Oxybasis glauca (Syn. *Chenopodium glaucum*) OAK-LEAVED GOOSEFOOT A branched annual to 50 cm, similar to *Chenopodium* (particularly *C. polyspermum* and *C. urbicum*) but distinguished by its leaves which are *mealy underneath* and green above; shallowly and regularly toothed. Perianth mealy. Seeds red-brown. Disturbed arable and maritime habitats. Throughout.

Dysphania

D. botrys

Very similar to *Chenopodium* but markedly *glandular-hairy*, aromatic and sticky.

Dysphania botrys (Syn. *Chenopodium botrys*) STICKY GOOSEFOOT An erect, long-stemmed, *sticky* glandular-hairy aromatic annual with alternate, oblong, *pinnately divided, rather oak-like leaves*. Flowers small and green, in long, narrow clusters. Throughout.

Blitum

Sparsely mealy perennials. Flowers arranged in terminal inflorescences; stamens 4–5.

Blitum bonus-henricus (Syn. *Chenopodium bonus-henricus*) GOOD KING-HENRY An erect, branched, rhizomatous *perennial* to 50 cm with stiff stems that are often streaked red. Leaves up to 10 cm long and *broadly triangular* with basal lobes, otherwise entire or shallowly lobed. Inflorescence a terminal, narrowly pyramidal raceme. Frequent in damp or fertilised waste places and other disturbed habitats. Throughout.

Beta BEET

Erect or ascending annuals or perennials with swollen roots and with alternate, not mealy leaves and cosexual flowers with 5 tepals.

Beta vulgaris SEA BEET A variable, hairless, *reddish* and rather fleshy annual or perennial to 1.5 m, but often prostrate. Leaves oval-lanceolate, dark green, leathery and shiny, and more or less untoothed, to 20 cm. Flowers small and green or purplish, borne at the top of dense, long spike; stigmas mostly 2; lower bracts 2–20 mm; bracts absent above. Common on maritime sands, cliffs and shingle. Probably throughout. *Beta maritima* (Syn. *B. vulgaris* subsp. *maritima*) is treated by some authors a distinct species, differing mainly in its sprawling (sometimes erect) habit, and uppermost bracts 10–20(35) mm, or absent. Probably throughout. *Beta macrocarpa* is similar to *B. vulgaris* but with flowering stems with flowers practically to the base, and *bracts to the top*. Rare and local. ES (incl. Balearic Islands), IT (incl. Sardinia), MA, PT.

Salicornia GLASSWORT

Salt-tolerant succulent herbs with slender, finger-like branches and minute leaves and flowers. Flowers in groups of 1–3, the central extending to make the group triangular. Many very similar species occur; only the most widespread and common are described here.

Salicornia ramosissima A spreading, short, succulent annual to 40 cm with *dark green, later red-purple stems*, the fruiting stems *strongly beaded with prominent 'waisted' joints*; each segment with a membranous margin 0.1–0.2 mm wide. Flowers borne in groups of 3, the central the largest. Common and widespread in salt-marshes and mudflats. ES, FR, IT, MA, PT. *Salicornia europaea* is a very similar annual to 35 cm (often grouped with *S. ramosissima*), differing in having yellowish green or reddish stems which are *scarcely beaded or waisted in fruit*, and with inconspicuous membranous margins (<0.1 mm) at the tips of the segments. Distribution poorly known due to confusion with the previous species. ES, FR, IT, MA, PT.

1. *Beta vulgaris*
2. *Salicornia ramosissima* (inset: flowers)
3. *Sarcocornia perennis*
4. *Arthrocnemum macrostachyum*

Sarcocornia (including *Arthrocnemum*)

Similar to *Salicornia* but *woody-based, shrubby* perennials; flowers minute, in groups of 3 in a row.

Sarcocornia perennis (Syn. *Arthrocnemum perenne*) A small, very succulent rather *short, spreading shrub to 30(70) cm with creeping, subterranean stems, at first dark green*, ageing red-brown or yellowish. Leaves opposite, scale-like and fused in pairs, appearing as segments. Flowers tiny, borne groups of 3, each with 2 stamens; flowers falling to leave 3 scars in the segment. Common in salt marshes, maritime sands and shingle. Throughout. *Sarcocornia fruticosa* (Syn. *Arthrocnemum fruticosum*) is very similar but with stems *erect to 1 m*, stout and non-rooting, and *blue-green*. Salt-marshes. Probably throughout. **Arthrocnemum macrostachyum** (Syn. *A. glaucum*) is similar to the previous species, rather erect, with many very woody, spreading to prostrate branches, bare below, glaucus and often partly *yellowish or reddish*. Throughout.

Suaeda SEABLITE

Herbs or small shrubs in saline habitats with fleshy, alternate, linear leaves and minute flowers with 5 tepals and 2–3 secondary bracts; flowers cosexual and female.

A. Plant a woody-based shrub.

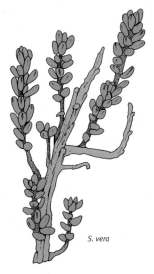

S. vera

Suaeda vera SHRUBBY SEABLITE A succulent, woody shrub to 1.2 m, leaves *densely crowded, blunt and fleshy*, alternate, small, 5–18 mm, semi-cylindrical, sessile and glaucous, becoming reddish or purplish. Flowers small, greenish, borne 1–3 in the axils of the upper leaves; stigmas 3. Common in the European part of the region on sea cliffs, maritime slopes and salt-marshes. ES, FR, IT, PT. *Suaeda ifniensis* is a similar, spreading subshrub to 40(50) cm, often strongly purple-tinged with cylindrical, succulent leaves to 20 mm long with a waxy bloom. Coastal and desert fringe habitats in northwest Africa. MA.

1. *Suaeda vera*	4. *Salsola kali*
2. *Suaeda maritima*	5. *Salsola brevifolia*
3. *Suaeda ifniensis*	6. *Traganum moquinii*

B. Plant a succulent annual herb.

Suaeda maritima ANNUAL SEABLITE A succulent annual to 30(75) cm with prostrate or ascending stems and distinct, *pointed, alternate, fleshy leaves* 3–25 mm long, ranging from blue-green to reddish or purple; *leaves slightly concave above in cross-section*. Flowers minute, green and borne in the leaf axils; stigmas 2. Seeds >1.5 mm long. Muddy coastal habitats. ES, FR, IT, PT. Prostrate forms with leaves with a distinct *whitish-grey bloom*; tepals *not* keeled and smaller seeds (1–1.5 mm) on coasts of the southern Iberian Peninsula are sometimes refered to as *S. albescens*. ES, PT. *Suaeda spicata* is very similar to the previous species but has *keeled tepals* and tiny seeds <1 mm long. ES, PT.

Salsola SALTWORT

Herbs or shrubs with fleshy, cylindrical leaves, and small cosexual flowers borne in the leaf axils with 5 tepals.

A. Plant herbaceous, with leaves and bracts sharp-pointed or spiny.

S. kali

Salsola kali PRICKLY SALTWORT A variable, hairy annual with a prostrate, spreading or erect habit to 50 cm (1 m). Leaves linear and narrow, 10–40(70) mm long, blue-green, *spine-tipped* and broadest at the base; opposite below. Flowers small and 5-lobed, solitary in the leaf axils with a pair of stiff, spiny bracts which exceed the flowers. Coastal environments. Throughout. *Salsola soda* is similar but hairless, with $^1/_2$-cylindrical, clasping, *not spine-tipped* leaves 15–75 mm long, and flower bracts equalling the flowers. Similar habitats. Throughout.

B. Plant a shrub or subshrub, with leaves and bracts *not* spiny.

S. vermiculata

Salsola vermiculata A succulent *shrub* to 1 m with *very small, alternate, crowded leaves* 5–12(25) mm long, semi-cylindrical, oval at the base and semi-clasping the stem, *often pubescent*. Flowers small and green or pink-tinged, the perianth to 12 mm across in fruit, stigmas shorter than the style. Common on sea cliffs and maritime slopes, and other coastal habitats. Throughout. *Salsola brevifolia* is very similar (sometimes grouped with the previous species), but has *short, oval to 3-sided basal leaves <3 mm long*. Throughout. *Salsola oppositifolia* is distinguished by its rather lax, linear leaves with a central groove that have an *opposite* arrangement. ES, MA.

Traganum

Succulent, woody shrubs with woolly nodes and fleshy, cylindrical leaves. Fruiting perianth wingless, with *2 horn-like teeth*.

Traganum moquinii A robust, spreading, woody-based, much-branched, succulent shrub to 1.5 m. Leaves 10 mm, succulent, yellowish with a whitish bloom, broadly cylindrical and bluntly pointed and angled. Flowers inconspicuous, dull yellow, borne solitary in the leaf axils, each with 2 secondary bracts. Fruiting perianth with *2 horn-like teeth*. Maritime dunes and desert fringe habitats on Atlantic North African coasts. MA.

PORTULACACEAE | BLINKS FAMILY

Annual or perennial herbs, often fleshy, with cosexual flowers with only 2 opposite sepals and 4–6 petals; stamens 3–14; styles 1–3(6). Fruit a capsule.

Montia

Stem leaves in several pairs, opposite or alternate. Flowers in terminal or axillay groups, white or pink; stamens 3–5; style 1, ovary superior. Fruit normally a 3-seeded capsule.

Montia fontana BLINKS A small, bright green, dense patch-forming annual or perennial 10 mm–20(50) cm with oval, opposite, spatula-shaped stem leaves (basal leaves absent). Flowers white, tiny, borne in small loose clusters with stems that lengthen in fruit; petals <2 mm, scarcely longer than the 2 sepals. Local in wet pastures. Throughout.

M. fontana

Claytonia

Stem leaves in 1 pair, opposite. Flowers in terminal cymes; petals 5; ovary superior. Fruit a 1-seeded capsule.

Claytonia perfoliata (Syn. *Montia perfoliata*) PERFOLIATE CLAYTONIA An annual to 30 cm with stem leaves fused to form a cup-like structure at the base of the inflorescence. Petals <5 mm, white and entire or notched. Waste ground. Local. ES, FR, PT.

Portulaca PURSLANE

Herbaceous annuals with 1-few, stalkless flowers; stamens numerous and ovary $^1/_2$ inferior. Fruit a many-seeded capsule.

Portulaca oleracea PURSLANE A fleshy, prostrate or ascending, patch-forming, leafy annual with stems to 30(50) cm. Leaves to 30 mm long, spoon-shaped and more or less opposite and densely crowded beneath the flowers. Flowers yellow with 4–5(6) petals 4–8 mm long which soon fall, revealing blunt sepals. Very common in bare or exposed habitats. Throughout.

1. *Portulaca oleracea*
2. *Opuntia maxima*

CACTACEAE | CACTUS FAMILY

A family of, typically, spiny succulents from the Americas, often with showy flowers. Cultivated and naturalised in the region.

Opuntia PRICKLY PEAR

Woody cacti with jointed stems and soon-falling leaves. A confused genus with conflicting names and descriptions in the literature (morphological traits are often continuous, or not clear or useful in the field). Native to North and South America. Further work is required to accurately resolve the genus in the region.

O. maxima

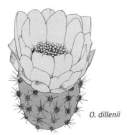

O. dillenii

O. stricta

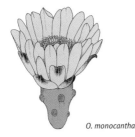

O. monocantha

A. Stem segments usually *without stout spines*. Flowers yellow.

Opuntia maxima PRICKLY PEAR A robust, blue-green, typically cactus-like plant *5–6 m tall*; woody and trunk-like at the base (rather tree-like) with *large*, flattened jointed stem segments 30–50 cm long and soon-falling, inconspicuous leaves. Bristles hooked and yellowish, *straight spines usually absent*. Flowers bright yellow. Fruit egg-shaped, 50–60(90) mm long, yellow or reddish when ripe; edible. The most common species to have naturalised in the region. Throughout.

Opuntia ficus-indica A *tree-like cactus* with a distinct trunk, to 5 m, with pale green or glaucous oblong segments 30–50 cm long *with no spines* but small yellow clusters of bristles. Flowers yellow-orange, large (>70 mm across). Fruit >50 mm, normally red. ES, FR, IT.

B. Stem segments generally with conspicuous, *stout* spines. Flowers yellow.

Opuntia ammophila A shrub with a sprawling habit, to 2 m tall; segments 80 mm–12 cm long, spiny (sometimes spineless). ES (Balearic Islands); possibly elsewhere.

Opuntia dillenii (Syn. *O. tuna*) A shrub to 1.5 m with blue-green stem segments 70 mm–40 cm long, *with 1–10 conspicuous, stout, yellow spines of various length*. Flowers yellow. *Fruits 50–75 mm, purple-red*. Naturalised in Iberian Peninsula. ES, PT. *Opuntia stricta* has a *spreading habit* with erect branches, and is *shorter* than the previous species, to 1 m high; stem joints 20–30 cm; spines normally present. Fruits reddish. Naturalised. ES (northeast), FR.

Opuntia monacantha A tall shrub 2–4(6) m with *bright, shiny green* joints 10–30 cm, typically with just 1–2 greyish or reddish spines. Flowers yellow. Fruits 50–75 mm, red-purple. FR, IT.

C. Flowers pink.

Opuntia pilifera A robust shrub with broad, flat, spiny segments, distinguished from the above species by its *pink flowers*. IT.

Cylindropuntia

Woody cacti similar to *Opuntia* (and previously classified in this genus) but with *cylindrical stem joints* and papery sheaths around the spines. Native to South America.

Cylindropuntia imbricata (Syn. *Opuntia imbricata*) A robust succulent 1–3(4) m high with *cylindrical* stem segments 20–50 cm long with very promiment ribs and clusters of 2–12 cream-white spines 8–30 mm. Flowers *dark pink-magenta*. Fruit grey-green, 40 mm. Native to Mexico, naturalised. ES.

Austrocylindropuntia

Woody cacti similar to *Cylindropuntia* but with spines *lacking papery sheaths*. Native to South America.

Austrocylindropuntia subulata (Syn. *Opuntia subulata*) A spreading, robust shrub to 3 m, similar to *Cylindropuntia imbricata* but with less prominently ribbed segments 35–50 cm long with prominent clusters of 1–4 spines to 70 mm long that lack papery sheaths. *Flowers pale pink*. Fruit green, 10 cm long. A native of Peru, naturalised. ES.

1. *Opuntia dillenii*	5. *Cylindropuntia imbricata*
2. *Opuntia ficus-indica*	6. *Aizoon canariense*
3. *Opuntia ammophila*	7. *Malephora purpureocrocea*
4. *Opuntia pilifera*	

AIZOACEAE

Herbaceous annuals to perennials with simple, opposite or alternate, succulent leaves. Flowers regular with numerous linear 'petals' (derived from staminodes, described as petals below); sepals 4–5(6); stamens 3-numerous. Fruit a capsule with (1)3-numerous seeds.

Aizoon

Annual or perennial herbs with a crystalline (papilose) surface. Leaves (mostly) alternate. Flowers with 4–5 tepals, often fused below into a short tube; ovary superior. Fruit a capsule with numerous seeds.

Aizoon canariense A short, slightly succulent annual to 30 cm, sprawling when mature. Leaves 40 mm wide, opposite below, alternate above, oblong-lanceolate, blunt and minutely hairy. Flowers solitary and unstalked with 5 yellow tepals 1–3 mm long and yellow stamens. Strongly southern. DZ, ES, MA, TN. *Aizoon hispanicum* has leaves just 10 mm across and tepals yellow or whitish, 7–15 mm long. ES (incl. Balearic Islands).

Carpobrotus HOTTENTOT FIG

Fleshy perennials with leaves triangular in cross section. Flowers with numerous petals and 8–20 stigmas. Fruit succulent, the seeds embedded in mucilage.

Carpobrotus edulis HOTTENTOT FIG A low succulent, trailing, mat-forming perennial with woody stems to 2 m long. Leaves opposite, 3-angled and upwardly curving, finely toothed along the edge and tapered towards the apex, 40 mm–9(13) cm long. Flowers large, 8–10 cm across, solitary, virtually stalkless, very showy, bright pink or yellow with yellow stamens. Fleshy, edible fruit. Native of South Africa, naturalised and invasive in some coastal areas; commonest in the west. ES, FR, IT, PT. *Carpobrotus acinaciformis* is very similar but with *leaves blue-green and broadest above the middle,* 40 mm–10 cm long; flowers magenta with *purple stamens,* 80 mm–10(12) cm across. Less common than *C. edulis.* ES, PT. *Carpobrotus chiliensis* has smaller leaves 20–40 mm long and flowers just 20–40(50) mm across. ES.

Ruschia

Dwarf shrubs with woody stems and succulent, triangular, blunt-angled leaves in opposite pairs. Flowers with 5 sepals and numerous petals; stigmas 4–5. Capsule opening by 4–5 valves.

Ruschia caroli A bushy, erect-spreading shrub to 80 cm with succulent leaves 15–70 mm, 2–8 mm wide, bluish with darker dots. Flowers bright fuschia-pink, 18–30 mm across. Native to South Africa; planted for ornament in pots, beds and in coastal resorts. Naturalised locally in the Algarve, possibly elsewhere. PT.

Malephora

Succulent perennials native to Africa. Leaves opposite and 3-angled. Flowers showy with numerous (65) petals and (150) stamens. Fruit an 8–12-parted capsule.

Malephora purpureocrocea A prostrate, spreading succulent perennial with stems with long internodes and *a white-grey bloom.* Leaves up to 50(60) mm long, opposite, cylindrical and bluntly 3-angled. Flowers very showy, typically 50 mm across, with numerous (to 65) bright pink petals and yellow-orange stamens. Planted in coastal areas, very locally naturalised.

Mesembryanthemum

Succulent annuals or biennials. Flowers with numerous stamens. Capsule with 4(5) valves. Easily recognisible in that the whole plant is densely covered in crystal-like vesicles.

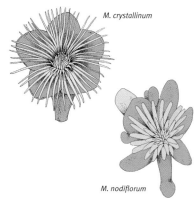

M. crystallinum

M. nodiflorum

A. Fleshy annuals with flowers with numerous petals and 5 stigmas, and leaves rounded or flattened in cross section.

Mesembryanthemum crystallinum ICE PLANT A spreading annual covered in glistening, frost-like *crystalline hairs*. Leaves stalked, flat, fleshy, oval, alternate and untoothed to 11.5 cm long. Flowers solitary, 20–30 mm across, virtually stalkless, yellowish or whitish. In sandy, rocky and saline environments. Throughout. *Mesembryanthemum nodiflorum* is similar but less crystalline and with *cylindrical-linear leaves* 10–25(30) mm long that are red-tinged, and smaller flowers to 15 mm across. Similar habitats. Throughout.

B. Fleshy perennials with small, brightly coloured flowers with numerous petals and 4 stigmas, and *flat leaves*.

Mesembryanthemum cordifolium (Syn. *Aptenia cordifolia*) A mat-forming perennial to 60 cm high with stems to 3 m long, with slightly fleshy, bright green, oval leaves which are pointed at the apex and heart-shaped at the base, 13–56 mm long. Flowers bright magenta and *Mesembryanthemum*-like, 10–18 mm across. An ornamental native to South Africa but widely planted and often naturalised. Throughout.

Lampranthus ICE PLANT

Sprawling dwarf succulent shrubs, woody below, with leaves in pairs and with brightly coloured flowers with numerous petals and 5 stigmas. Capsule woody.

Lampranthus multiradiatus (Syn. *L. roseus*) ROSY DEW-PLANT A spreading succulent shrub to 25(60) cm with opposite, angled leaves (8)20(40) mm long, scarcely pointed at the tips. Flowers bright pink, 30–50 mm across with numerous tepals and whitish stamens. Native to South Africa, naturalised. ES (incl. Balearic Islands), PT.

Drosanthemum

Crystalline, creeping, succulent perennials with showy (often pink), rather small flowers with numerous stamens; sepals 5; petals numerous; stigmas mostly 5. Capsule woody.

Drosanthemum floribundum A crystalline, creeping, all-green succulent perennial to 12 cm high with opposite, fleshy, stalkless leaves 5–16 mm long. Flowers 20–25 mm across, *bright pink* with many narrow petals and numerous anthers; perianth tube semi-globose. Native to South Africa but planted and naturalised on coasts. ES (incl. Balearic Islands), PT. Forms with paler flowers and conical perianth tube have traditionally been described as *D. candens*.

1. *Carpobrotus edulis*
2. *Ruschia caroli*
3. *Mesembryanthemum crystallinum*
4. *Mesembryanthemum nodiflorum*
5. *Mesembryanthemum cordifolium*
6. *Lampranthus multiradiatus*
7. *Drosanthemum floribundum*

Tetragonia

Prostrate or ascending, leafy annuals or shrubs with alternate, flat leaves with long stalks. Flowers mostly solitary with 4–5 sepals, 0 petals; stigmas 3–8. Fruit woody, ridged with 1 seed in each of 3–8 cells.

Tetragonia tetragonioides New Zealand spinach A robust trailing or scrambling, leafy plant to 1 m with triangular leaves 15 mm–11 cm long, superficially like a member of the Amaranthaceae. Flowers small, borne in the axils, virtually stalkess and yellow with 5 petals. Fruit horned. Native to New Zealand and Australia but cultivated as a source of 'spinach' and possibly naturalised in the area. Locally naturalised on dunes. PT.

NYCTAGINACEAE

Trees, shrubs and woody climbers native to South America with opposite leaves. Flowers with 5 showy bracts, cosexual; stamens 5; style 1. Fruit surrounded by the perianth tube forming a false fruit.

Bougainvillea

South American vigorous climbing shrubs with flowers with showy bracts; stamens 8. Fruit a 5-lobed achene.

B. glabra

Bougainvillea glabra BOUGAINVILLEA A vigorous woody, virtually hairless climber to 10 m. Leaves opposite or alternate, to 60 mm long, untoothed and stalked. Flowers in groups of 3, inconspicuous, whitish-yellow and funnel-shaped, surrounded by large, flower-like crimson bracts, 1 to each flower. Widely planted but rarely naturalised. Throughout. ***Bougainvillea spectabilis*** is very similar but with leaves with a densely hairy underside, and flowers with straight hairs and a purplish, not greenish outer corolla. Bracts red, orange or violet. Throughout.

Mirabilis

South American tuberous perennials typically with fragrant, deep-throated flowers; stamens 3–5. Fruit an 0–5-angled achene.

Mirabilis jalapa MARVEL OF PERU A hairless or shortly hairy perennial to 1.5 m with large, oval, untoothed, leaves narrowing to a point; often rather wrinkled, stalked below and short-stalked to sessile above. Flowers borne in terminal clusters and variably white, yellow, pink, red or variegated; tube 25–35 mm long, and with 5 lobes. Native to tropical America but widely planted and naturalised in southern and eastern Spain, possibly elsewhere. ES.

1. *Bougainvillea spectabilis*
2. *Bougainvillea glabra*
3. *Mirabilis jalapa*
4. *Phytolacca acinosa* (in flower)
5. *Phytolacca acinosa* (in fruit)

PHYTOLACCACEAE

Herbaceous plants, shrubs and trees with alternate, entire leaves. Flowers borne in leaf-opposed racemes; perianth 5(4–9)-parted and persistent in fruit; stamens 5–25(30); ovary composed of 5–15 1-seeded carpels. Fruit succulent, berry-like and lobed.

Phytolacca

Herbs or shrubs. Flowers unisexual or cosexual; stamens 5–25(30); carpels 5–15. Fruit a berry.

Phytolacca acinosa POKEWEED A tall and leafy herbaceous perennial to 1.5(3) m tall with thick, ribbed stems, often reddish. Leaves large and oval, 12–20(25) cm long. Flowers green or pink, borne in erect, cylindrical spikes; stamens and styles 7–10, tepals more or less equal. Fruit a glubular cluster of red, later black berries 4 mm long borne in dense cylindrical spikes. Native to North America, widely naturalised. Throughout. (*P. americana* is often recorded in error, which has narrower leaves and long, arching racemes). *Phytolacca heterotepala* is similar but with *unequal tepals*, 9–21 stamens, and 8–9 carpels. Natualised near the coast of southern and eastern Spain, possibly elsewhere. ES.

SANTALACEAE

Woody or herbaceous perennials, hemi-parasitic on the roots of surrounding vegetation. Flowers small, cosexual or unisexual; stamens 3–5; style 1. Fruit a berry or nut.

Osyris

Dioecious shrubs with angled stems and entire leaves. Flowers with 3–4 tepals and 3–4 stamens or 1 style. Fruit a berry.

A. Bracts persistent in fruit.

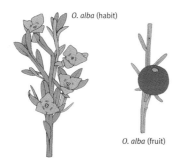

O. alba (habit)

O. alba (fruit)

Osyris alba A superficially broom-like, small yellowish shrub 50 cm–1.5(2) m. Leaves *small*, 15–20(40) mm long, and narrow, just 2–3(4) mm wide, alternate, linear and entire, with a single mid-vein. *Bracts green, leaf-like and persistent in fruit.* Flowers sweetly scented, yellow, with 3(4) tepals, the male in small clusters, the female solitary, and borne on separate shrubs; stamens with a cluster of hairs. Fruit a red berry 6–7 mm long. Woods and scrub; common. Throughout

B. Bracts soon-falling in fruit.

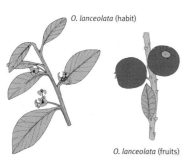

O. lanceolata (habit)

O. lanceolata (fruits)

Osyris lanceolata (Syn. *O. quadripartita*) Similar to *O. alba* (virtually indistinguishable in some areas where intermediates occur) but typically taller, 1–2.5(3) m, with leaves yellowish and *larger*, 18–30(45) mm, leathery often with pronounced pinnate veins (not a reliable diagnostic in all populations) and *small, papery bracts that are shed before fruiting*. Tepals 4 (sometimes 3), and flowers male or cosexual, borne on separate plants (androdioecious); stamens hairless. Fruit a red berry 7–9 mm. Frequent on coastal garrigue in the west. ES, MA, PT.

Viscum MISTLETOES

Hemi-parasitic herbs with opposite leaves. Flowers 3–5, virtually stalkless; tepals 4; stamens 4; style absent. Fruit a sticky 1-seeded berry.

Viscum album MISTLETOE A woody, hemi-parasitic and rootless shrub which forms rounded clumps to 2 m across on the branches of trees (often apple, poplar and lime). Leaves 20–80 mm long, leathery, borne in pairs, widest towards the tip. Flowers yellowish and inconspicuous. Fruit a sticky white berry 6–10 mm. Scattered across the region though rather rare, especially in hot, dry areas. Throughout. *Viscum cruciatum* is very similar but has *red berries*. ES, MA, PT.

SAXIFRAGACEAE | SAXIFRAGE FAMILY

Annual to perennial herbs with alternate or all basal leaves (rarely opposite). Flowers regular, cosexual with 4–5 petals and 5–10 stamens; carpels 2(3). Fruit 2 follicles fused to form a capsule. Many species occur at higher altitude in the region which are not described here.

Saxifraga

Hairy, often rosette-forming annuals or perennials with simple or almost compound leaves. Flowers with 5 petals and 10 stamens; carpels 2. Fruit a capsule.

Saxifraga granulata MEADOW SAXIFRAGE A perennial to 50 cm high, with most leaves in a basal rosette; leaves stalked, kidney-shaped, to 40 mm long and hairy. Flowers white, with petals 9–16(20) mm long, borne singly or in a loose inflorescence on erect, slender stems. Cool, damp woods and rocky habitats. ES, FR, IT, PT. *Saxifraga corsica* differs in having 3-lobed basal leaves and inflorescences branched from the ground with spreading branches. ES (east), FR (Corsica), IT (Sardinia).

1. *Osyris lanceolata*
2. *Osyris alba*
3. *Viscum album*
4. *Saxifraga granulata*
5. *Paeonia broteri*
6. *Aeonium arboreum*

PAEONIACEAE | PEONY FAMILY

Perennial herbs with alternate, divided leaves. Flowers solitary, large and showy, with usually 5 sepals and 5–8(13) petals; stamens numerous. Fruit consisting of 2–8(9) fleshy follicles, with large black or brown seeds.

Paeonia

Robust, leafy perennials. Flowers large and conspicuous, regular, solitary; stamens ~140. Follicles variably hairy.

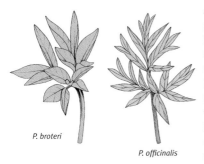

P. broteri

P. officinalis

Paeonia broteri PEONY A medium, almost hairless perennial with untoothed lower leaves with 10–17(30) hairless, narrowly elliptic leaflets 30–45 mm wide, that are grey-green below. Flowers solitary and large, 9–15 cm across, purplish red, with numerous yellow stamens. Follicles 2–5(6), *softly hairy*. Local on rocky, forested hill slopes. ES, PT. *Paeonia officinalis* is similar but with leaves hairless above, and bluish-green and hairy beneath, the divisions 10–25(30) mm wide and a deeply channelled, pubescent petiole. Flowers red with red filaments. Follicles 2–4, *more or less hairless*. Meadows and bushy places. ES, PT.

Paeonia mascula An erect, bushy herb to 80 cm with large leaves with *leaflets usually not further divided*; lower leaves normally with 6–13(16) elliptic to oval leaflets 25 mm–12 cm cm wide with untoothed margins; hairless beneath. Flowers solitary, red and opening widely, 10–15 cm across with yellow anthers and 5–7 petals. Follicles in groups of 3–4(5), usually hairy. ES, FR, IT. Subsp. *russoii* has just 9–10 leaflets that are broadly oval. FR (Corsica), IT (Sardinia and Sicily). *Paeonia arietina* is similar to *P. mascula* but with narrow, elliptic leaflets that are *hairy beneath*. IT. *Paeonia coriacea* is distinguished from the above species by having leaflets with wavy margins that are *blue-grey beneath*, and just 2(3–4), hairless follicles. DZ, ES, MA. *Paeonia cambessedesii* has leaves flushed purple, *especially underneath*, and 5–8(9) hairless, purplish follicles. An island endemic. ES (Balearic Islands).

CRASSULACEAE

Typically succulent annuals and perennials with alternate, opposite or whorled leaves. Flowers regular, star- or bell-shaped with 3–18(20) sepals and petals and an equal number of stamens, or 2 x as many. Fruit a cluster of follicles.

Aeonium

Woody perennials *with branches bearing terminal rosettes of overlapping leaves*. Flowers with 8–11-numerous petals and 2 x as many stamens.

Aeonium arboreum AEONIUM A stout, succulent, branched perennial to 1 m with stems with conspicuous leaf scars, ending in rosettes of overlapping green, shiny, hairless, bristle-edged leaves. Flowers bright yellow, to 20 mm across, borne in dense, rounded clusters. Native to Morocco (and Canary Islands); locally naturalised elsewhere. ES, FR, IT, MA, PT.

Crassula

C. tillaea

Aquatic or terrestrial annuals and perennials, usually hairless with succulent leaves, often fused in pairs. Flowers with 3–5-numerous sepals, petals and stamens.

Crassula tillaea MOSSY STONECROP A tiny, often *dark red* moss-like annual with prostrate stems. Leaves 1–2 mm, oval and crowded. Flowers borne in small groups in the leaf axis, 1–2 mm across with petals shorter than the sepals; white or pale pink; *sepals and petals 3* (rarely 4). ES, FR, PT. *Crassula vaillantii* is similar but with rather sparser leaves and with *flowers with 4 sepals and petals*. Throughout.

Sedum STONECROPS

Succulent annuals or perennials with flat or cylindrical leaves. Flowers with 4–9 (numerous) petals and 2 x as many stamens.

A. Leaves not flattened; flowers greenish or yellow with (5)*6–9 petals.*

Sedum sediforme (Syn. *S. nicaeense*) A very fleshy, short to medium perennial to 60 cm with flowering and non-flowering shoots, woody at the base, and reddish throughout. Leaves grey green, later red and borne in spiralled rows. Flowers yellowish-white to bright yellow with 5–8 petals, borne on tall stalks with a dense sub-spherical inflorescence, then with recurved branches when mature, concave in fruit. Common in dry rocky or sandy habitats. Throughout. *Sedum amplexicaule* (Syn. *S. tenuifolium*) is similar but with small grey-green leaves closely overlapping and clasping the stem at the base, and *few flowers borne in 1-sided, lax clusters*; petals yellow with reddish mid-veins. Throughout. **Sedum forsterianum** is similar to *S. sediforme* but has denser, more slender, linear *blue-grey* leaves which are *flat on top, forming distinctive terminal, tassel-like clusters on non-flowering shoots*. Rocky slopes above sea level. ES, FR, PT. **Sedum ochroleucum** is similar to *S. sediforme* but shorter, to 30 cm tall, and less robust. Leaves rounded and spurred at the base. Flowers *larger, to 20 mm across* and pale yellow; sepals 5–7 mm long, narrowly-triangular and glandular-hairy; petals to 10 mm long; inflorescence flat-topped in bud. Rare or absent in much of the south, west and the islands. FR, IT.

B. Leaves not, or scarcely, flattened; flowers greenish or yellow, normally with *5 petals*.

Sedum acre WALLPEPPER A low, succulent, tufted, bright yellow-green hairless perennial sending up ascending or erect flowering shoots to 29 cm. Leaves small, 3–5 mm, oval and broadest at the base, blunt and overlapping. Flowers *bright yellow*, borne in small clusters; petals 5, 6–8 mm. In rocky and sandy habitats. Widespread but absent from most islands and the east. DZ, ES, FR, IT, MA, PT.

Sedum litoreum A low to short annual similar in general appearance to *S. rubens* with short, reddish stems, but with *slightly flattened, broadly oblong to spatula-shaped leaves* and *yellow flowers*. Rocky outcrops. FR, IT.

1. *Crassula tillaea*	5. *Sedum dasyphyllum*
2. *Sedum sediforme*	6. *Sedum album*
3. *Sedum forsterianum*	7. *Sedum acre*
4. *Sedum ochroleucum*	

C. Leaves not flattened; *flowers white* (or flushed with pink, rarely blue) normally with 5(6) petals.

Sedum album WHITE STONECROP A low to short, hairless, tufted and loosely *mat-forming* perennial with creeping stems, sending up flowering stems to 20 cm. *Leaves alternate*, linear-cylindrical to egg-shaped, 4–12 mm long, *rather shiny and often reddish*. *Flowers white* (sometimes pink-tinged) borne in flat-topped clusters of >20; with 5 petals 2–4 mm long and 10 stamens. Rocky habitats and old walls. Throughout.

Sedum dasyphyllum THICK-LEAVED STONECROP A fleshy, grey-pink-tinged perennial with *glandular-hairy* stems to 10 cm. Leaves ovoid to almost spherical, 3–6(10) mm long, *mostly opposite*, and slightly flattened above. Flowers pink-white, with 5–6 petals 2.5–4 mm long. Rocks and walls. Rare in the far west and east. ES, FR, IT, MA. *Sedum brevifolium* is very similar but *completely hairless*, with globose (bead-like) leaves. ES, FR, PT.

Sedum andegavense A low, *strongly red-flushed*, mat-forming, glandular-hairy annual to just 10 cm with ovoid to sub-globose, *bead-like leaves*. Flowers white or pinkish, borne on *short stalks* to 1.5 mm. Rocky slopes and sandy habitats. ES, FR, PT.

Sedum rubens RED STONECROP A *very small*, erect annual, usually to just 30 mm–9(15) cm tall, often sticky and glandular above. *Leaves linear*, alternate, 7–16(20) mm long, greyish-green, tinged with red. Flowers whitish or pink with darker mid-veins, borne in small clusters; *petals 5, 4–5(6) mm long, 2 x the length of the sepals*. Rocky habitats and walls. Throughout. *Sedum caespitosum* is similar, also short (20–40 mm), but *hairless* and with egg-shaped (not linear) leaves. Petals 2.5–3(4) mm long; *sepals fused*. ES, MA. *Sedum caeruleum* like the previous species is short, reddish and mat-forming with erect stems, with linear-oblong leaves, and *flowers white flushed with light blue*, borne in lax clusters. FR (Corsica), IT (Sardinia and Sicily), MA.

D. Leaves flat and fleshy.

Sedum stellatum STARRY STONECROP A small, *hairless*, fleshy annual 50 mm–15 cm with stout, erect or ascending stems. *Leaves short-stalked, rounded and flat*, 6–20(32) mm long, *blunt-toothed* some opposite below but *alternate above*. Flowers pink, borne in lax, outward-spreading, leafy clusters; petals 5, each 4.5 mm long (equal or shorter than the sepals); stamens 8–10. Rocky habitats and old walls. ES, FR, IT. *Sedum cepaea* is similar in form but much taller, 15–28(40) cm, all leaves opposite or whorled, with white or pink flowers borne in long, lax, leafy inflorescences; petals 3–4(5) mm. Shady habitats on higher ground. ES, FR, IT.

Hylotelephium

Flat-leaved succulents still widely included in the genus *Sedum*. Flowers with 4(5) petals.

Hylotelephium telephium (Syn. *Sedum telephium*) ORPINE A fleshy perennial to 60 cm with several erect, red-tinged stems in clusters. Leaves flat, oval, alternate and irregularly-toothed, 20–80 mm long. Flowers *red, purplish or lilac,* borne numerously in dense terminal clusters; petals 3–5 mm. Shady woods and rocky habitats at higher altitude. Rare in the region generally, and absent from the far south. ES, FR, IT. *Hylotelephium maximum* is similar but with *whitish or yellowish flowers.* Similar habitats and distribution. ES, FR.

Umbilicus NAVELWORTS

Perennial herbs with *round leaves joining their stalks at the centre of the blade* (peltate). Flowers borne in spike-like racemes; petals 5, stamens 2 x as many, fused to the corolla.

Umbilicus rupestris NAVELWORT A fleshy, hairless perennial 15–30(60) cm high with distinctive basal circular leaves with a central hollow, borne on long stalks; upper leaves smaller, more kidney-shaped and with rounded teeth. Flowers with 5 petals and 2 x as many stamens, whitish-green or yellowish, sometimes pink, tubular, and *drooping,* borne in long, tapered racemes in which the flowers occupy $^2/_3$ the stem. Common on rocks, cliffs and old walls and dunes Throughout. *Umbilicus gaditanus* (Syn. *U. horizontalis*) is similar but with flowering stems with numerous crowded, linear leaves, and flowers held *horizontally* (not drooping) on stalks 13–70 cm, the flowers occupying $^1/_3$ (or $^1/_2$) the stem. Absent from some islands. ES, FR, IT, MA. *Umbilicus heylandianus* has *very long, dense flower spikes 40–80(90) cm* with *pale yellow flowers.* Strongly western. ES, MA, PT.

Sempervivum HOUSE-LEEKS

Glandular perennials with leaves in basal rosettes. Flowers borne in terminal clusters, 8–18 (mostly 13)-parted with 2 x as many stamens.

Sempervivum tectorum HOUSE-LEEK A distinctive perennial to 40(60) cm with dense basal rosettes of spine-pointed, dull grey-green leaves, often flushed reddish above, 20–40(60) mm long. Flowers dull red with 8–18 petals; petals narrow and 8–12 mm long, borne in rather rounded, dense clusters on erect, leafy, long-persisting stems. A mountain-dwelling species, widely cultivated; absent from hot, dry areas. FR.

CYNOMORIACEAE

C. coccineum

A family (with 1 genus) of root parasites of members of the Amaranthaceae. Flowers unisexual or cosexual, borne in a dense, brush-like inflorescence; tepals (1)3–6(8); stamen 1. Fruit a small, 1-seeded nut.

Cynomorium coccineum MALTESE FUNGUS A highly distinctive blackish-red plant sprouting as a club-shaped structure up to 25(30) cm high from an extensive underground rhizome system with lanceolate scale leaves. Inflorescence spike-like: cylindrical with very dense, tiny flowers; tepals 3–6(8); stamens solitary, exserted. Parasitic on halophytic shrubs in saline habitats such as salt marshes and sea cliffs. Widespread but strongly southern and very local. DZ, ES, IT (incl. Sardinia and Sicily), MA, MT, PT, TN.

1. *Hylotelephium maximum*
2. *Umbilicus rupestris*
3. *Umbilicus gaditanus*
4. *Sempervivum tectorum*

VITACEAE | GRAPE FAMILY

A large, primarily tropical family of climbers with leaf-opposed tendrils. Flowers small, borne in clusters; petals and stamens 5; style 1. Fruit a 1–4-seeded berry.

Vitis

Leaves simple, palmately lobed. Flowers with fused petals, falling as the flowers open.

1. *Cynomorium coccineum*
2. *Vitis vinifera*
3. *Geranium dissectum*
4. *Geranium molle*
5. *Geranium robertianum*
6. *Geranium lucidum*
7. *Geranium purpureum*

Vitis vinifera GRAPE A climbing shrub to >10 m with alternate leaves and tendrils opposite; tendrils branched. Leaves long-stalked and palmately 5–7 lobed, coarsely toothed. Flowers small, greenish, cosexual, borne in clusters hanging when mature. Fruit a berry (grape). Cultivated Throughout. Subsp. *sylvestris* is the wild form which differs in being dioecious (also hybridising with cultivated vines). ES, FR, IT.

GERANIACEAE | GERANIUM FAMILY

Herbs with alternate palmately or pinnately lobed leaves. Flowers borne in cymes, umbels or solitary, usually more or less regular with 5 sepals and petals; stamens (3)5 or 10; style 1. Fruit with 5 1-seeded portions united into a prominent beak.

Geranium

Annuals to perennials with simple, palmately lobed leaves. Flowers regular with 10 stamens; style with 5 branches. Fruit beaked.

A. Perennials with large flowers; petals long, 7–10 mm, and *notched*.

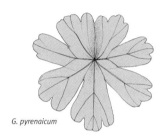

G. pyrenaicum

Geranium pyrenaicum An erect, hairy *perennial* to 60 cm with palmately lobed leaves to 50 mm across, rather deeply divided into 5–7(9) lobes ($^2/_3$ to the base) with straight, entire margins, wavy towards the tip. Flowers large, borne in pairs, and pink-purple or lilac with *deeply notched* petals 7–10 mm long, all 10 stamens with anthers. Local in dry, grassy habitats. ES, FR, IT, MA, PT.

B. Annuals with small flowers; petals usually <7(10) mm; *sepals spreading or ascending*.

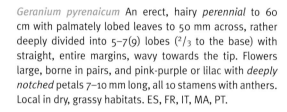

G. molle

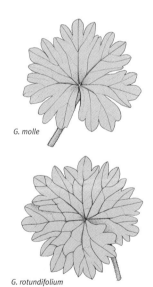

G. rotundifolium

Geranium molle DOVE'S-FOOT CRANE'S-BILL A low to short, sprawling annual to 40 cm with stems branched from the base, *grey-green and softly hairy*. Basal leaves long-stalked, rounded or kidney-shaped, divided >70 % of the radius into 5–7(9) wedge-shaped, 3-lobed segments ($^1/_2$–$^2/_3$ to the base); upper leaves more deeply divided and short-stalked or unstalked. Flowers pink-purple, the petals 4–6 mm long, borne in lax clusters, the *petals deeply notched*; outer stamens lacking anthers; flower stalks with short and long hairs. Field and grassy habitats; common. Throughout. Less hairy, taller forms in Italy (incl. Sicily) have traditionally been described as *G. brutium*. *Geranium pusillum* SMALL-LEAVED CRANE'S-BILL is similar to *G. molle* but flower stalks with hairs all short (not some long); flowers small, the petals 2.5–4 mm long. Garrigue. Widespread but local and scattered. Throughout. *Geranium rotundifolium* is similar but with shallowly 5–9-lobed leaves (<$^1/_2$ to the base) and bright pink flowers with *unnotched or slightly notched* petals, rounded at the tips, 5–7 mm long. Similar habitats. Throughout.

Geranium dissectum CUT-LEAVED CRANE'S-BILL A short to medium, spreading, hairy annual to 60 cm with ascending flowering stems. Leaves rounded in outline but *deeply dissected into 5–7 lobes almost to the base,* with sub-lobes. Flowers bright pink with shallowly notched petals 4.5–6 mm long; *flower stalks <15 mm long*; sepals spreading and with pointed tips. Fruit ridged and hairy. Common in a range of habitats, especially damp, disturbed, grassy places. ES, FR, IT, MA, PT. *Geranium columbinum* LONG-STALKED CRANE'S-BILL is similar but with larger flowers on *long stalks, mostly 25–60 mm long*; petals not notched, 7–10 mm long. ES, FR, IT, MA, PT.

C. Annuals with, normally, small flowers; petals 5–10(14) mm; *sepals erect and curved at the tips.*

G. robertianum

G. purpureum

Geranium robertianum HERB ROBERT A short to medium, hairy, *very aromatic* annual or biennial to 50 cm; usually *strongly flushed with red* or purple. Leaves palmate, the lower leaves with 3–5 pinnately lobed segments. Flowers pink or sometimes white, the petals slightly notched or rounded, 8–10(14) mm long; pollen orange. Fruit hairy and ridged. Very common in cool, damp shady places and woods. Throughout. *Geranium purpureum* LITTLE ROBIN is similar to *G. robertianum* but less flushed with red, flowers purplish-pink and smaller, the petals 5–9 mm, with yellow pollen. Similar habitats. ES, FR, IT, MA, PT. *Geranium lucidum* SHINY CRANE'S-BILL is similar to *G. robertianum* but readily distinguished by its *hairless* (or sparsely hairy), *shiny leaves* which are circular in outline and divided into 5(7) lobes. Petals deep pink, with rounded tips, 8–10 mm; sepals keeled on the back. Throughout.

Erodium STORK'S-BILL

Annuals to perennials like *Geranium* but generally with pinnately lobed leaves and often slightly zygomorphic flowers with 2 petals larger than the other 3; stamens 5. Fruit beaked.

A. Some or all leaves *shallowly-lobed* (to <¹/₂ the width of the blade).

Erodium reichardii HERON'S-BILL A low, greenish, hairy, spreading or small clump-forming perennial with leaves oval-triangular in outline, shallowly dissected into rounded lobes, often with reddish stalks. Flowers *white or very pale pink, with darker veins*; rather large relative to the plant. Coastal rocks and cliffs; a rare island endemic. ES (Balearic Islands).

E. chium

Erodium chium THREE-LOBED STORK'S-BILL A robust, low to medium, hairy perennial or biennial to 40 cm. Leaves oval, those below divided into 3 toothed, blunt lobes. Flowers pink-purple, 10–18 mm across, borne in 2–8 flowered clusters on non-glandular flower stalks; 2 petals slightly larger than the remaining 3; sepals with hairs *not glandular*. Fruit with short white hairs, the beak 20–40 mm long. Open, rocky places and roadsides. Throughout.

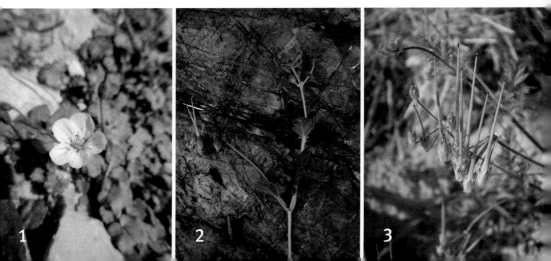

1 2 3

E. malacoides

E. laciniatum

E. botrys

E. cicutarium

Erodium malacoides MALLOW-LEAVED STORK'S-BILL An erect to sprawling short to medium, glandular-hairy biennial to 40 cm. Leaves oblong, heart-shaped at the base, those below toothed and sometimes 3–several-lobed, covered in shiny glands. Flowers purplish pink, 11–18 mm across, borne in 3–7-flowered clusters with at least 3 bracts at the base, borne on glandular-hairy stalks; *sepals with glandular hairs*. Fruit beak 20–35 mm long. Very common on bare or cultivated ground. Throughout. *Erodium laciniatum* is similar to *E. malacoides* but with leaves more deeply cut and not markedly glandular-hairy beneath, and flowers in clusters of 4–9 with just 2 bracts at the base. Fruit beak longer, to 90 mm. On coastal sands. Throughout, though absent from some islands. *Erodium maritimum* SEA STORK'S-BILL is similar to *E. malacoides* but *smaller* to 10(20) cm, with leaves <30 mm wide and flowers to 6 mm across. *Fruit with beak just 0.8–1 mm long.* FR, IT. *Erodium corsicum* is similar to *E. malacoides* but a perennial, with *grey-green leaves*. Coastal rocks. FR (Corsica), IT (Sardinia).

B. Leaves 1–2-*pinnately divided*.

Erodium botrys MEDITERRANEAN STORK'S-BILL A short, hairy annual to 50 cm with an obvious stem above ground. Leaves bristly, to 50 mm across, oval and deeply pinnately lobed and toothed, at least on the upper leaves. Flowers to 30 mm across, bluish with darker veins, borne in clusters of up to 4; bracts brown. Fruit with a long beak, 60 mm-(9)11 cm. Dry rocky habitats and roadsides. Throughout.

Erodium cicutarium COMMON STORK'S-BILL A very variable, low to medium, erect or prostrate, hairy (sometimes sticky), aromatic annual to 60 cm (often less). Leaves *deeply pinnately divided* without smaller lobes between the larger ones. *Stipules pointed* and whitish. Flowers purplish, pink or white, (7)10–18 mm across with 3–7(12) in a cluster; petals 4–9 mm, the upper 2 petals normally larger and with a blackish patch. Bracts brownish. Fruit hairy, with a beak 15–40 mm long. Very common in a range of open and disturbed habitats. Throughout. Many variants have been described. *Erodium lebelii* (Syn. *E. cicutarium* subsp. *bipinnatum*) is *smaller* with stems to 15(25) cm with *few*, just 2–4(5) pale pink flowers per cluster, *small,* <10 mm across. Fruit with beak to 22 mm. Fixed dunes. ES, PT. *Erodium aethiopicum* is a similar annual to 10(50) cm, generally prostrate to spreading. Leaves 10 mm–17 cm, pinnately divided into obtuse lobes, often purple-tinged. Flowers borne in clusters of 3–8, *larger,* with petals 5–10 mm, unequal and violet. Fruit 4.3–5.5 mm with a beak 30–55 mm. Coastal sand dunes. Throughout.

1. *Erodium reichardii*
2. *Erodium malacoides*
3. *Erodium cicutarium*

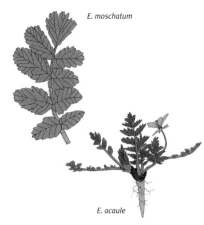

E. moschatum

E. acaule

Erodium moschatum MUSK STORK'S-BILL A spreading annual to 60 cm, similar to *E. cicutarium* but always stickily hairy and *smelling of musk*, leaflets only shallowly lobed (<¹/₂ to the midrib). *Stipules blunt.* Flowers larger, to 28 mm across, violet or pinkish purple. Fruit with beak 20–45 mm long. On cultivated and waste ground. Throughout.

Erodium acaule STEMLESS STORK'S-BILL Similar to *E. cicutarium* but *stemless* (though some populations of *E. cicutatium* appear stemless in exposed habitats). Leaves with petioles shorter than the leaf-blades. Flowers lilac without darker patches, to 22 mm across. Fruit with white hairs and a beak to 50 mm long. In dry habitats, less frequent than *E. cicutarium*. Throughout, though absent from some islands.

LYTHRACEAE | LOOSETRIFE FAMILY

Annual to perennial herbs with leaves simple and opposite or in whorls of 3. Flowers cosexual, regular, usually with (4)6 sepals and *6 petals* (0–5) often pink or purple; stamens (2)6–12; style 1. Fruit a 2-valved capsule.

Lythrum LOOSETRIFE

Herbs with tubular or bell-shaped calyx with 4–6 teeth and 4–6 petals <8 mm long.

A. Erect plants with small, normally *solitary* (or 2) pink-purple flowers.

L. junceum

L. hyssopifolia

Lythrum junceum A hairless, short to medium perennial to 70 cm with much-branched, sparse stems. Leaves mostly alternate, elliptic and stalkless. *Flowers small*, borne 1(2) in each leaf axil, purple, rarely white, solitary; petals 6, 5–6 mm long; *stamens 12, some or all protruding*. Damp habitats. Throughout.

Lythrum hyssopifolia GRASS POLY An erect, hairless annual to 25 cm, similar to *L. junceum*, with linear-lanceolate, rough-margined leaves. Flowers pink, borne 1(2) in each leaf axil with *4–6 stamens, not protruding;* petals 2–3 mm. Seasonally flooded areas and damp places. Throughout. *Lythrum tribracteatum* is similar to *L. hyssopifolia* but with *broadly* linear to oval leaves, and with appendages on the inner sepals *longer (not shorter) than the sepals themselves;* petals purple, just 2–3 mm long. Scattered in distribution. Throughout, though absent from most islands. *Lythrum thymifolia* is similar to *L. hyssopifolia* but with *tiny, narrow leaves 0.75–1(2.5) mm wide,* and flowers with 4 petals 1.5–3 mm long and 2–3 stamens. Throughout except Corsica and Siciy.

1. *Lythrum salicaria*
2. *Punica granatum*

B. Erect plants with conspicuous pink-purple flowers clustered in whorls; petals >8 mm long.

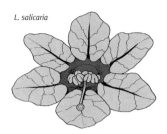

L. salicaria

Lythrum salicaria PURPLE LOOSETRIFE An erect, more or less hairless or shortly grey-hairy perennial to 1.5 m; stems sparingly branched with stalkless leaves in whorls of 3 or opposite; lanceolate. *Flowers clustered in whorls, large, conspicuous, and pink-purple,* borne in long, dense terminal spikes; petals 8–10 mm; stamens 12, some or all protruding. Common in damp places and on riverbanks. Throughout.

C. Creeping annuals with prostrate or ascending (rarely erect) stems.

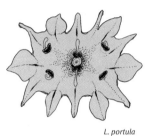

Lythrum portula WATER PURSLANE A low to *prostrate creeping, hairless annual with stems rooting at the internodes,* to 25 cm high. Leaves opposite, fleshy, often reddish and tapered into a short stalk, oval-spatula-shaped. Flowers purple, very small, borne solitary in the leaf bases; petals 6 or absent, 1 mm long; stamens normally 6. Seasonally flooded and waterlogged muddy ground. Throughout. *Lythrum borysthenicum* is similar but *bristly*, at least when young, and with *stems not rooting at the internodes*. Throughout, through absent from islands except for Corsica and Sardinia.

L. portula

1

2

Punica POMEGRANATE

Fruit-bearing deciduous shrubs or small trees, best known for the pomegranate. Flowers with 4–6(9) sepals and petals and as many or 2 x as many stamens. Fruit a capsule or berry. Recently established to be part of the family Lythraceae.

P. granatum

Punica granatum POMEGRANATE A deciduous shrub or small tree to 5 m with spiny, 4-angled young stems. Leaves opposite, shiny, bright green, roughly oblong, untoothed and virtually unstalked. Flowers to 40 mm across, with 5–9 scarlet, crumpled petals and a fleshy calyx (hypanthium), borne in clusters of 1–3 near the ends of the branches. Fruit rounded, to 90 mm across. Widely cultivated in the region, and naturalised. Throughout.

Lagerstroemia

Deciduous or evergreen shrubs and trees native to Asia and Australasia. Leaves mostly opposite and simple. Flowers with 6(12) petals; stamens numerous and in 2 forms (6 with large anthers + 12–100 with small anthers); style 1. Fruit a 3–6-valved capsule.

Lagerstroemia indica CAPE MYRTLE A deciduous shrub or tree to 10 m with 4-angled twigs. Leaves 25–70 mm (10 cm), opposite, sub-opposite or in groups of 3, round or oblong to elliptic, untoothed and short-stalked. Flowers showy, pink-purple (or white), 25–35 mm across; petals 6, with undulate margins; stamens numerous. Planted on roadsides. ES, FR, IT.

1. *Epilobium hirsutum*
2. *Epilobium parviflorum*
3. *Oenothera lindheimeri*

1   2   3

ONAGRACEAE | WILLOWHERB FAMILY

Annual to perennial herbs with simple, alternate or opposite leaves. Flowers mostly regular, with 2–4 free sepals and petals and 2, 4 or 8(10–12) stamens; style 1. Fruit a capsule or berry splitting lengthways with distinctive cottony seeds, or a 1–2-seeded nut.

Epilobium WILLOWHERB

E. hirsutum

E. parviflorum

Perennial herbs with lower leaves opposite. Flowers pink or purple with 4 sepals and petals; stamens 8; ovary 4-parted. *Fruit a linear capsule splitting into 4 valves to reveal seeds with long plumes of hairs.*

A. Stigma with 4 lobes in a cross.

Epilobium hirsutum GREAT WILLOWHERB A robust, densely and softly hairy perennial to 1.8 m with spreading non-glandular and glandular hairs. Leaves opposite, lanceolate, unstalked and *partially clasping the stem below*; markedly toothed. Flowers with bright pink, notched petals 10–16(18) mm long, borne in a leafy raceme; stigma 4-lobed. Common in damp places and near rivers. Throughout. *Epilobium parviflorum* SMALL-FLOWERED HAIRY WILLOWHERB is similar to (and hybridising with) the previous species but *smaller* in all parts (to 75 cm), with leaves not clasping the stem and *small, pale pink flowers with petals 5–9 mm long*. Local in damp places. Throughout.

B. Stigma club-shaped (clavate).

Epilobium palustre MARSH WILLOWHERB A perennial to 60 cm with sparse, adpressed, non-glandular hairs (some glandular hairs above) *rounded stems without lines or ridges, and with untoothed, virtually stalkless leaves* (leaf stalks short, <4 mm). Flowers pale pink to white, borne in lax, coarsely hairy racemes; petals 4–7 mm; *stigmas club-shaped*. Marshes and fens; local. Throughout. *Epilobium roseum* SMALL-FLOWERED WILLOWHERB is similar but has *distinctly stalked leaves* (4–15 mm long) *with toothed blades; stem with 2 raised lines*. Petals 4–7 mm. Damp habitats on higher ground. ES, FR, IT.

Epilobium tetragonum SQUARE-STALKED WILLOWHERB An erect perennial to 75 cm, similar to the previous species but with inflorescences with dense, white, adpressed hairs (no glandular hairs) and distinguished by its *clearly 4-ridged, often winged stems*. Petals pink-purple, 5–7 mm. Fruits 65–80 mm (10 cm) long. Throughout. *Epilobium obscurum* SHORT-FRUITED WILLOWHERB is similar to *E. tetragonum,* also has has 4-ridged stems, petals 4–7 mm, and *shorter fruits, (30)40–60(56) mm long*. Throughout.

Oenothera EVENING PRIMROSE

Annual to perennial herbs with alternate leaves. Flowers regular or zygormorphic (species formerly classified as *Gaura*). Petals (3)4; stamens 8. Fruit a capsule. A genus recently expanded to include the long-recognised genus *Gaura* in the light of DNA sequence data.

Oenothera lindheimeri (Syn. *Gaura lindheimeri*) A clumped perennial with numerous ascending stems to 1(2) m and oblong to spatula-shaped leaves to 75 mm long; hairy throughout and glandular above. Flowers *zygomorphic*, white, flushed pink, with 4 petals to about 25 mm across borne in lax, slender racemes. Fruit a non-splitting, 5-angled nut. Native to North America; cultivated and now naturalised fairly frequently with the potential to increase in abundance (though still rarely recorded). ES, PT.

MYRTACEAE | MYRTLE FAMILY

A mainly tropical family of shrubs and trees with normally opposite, simple leaves. Flowers with 4–5 sepals and petals and numerous stamens; style 1. Fruit a many-seeded capsule or berry.

Myrtus MYRTLE

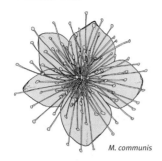

M. communis

Trees or shrubs with opposite leaves. Flowers with 5 free sepals and petals; stamens numerous. Fruit a berry.

Myrtus communis COMMON MYRTLE An erect, much-branched evergreen shrub to 1.5 m, glandular-hairy when young. Leaves opposite, shiny deep green, lanceolate and pointed, *aromatic* when crushed, 20–50 mm long. Flowers white, to 30 mm with rounded petals and numerous, conspicuous protruding stamens. Berry blue-mauve then bluish-black when ripe, 7–10 mm long. Common on maquis and coastal garrigue, also planted. Throughout.

Melaleuca BOTTLEBRUSH

M. citrina

Shrubs native to Australia. Leaves alternate. Flowers borne in dense inflorescences, petals 5; stamens numerous. Fruit a capsule.

Melaleuca citrina (Syn. *Callistemon citrinus*) BOTTLEBRUSH An evergreen shrub to 2 m with stiff, arching stems. Leaves leathery green, narrowly elliptic, entire, and lemon-scented when crushed. Flowers crimson, borne in brush-like heads with numerous prominent stamens. Fruit a capsule. Native to Australia; widely planted. Throughout.

Eucalyptus

Evergreen trees native to Australasia with peeling bark and broad, erect leaves when juvenile; leaves on mature trees narrow and pendent. Flowers borne in open clusters, with >100, prominent stamens. Fruit a woody capsule concealed by a fleshy cap (transformed sepals) when in bud. Widely planted and important for timber production in the region, particularly in the Iberian Peninsula.

A. Flowers white or cream, borne on *stalked inflorescences*.

E. robustus

Eucalyptus robustus SWAMP MAHOGANY A large tree to 30 m with *persistent* (not peeling) bark. Juvenile leaves lanceolate, mature leaves narrower and tapered; shiny green, *with stalks 20–35 mm*. Flowers white, to 25 mm in *clusters of 9–12(15)*. Fruit bell-shaped, 10–15(18) mm across. Planted in swampy and slightly saline areas. PT (south). *Eucalyptus resinifera* RED MAHOGANY is taller to 40 m with *reddish* bark and egg-shaped fruits to 8 mm. ES, FR, IT, PT.

1. *Myrtus communis*
2. *Melaleuca citrina*
3. *Eucalyptus globulus*

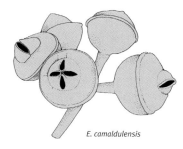

E. camaldulensis

Eucalyptus camaldulensis RIVER RED GUM A tree to 15(50) m with smooth, white, peeling bark. Juvenile leaves oval-lanceolate, grey-green, mature leaves much narrower, 80 mm–25(30) cm long. Flowers yellowish-white, borne in clusters of 5–12, and *distinctly stalked; stalks 6–15(20) mm long*. Fruit *hemispherical*; longer than wide, to 6 mm and with a broad, raised rim. Commonly planted for timber southwest Iberian Peninsula. ES, PT. *Eucalyptus tereticornis* is very similar to *E. camaldulensis* but has larger, cone-shaped fruits to 10 mm. Local. PT.

B. Flowers white or cream and stalkess or virtually stalkless.

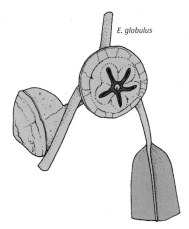

E. globulus

Eucalyptus globulus BLUE GUM A large tree to 30(40) m with smooth, peeling bark. Juvenile leaves oval-lanceolate with a heart-shaped base, mature leaves narrower and tapered, 10–15(40) cm long; grey, and aromatic when crushed. Flowers white or pink, to 35 mm, and *unstalked*. Fruit rounded and somewhat tapered towards the base, *large* to 10–15(18) mm across; wider than long. A native of Tasmania, very widely planted on hillsides in the region. Throughout. Subsp. *globulus* has *solitary flowers*. Subsp. *maidenii* (Syn. *E. maidenii*) is like *E. vimnalis* but with 3–7-flowered umbels, *much larger leaves* 12–30(40) cm long, and bluish fruits with 1–2 ribs. *Eucalyptus vimnalis* MANNA GUM is similar to *E. globulus,* with mature leaves 11–18 cm long, bark that peels in long ribbons, flowers in *umbels or long racemes*, usually of 3 (not solitary), and short-stalked. *Fruit small;* <10 mm across. Planted for timber and as an ornamental. ES, FR, IT, PT.

C. Flowers reddish, pinkish or orange.

Eucalyptus torquata CORAL GUM A small tree to 12 m and with flower buds with horned appendages; *flowers red, pink or orange*. Planted as an ornamental in hot, dry areas. FR. PT.

SALICACEAE | WILLOW FAMILY

Deciduous trees and shrubs usually with alternate, simple, toothed leaves. Flowers reduced, borne in catkins; male flowers with 1-many stamens; female flowers with 1(–2) short styles. Fruit a 2-valved capsule.

Salix

Trees or shrubs with winter buds with 1 outer scale. Flowers borne before or after the leaves. Stamens 1–5(12). Few species in the Mediterranean (many species in cooler parts of the region not included here).

Salix pedicellata MEDITERRANEAN WILLOW A deciduous shrub or tree to 8 m with flaking bark and grey-downy twigs. Leaves 50 mm–10 cm, oblong to lanceolate, toothed or scarcely toothed; stipules large and heart-shaped, soon falling. Dioecious; catkins 30–60(70) mm long. Streams, rivers and ravines, uncommon. DZ, ES, FR (Corsica), IT (incl. Sardinia and Sicily), TN.

VIOLACEAE | VIOLET FAMILY

A widespread family of herbs and shrubs. Flowers zygomorphic, solitary; sepals 5, separate; petals 5, the lowermost forming a lip, and extended behind into a spur; stamens 5; style 1. Fruit a 3-valved capsule. Alpine species not described here.

Viola VIOLETS

Herbs or shrubs with alternate stalked leaves with stipules at the base. Flowers with 5 sepals and 5 petals; stamens 5; carpels 3. Fruit a capsule.

1. *Viola arborescens*

A. Plant herbaceous.

Viola alba MEDITERRANEAN WHITE VIOLET A low, slightly hairy perennial 50 mm–15(20) cm with short, non-rooting runners. Leaves borne in basal tufts, slightly hairy or hairless, oval to triangular, pointed, dark green; *stipules linear-lanceolate and deeply fringed.* Flowers white or blue-violet, to 20 mm across, fragrant, with an upturned spur 3.5–6 mm. Rare in North Africa. DZ, ES, FR, IT, MA. *Viola odorata* SWEET VIOLET is similar in form, 80 mm–20 cm high, with *creeping, rooting stems above ground* and stipules broad and shortly fringed. Flowers usually dark violet-purple (rarely white) and fragrant; spur 4–5(7) mm. ES, FR, IT. *Viola hirta* HAIRY VIOLET lacks creeping stems and has narrower, triangular-oval leaves hairy on both sides. *Flowers not fragrant; spur 3–4(5) mm. ES, FR, IT.*

V. kitaibeliana

V. arborescens

Viola kitaibeliana (Syn. *V. hymettia*) A low, bristly hairy annual 40 mm–25 cm, with tufted leaves, oval below, narrower above, and somewhat toothed. Stipules pinnately divided with a large terminal lobe. Flowers violet or white *with a yellowish centre* and with darker veins; spur short, 1.5–3 mm long. Grassy habitats; widespread but uncommon. ES, FR, IT, MA, PT.

B. Plant with woody stems.

Viola arborescens TREE VIOLET A short, scrambling, shrubby, ascending perennial that is *woody* and often rather corky beneath, 10–20(30) cm high. Leaves oval to linear-lanceolate, pointed and slightly toothed. Stipules small and narrow, pinnately lobed. Flowers white flushed pale violet, to 15 mm, the spur short and blunt, 2 mm long. Fruit erect and hairless when ripe. Local in shrubby thickets and rock fissures near the sea. DZ, ES, FR, IT, PT.

Viola cazorlensis A distinctive, *spreading, mat-forming perennial* with short, woody stems 10–20(25) cm high and crowded, narrow linear-lanceolate, dark green leaves. Flowers borne prolifically, 1–3 per stem, *bright pink-purple; spur long, 17–30 mm.* Fruit an ovoid capsule. Rocky cliffs and fissures in southern Spain only; rare. ES.

1

ELATINACEAE | WATERWORT FAMILY

Small aquatic annuals with simple, opposite leaves and leafy stipules. Flowers minute, solitary or 2–5 in the leaf axils, cosexual and regular; sepals and petals 3–4; stamens 2 x as many as petals; styles 3–4, very short. Fruit a capsule.

Elatine

Aquatic or water-margin (sometimes terrestrial) annuals, superficially similar to Portulacaceae or Caryophyllaceae but with distinctive floral characteristics and leafy (not papery) stipules unlike the latter family.

Elatine alsinastrum A short, hairless annual to 50 cm high with leaves in *whorls* of 6–18, linear to lanceolate-oval, to 25 mm if submerged, 5–13 mm if exposed. Flowers solitary, minute, reddish or greenish, unstalked; stamens 8. Fruit a tiny 4-parted capsule. Wet, muddy and aquatic habitats. Throughout, except for many islands. *Elatine macropoda* is a similar, smaller annual 20 mm–10 cm, often patch-forming and with *opposite leaves* and stalked flowers. Saline damp habitats. Throughout. *Elatine hexandra* is similar with flowers in clusters of 2–5; stamens 6. ES, FR, IT, PT.

EUPHORBIACEAE | EUPHORBIA FAMILY

A large family of herbs, shrubs, trees and lianas, often with a latex. Flowers monoecious or dioecious; perianth absent or 3(5)-lobed; male flowers with 1–many stamens; female flowers with 2–3 styles. Fruit a capsule.

Floral arrangement of an *Euphorbia* inflorescence

Bract

Developing fruit

Horned glands

Cyathium

Female flower

Male flowers

Euphorbia (Spurges)

Leaves normally alternate and entire, with a sticky, milky latex. Flowers in small groups surrounded by a cup-shaped structure with 4–5 round or crescent-shaped glands, a solitary female flower and male flowers with 1–several stamens, the whole structure forming a cyathium. Often difficult to identify in the field; many species occur in the region so mainly widespread and common species are described.

A. Mature plant more or less *leafless and cactus-like* (distinctly succulent).

Euphorbia officinarum A more or less hemispherical *cactus-like shrub* with erect, swollen, grooved stems with many spines arranged in pairs along the angles; leaves reduced to tiny tubercles. Cyathium reddish, yellow or purple, borne at the tips of the branches, borne in summer. Various bare and open habitats from sea level to 1,900 m. A rare endemic. MA.

Euphorbia ingens CANDELABRA TREE A large, imposing, *Cactus-like* perennial to 9 m tall. Stems bright green and 4-angled with irregular, spiny margins. Native to South Africa, occasionally planted in landscaped areas or cultivated in parks and gardens. ES, PT.

B. Plant a robust *shrub* (1–3 m), often woody at the base, and spineless.

Euphorbia dendroides TREE SPURGE *A stout, shrubby and woody shrub to 2(3) m tall.* Leaves oblong-lanceolate, 35–70 mm long, borne in the autumn and winter and falling in late spring to leave bare branches. Umbel with 4–8(10) rays; glands almost round and irregularly lobed; bracts yellowish and broader than the leaves. Leaves turn red with age. Rocky coasts, common on coasts of North Africa and Sicily. DZ, ES (incl. Balearic Islands), FR, IT (incl. Sicily), MA, TN.

Euphorbia characias LARGE MEDITERENNEAN SPURGE *A large*, robust, *hairy* perennial with very thick, *unbranched* stems, to 1.5 m. Leaves bluish or greyish green, linear-lanceolate, untoothed and crowded towards the upper part of the stems, 25–90 mm (13 cm) long. Umbels with 9–20 short rays forming dense rounded or oblong heads at the apex of the stems. *Bracts fused in pairs around the flowers*, glands dark red-brown, notched or with short horns. Capsule smooth and softly hairy; seeds greyish. Exposed or shady, dry places; absent from the far southeast. ES, FR, IT, MA, PT.

1. *Euphorbia officinarum* PHOTOGRAPH: SERGE D. MULLER
2. *Euphorbia ingens*
3. *Euphorbia dendroides*
4. *Euphorbia characias*

E. squamigera

Euphorbia squamigera A woody *shrub* to 1.5 m with much-branched stems always *bare below* and densely leafy above. Leaves *narrow*, linear-lanceolate, *pointed* and hairless or hairy beneath, entire or minutely toothed, 25–60 mm long. Ray leaves rounded or diamond-shaped, also pointed. Rays 5(6), capsule hairless and grooved, with tubercles. Seeds smooth and brown. On sunny, rocky hillsides and sea cliffs; distribution poorly known due to confusion with the annual species *E. clementei*. DZ, ES (southeast + Balearic Islands), MA (also recorded PT).

1. *Euphorbia paniculata* subsp. *monchiquensis*
2. *Euphorbia regis-jubae*
3. *Euphorbia spinosa* (inset: flowers above, fruits below)
4. *Euphorbia helioscopa*

Euphorbia paniculata A robust, tuberous, virtually hairless perennial to 1.5 m with narrow, lanceolate to oblong leaves *abrubtly contracted at the base, or clasping the stem at the base,* 25 mm–10 cm long. Rays 5(6); ray leaves broadly oval to diamond-shaped. Iberian Peninsula and North Africa. DZ, ES, MA, PT, TN. Subsp. *paniculata* has *broad, oblong-lanceolate leaves* to 70 mm long. DZ, ES, MA, PT, TN. Subsp. *monchiquensis* (Syn. *E. monchiquensis*) is also regarded as a distinct species and is more robust and has lanceolate leaves 40 mm–10 cm long, gradually narrowed at the base. Hill forests, rare. PT. Subsp. *welwitschii* (Syn. *E. welwitschii*) is more slender and has elliptic leaves 30–40 mm long and smaller, ovoid seeds to 2.8 mm (not 4 mm) long. PT.

Euphorbia pulcherrima POINSETTIA A shrub to 2 m tall with dark green leaves to 16 cm long and *leaf-like, showy red bracts*. A commonly planted ornamental native to Central America in parks and gardens; rarely, if ever, naturalised. Throughout.

Euphorbia regis-jubae A robust, much-branched, rather rounded shrub to 1.2(3) m with thick, succulent, spineless stems. Leaves 15 mm–10 cm long, light yellow-green and linear. Rays 4–10; bracts 7–15 mm, *bright pale yellow*. Capsules 5.2–7.3 mm. An important component of the succulent thicket vegetation on the Atlantic littoral of northwest Africa. MA.

C. Plant a small shrub (to 40 cm), woody at the base and *spiny*.

Euphorbia spinosa SPINY SPURGE A domed, *cushion-like* shrub forming intricately branched mounds to 40 cm high; branch-tips *spiny*. Leaves blue-green and lanceolate. Bracts broader than the stem leaves; umbels with 1–5 rays. A range of habitats, usually dry and rocky at a range of altitudes. FR, IT.

D. Plant herbaceous with oval to rounded glands in cyathium *without horn-like projections*.

E. helioscopa

E. clementei

Euphorbia helioscopia SUN SPURGE A short, erect hairless annual, normally with a single stem, to 30(50) cm. Leaves oval or spoon-shaped, broadest above the middle and *toothed* in the upper $^1/_2$, 4–35(60) mm long. *Umbel 5-rayed (small plants with 2–3 rays), with 5 distinctive bracts* at the base, yellowish and similar in shape to the leaves. Glands oval-shaped and untoothed. Capsule smooth and unwinged; seeds brown and wrinkled. A common weed on waste ground and bare soil. Throughout.

Euphorbia clementei An *annual* to 80 cm, similar to, and often confused with the woody perennial species *E. squamigera* but with leaves *widest above the middle,* minutely toothed along the margins, *not pointed*, and *gradually narrowed into a very short petiole;* leaves 15–40(70) mm long. Rays (3)4–5; ray leaves oval or diamond-shaped; glands without horns. Local on sea cliffs in southwest Iberian Peninsula and North Africa. DZ, ES, MA, PT, TN.

E. Herbaceous, little-branched annuals with glands in cyathium sickle-shaped or with horn-like projections.

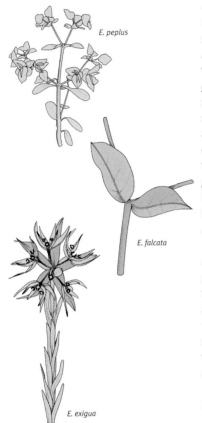

E. peplus

E. falcata

E. exigua

Euphorbia peplus PETTY SPURGE An erect, hairless annual 20 mm–20(40) cm, branched at the base. Leaves green, oval to rounded, untoothed and short-stalked, 4–30 mm long. Umbels with *3 main rays* (rarely 2–5), with 3 bracts that are triangular-oval to spoon-shaped, *green* and unstalked. Glands kidney-shaped with long, slender horns. Capsule smooth but with 2 ridges; seeds pale grey and pitted. Common on disturbed and waste ground. Throughout. Forms with poorly developed umbels and scarcely pitted seeds have traditionally been described as *E. peploides*. *Euphorbia falcata* SICKLE SPURGE is similar to *E. peplus* but with few, lanceolate, 3-veined unstalked leaves 2–25(40) mm long with a waxy bloom and an *unridged* capsule with *unpitted* seed. Similar habitats. Throughout.

Euphorbia exigua DWARF SPURGE A very small, hairless, grey-green annual, 20 mm–20(30) cm high, branched from the base. Leaves *very narrow*-lanceolate, untoothed and unstalked, 2–25 mm long. Rays 2–5(7), branched x 1–3(5), with narrowly triangular bracts. Glands crescent-shaped with 2 horns. Capsule shallowly grooved; seeds wrinkled and grey. Very common on cultivated, grassy and fallow habitats. Throughout.

F. Herbaceous, *many-stemmed*, *thin-leaved perennials* with glands in cyathium sickle-shaped or with horn-like projections.

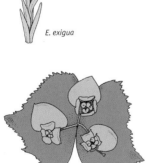

E. serrata

Euphorbia serrata SAW-LEAVED SPURGE A short-medium greyish or bluish perennial 20–60 cm tall with a woody stock. Leaves narrow-oblong with finely *toothed margins*, 20–70 mm long. Umbels with 3–5 rays, branched x 1–3(5) with lanceolate to rounded, *yellow* bracts. Glands oval-shaped, squared at 1 end. Seeds *pitted*. Dry habitats and garrigue. ES, FR, IT, PT.

Euphorbia cyparissias CYPRESS SPURGE A short to medium, hairless, rhizomatous perennial with erect stems to 50 cm, usually unbranched at the base and branched above to form numerous non-flowering shoots. *Leaves crowded, long and narrow, 0.5–2(3)mm wide,* stalkless, turning yellowish with age. Umbels with *many rays* (11–18); glands horned. Waste places and grassy habitats, usually above sea level; absent from many islands and the far south. ES, FR, IT.

E. segetalis

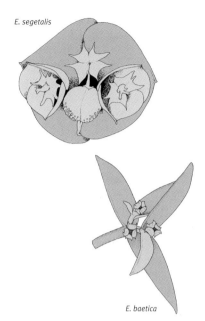

Euphorbia segetalis (Syn. *E. pinea*) A hairless annual or perennial, simple or with some branches at the base, to 80 cm (1 m). Leaves alternate and *narrow* (2–4 mm), linear to linear-lanceolate, 27–40 mm long. Rays 4–6(8), *bracts diamond-shaped* and yellow-green, glands with 2–4 horns. Capsule rough and glandular; seeds pale grey. In open, maritime habitats. Throughout. Perennial forms have traditionally been described as *E. pinea*.

Euphorbia baetica A hairless, bluish or yellowish, tuft-forming perennial 16–40 cm with a far-creeping rhizome. Leaves linear to linear-lanceolate, untoothed, and with 2–5 prominent veins beneath, 20–40 mm long. Rays 5(3–7), branched x 1–2(3). Glands more or less semi-circular and with horn-like projections. Cyathium lobes *densely fringed with fine hairs*. Capsule scarcely grooved; seeds grey and pitted. Various habitats. Southwest Iberian Peninsula. ES, PT.

E. baetica

Euphorbia nicaeensis An erect medium, hairless *white-grey* perennial, normally with *red stems* 20–80 cm tall. Leaves succulent, lanceolate to oblong and 3-veined, 25–50 mm long. Umbels with 8–16 rays, rays simple or branched x 2–3; bracts yellowish, oval to kidney-shaped, glands notched, sometimes with 2 short horns. Capsule 3.5–4.5 mm, rough; seeds pale grey and pitted. Garrigue and maquis. Widespread but absent from most islands; common in northeast Spain. ES, FR, IT.

1. *Euphorbia exigua*
2. *Euphorbia cyparissias*
3. *Euphorbia nicaeensis*

Euphorbia transtagana A rather short, ascending, slightly glaucous and sometimes red-flushed annual 50 mm–20(30) cm tall. Leaves more dense at the base lax further up the stem; 7–17 mm long, variably lanceolate to diamond-shaped and narrowed at base. Umbels with 3–4(5) rays, simple or branched x 2; bracts broadly oval to diamond-shaped, sometimes indistinctly toothed along the margin. Cyathium short-stalked; nectaries with 2 appendages. Seeds grey, ovoid and ribbed. Rather uncommon; dry habitats in southern Iberian Peninsula. ES, PT.

1. *Euphorbia myrsinites*
2. *Euphorbia myrsinites*
3. *Euphorbia rigida*
4. *Euphorbia paralias*

G. Herbaceous, many-stemmed, *fleshy-leaved perennials* with glands in cyathium sickle-shaped or with horn-like projections.

Euphorbia myrsinites BROAD-LEAVED GLAUCOUS SPURGE A low to short, *prostrate* perennial with unbranched, fleshy, densely leafy stems radiating from the centre 20–40 cm long. *Leaves succulent, grey and borne in distinctly regular arrangement around the stem;* oval to rounded and pointed at the tips. Rays 5–12, bracts bright yellow-green and heart or kidney-shaped; glands weakly horned. Capsule 5–7 mm; seeds smooth and grey-brown. East of the region though also cultivated. IT. *Euphorbia oxyphylla* (Syn. *E. broteri*) is similar, also bluish and with fleshy leaves and stems, but with *stems erect to ascending.* Mainly central Iberian Peninsula. ES, PT.

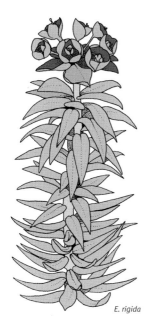

E. rigida

Euphorbia rigida NARROW-LEAVED GLAUCOUS SPURGE A hairless and distinctly bluish perennial with numerous stout, erect to ascending stems arising from a *woody stock*, to 30 cm. Leaves fleshy, lanceolate and pointed, often flushed with purple. Rays 6–12, forked x 1 or x 2, ray leaves oblong. Glands minutely *horned*. Capsule strongly *3-sided*; seeds smooth and whitish. Local on garrigue and maquis; cultivated in gardens. IT, MA.

Euphorbia paralias SEA SPURGE A short to medium, hairless, stiffly erect, fleshy perennial to 60(80) cm forming clumps. Leaves grey-green, regularly and closely set on the stem and *overlapping*, oval, *broadest towards the base* and concave above, *midrib obscure below*, 8–20 mm long. Umbels with 3–6 rays, bracts oval and concave. Glands kidney-shaped with long horns. Capsule rough along the back; seeds pale grey and *smooth*. Common in rocky and sandy maritime habitats and on dunes. Throughout.

Euphorbia portlandica PORTLAND SPURGE A hairless perennial 10–40(50) cm, similar to *E. paralias* with stems prostrate to ascending (rather than stiffly erect), but less succulent. Leaves oval, pointed and *broadest towards the tip*, tapered at the base, grey-green with a *prominent midrib below*, 6–18 mm. Umbels with 4–5(6) rays, bracts oval below, triangular to diamond-shaped above. Glands yellow and kidney-shaped with prominent horns. Capsule deeply furrowed; seeds *pitted*. Sea cliffs and other maritime habitats; an Atlantic species extending to southwest Iberian Peninsula. PT.

E. terracina

Euphorbia terracina A medium, hairless, *succulent* perennial 40 mm–90 cm with erect to ascending stems and non-flowering lateral branches. Leaves oblong to linear-lanceolate, minutely toothed, regularly and closely set on the stem and overlapping but *flat*, 4–60 mm long. Umbels with 2–5(6) rays with as many oblong to diamond-shaped, green bracts. Glands with 2, long, slender horns. Capsule smooth; seeds pale grey and smooth with a boat-shaped, fleshy structure (caruncle) attached. Open habitats. Throughout.

H. *Chamaesyce* group: small-flowered, often low, *prostrate* annuals; closely related, very similar species, now re-classified in the genus *Euphorbia*.

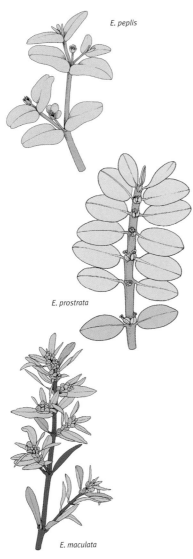

E. peplis

E. prostrata

E. maculata

Euphorbia peplis (Syn. *Chamaesyce peplis*) PURPLE SPURGE A flat, prostrate, *ground-hugging*, hairless annual with *4 (sometimes 3 or 5) main branches* at the base. Stems red or purple, *leaves fleshy*, grey-green and small to 11 mm, opposite, oblong, and with a single rounded lobe at the base. Flowers tiny, greenish with semicircular red-brown glands, borne laterally or in clusters but not in umbels. Capsule nearly smooth and purplish. On sandy and shingly shores. Throughout. *Euphorbia serpens* (Syn. *Chamaesyce serpens*) is very similar (often confused with *C. peplis*) but has up to 16 branches and leaves symmetrical rather than with rounded lobes at the base. Native to Tropical America. Throughout.

Euphorbia prostrata (Syn. *Chamaesyce prostrata*) A spreading herb, similar to *C. peplis* but often hairy, and with *up to 10 branches at the base* and stems that are hairy above, *leaves slightly serrated on the margins,* and *capsules hairy on the keels.* A North American weed naturalised locally in sandy places. Throughout. ***Euphorbia maculata*** (Syn. *Chamaesyce maculata*) is similar to *C. prostrata,* with up to 8 branches and *leaves often with a dark central spot* and with a capsule either virtually hairless, or more often *entirely covered with closely adpressed hairs.* A North American weed naturalised in ruderal places. Throughout. *Euphorbia canescens* (Syn. *Chamaesyce canescens*) is similar to the previous species, *with numerous (up to 25 stems) branching from the base*, with plain or purple-spotted leaves, but with capsules with *spreading (not adpressed) hairs.* Throughout. *Euphorbia humistrata* (Syn. *Chamaesyce humistrata*) has rather narrow, often spotted leaves and *hairless capsules.* Native to Atlantic American coasts. ES (south).

Chrozophora

Monoecious, hairy annual or perennial herbs or shrubs with simple, alternate leaves. Male flowers with 5–15 stamens; female flowers with 3 stigmas. Fruit a capsule.

Chrozophora tinctoria A grey-hairy annual 10–40 cm high with simple, wavy-margined leaves 20–60(80) mm long. Flowers unisexual, the male with yellow triangular corolla lobes 1.5–2 mm and 10 stamens; female flowers with greenish, linear corolla lobes 3–3.5 mm. Fruit sub-spherical, 5–8 mm across. Local in disturbed habitats. Throughout.

Mercurialis MERCURY

Annual or perennial herbs (or shrubs) with opposite leaves. Flowers unisexual, green and inconspicuous; sepals 3; stamens 8–25; carpels 2(3–4). Fruit a 2(3–4)-parted capsule.

A. Annuals.

M. annua

Mercurialis annua ANNUAL MERCURY A dioecious, branched, erect annual 30–70 cm high, more or less hairless. Leaves 20–70 mm long, opposite, oval to elliptic, toothed and long-stalked, shiny green, with minute hairs along the margins (<0.4 mm). Male flowers borne on dense, long, erect greenish spikes, female flowers few, borne in lateral clusters. Fruit 2.4–2.6 mm, 2-lobed and bristly. A common weed on disturbed ground. Common and widespread. DZ, ES, FR, IT, MA, PT, TN. The following are variably treated as true species, or as subspecies: *Mercurialis ambigua* (Syn. *M. annua* subsp. *ambigua*) is shorter, often to just 10 cm tall (80 mm–50 cm), with oval leaves with *hairs >0.4 mm along the margins*. Fruit with some bristles. Most of Iberian Peninsula. ES, PT. *Mercurialis elliptica* (Syn. *M. annua* subsp. *elliptica*) is *completely hairless* with regularly toothed leaves. ES, MA, PT.

1. *Euphorbia peplis*
2. *Euphorbia prostrata*
3. *Euphorbia canescens*
4. *Euphorbia maculata*
5. *Chrozophora tinctoria*
6. *Mercurialis annua*
7. *Mercurialis ambigua*

M. tomentosa

B. Perennials.

Mercurialis tomentosa A densely woolly hairy, woody-based, dioecious perennial 30–70 cm tall; very distinct from the previous species and superficially rather like a member of the Lamiaceae. Leaves 5–40 mm long, more or less stalkless and elliptic-lanceolate, entire or scarcely toothed; grey-woolly. Male flowers borne in rounded clusters in interrupted spikes; female flowers almost stalkless. Fruit woolly-hairy. ES, FR, PT.

Ricinus CCSTOR OIL PLANT

Shrubs without a milky latex and with *palmately lobed leaves*. Flowers unisexual; perianth with 3–5 tepals; *stamens conspicuous and in fascicles* (branched); ovary 3-parted. Fruit a capsule.

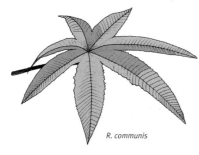

R. communis

Ricinus communis CASTOR OIL PLANT An imposing monoecious annual or shrub to 5(7) m, flushed red, bronze or purple. Leaves shiny, large, 10–36(60) cm across and palmate, with 5–9 coarsely toothed lobes. Flowers in large terminal panicles with the male below with yellowish stamens, and the female above and with bright red stigmas. Fruit a 3-parted, spiny capsule 18–20 mm across; seeds bean-like, 10–15 mm long. Native to tropical Africa, widely planted in towns and naturalised. Throughout.

1. *Ricinus communis*
2. *Hypericum perforatum*
3. *Hypericum triquetrifolium*
4. *Hypericum hircinum*
5. *Hypericum balearicum*

HYPERICACEAE

Shrubs or herbs, often with numerous glands and simple, opposite or whorled leaves. Flowers regular, yellow, with 5 free petals and sepals; stamens numerous; styles 3–5. Fruit a capsule or succulent and berry-like.

Hypericum

Herbs and shrubs easily recognised by the opposite leaves and flowers with 5 yellow petals and numerous stamens.

A. Plant a herbaceous perennial.

Hypericum perforatum PERFOLIATE ST. JOHN'S-WORT An erect herb to 50(80) cm with a cylindrical stem with *2 lines running down its length*. Leaves oval to linear, to 30 mm long, *scarcely stalked*, hairless and blunt with numerous transparent glands. Flowers bright yellow, with petals 9–15 mm long, often with black dots along the edges; sepal margins entire. One of the most common species in a range of habitats and altitudes in the region. Throughout. *Hypericum triquetrifolium* is similar but shorter (below 50 cm), and bushier, often with much-branched, spreading stems and *small leaves clasping the stem and with markedly wavy margins*. Various habitats including garrigue, car parks and path margins. DZ, IT, TN.

Hypericum perfoliatum A medium to tall, erect to spreading, normally hairless perennial to 70(80) cm with stems with 2 lines. Leaves blue-green, opposite, lanceolate and *clasping the stem at the base*, without wavy margins, 8–50 mm long. Flowers yellow with petals 8–12 mm long with blunt sepals with black markings. Fruit 8–10 mm with raised orange warts. Various habitats; common. Throughout. *Hypericum australe* is similar but shorter, 10–40 cm with *semi-prostrate stems rooting at the base* and leaves ascending or pressed to the stem, 5–12 mm long. Petals 8–11 mm. Damp habitats in the islands, France and Italy. ES, FR, IT.

Hypericum tetrapterum SQUARE-STALKED ST. JOHN'S-WORT An erect, herbaceous perennial to 60 cm (1.2 m) with *square stems with 4 winged angles* (0.25–0.5 mm). Leaves small and oval and unstalked, 9–35 mm. Flowers pale yellow with petals 5–6.5(7) mm long; sepals narrow and pointed, without black dots. Water-logged habitats and woods, widespread but absent from the far west and some islands. DZ, ES, FR, IT, MA, TN. *Hypericum undulatum* is similar, and also has 4-angled stems but with *wavy-edged leaves* (7–35 mm) long which are red-tinged with conspicuous black dots. Petals 7–9 mm. Probably throughout; rare or absent in much of the east.

B. Plant a woody-based shrub.

Hypericum hircinum STINKING TUTSAN An erect, semi-evergreen shrub to 1.5(3) m with (2)4-angled stems, *smelling of goats when crushed*. Leaves narrowly lanceolate to broadly oval and unstalked, 12–50 mm long. Flowers yellow with petals 10–18(20) mm long borne in few-flowered, branched terminal inflorescences; *sepals shorter than the petals* (4–7 mm); *stamens longer than the petals*. Fruit ellipsoidal, 6–13(16) mm long. Damp and shady habitats. ES, FR, IT, MA. *Hypericum androsaemum* is similar but scarcely aromatic and with small flowers with petals 4–10 mm long, and with *sepals almost as long or exceeding the petals* (4–12 mm). Local, in damp woods. DZ, ES, PT.

Hypericum aegypticum Eɢʏᴘᴛɪᴀɴ Sᴛ. Jᴏʜɴ's-ᴡᴏʀᴛ A low, spreading shrub to 30 cm. Leaves small, green and rather crowded, narrowly oblong and leathery and slightly concave. *Flowers borne singly*, but many to a stem, in long spikes (unless on stunted specimens); bright yellow, rather flax-like. Coastal cliffs and scree. DZ, IT (incl. Sardinia and Sicily), MA, MT, TN.

Hypericum balearicum A, small and low shrub to 1(2) m lacking black dots but with *prominent resinous warts on the leaves and stems*. Leaves oval and leathery with crinkly edges, 4–13 mm long. Flowers terminal, yellow and solitary, with petals 13–22 mm long; sepals 5–7 mm. Dry woods and rocky habitats; an island endemic. ES (Balearic Islands).

LINACEAE | FLAX FAMILY

Annuals or perennials with simple, opposite or alternate leaves. Flowers in branched inflorescences; sepals and petals 4–5, free; stamens 4; styles 4–5. Fruit a 8–10-valved capsule.

Linum ꜰʟᴀx

Hairless annuals to perennials. Flowers 5-parted with white or blue petals. Capsule 10-valved.

L. strictum

L. maritimum

A. Annuals with yellow flowers.

Linum strictum ᴜᴘʀɪɢʜᴛ ʏᴇʟʟᴏw ꜰʟᴀx A short, erect annual 10–45 cm with narrowly lanceolate leaves with inrolled margins; margins very rough. Flowers small with yellow petals 6–12 mm long which exceed the long-pointed sepals, borne in branched, spreading clusters or short lateral clusters in a rigid, spike-like inflorescence. Coastal sands and other dry places; common. Throughout. *Linum setaceum* is similar to *L. strictum* but with linear leaves <0.5 mm wide, *densely crowded in the middle of the stem*, and the inflorescence lax and much-branched; flowers pale yellow with dark veins. ES, MA, PT.

B. Woody-based perennials with yellow flowers.

Linum maritimum A slightly hairy or hairless perennial *with a woody stock* and long, erect or ascending stems. *Leaves small* and upward-pointing, greyish, lanceolate to narrowly elliptic, the *lower leaves opposite*, 3-veined, the upper leaves alternate and 1-veined. *Flowers yellow*, borne in lax clusters, the petals to 13 mm long. Damp, saline soils. ES, MA, PT.

1. *Linum strictum*
2. *Linum bienne*
3. *Linum suffruticosum*

Linum flavum YELLOW FLAX A variable, often tall, hairless, erect, branched perennial to 60 cm with a woody stock and *rather large, dark green, 3(5)-veined leaves 20–35 mm long and 3–12 mm wide,* lanceolate above and spatula-shaped below. Flowers *large and bright yellow;* petals 20 mm long. Dry, grassy habitats in the east of the region. IT. *Linum campanulatum* is similar but with upper leaves narrowly lanceolate and *with whitish margins,* and 2 small glands at the base. Absent from most islands and the far south. ES, FR, IT.

C. Flowers blue; sepals without glandular hairs.

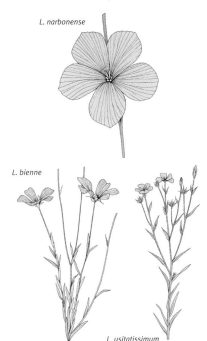

L. narbonense

L. bienne

L. usitatissimum

Linum narbonense BEAUTIFUL FLAX A short to medium, hairless, greyish perennial 20–50 cm high with erect to ascending stems. Leaves alternate, greyish, linear and long-pointed, mostly 1- or 3-veined. *Bracts with membranous margins.* Flowers few, bright blue, and *large* with petals 25–35(40) mm long. Grassy places. Absent from much of the east. ES, FR, IT, MA, PT.

Linum bienne PALE FLAX An annual to perennial herb with slender, erect to spreading stems, often branched below (not in small specimens), 10–60 cm. Leaves alternate, linear and long-pointed, mostly 3-veined, 0.5–1.5 mm wide. Flowers *pale blue,* borne on slender stalks in loose clusters or singly, petals 8–12 mm long, exceeding the oval, long-pointed, papery-margined sepals (4–6 mm long). Capsule 4–6 mm. Common on garrigue. Throughout. *Linum usitatissimum* CULTIVATED FLAX is probably derived from *L. bienne* and similar, but *usually an unbranched annual* to 85 cm with *larger, darker blue* or white flowers with petals 12–20 mm long. Capsule 6–9 mm. Disturbed habitats. Throughout.

1   2   3

D. Flowers pink; sepals glandular-hairy.

Linum viscosum STICKY FLAX A hairy-stemmed perennial to 60 cm with hairy-margined leaves to 8 mm wide, 3–5-veined. Flowers virtually stalkless, the petals *pink with darker veins* (rarely white or blue), 18–22 mm long, borne in elongated flower clusters. Grassy habitats. ES. FR, IT.

E. Flowers white.

Linum suffruticosum WHITE FLAX A stiff-branched shrubby hairless perennial *with numerous non-flowering leafy shoots* to 50 cm tall with numerous stiff, linear leaves to 1 mm wide. Flowers conspicuous, white with a purplish centre (yellow in bud), with petals 15–30 mm long, much-exceeding the sepals; stamens virtually included in the corolla tube; sepals 3-veined. Absent from most islands. ES, FR, IT. *Linum tenuifolium* is similar but with 1-veined sepals and (often) clearly protruding stamens. Absent from most islands. ES, FR, IT.

Radiola

Annuals with opposite leaves. Flowers 4-parted with white petals. Capsule 8-valved.

Radiola linoides A much-branched, *extremely slender* and small annual to 10 cm with 1-veined leaves. Flowers tiny with 4 sepals and petals to 1 mm. Capsule 0.7–1 mm. Open, sandy ground. Throughout.

PASSIFLORACEAE | PASSION FLOWER FAMILY

A pantropical family of vines and shrubs. Flowers regular, unisexual or cosexual, with 3–5 sepals, petals and stamens. Fruit a capsule or berry.

Passiflora

Mostly vines with tendrils borne in the leaf axils and spirally arranged leaves. Flowers complex, with a tubular calyx, 5 petals and 5 sepals and a corona of thread-like elements, as well as 5 conspicuous stamens on a column with the ovary and 3 stigmas.

Passiflora caerulea COMMON PASSION FLOWER A vigorous, tropical-looking vine with stems which climb by means of coiled tendrils. Leaves alternate, palmately lobed with 5–7 oblong lobes, untoothed and dark green above, paler below. Flowers very distinctive, consisting of a tubular calyx, 5 greenish-white petals and 5 sepals and a corona of thread-like elements banded purple, white and blue, and 5 conspicuous stamens on a column which carries the ovary and 3 stigmas. Fruits ovoid, orange, and rather large, to 80 mm long. Very commonly cultivated in towns and gardens. Throughout.

OXALIDACEAE | OXALIS FAMILY

Perennial herbs, often with a bulbous stock or rhizomes, with clover-like leaves. Flowers with 5 petals and sepals; stamens 10; styles 5. Fruit a capsule.

Oxalis

Perennial herbs, often on disturbed ground, with clover-like leaves with 3 leaflets (ternate). Flowers regular with 5 petals and sepals. Root morphology an important diagnostic.

A. Flowers yellow.

Oxalis pes-caprae BERMUDA BUTTERCUP A low, tufted perennial with numerous leaves arising from a bulbous stock; far spreading; leaves withering soon after flowering. Leaves trifoliate and clover-like. *Flowers bright yellow and tubular* with petals 13–26 mm, borne in loose umbels on long stalks to 30 cm. An abundant and highly invasive weed in some coastal regions and disturbed land; increasing. Throughout. *Oxalis corniculata* PROCUMBENT YELLOW SORREL also has yellow flowers but has *stems rooting at the nodes*, and often dull-purple leaves. Petals 5–9 mm long. Cultivated land and coasts. Throughout.

B. Flowers bright pink (species virtually identical and observation of the roots necessary).

Oxalis articulata PINK OXALIS A tufted perennial with *brown scaly rhizomes* and long-stalked, all basal, clover-like leaves with heart-shaped leaflets, often covered with orange or brown dots. *Flowers pink,* borne in broad, umbel-like clusters; on stalks to 35 cm; petals slightly hairy, 12–19 mm long. Waste ground and gardens. ES, FR, PT. *Oxalis debilis* is virtually identical (often grouped with *O. articulata*), also producing leaves with orange dots and pink flowers with petals 10–18 mm long, and differeng chiefly in producing *underground bulbils*. Throughout (distribution poorly known due to confusion with *O. articulata*). *Oxalis latifolia* is virtually identical to *O. debilis* with bulblets formed at the end of short rhizomes, and leaflets without orange dots. Petals 9–13 mm. Throughout (distribution poorly known due to confusion with *O. articulata*).

C. Flowers pale whitish-pink.

Oxalis incarnata A virtually hairless, tufted perennial with a bulb from which arise *branching stems with aerial bulbils in the leaf axils.* Flowers *solitary, pale white-lilac* with darker veins; petals 12–20 mm long. Native to South Africa, locally naturalised in recently disturbed places in southern France and south-central Italy, possibly elsewhere. FR, IT.

---

1. *Passiflora caerulea* (fruit) 4. *Oxalis incarnata*
2. *Passiflora caerulea* (flower) 5. *Oxalis debilis*
3. *Oxalis pes-caprae*

LEGUMINOSAE (FABACEAE) | PEA FAMILY

The third largest family of flowering plants, consisting of herbs or trees with trifoliate or pinnately compound leaves. Flowers zygomorphic, with an upper petal (standard), 2 lateral wings which lie on the side of the 2 lower, typically fused petals (keel), concealing the 10 stamens and single style; stamens 9 (sometimes all 10), fused into a basal tube. Fruit (legume), highly variable, often splitting and pod-like, or a nut.

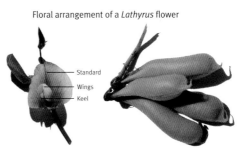

Floral arrangement of a *Lathyrus* flower

Standard
Wings
Keel

Fruits in the Leguminosae are highly variable but often pod-like, such as these *Erophaca* fruits

Bauhinia

A tropical genus of trees or climbers with simple, *entire* (or bilobed leaves). Flowers showy with 5, equal or unequal petals; stamens 2, 3, 5 or 10. Fruit linear and pod-like.

Bauhinia variegata ORCHID TREE A medium-sized, semi-deciduous tree to about 10 m tall. Leaves entire, rather broadly oval and heart-shaped at the base, to 10 cm long. Flowers conspicous: pink-white with dark pink markings, 5 petals, to 12 cm across; stamens 5. Fruit pod-like, to 30 cm long. Native to southern Asia but planted as an ornamental in southern Spain, possibly elsewhere. ES.

Senna

A mainly tropical genus of herbs, shrubs and tress with pinnately divided leaves. Flowers typically yellow with 5 sepals and petals, petals not fused in a tube; stamens 10, (7 fertile, all free). Fruit pod-like, several-seeded.

Senna didymobotrya (Syn. *Cassia didymobotrya*) CASSIA An evergreen, vigorous and robust shrub to 5(9) m with leaves to about 50 cm long with *14–18 pairs* of broadly elliptic leaflets. Flowers bright yellow, borne in large, dense, terminal *racemes with numerous blackish-brown buds* at the apex. Native to Tropical America but widely planted. Throughout.

Senna corymbosa (Syn. *Cassia corymbosa*) POPCORN CASSIA An evergreen or semi-evergreen, bushy and leafy shrub to 1(2) m with pinnately divided leaves with *2–3 pairs* of broadly elliptic leaflets which are rounded at the base. Flowers numerous, bright-yellow, borne in loose terminal clusters, each to 30 mm across with 5 unequal petals; stamens curved, and exceeded by their petals. Native to Tropical America but widely planted. Throughout.

Cercis

Shrubs and trees with *entire leaves*. Flowers with 10 free stamens. Fruit flattened with a dorsal *wing*.

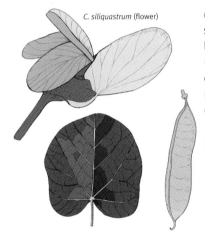

C. siliquastrum (flower)

Cercis siliquastrum JUDAS TREE A deciduous shrub or small tree to 10(12) m. Leaves heart-shaped 10–12 cm long, blunt, long-stalked and hairless. Flowers pink (rarely white), 14–20 mm long, borne in clusters arising directly from older branches, among previous blooms' pods, before the leaves appear, with a bell-shaped calyx. Fruit pendent, linear-oblong, laterally flattened, 50 mm–10 cm, with a narrow wing along 1 edge. Widely planted as an ornamental. Throughout.

1. *Bauhinia variegata*   5. *Cercis siliquastrum* (fruits)
2. *Senna didymobotrya*   6. *Ceratonia siliqua* (female flowers)
3. *Senna corymbosa*   7. *Ceratonia siliqua* (fruits)
4. *Cercis siliquastrum*

Ceratonia CAROB

Shrubs and trees. Normally dioeceous. Flowers inconspicuous (greenish), *regular;* stamens 2–8. Fruit a non-splitting, pod-like legume.

C. siliqua

Ceratonia siliqua CAROB TREE An evergreen shrub or tree to 10 m with leaves to 24 cm long, pinnately divided into 1–5 pairs of dark green, rounded, leathery, untoothed leaflets; terminal leaflet absent. Flowers green or reddish and small with 5 sepals but without petals, borne in lateral racemes *directly from the trunk* in early autumn. Fruit large, 45 mm–20 cm, linear-oblong and laterally flattened, bluish-brown when ripe and pendent. Very common maquis, roadsides and field boundaries, a relic of cultivation. Throughout.

Albizia

Trees with pinnately divided leaves. Flowers *regular* with *numerous, free, conspicuously protruding stamens*. Fruit a pod-like legume.

A. julibrissin

Albizia julibrissin ALBIZIA A deciduous tree 5–12(15) m tall with a broad crown. Leaves large, 20–45 cm long, and 2-pinnately divided with numerous (20–30) oblong leaflets, hairy beneath. Flowers borne in large spherical heads of up to 50, long stalked; corolla tubular with 5 even teeth; greenish-white with a *conspicuous fringe of long, pink stamens* 20–30 mm long. Fruit oblong, 40 mm– 20 cm long and with prominent seeds. Native to Asia; commonly planted as an ornamental in towns. Throughout.

Robinia

Trees with pinnately divided leaves. Flowers borne in pendent racemes; stamens 10, inserted (9 inferior, united into a tube, 1 free). Fruit a 2-valved, splitting, pod-like legume.

R. pseudacacia

Robinia pseudacacia FALSE ACACIA A deciduous tree to 25(52) m with irregularly fissured, grey-brown bark, and spine-like stipules on younger branches. Leaves 10–25 cm long, pinnately divided with 9–19 leaflets, more or less hairless. *Flowers white*, scented, with a standard yellowish at the base, borne in *pendent* racemes. Fruit linear-oval, 50 mm–10 cm long, flattened. Native to North America, commonly planted as an ornamental. Throughout. *Robinia viscosa* is similar but has pale pink to mauve flowers. Planted locally throughout.

Acacia ACACIA

Trees or shrubs with either 2-pinnately divided true leaves, or leaves reduced to a single blade (phyllode) and flowers borne in dense clusters; stamens numerous, free. Fruit a splitting or non-splitting pod-like legume. Species in the region are native to South Africa and Australia and the cultivated forms can be difficult to distinguish.

A. Leaves 2-pinnately divided.

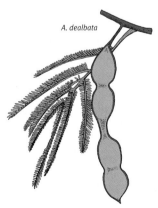

A. dealbata

Acacia dealbata SILVER WATTLE A bushy tree 12–15(30) m tall with smooth grey bark and silvery-hairy twigs and young leaves. Leaves *2-pinnately divided* with 10–26 pairs of primary divisions; leaflets 2–5 mm long, stipules rudimentary. Flowers *pale, bright yellow* and fragrant, borne in heads 5–6 mm across that form large terminal panicles that exceed the leaves. Fruit to 10 cm, linear-oblong and laterally flattened, bluish-brown. Native to Australia, widely planted as an ornamental and on roadsides. ES, PT. *Acacia mearnsii* is similar to the previous species but shorter 7–10(15) m, with *yellow hairy* young leaves, dull *pale yellow flowers* and blackish pods distinctly constricted between the seeds. PT. *Acacia karroo* is superficially similar to the previous species, but with *prominent spines present*; native to South Africa but planted as an ornamental. ES, FR (Corsica), IT (Sicily), PT.

B. Both 2-pinnately divided leaves and phyllodes (leaf-like blades) present.

Acacia melanoxylon bLACKWOOD ACACIA A dense, pyramidal tree to 40 m *with both true compound leaves and phyllodes present;* phyllodes elliptic-lanceolate (not long and linear), greyish. Flowers borne in small clusters in the leaf axils; *creamy-white.* ES, FR, IT, PT.

C. *Pinnately divided leaves absent,* reduced to phyllodes (leaf-like blades); *flowers borne in spherical heads.*

A. retinodes

Acacia retinodes SWAMP WATTLE A shrub or small tree 8–15(30) m tall with *spreading or upward-facing branches.* Phyllodes lanceolate, wavy or straight, light green and leathery with a single vein, 60 mm–14 cm long and *narrow,* 2–15 mm wide. Flowers pale yellow, borne in dense spherical heads 10–12 mm across, borne up to 10 in a raceme. Fruit 40 mm–12 cm, scarcely constricted between the seeds. Native to Australia, widely planted on roadsides throughout; common. ES, FR, IT, PT. *Acacia saligna* (Syn. *Acacia cyanophylla*) BLUE-LEAVED WATTLE is similar to *A. retinodes* but shorter (<8 m) and with slightly broader phyllodes 5–50 mm wide, and flowers dark yellow, borne in *long, drooping branches;* heads 6–8 mm. Fruit distinctly constricted between the seeds, 50 mm–14 cm long. Planted to stabilise dunes and as an ornamental but less common than the former species. European Mediterranean and islands. ES, FR, IT, PT. *Acacia pycnantha* is similar to the previous species and has similar sized but *sickle-shaped,* assymetrical phyllodes. ES, FR, IT, PT.

1. *Albizia julibrissin*
2. *Robinia pseudacacia*
3. *Acacia mearnsii*

1
2
3

D. *Pinnately divided leaves absent*, reduced to phyllodes (leaf-like blades); *flowers borne in elongated heads.*

Acacia longifolia WHITE SALLOW A tall shrub or tree 1–8(10) m with lanceolate phyllodes 6–20 cm long and 3–15 mm wide, rather like willow leaves. Flowers bright yellow, borne in *elongated, spike-like heads* in the axils of the phyllodes, 20–60 mm long and 5 mm wide. ES, FR, IT, PT.

Anagyris

Large shrubs with trifoliate leaves. Yellow flowers borne in short racemes; stamens 10. Fruit a pod-like legume.

Anagyris foetida BEAN TREFOIL A poisonous, deciduous, unpleasant-smelling shrub 2–4 m high. Leaves trifoliate with narrowly elliptic leaflets 6–40 mm long, silvery-hairy below and with papery stipules 5–10 mm. Flowers yellow, borne in short clusters; calyx bell-shaped, bluish. Fruit 60 mm–20 cm long. In dry, rocky places. Throughout.

1. *Acacia saligna* (inset: fruits)
2. *Acacia melanoxylon* (inset: fruits)
3. *Anagyris foetida* (inset: fruits)
4. *Calicotome villosa*
5. *Calicotome spinosa* (fruit)
6. *Calicotome spinosa*

Calicotome

Spiny shrubs with simple, inconspicuous leaves. Flowers with tubular calyx with 5 teeth; stamens 10, fused in a tube. Fruit a pod-like legume.

C. villosa

Calicotome villosa HAIRY THORNY BROOM A superficially gorse-like, erect shrub 1.5–3 m tall with slender spines on the branches. Leaves trifoliate, silver-hairy below, leaflets 4.5–15 mm. Flowers yellow and large, to 18 mm long, borne in umbrella-shaped clusters of 3–5; calyx tubular with 5 teeth, the upper part falling in flower to leave a cup-shaped structure. *Fruit woolly-hairy* (hairs to 2 mm long), 21–41 mm long. Locally common on maquis and garrigue. Throughout (absent from some islands). *Calicotome infesta* THORNY BROOM is similar but with short-hairy pods 22–40 mm (all hairs <1 mm). Nothern and western and absent from some islands. ES, FR, IT. *Calicotome spinosa* is similar to the previous species but with flowers often solitary (sometimes 3–8) and *hairless fruits* 21–52 mm. Local and absent from many areas. DZ, ES, FR, IT.

Cytisus BROOM

Spineless shrubs with leaves with 1 or 3 leaflets. Flowers yellow (sometimes white), borne in racemes; all 10 stamens in a tube. Fruit several–many-seeded.

A. Plant >1 m, *leaves mostly trifoliate; flowers yellow.*

Cytisus scoparius BROOM A spineless, much-branched shrub 1–2.5 m, with long, slender and flexible stems, normally *5-angled and ridged*, hairy or not. Leaves small, trifoliate (1-foliate and stalkless on young branches); oval-elliptic, 11–14 mm *with short stalks* (13 mm). Flowers large, 15–20 mm, golden yellow, *solitary or paired*. Fruit oblong, compressed, 25–50 mm and *hairy along the margins only*. Very common in wooded areas on acid soils. Throughout. *Cytisus arboreus* TREE BROOM is similar, has all trifoliate, *stalked leaves* (32 mm), *and (5)7–8-angled stems*. Subsp. *baeticus* (Syn. *C. baeticus*) is distinguished by its *fruits covered in long, white, woolly hairs*. Wooded hills. DZ, ES, MA, PT. Subsp. *catalaunicus* has hairless fruits, except along the margins. Hills and mountains. ES, FR, MA. *Cytisus grandiflorus* is similar to the previous species, with stems with 5(7) ridges, with ovary and fruits hairy all over (not just at the margins), and *virtually stalkless leaves*. ES, MA, PT.

Cytisus villosus A shrub 1.5–3 m similar to *C. scoparius* with 5-angled, hairy branches, stalked leaves (stalks 10–17 mm) with all trifoliate leaves with elliptic leaflets 20–47 mm with adpressed hairs, and *large yellow flowers* (17–22 mm) *with darker red markings on the base of the standard petal,* borne solitary or 2–3(4) *in each leaf axil,* forming lax, leafy clusters. Fruit 35–50 mm. Woods and rocky slopes. ES, FR, IT.

B. Plant 80 cm–1 m tall, leaves mostly trifoliate; *flowers white.*

Cytisus multiflorus WHITE SPANISH BROOM An erect, much-branched shrub 80 cm–1 m, similar to *Retama* (but with differing calyx and fruit) with stems with 6(8) ridges and very short-stalked (1.5–10 mm), trifoliate leaves with silver hairs; leaflets to 4.5 mm. *Flowers white, borne 2–3(4) per group in long interrupted racemes.* Fruit 15–31 mm with 4–5 seeds. Maquis, hills and roadsides in Iberian Peninsula; introduced further east. ES, FR, IT, PT.

C. Plant short, <1 m; *leaves simple; flowers yellow.*

Cytisus decumbens A spreading to ascending, low shrub just 80 mm–20 cm with 5-angled, hairy stems and *simple leaves;* leaves virtually hairless above, hairy below, 3–11 mm. Flowers 12–15 mm, yellow, borne numerously in the upper leaf axils. Fruit 18–28 mm, with spreading hairs. Absent from hot dry areas, and the far west, east and south. ES, FR, IT.

Spartium

S. junceum

Spineless, broom-like shrubs with stiff, rush-like branches; leaves simple, often absent or inconspicuous. Flowers with stamens 10, fused in a tube. Fruit many-seeded.

Spartium junceum SPANISH BROOM A large, spineless, broom-like shrub to 3 m with cylindrical, blue-green, rush-like stems. Leaves sparse, linear-oval and soon-falling, 15–30 mm. Flowers large, 20–28 mm, and bright yellow, solitary but in large numbers; sweetly scented; *calyx spathe-like.* Fruit 40 mm–12 cm, flattened. Common on dry slopes and in woods; also planted. Throughout.

1. *Cytisus scoparius*
2. *Cytisus scoparius* (fruiting plant)
3. *Spartium junceum*
4. *Genista tinctoria*
5. *Genista hirsuta*
6. *Genista tricuspidata*
7. *Genista tricuspidata* (fruits)

Genista GREENWEED

Spiny or non-spiny shrubs with simple leaves and yellow flowers; *calyx tube as long as or longer than the lips;* stamens 10, fused in a tube. Fruit 1–several seeded. Many closely related species which are not included here, occur in the hills of the Iberian Peninsula.

A. Shrubs without spines; flowers borne in lax, elongated clusters.

Genista tinctoria DYER'S GREENWEED A tufted, erect or ascending small shrub 60 cm (1 m) with simple, linear-lanceolate, stalkless, hairless leaves to 30 mm long. Flowers 10–15 mm, yellow, borne in rather lax, *long racemes,* terminal or in the axils; flowers hairless. Fruit oblong, flat and hairless, 15–30 mm. Absent from most islands, and the far west and south. ES, FR, IT.

B. Spiny, gorse-like shrubs <2 m.

Genista hirsuta A low, intricately spiny, gorse-like shrub to 50 cm (1.5 m) with simple and branched spines on older stems. Leaves simple, elliptic, with long hairs on the margins and lower surface, hairless above, 8–14 mm. Flowers 10–16.5 mm, yellow, borne in crowded terminal, *silvery-hairy and spiny racemes.* Fruit 5–6 mm and hairy. Southwest Iberian Peninsula and northwest Africa; common in the Algarve where plants have historically also been described as *G. hirsuta* subsp. *algarbiensis* and *G. algarbiensis.* ES, MA, PT. *Genista tournefortii* is similar to *G. hirsuta* but with *flowering racemes with few spines.* Fruit longer, 7–8(10) mm. Central-west Iberian Peninsula. ES, PT.

Genista tricuspidata A low, spiny shrub 30 cm–2 m, with *leathery young branches*; branches later woody and hairless. Leaves with a central developed leaflet 4–19 mm long and lateral appendages. Flowers yellow 10–13.4 mm, borne numerously in terminal heads; *calyx and flower-stalks leathery.* Fruit ovoid, 6.5–8(9) mm. DZ, ES (incl. Balearic Islands), MA, TN.

Genista hispanica SPANISH GORSE A low, intricately spiny, gorse-like shrub to 50(70) cm with spineless, leafy young stems; older stems with numerous *branched lateral spines.* Leaves stalkless, simple, elliptic and hairy beneath. Flowers 8–14 mm, yellow, borne in terminal clusters; *bracts tiny, <1 mm* (longer in the above species). Fruit 8–10(11) mm with 2–3 seeds. Rocky scrub. ES, FR.

Genista scorpius An erect or spareading, *intricately branched and spiny shrub* 30 cm–2 m high; spines stout, lateral and spreading, yellow at the tips. Leaves inconspicuous, alternate, *simple,* elliptic and slightly hairy beneath, 1.5–9 mm. Flowers bright yellow and hairless, 9–12 mm long, borne *directly on the spines.* Fruit narrowly oblong, 14–35 mm long and constricted beneath the seeds; *hairless.* Locally common in dry, bare habitats at various altltudes. ES, FR. *Genista corsica* CORSICAN GORSE is similar but with *flowers not borne directly on the spines*, and fruit to just 20 mm long. Garrigue. FR (Corsica), IT (Sardinia).

Genista triacanthos A gorse-like shrub 30 cm–2 m high with robust, diffuse stems, hairless or nearly so, bearing *tripartite* spines, leaves with 3–5 developed leaflets 3.5–10 mm long. Flowers 7–9.5 mm, sulphur-yellow, borne in racemes that *do not end in a spine-tip*, and which lengthen after flowering. Fruit 6.5–7.5 mm, with 3–8 aborted seeds. Southwest Iberian Peninsula and northwest Africa. ES, MA, PT. *Genista tridens* is similar but has *racemes ending in a spine-tip.* Fruit 5–6 mm, with 2 aborted seeds. ES, MA.

C. Spineless shrubs <2(3) m, with broom-like or rush-like stems.

Genista umbellata A distinctive, rounded, tufted shrub 30 cm–1.5 m high with numerous erect, almost leafless, *rush-like branches* bearing simple, hairy, narrowly elliptic leaves 3–16 mm long (those above 1-foliate) and *terminal dense, rounded heads of 5–30 yellow flowers* 9–14 mm long; calyx conspicuously silvery-hairy, 4.5–7 mm long. Fruit 8–24 mm. Locally abundant on dry slopes and banks in southern Spain; common in southern Andalucía. DZ, ES, MA.

Genista ephedroides A medium, tufted shrub to 1 m with slightly hairy, trifoliate leaves and *numerous broom-like, greyish stems*, superficially similar to an *Ephedra* bush. Flowers yellow and numerous, scattered along the terminal branches; flowers virtually hairless. IT (south + Sardinia and Sicily). *Genista spartioides* is similar, also *broom-like*, though taller, to 1.5(3) m, with leaves *hairy on both surfaces* and small flowers, 9.5–13 mm, borne in interrupted racemes; keel longer than the wings and standard petals. DZ, ES, MA.

Genista pulchella A non-spiny, domed, dark green, *dense, spreading shrub* to 15(75) cm, almost conifer-like when not in flower; branches alternate, rooting at the nodes and hairy when young. Leaves with solitary leaflets 2–4 mm. Flowers 10–12 mm, yellow, borne solitary and laterally. Fruit 9–17 mm. Rocky limestone scree. ES, FR.

Genista monspessulana (Syn. *Teline monspessulana*) A broom-like, leafy shrub to 3 m. Stems with 6–8 ridges. Leaves stalked and trifoliate with rather large, oval, variably hairy leaflets 6–17 mm long, 4–10 mm wide. Flowers yellow, 13–16.5 mm long, borne in lateral clusters of 3–7(9); calyx silvery-hairy and 2-lipped. Fruit oblong, flattened, 15–27 mm and white-woolly. Common on garrigue. Throughout, except for the Balearic Islands and the far east of region. ES, FR, IT, MA, PT. *Genista linifolia* (Syn. *Teline linifolia*) is similar but shorter, to 1.5(2) m, with *long, narrow leaflets*, shiny dark green above and noticeably paler beneath, 8–30 mm long and *just 1–5 mm wide*. ES (incl. Balearic Islands), FR.

G. monspessulana

D. Spineless, tall, *tree-like* shrubs to 5 m.

Genista aetnensis MOUNT ETNA BROOM A large shrub or small tree 3(5) m high with a substantial trunk when mature; young stems slender and much-branched; not spiny. Leaves small, simple and elliptic, soon-falling. Flowers yellow, 8–13 mm long, borne in lax terminal racemes. Fruit oval, flattened, 8–14 mm long and hairless. Rocky maquis, often in large numbers. IT (Sardinia and Sicily).

Pterospartum

Mostly spineless shrubs with markedly *winged* stems. Leaves 1-foliate. Flowers yellow, with standard equalling the keel; stamens 10, fused in a tube.

Pterospartum tridentatum (Syn. *Chamaespartium tridentatum, Genista tridentata*) A dwarf, much-branched shrub 15–50 cm (1 m) with distinctly *winged* young stems; wings undulate and leathery, elongated at each node to form 3 small, tooth-like lobes. Flowers 9.5–15 mm, yellow, borne in crowded terminal clusters of 3–10. Fruit 8.5–14 mm. Woods and heaths. ES, MA, PT.

Ulex GORSE

Very spiny, densely-branched shrubs with adult leaves reduced to weak spines. Flowers yellow; calyx yellowish and divided to the base into 2 lips; stamens 10, fused in a tube. Fruit (1)2–6(8)-seeded. Other species besides those described, which are difficult to distinguish, occur in the region.

A. Flowers with a standard petal (12)14–18 mm long.

Ulex europaeus GORSE A stout, erect, very spiny shrub to 2(2.5) m (shorter in exposed habitats), densely branched above and bare below. Young twigs and spines glaucous; twigs somewhat hairy. Leaves highly reduced, to 8 mm; *spines stout, straight, deeply furrowed and hairless*; calyx yellowish, 2-lipped and persistent, 10–16(20) mm, flowers large, the standard (12)14–18 mm, *pale yellow* with straight wings longer than the keel. *Secondary bracts at least 2 mm wide.* On acid heaths and in hedges; planted in west of range. ES, FR, PT.

1. *Genista scorpius*
2. *Genista triacanthos*
3. *Genista umbellata*
4. *Genista umbellata*
5. *Genista pulchella*
6. *Genista aetnensis*
   PHOTOGRAPH: GIANNIANTONIO DOMINA
7. *Pterospartum tridentatum*
8. *Ulex europaeus*

B. Flowers with a standard petal <12(13.5) mm long.

Ulex minor DWARF GORSE A spiny shrub to 1(1.5) m, similar in general appearance to *U. europaeus*, often spreading, with spines to 10 mm long, not rigid and weakly furrowed or striped. *Flowers deep yellow,* and small, the standard 7–12(13) mm long; calyx 5–10 mm, the *teeth divergent.* Rocky scrub; a northern European species that extends to southern Portugal. PT.

U. parviflorus

Ulex parviflorus SMALL-FLOWERED GORSE A densely spiny shrub to 1.5(2) m, generally with *blue-grey, virtually hairless or finely hairy stems and long, stiff, arched spines to 30 mm which are widely spaced and greatly exceeding the flowers.* Flowers yellow, *small, the standard 8.5–11 mm long*, the wings shorter than the keel; calyx 7–10 mm, yellowish and 2-lipped. Rocky scrub. ES, FR, PT. *Ulex argenteus* is similar but the young shoots are *covered in short, silky, straight, white hairs.* Standard 7–13.5 mm; *calyx with narrow lips* (2 mm wide). Southwest Iberian Peninsula. ES, PT. Subsp. *argenteus* has straight, slender spines; calyx 7.5–9.5 mm. Subsp. *subsericeus* has robust, curved spines; calyx larger, 9–12 mm. Subsp. *erinaceus* (Syn. *U. erinaceus*), also widely regarded as a distinct species, has a low, compact habit to 40 cm and is *silvery* and with a distinctly *hairy calyx* 10.5–13.5 mm; standard 11.5–13.5 mm. Rare. PT (southwest).

Stauracanthus

Very intricately spiny gorse-like shrubs with few leaves and yellow flowers. Very closely related to *Ulex* (and now included under this genus by some authors) but with a short tube at the base of the calyx (not divided to the base into 2 lips); stamens 10, fused in a tube.

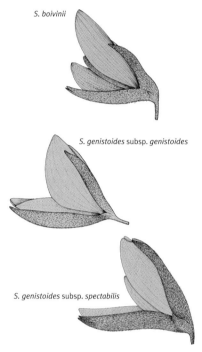

S. boivinii

S. genistoides subsp. *genistoides*

S. genistoides subsp. *spectabilis*

Stauracanthus boivinii STAURACANTHUS A dense and *very intricately spiny shrub* to 50 cm, similar to a gorse, but with mostly opposite leaves (leaves few). *Stems more or less hairless* or with some brown hairs, spines straight, to 4 mm, leaves tiny, to 3 mm; *calyx small (5–8 mm) with a 2-toothed upper lip about* $^1/_2$ *the size of the lower lip*; flowers yellow. Fruit oblong, *small* (8–9(12) mm), greatly exceeding the calyx. Garrigue and pine forests (and mountains in North Africa). DZ, ES, MA, PT. ***Stauracanthus genistoides*** is very similar though normally has a more robust and compact habit (to 1.5 m) and a preference for coastal dunes, and is most easily differentiated by the *upper calyx lip which equals the lower*, 9–11 mm. Fruit 15–25 mm. ES, PT. The following subspecies are widely recognised but show continuous variation and may not be distinct: Subsp. *genistoides* has slender, flexible spines and a slightly hairy standard. Subsp. *spectabilis* (Syn. *S. spectabilis*) is treated by some authors as a separate species, but differs only marginally in having a very hairy standard and larger flowers with a *longer calyx*, 12–15 mm long. ES, MA, PT.

Retama

Rush-like, branched, spineless shrubs with slender, alternate branches and simple leaves. Flowers with stamens 10, fused in a tube. Fruit 1–2-seeded, (normally) non-splitting.

Retama raetam WHITE BROOM An erect, much-branched, spineless shrub to 2 m with pendent lateral branches, silvery when young. Leaves sparse, linear and silvery-hairy, soon-falling. *Flowers borne abundantly along drooping branches, white,* to 17 mm long, borne in dense racemes; sweetly scented; calyx soon-falling. Fruit club-shaped, to 20 mm long and beaked. Coastal sands. DZ, IT (Sicily), TN.

Retama sphaerocarpa LYGOS A much-branched spineless shrub to 2(3) m with erect to spreading branches which are hairless, ribbed and with a *silvery sheen*. Leaves small and linear, 6–11 mm, silvery-hairy and soon falling. Flowers yellow, the standard 4–5 mm long; borne in lax, lateral clusters; calyx persistent in fruit. Fruit smooth and egg-shapd, 7.5–12 mm long. Dry slopes in the southern Iberian Peninsula. ES, PT. *Retama monosperma* has *white flowers* with a standard petal 9–10.5 mm long and *fruit rough,* at least partially, 10–22 mm. Mainly near the coast. ES, PT.

---

1. *Ulex minor*
2. *Stauracanthus boivinii*
3. *Stauracanthus boivinii*
4. *Stauracanthus genistoides*
5. *Retama sphaerocarpa*
6. *Retama monosperma* (fruits)

Adenocarpus

Spineless shrubs with trifoliate leaves. Flowers orange-yellow; stamens 10, fused in a tube. Fruit many-seeded, splitting.

A. complicatus

1. *Lupinus luteus*
2. *Lupinus micranthus*
3. *Lupinus angustifolius*

Adenocarpus complicatus ADENOCARPUS An erect, sparsely hairy, spineless shrub to 3 m, rather broom-like but leafy, with leaves trifoliate with narrow elliptic, silkily hairy leaflets 6–13 mm, and flowers *orange-yellow*, the standard 10–15 mm in *long, lax terminal clusters* of >7; calyx tubular, with a bilobed upper lip, and trilobed lower lip. Fruit oblong-lanceolate, to 30 mm. Local, and absent from many islands. ES, FR, IT, PT. *Adenocarpus anisochilus* (Syn. *A. complicatus* subsp. *anisochilus*) is treated by some authors as a distinct species, and has very *congested racemes; standard 16–20(22) mm.* Algarve only. PT. *Adenocarpus telonensis* has *larger flowers* to 20 mm long in terminal umbel-like heads. ES, FR, MA, PT. *Adenocarpus hispanicus* is similar but with leaflets with *inwardly rolled margins*. Standard 12–19(25) mm. Fruit to 60 mm. Central Iberian Peninsula. ES. Subsp. *argyrophyllus* (also treated as a distinct species) has leaves densely silvery-hairy above. Central Iberian Peninsula, possibly southern Spain and Portugal. ES, (PT).

Lupinus LUPIN

Herbs with palmate, lobed leaves. Flowers borne in long, conspicuous terminal racemes; stamens 10, fused in a tube. Fruit with 2–many-seeds, splitting.

A. Flowers yellow.

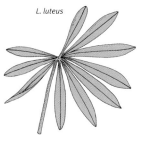

L. luteus

Lupinus luteus YELLOW LUPIN A robust, medium-tall hairy annual to 1 m. Leaves palmately lobed with 5–9 oblong leaflets, flowers bright yellow, 15–18 mm long borne in long whorls along the raceme, scented. Fruit 40–60 mm long, densely hairy, black when ripe. Sandy places, and cultivated for fodder; common in southwest Iberian Peninsula. ES, FR, IT, PT. *Lupinus gredensis* is similar but has *pale yellow-cream flowers, ageing violet*. Iberian Peninsula. ES, PT.

B. Flowers blue.

L. micranthus

L. cosentinii

Lupinus micranthus HAIRY LUPIN A short, *short-brown-hairy* annual to 50 cm with palmate leaves of 5–9 sparsely hairy oblong leaflets. Flowers blue, 10–14 mm long, the standard with a whitish central region, and the keel tipped with black-violet, borne in irregular whorls along the raceme. Fruit 30–50 mm, *hairy* and red-brown when ripe. Cliff-tops and verges. Throughout, though absent from some islands. *Lupinus cosentinii* (Syn. *L. varius*) is similar to, and often confused with the above species but much taller, to 1.3 m with *long, silky hairs* and *larger flowers 15–17 mm long*, the standard with a paler, white or yellow region, borne in irregular whorls along the raceme. Fruit reddish-brown when ripe, 40–60 mm. Similar habitats though local. Probably locally throughout.

Lupinus angustifolius NARROW-LEAVED LUPIN An annual to 1 m similar to *L. micranthus* but with 5–9 *narrower* leaflets (2–8 mm wide), linear to spatula-shaped, slightly hairy beneath. Flowers dark blue, 12–16 mm long, borne alternately along the raceme. Fruit 40–70 mm with short hairs, yellow to black when ripe. Common in waste places and on cultivated ground. Throughout.

C. Flowers white.

L. angustifolius

Lupinus albus WHITE LUPIN An erect annual to 1(2) m with palmate leaves with 5–7(9) leaflets, hairless above. *Flowers white, tipped with blue*, 18–20 mm long, borne in lax, terminal spikes. Fruits long-hairy, 60 mm–13 cm. Mainly eastern but widely naturalised as a casual. Throughout.

Colutea BLADDER SENNA

Deciduous shrubs. Flowers yellow or orange, borne in axillary clusters with a broad standard and blunt keel; stamens 10, 1 free. Fruit *strongly inflated* and later papery.

Colutea arborescens BLADDER SENNA A lax, much-branched deciduous shrub 2–3(5) m high with pinnately divided leaves with 3–6 pairs of oval to elliptic leaflets 17–23 mm long. Flowers yellow, borne in lax axillary racemes of 3–7, each 16–20 mm long; calyx 6.5–8 mm long, the lower teeth to 0.7–2 mm long; *ovary hairless*. Fruits very distinctive: 52–55 mm long, *strongly inflated* and pink to red-flushed, later brown and papery; hairless except for along the keel. Widespread in light woodland and maquis; common in southern France and northeast Spain, local elsewhere. ES, FR, IT. Subsp. *hispanica* (Syn. *C. hispanica*) has an *ovary shortly hairy throughout*. Calcareous woods in central, southern and eastern Spain. ES. Subsp. *gallica* (Syn. *C. brevialata*) has smaller flowers and calyx just 4.5–6.5 mm long but with lower teeth long: 2–3 mm. Distribution improperly known. ES, FR, IT.

Astragalus

Annual or perennial herbs with pinnately divided leaves terminating in a single leaflet. Flowers borne in lateral clusters; keel blunt (not toothed or pointed); stamens 10, 1 free. Fruit variable (sometimes inflated). A very large genus of which only the most widespread or conspicuous are described here.

A. Leafy stems absent, flowers borne on long, leafless stems arising directly from a rosette of leaves.

Astragalus monspessulanus MONTPELLIER MILK-VETCH A perennial herb to 20 cm with more or less prostrate *rosettes of leaves*; leaves long, with 7–21 oval leaflets 3–11 mm long, adpressed-hairy beneath. Flowers 17–25 mm, mauve, borne in dense, ovoid clusters on long, slender, spreading stems. Fruit long and cylindrical, 20–30(50) mm. ES, FR, IT.

B. Annuals or perennials with flowers borne on leafy stems; flowers yellowish or whitish.

Astragalus cicer WILD LENTIL A robust and vigorous, ascending perennial to 60 cm (1 m). Leaves to 13 cm long with 17–31 pairs of oval, short-hairy leaflets. Flowers 12–16 mm, dull pale yellow-white, borne in dense, oblong racemes. Fruit 10–15 mm, inflated, papery and covered with *short black and white hairs*. Grassy habitats. Common in southeast Europe, extending locally into cool, grassy habitats in the western Mediterranean. IT, FR.

Astragalus odoratus LESSER MILK-VETCH An erect to ascending perennial to 30 cm. Leaves pinnately divided with numerous (19–29) leaflets. Flowers 9–12 mm long, white-cream to yellow, sometimes flushed with dull violet in bud, borne in rather dense racemes. Fruit 8–10 mm, laterally compressed and sparsely short-hairy. Native to the eastern Mediterranean, locally naturalised in grassy habitats in southeast France and central-east Italy, possibly elsewhere. FR, IT.

Astragalus boeticus A hairy, short-medium, erect annual with pinnately divided leaves of 6–13 pairs of oval leaflets, notched at the tip, hairless above, and slightly hairy below. Flowers pale sulphur yellow-white, 8–11(14) mm, borne on dense racemes on stalks as long as the leaves or slightly shorter; wings longer than keel. Fruit 20–45(60) mm, oblong, triangular in section, shortly hairy, borne in dense clusters erect at first, then drooping. Sandy places and cliff-tops; common. Throughout.

1. *Colutea arborescens*   5. *Astragalus boeticus*
2. *Astragalus cicer*   6. *Astragalus epiglottis* (fruits)
3. *Astragalus odoratus*   7. *Astragalus epiglottis*
4. *Astragalus solandri*

Astragalus solandri A *prostrate, spreading* annual, branched from the base with white-hairy stems 50 mm–45 cm. Leaves pinnately divided. Flowers pale greenish-yellow, borne in dense heads on long stalks. *Pods strongly arched*, compressed and tinged red-brown, rather shiny with minute white hairs. Dunes and bare ground, mainly on North African Atlantic coasts. MA.

C. Annuals or perennials with flowers borne on leafy stems; flowers typically violet, bluish or reddish.

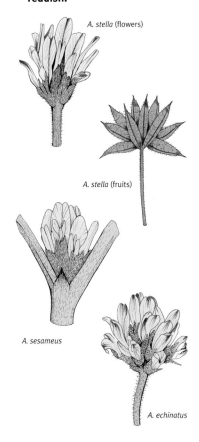

A. stella (flowers)

A. stella (fruits)

A. sesameus

A. echinatus

Astragalus stella A low to short, *almost prostrate* annual to 35 cm *with long-spreading or ascending stems*. Leaves with 5–11 pairs of oval leaflets, hairy on both surfaces. Flowers yellowish, or reddish-violet, 5.5–11 mm long, borne on dense racemes $^1/_2$ to the same length as the leaves. Fruit 10–15 mm, more or less erect, almost straight, laterally compressed, *borne in star-shaped clusters*. Common on disturbed ground, particularly in the west. Throughout. *Astragalus epiglottis* is similar to *A. stella* but with just 3–7(10) leaflets, very small flowers just 3–4 mm long, with *5 stamens, not 10*. Fruit *5–9 mm, distinctly triangular-heart-shaped*, borne in spiky clusters. Local. Throughout. *Astragalus sesameus* is a similar, erect annual to 36 cm, with hairy stems and rather hairy leaves, similar in most respects to *A. stella* but *with racemes scarcely stalked* and *more erect* and flowers 7–9 mm, bluish-violet (not reddish-violet). Fruit 9–17 mm. Throughout. *Astragalus echinatus* is similar to the previous species, somewhat hairy, to 40 cm, with leaflets in 5–9(12) pairs, notched at the tip, hairy beneath and hairless above. Purplish flowers 8 mm long are borne in very dense racemes (20 flowers) with stalks equalling or longer than the leaves. Fruit 8 mm long, oval-triangular, flattened and *covered in bristles*, with a hooked beak. Throughout.

Astragalus algarbiensis A perennial with stems to 25 cm, erect or ascending. Leaves 7–12 cm, leaflets 8–12(13) in pairs, *wedge-shaped*, hairy beneath, hairless above. Flower stalks more or less equalling the leaves; flowers 9–11 mm, violet or dull yellow, borne in dense racemes of pendent flowers. Fruit 7–10 mm, *crescent-shaped, wrinkled, and hairless*. ES, MA, PT.

Astragalus mareoticus A *prostrate, spreading annual*, branched from the base. Leaves with 6–8 pairs of leaflets, with short white hairs above and long white hairs below. Flowers few; corolla mauve. Fruits spreading, thick, strongly arched and grooved, purple-tinged and with minute, adpressed hairs. Dunes and desert fringe habitats in North Africa. DZ, MA, TN.

---

1. *Astragalus echinatus*   4. *Astragalus tragacantha* subsp. *vicentinus*
2. *Astragalus mareoticus*   5. *Astragalus sicula* PHOTOGRAPH: GIANNIANTONIO DOMINA
3. *Astragalus balearicus*

D. Densely spiny, cushion-like shrubs.

Astragalus tragacantha (Syn. *A. massiliensis*) An extremely *spiny, intricately branched and cushion-like low shrub* to 30 cm. Leaves with 5–10(12) pairs of elliptic leaflets ending in a spine, densely hairy beneath. Flowers white, borne in short racemes. Fruit oblong, shortly hairy, 7–13 mm. Locally common in exposed sea cliffs. ES, FR (Corsica), IT (Sardinia), PT, TN. Subsp. *vicentinus* is regarded by some authors as distinct; cliff-tops in the Algarve. PT. *Astragalus balearicus* is a similar cushion-like shrub to 25 cm high, but with leaves with *just 3–5 pairs of leaflets* which are also less hairy. Common on sea cliffs on the Balearic Islands, where it is endemic. *Astragalus sirinicus* is very similar to the previous species, with a dense, flat, spreading habit and white flowers *flushed pale violet or yellow*. FR (incl. Corsica), IT (incl. Sardinia).

Astragalus sicula Sicilian milk-vetch A distinctive, *compact, flattened, spiny cushion-like subshrub*. Leaves blue-green, pinnately divided and hairy. Flowers pink-purple, borne in dense clusters in the leaf axils; calyx white-woolly. Fruit pod-like. An endemic, sub-dominant species on bare slopes on Mount Etna. IT (Sicily).

Erophaca

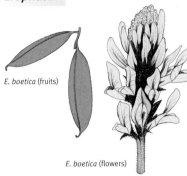

E. boetica (fruits)

E. boetica (flowers)

Similar to *Atragalus* (not recognised as distinct by all authors); *pods greatly inflated*.

Erophaca boetica (Syn. *Astragalus lusitanicus*) An erect, robust perennial to 50 cm (1 m) with grey-green leaves with 7–12 elliptic leaflets, hairless above. Flowers 25–31 mm long, cream or greenish, borne in lateral, long-stalked, dense and many-flowered racemes. *Fruit oblong, large and inflated*, 50–70 mm long and pendulous in clusters. Common on maquis and garrigue in the southwest. ES, MA, PT.

Glycyrrhiza

Similar to *Astragalus* but with short, linear or curved fruits and one-celled, not splitting. Leaves pinnately divided with a terminal leaflet; stipules very small.

Glycyrrhiza glabra LIQUORICE An erect, robust, hairless perennial to 1 m with pinnately divided leaves with 4–8(17) elliptic leaflets, sticky beneath, and with *minute or absent stipules*, and *short-stalked* clusters of bluish flowers, 10–13(15.5) mm long. Fruit 10–25 mm long. Rare or absent in much of the west and south. ES, FR, IT. ***Glycyrrhiza echinata*** SPINY-FRUITED LIQUORICE is very similar but with *fruits covered in long, brown-red spines*. Southern Italy; rare or absent elsewhere. IT.

Bituminaria

Lax, spineless herbs smelling of tar when crushed, with trifoliate leaves. Flowers bluish, borne in dense, globular heads; stamens 10, fused in a tube. Fruit 1-seeded, non-splitting.

Bituminaria bituminosa (Syn. *Psoralea bituminosa*) An erect, branched perennial 20 cm–1 m high. Leaves long-stalked, trifoliate with narrow to broad untoothed leaflets 12–90 mm long. Flowers 11–18 mm long, blue-violet borne on long-stalked clover-like clusters. Fruit ovoid, 13–19 mm, flattened with a sickle-like beak. Common on wasteground and in disturbed habitats.

1

2

3

4

5

Cullen

Lax, spineless herbs smelling of tar when crushed, similar to *Bituminaria* but identified as genetically distinct. Flowers with 10 stamens fused in a tube. Fruit 1-seeded, *fleshy and hairless,* non-splitting.

Cullen americana (Syn. *Bituminaria americana; Psoralea americana*) is an erect perennial to 50 cm (1 m) vaguely similar to *B. bituminosa* but with oval-round *toothed leaflets* 8–50 mm and flowers 6–7(8) mm borne fairly densely in longer racemes on stalks equalling the leaves; white, violet-tipped flowers. Similar habitats. Throughout.

Vicia VETCH

Climbing or scrambling annual or perennial herbs with unwinged stems, often bearing leaf tendrils; leaves pinnately divided. Flowers with 10 stamens, 9 in a tube, 1 free. Fruit flattened and splitting.

A. *Flowers small,* 2–10(15) mm, borne in clusters of 1–9(20) on stalks equalling or longer than the leaves.

V. hirsuta

V. disperma

Vicia hirsuta HAIRY TARE A short to medium, delicate, hairy to nearly hairless annual to 80 cm with leaves with 4–10 pairs of linear to oblong leaflets, each to 12 mm, notched at the tip and with a fine point; tendrils branched. Flowers white, tinged purple and small, 3–5 mm long in 1–9-flowered racemes; *calyx teeth equal. Fruit hairy,* 6–11 mm, black when ripe; seeds mostly 2. Grassy habitats. ES, FR, IT, MA, PT. The following species are similar: *Vicia disperma* is sparsely hairy with pale purple flowers 3–5 mm and broad, *hairless fruits* (except sometimes the margins) 14–19 mm, with 1–2 seeds. ES, FR, IT, MA, PT. *Vicia parviflora* (Syn. *V. tenuissima*) has leaves with *2–4(5) pairs of* leaflets, and 1–4(5)-flowered racemes that greatly exceed the leaves; flowers 6–9 mm with *limb of standard longer than its claw.* Fruit hairless, 12–17 mm; seeds 4–6(8). ES, FR, IT, MA, PT. *Vicia ervilia* has 10–16 pairs of linear leaflets with a short terminal point in place of a tendril. Flowers 6–8 mm, whitish or pale pink with darker veins. *Fruit hairless and constricted between the seeds;* 14–22 mm; seeds 2–4. Local, throughout. *Vicia leucantha* is similar to *V. ervilia* but has leaves with just 5–6 pairs of leaflets and tendrils. Fruits to 30 mm, hairy and *not constricted between the seeds;* seeds 3–4. ES (Balearic Islands), FR (Corsica), IT (incl. Sardinia and Sicily), MA, MT. *Vicia vicioides* has 5–7 leaflets, rather dark pink to purple flowers 5–7(8) mm long, borne in dense clusters of 5–20, with *hairy calyces.* Fruit semi-circular, 14–21 mm with 1–3 seeds. Far west only. ES, MA, PT.

1. *Erophaca boetica*
2. *Glycyrrhiza glabra*
3. *Glycyrrhiza echinata* (fruits)
4. *Glycyrrhiza echinata* (flowers)
5. *Bituminaria bituminosa*

V. tetrasperma

Vicia tetrasperma A more or less hairless annual with 3–6(8) pairs of linear leaflets and entire stipules. Racemes with *1 or 2(6) flowers, about equalling (not exceeding) the leaves*; corolla pale purple; 12–15 mm, the calyx teeth unequal. Fruit hairless, 20–35 mm, *with 3–10 seeds*. Distribution poorly known due to confusion with other species. ES, FR, IT, MA, PT. *Vicia pubescens* is similar but *slightly hairy throughout*, with up to 6-flowered racemes and hairy fruits. ES, FR, IT, MA, PT.

B. Flowers 10–20(24) mm, borne in clusters of 6–30 on stalks longer than the leaves.

Vicia onobrychioides FALSE SAINFOIN A hairy or hairless perennial to 60 cm. Leaflets in 4–9 pairs, linear to oblong and *narrow*, just 1–4(5) mm wide, with tendrils present; stipules toothed. Flowers large, 14–24 mm, borne in rather 1-sided, long-stalked racemes of 4–12, bright *blue-violet with a paler, whitish keel*. Fruit 27–35 mm, reddish and hairless with 2–8 seeds. Disturbed and sandy waste places. ES, FR, IT, MA, PT. *Vicia altissima* is similar but has broader, sometimes toothed leaflets 6–8 mm wide and *paler, whitish-violet flowers* 15–18 mm long. Fruit brown and hairless, to 35 mm with 4–6 seeds. ES, FR, IT, MA.

Vicia cracca TUFTED VETCH A variable, often hairy, vigorous, scrambling perennial to 2 m with leaves with 5–15 pairs of leaflets and branched tendrils, and entire, $^1/_2$-arrow-shaped stipules. Flowers 8–12(13) mm, *borne in dense, long narrow racemes of 10–40 blue-violet flowers* on stout stalks, drooping; calyx teeth very unequal (the upper minute). Fruit hairless, 10–25 mm with 2–6(8) seeds. Grassy habitats on higher ground, absent from much of the far south. ES, FR, IT, MA.

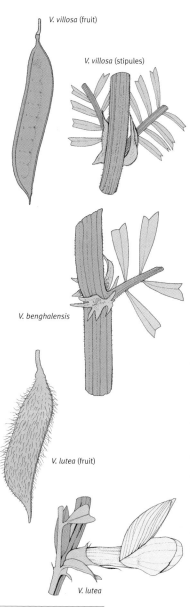

V. villosa (fruit)

V. villosa (stipules)

V. benghalensis

V. lutea (fruit)

V. lutea

1. *Vicia hirsuta*
2. *Vicia tetrasperma*
3. *Vicia cracca*
4. *Vicia benghalensis*
5. *Vicia lutea*

Vicia villosa FODDER VETCH A variable annual herb to 2 m, usually scrambling or climbing, similar to *V. cracca*. Leaves pinnately divided with 4–12 pairs of linear-elliptic leaflets, and *unlobed*, hairy stipules and branched tendrils. Flowers 10–20 mm, violet to purple, often with creamy wings borne in conspicuous racemes of 10–30 flowers; stalk shorter than the subtending leaf; *calyx with a swelling at the base; standard petal with a basal stalk-like part 2 x the length of the blade*. Fruit brown and hairless, 20–40 mm with 2–8 seeds. Scattered as a native but widely naturalised in Europe. ES, FR, IT, MA, PT.

Vicia benghalensis A short to tall, hairy annual to 80 cm with leaves with 5–9 pairs of linear to elliptic leaflets; tendrils branched; stipules toothed or not. Flowers 2–20, *reddish to burgundy-coloured with blackish tips*, 15–18 mm long, with calyx convex at the base, borne in racemes longer than the leaves. Fruit hairy, 21–40 mm with 2–5 seeds. Garrigue, pathsides. Rare or absent in much of the east. ES, FR, MA, PT.

C. Flowers *yellow or white*, <6 per cluster, on stalks *much shorter than the leaves.*

Vicia lutea YELLOW VETCH A tufted, prostrate, hairless to hairy annual to 50 cm with leaves with 3–9 pairs of linear to oblong, bristle-tipped leaflets; tendrils simple or branched. Flowers 17–25 mm, hairless, borne in groups of 1–3, dull *yellowish-white* (sometimes flushed purple) with a hairless standard; calyx teeth unequal. Fruit yellow-brown and densely hairy, 25–43 mm with 2–4 seeds. Coastal, sandy habitats. Rare or absent in the east. DZ, ES, FR, MA, PT. *Vicia hybrida* is similar to *V. lutea* but with *solitary* flowers 18–25 mm with a *hairy standard*, borne on short stalks, and unequal calyx teeth. Fruit 27–32 mm with 2–5 seeds. Damp habitats. Distribution poorly known due to confusion with the previous species. DZ, ES, FR, IT, MA, PT. *Vicia grandiflora* LARGE YELLOW VETCH is similar to *V. lutea* but with *large flowers* 22–35 mm long, yellowish or white, sometimes purple-tinged. Fruit hairy. IT.

Vicia faba BROADBEAN An erect, hairless, square-stemmed annual to 80 cm (1 m) with 2–3 pairs of large, oval, grey-green leaflets without tendrils. Flowers white with blackish-purple blotches, borne in virtually stalkless clusters of 1–6. Fruit long, to 30 cm with 4–8 seeds. Native to Asia, cultivated virtually throughout.

D. Flowers *bi- or tri-coloured*, <6 per cluster, on stalks *much shorter than the leaves.*

Vicia melanops BLACK VETCH A short to medium, sprawling, hairy annual to 50 cm with leaves with 4–8 pairs of leaflets. *Flowers tricoloured,* 17–21 mm long, solitary or in clusters of 2–3, with a *yellow-green standard, dark black-purple wings and purplish keel*; calyx hairy. Fruit hairy on the margins only, to 32 mm with 2–4 seeds. Garrigue. Mainly eastern; absent from most islands. ES, FR, IT. *Vicia barbazitae* is similar, but with normally solitary flowers to 22 mm which are *pale yellow with blue-purple wings*. Fruit hairless at both ends. Central in distribution. FR (incl. Corsica), IT.

E. Flowers a shade of *purple*, ≤6 per cluster, on stalks *much shorter than the leaves*.

Vicia sepium BUSH VETCH A scrambling, spreading perennial to 1 m, downy or hairless with leaves with 3–9 pairs of almost rounded leaflets ending in bristle-tips; stipules ½-arrow-shaped and spotted. Flowers 12–15 mm, *dull bluish-purple with red-purple calyces* with unequal teeth, borne in short-stalked clusters of 2–6. Fruit black and hairless, 20–35 mm with 3–10 seeds. Grassy places on higher ground; absent from hot, dry areas and the far south. ES, FR, IT. *Vicia pannonica* is similar, with 4–19 leaflet pairs, clusters of 1–4 dull purple (sometimes cream) flowers 20–35 mm borne on short stalks; *the standard petal hairy on the back*. Fruit yellowish and hairy, 20–35 mm with 2–8 seeds. ES, FR, IT, MA.

Vicia sativa COMMON VETCH A variable, vigorous, sprawling, hairy annual to 1.5 m. Leaves with 3–8 pairs of linear to heart-shaped leaflets, *tendrils branched*; *stipules toothed*. Flowers purplish-red, *solitary or paired* (up to 4), and often large, *20–30 mm long;* calyx teeth equal. Fruit yellowish or blackish, long, 35–68 mm with 4–9 seeds. The commonest *Vicia* species; frequent in all grassy and disturbed habitats. Throughout. *Vicia lathyroides* SPRING VETCH is similar but small and prostrate to 20 cm with leaves with 2–4 pairs of bristle-tipped leaflets, *unbranched or undeveloped tendrils,* and *small, solitary flowers just 6–9 mm long*. Garrigue and bare ground. ES, FR, IT, MA. *Vicia peregrina* is similar to *V. sativa* but has *untoothed stipules*, smaller flowers 11–19 mm long with a deeply notched standard petal. Throughout.

Vicia narbonensis PURPLE BROADBEAN A downy, erect, square-stemmed annual to 60 cm with *large, grey-green leaves with 1–3 pairs of leaflets 11–39 mm wide* (rather like a broadbean plant). Flowers 10–30 mm, borne in clusters of 1–3(6), *dark purple*. Fruit short-hairy, 30–50(70) mm with 4–7 seeds. Damp fields and ditches. Rare or absent from some islands, but scattered more or less throughout.

Lathyrus PEAS

Similar to *Vicia* but typically with angled or *winged stems*, and often fewer, parallel-veined leaflets, sometimes reduced to a simple blade or tendril. Flowers with 10 stamens, 9 in a tube, 1 free. Fruit a pod-like legume.

A. Flowers yellow.

Lathyrus aphaca YELLOW VETCHLING A hairless, waxy grey-green scrambling annual to 40 cm (1 m) with unwinged (angled) stems. Mature plants with *leaves reduced to tendrils, but stipules large and leaf-like, broadly triangular and paired,* 6–50 mm long. Flowers borne solitary on erect stalks, yellow, 10–13 mm long. Fruit hairless, 17–35 mm long. Grassy and disturbed habitats. Throughout.

Lathyrus ochrus WINGED VETCHLING A medium to tall, hairless, pale greysh annual to 1.5 m with *very broadly winged stems*, and leaves borne on *leaf-like stalks*; lower leaves oval-oblong, upper leaves with 2–3(4) pairs of leaflets. Flowers *pale yellow*, usually solitary, 13–21 mm long. *Fruit 39–65 mm with 2 wings* along the upper edge. Garrigue. Throughout.

1. *Vicia faba*
2. *Vicia sepium*
3. *Vicia sativa*
4. *Lathyrus aphaca*
5. *Lathyrus ochrus*
6. *Lathyrus annuus*

L. annuus

Lathyrus annuus ANNUAL YELLOW VETCHLING A climbing or scrambling annual to 1 m with leaves with 1 pair of lanceolate leaflets, arrow-shaped stipules and branched tendrils. Flowers 12–18 mm, borne in erect, long-stalked racemes of 1–3, corolla *yellow to orange* and red-veined; calyx teeth equal. Fruit pale brown and glandular when young, 40–80 mm. Disturbed habitats. Throughout.

B. Flowers reddish, purple, pink or white. Leaves simple or with just 2 leaflets; stems with wings <1 mm or wingless.

Lathyrus nissolia GRASS VETCHLING A delicate, ascending, hairless to downy annual with *unwinged stems* to 90 cm. *Leaves grass-like,* lacking leaflets and tendrils, and with very small, narrow stipules to 2 mm. Flowers solitary or paired, long-stalked and crimson, 8–18 mm long. Fruit pale brown, 22–60 mm. Grassy habitats, often near the coast. Throughout.

Lathyrus cicera RED VETCHLING A slender, hairless annual to 80 cm (1 m) with *narrowly winged stems*. Leaves with 1 (rarely 2) pair(s) of lanceolate-oval leaflets, with simple or branched tendrils on the upper leaves. Flowers 12–18 mm, solitary on stalks to 30 mm, *red, orange or brownish*; *calyx teeth equal*. Fruit hairless, 25–50 mm, and with 2 keels on the upper edge. Garrigue and grassland. Throughout. *Lathyrus setifolius* NARROW-LEAVED RED VETCHLING is similar but has scarcely winged stems and smaller orange-red flowers 9–13 mm with slightly *unequal calyx teeth*. Fruit 19–33 mm. Garrigue; rather local. Throughout. *Lathyrus gorgoni* is similar to *L. cicera* but has *large, pale orange flowers* with spreading sepals. IT (Sardinia and Sicily). *Lathyrus hirsutus* HAIRY VETCHLING is similar to the previous species in form, but sparsely hairy and with *pale violet-blue flowers with pink wings,* 7–15 mm. Fruit 15–45 mm. Throughout.

1. *Lathyrus nissolia*
2. *Lathyrus cicera*
3. *Lathyrus tuberosus*
4. *Lathyrus angulatus*
5. *Lathyrus clymenum* (pale-flowered form)
6. *Lathyrus clymenum*

Lathyrus tuberosus FYFIELD PEA A scrambling, hairless perennial with *angled but mostly unwinged stems* to 1.2 m. Leaves with a single pair of elliptic leaflets and simple or branched tendrils and narrowly ¹/₂-arrow shaped tendrils. *Flowers bright reddish-pink*, borne on long stalks in groups of 2–7, 12–20 mm long; calyx teeth triangular. Fruit brown and hairless, 16–28 mm long. Local in grassy and waste habitats. ES, FR, IT, MA.

C. Flowers reddish, purple, pink or white. Leaves simple or with just 2 leaflets; *stems broadly winged.*

Lathyrus sylvestris NARROW-LEAVED EVERLASTING PEA A climbing, hairless or downy perennial herb with *broadly winged stems* to 2 m long. Leaves with a single pair of narrowly lanceolate, 3-veined leaflets 2–19 mm wide with branched tendrils; *stipules lanceolate, <¹/₂ the width of the stem*. Flowers 13–19 mm, borne on rather long stalks (30 mm–21 cm), pinkish-purple flushed yellow; calyx teeth shorter than their tube. Fruit 50–70 mm. Rare or absent in much of the extreme east, south and west. ES, FR, IT, MA. *Lathyrus latifolius* BROAD-LEAVED EVERLASTING PEA is similar, to 3 m tall, with *broader, more oval leaflets 2–40 mm wide, and stipules >¹/₂ the width of the stem*; flowers 3–12 and larger, 15–30 mm and magenta-pink. ES, FR, IT, MA, PT.

Lathyrus odoratus EVERLASTING PEA A clambering annual to 2.5 m, slightly hairy. Leaves with a single pair of oval to elliptic, slightly undulate leaflets. Flowers 1–3(4), bicoloured, pink or purple, 20–35 mm and *sweetly scented*, borne in long-stalked racemes (stalks 14–20 cm). Fruit 55–65 mm. Dunes. IT (incl. Sicily); naturalised elsewhere.

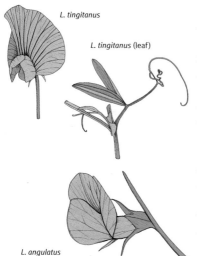

L. tingitanus

L. tingitanus (leaf)

L. angulatus

Lathyrus tingitanus TANGIER PEA A clambering, hairless annual herb with winged stems to 1.8 m. Leaves with *1 pair* of linear-lanceolate leaflets, tendrils branched; stipules arrow-shaped at the base; oval. *Flowers large and bright red-purple*, 25–35 mm long, borne in racemes of 2–3(4) on stalks 28 mm–16 cm. Fruit hairless and shiny, 80 mm–11 cm. Maquis and naturalised from cultivation. ES, IT (Sardinia), MA, PT.

Lathyrus angulatus A small, clambering, grey-green annual with winged stems and *long, paired, linear leaflets. Flowers small and pale blue or purple*, 9–14 mm, borne on flower stalks 25–85 mm, much longer than their leaves, and projecting in a point beyond the point of flower attachment. Fruit 25–45 mm, scarcely veined. Sandy and rocky habitats. Absent from many islands, the far south, and much of Italy. ES, FR, IT, MA, PT.

D. Flowers red, purple, pink or white. *Leaves with 4–12 leaflets; stems broadly winged.*

Lathyrus clymenum A medium to tall, scrambling, hairless annual to 1 m with *broadly winged stems and leaves with winged, leaf-like stalks;* leaves with 4–8 pairs of narrower, *linear-elliptic* leaflets, 0.5–14 mm wide. Flowers variably pink, dull yellow or crimson, 12–25 mm, borne in racemes of 1–2(3). Fruit *grooved* on the upper side. Roadsides, hedges and tracks. Throughout, though rarer in the east. Some authors recognise forms with *very narrow leaflets* (<5 mm), bicoloured flowers with white or pink wings and ungrooved pods as *L. articulatus*.

Pisum PEA

P. sativum

Herbs with unwinged stems, *large, leafy stipules* (larger than the leaflets) and branched tendrils. Flowers with 10 stamens, 9 in a tube, 1 free. Fruit a pod-like legume.

Pisum sativum WILD PEA A variable, medium-tall, clambering, hairless annual to 2 m (often less), stems *not winged*. Leaves with 1–3 pairs of rounded to elliptic leaflets, more or less heart-shaped at the base, tendrils branched; stipules large and leafy, often blotched. Flowers white to purple, 15–35 mm, borne in 1–3-flowered racemes, wing *petals fused to the keel*. Fruit hairless, net-veined, 30 mm–12 cm. Throughout.

1. *Ononis reclinata*   4. *Ononis natrix* subsp. *natrix*
2. *Ononis spinosa*   5. *Ononis natrix* subsp. *ramosissima*
3. *Ononis mitissima*

Ononis RESTHARROW

Herbs or subshrubs with simple or trifoliate leaves (often both), the veins of the leaflets ending in marginal teeth; tendrils absent. Flowers with calyx deeply toothed with nearly equal teeth; all 10 stamens forming a tube. Fruit straight, 1–many-seeded, splitting. Many closely related species occur in the region, of which just a few are decribed here.

A. Flowers white, pink or purple.

O. reclinata

O. spinosa

O. mitissima

Ononis reclinata A low, slender, *spreading* annual with *shaggy-hairy* stems <15 cm. Flowers *small* (just 5–10 mm), pink and solitary forming loose, leafy inflorescences; corolla equalling or shorter than the calyx; calyx with *entire teeth*. Fruit 8–14 mm with *10–20 seeds*. Widespread in coastal habitats. Throughout. Subsp. *dentata* (also treated as a true species, *O. dentata*) has small flowers 5–8 mm with 1–3-*toothed or lobed calyx teeth*. Fruit 6–10 mm with 9 seeds. ES, IT (Sardinia and Sicily), PT.

Ononis spinosa SPINY RESTHARROW A *very spiny* dwarf-shrub to 70(80) cm, with 2 opposite rows of hairs on young stems. Leaves trifoliate, or simple above. Flowers large, pink, 10–20 mm, usually borne singly at each node, forming a loose inflorescence. Fruit hairy, 6–10 mm, usually with a single (2–4), warty seed(s). Dry, rocky and waste places and grassland, very widespread. Throughout.

Ononis diffusa A short to medium, *very sticky, glandular-hairy* annual to 40(60) cm. Leaves trifoliate, with *toothed* margins. Flowers small, 8–14 mm, the corolla exceeding the calyx; pink with a whitish keel, borne singly at each node in a *dense, oblong spike*. Fruit 4.8–5.5 mm with 3–4 seeds, slightly curved. Coastal habitats. ES, PT. *Ononis serrata* is similar but shorter, to 30 cm, with a smaller corolla, just 6–8 mm long, not exceeding its calyx. Strongly southern in distribution. DZ, ES, MA, TN. ***Ononis mitissima*** is similar to *O. diffusa* but is a rather taller and more erect, scarcely hairy annual, to 60 cm, with trifoliate bracts with conspicuous white stipules. Flowers pink and exceeding their calyces, 10–12 mm. Fruit 5–6 mm with 2–3 seeds. Throughout.

Ononis alopecuroides A short to medium annual, distinguished by its *large leaves with a single leaflet, and winged stalk*. Flowers pale pink and white, rather large, 16–19 mm long, borne in very dense, hairy terminal racemes; all bracts with a single leaflet. Fruit to 10 mm long. Various habitats. Widespread, absent from the Balearic Islands. ES, IT.

B. Flowers yellow.

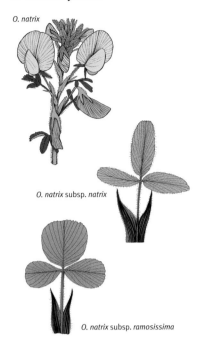

O. natrix

O. natrix subsp. *natrix*

O. natrix subsp. *ramosissima*

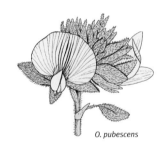

O. pubescens

1. *Ononis minutissima*
2. *Ononis pubescens*
3. *Ononis pusilla*

Ononis natrix LARGE YELLOW RESTHARROW much-branched, green, *sticky* subshrub to 40(60) cm, with stems woody below, and densely leafy. Leaves trifoliate with oval, toothed leaflets. Flowers yellow, the standard petal veined with red externally, (6)11–25 mm, *solitary* but borne in loose, leafy clusters. Fruit 11–25 mm, pendulous and hairy, with 2–27 seeds. Common and widespread across the region, especially near the coast. Throughout. The following are regarded by some as distinct species, but seem to be continuously variable: Subsp. *natrix* has *long sticky stem hairs* and calyx teeth much longer than their tubes; corolla 11–25 mm. Fruit 13–25 mm with 3–27 seeds. Throughout. Subsp. *ramosissima* is often compact and has very *short glandular stem hairs*, corolla 9–16(18) mm, and calyx teeth *just* exceeding the length of their tube. Fruit 11–22 mm with 2–8 seeds. ES, PT.

Ononis viscosa A variable species, similar to *O. natrix* but a *taller*, softly hairy annual to 1.3 m with flowers 5–16 mm with a pinkish, glandular-hairy standard, and *3-veined* calyx teeth. Fruit 10–25 mm with 3–10 seeds. Throughout. Subsp. *sicula* (also treated as a true species *O. sicula*) is short, to 15(25) cm tall with glandular-hairy flower-stalks that lack long spreading hairs; corolla 5–10 mm. Fruit 9–14 mm with 10–20 seeds. Dry habitats and cultivated land. Throughout. ***Ononis pubescens*** is a *short* annual to 75 cm, similar to *O. viscosa* and also extremely sticky, often with some leaves 1-foliate. Flowers larger, 12–22 mm, yellow, the standard red-veined, the calyx teeth *5-veined with broad, oblong-lanceolate teeth*. Fruit 8–12 mm with just 2–3 seeds. Scattered throughout, but absent from many islands.

1

2

3

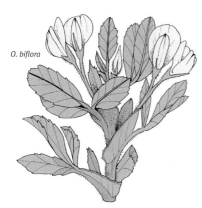

O. biflora

O. pusilla

O. variegata

Ononis ornithopodioides BIRD'S-FOOT REST HARROW A low to short, erect, much-branched annual with glandular-hairy stems to 30 cm. Leaves trifoliate with oval leaflets, broadest above the middle. Flowers yellow (occasionally with pink veins), 7–8.5 mm long, borne in sparse, leafy panicles which elongate in fruit. Fruit linear, 12–20 mm long and *constricted between the seeds;* seeds 7–10. Garrigue and pine woods. Throughout. *Ononis biflora* is a similar, erect annual to 50 cm distinguished from the previous species by its somewhat fleshy, sparsely hairy leaves, and 1–2-*flowered* primary branches of the inflorescence (not always 1-flowered), and pink-veined yellow, *hairless* flowers 13–15 mm long. Fruit 17–25 mm. Southern in distribution. DZ, ES, MA, PT, TN. *Ononis maweana* (Syn. *O. hackelii*) is very similar to *O. biflora* but with flowers 7.5–13 mm with a white, pink or purple and glandular-hairy standard, and yellow wings and keel. Fruit 7–10 mm. Sandy soils in southern Portugal only. PT.

Ononis pusilla A shortly hairy, sometimes rather straggling perennial with stems woody below to 35(55) cm. Leaves trifoliate with elliptic to rounded leaflets, often notched. Flowers yellow, 6–12 mm long, borne in lax spikes with leafy bracts; *calyx lobes hairy, long, equalling the corolla, and later opening to become star-like.* Fruit small, 4.5–10.5 mm long with 3–10 seeds. On rocks, scree and in pine woods. Throughout, though absent from the Balearic Islands. *Ononis minutissima* is very similar but with rooting, hairless stems to 40 cm and calyx lobes *hairless or scarcely hairy*; corolla 7–14 mm. Fruit 5–8 mm with 1–3 seeds. Local from Spain eastwards, fairly common in southern France. ES, FR, IT, TN (rare).

Ononis variegata A *low, spreading, mat-forming annual* to 40 cm, branched nearly from the base; densely hairy with *bluish green*, toothed leaves with distinct veins. Flowers yellow, 10–15 mm long, borne in lax, terminal racemes. Fruit oblong and slightly hairy, 8–11 mm long. Local on sand dunes. Throughout, though absent from some islands.

Melilotus MELILOT

Annuals or biennials with trifoliate leaves often with toothed leaflets. Flowers small and sweet-smelling, borne in elongated racemes; stamens 10, 9 in a tube, 1 free. Fruit straight, 1–2-seeded, (normally) non-splitting.

A. Tall biennial with yellow flowers; fruit hairy.

Melilotus altissimus TALL MELILOT A tall, branched biennial to 1.5 m with oblong leaflets, the uppermost almost parallel-sided, toothed, and with bristle-like stipules. Flowers yellow, 50–70 mm long, *the standard and keel equal*. Fruit net-veined and black, 5–7 mm, *hairy*. Various disturbed habitats. ES, FR, IT.

B. Short to medium annuals with yellow flowers; fruit hairless.

Melilotus indicus SMALL MELILOT An erect, branched or simple annual to 40 cm with trifoliate leaves with toothed leaflets and more or less untoothed stipules. *Flowers tiny, 2–3.5 mm*, pale yellow to whitish and borne in *dense, many-flowered spikes*; wings and keel equal, and shorter than the standard. Fruit more or less spherical, pale brown when ripe, 1.5–3 mm and hairless. Common on disturbed ground and on damp, sandy soils. Throughout. *Melilotus italicus* is similar to *M. indica* but to 80 cm, with larger flowers 7–8 mm long, and fruit 4–5 mm, wrinkled and with small depressions when ripe. Dry open habitats. Widespread though local. Throughout.

M. elegans

Melilotus elegans An annual to 2 m similar to *M. indicus* with a stem hairy above, and small flowers 4.5–5 mm with the standard and wing petals *shorter* than the keel, borne in 20–35-flowered racemes. Fruit ovoid, compressed and with very prominent transverse veins, 3–4.5 mm. Scattered throughout. *Melilotus segetalis* is similar to *M. elegans* but with *racemes (equalling or exceeding their leaves) of 15–100 larger flowers 5–7 mm*; standard shorter than the keel. Fruit *yellow, egg-shaped* and with concentric grooves; 4–5 mm. In damp habitats. Throughout. *Melilotus sulcatus* FURROWED MELILOT is very similar to *M. segetalis* but with small flowers 2.8–4.5 mm, and a more or less *stalkless*; *inflorescence shorter than its leaf*. Fruit globose, 3–4.5 mm. Throughout.

Melilotus messanensis (Syn. *M. siculus*) An annual to 60 cm similar to *M. segetalis* with yellow flowers 4–5 mm with more or less equal standard and keel that exceed the wings, borne on erect to sub-erect stalks, in 4–10(14)-flowered racemes much *shorter* than the subtending leaves. Fruit yellowish with concentric grooves, 7–8(9) mm. Cultivated and damp maritime places. Throughout.

M. messanensis

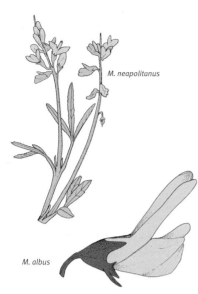

M. neapolitanus

M. albus

Melilotus neapolitanus (Syn. *M. spicatus*) A short annual 15–40(80) cm tall, hairy above with yellow flowers 4–5 mm with equal wings and keel, borne in *short, lax racemes* of 6–16(20) flowers. Fruit small, 3–4 mm, globose, not flattened and with a short beak. In dry, sandy places. Throughout.

C. Flowers white.

Melilotus albus WHITE MELILOT A rather vigorous annual to 1.5 m, similar to other *Melilotus* spp. and distinguished by its *white flowers*; corolla 4–5 mm. Fruit 3–5 mm, ridged, brown and hairless. common in a range of dry, disturbed habitats. Throughout.

1. *Melilotus altissimus*
2. *Melilotus indicus* (fruits)
3. *Melilotus indicus*
4. *Melilotus albus*
5. *Trigonella stellata*

Trigonella FENUGREEK

Annuals with trifoliate, toothed leaves. Flowers borne solitary in leaf axils or in lateral racemes; stamens 10, 9 in a tube and 1 free. Fruit straight or curved with 1–many seeds in 2 rows; eventually splitting.

A. Flowers yellow and borne numerously (to 15) in elongated clusters.

Trigonella corniculata SICKLE-FRUITED FENUGREEK *An erect or spreading, more or less hairless* annual to 50 cm, superficially similar to *Melilotus*. Leaves with 3 oval leaflets. Flowers yellow, 5–7 mm borne in *long-stalked cylindrical racemes* of 8–15; *calyx teeth unequal*. Fruit *linear and pendent, slightly curved*, 10–18 mm (without beak). Various dry, grassy habitats. ES, FR, IT, MA. *Trigonella maritima* is similar but with fewer flowers per cluster (5–10), borne in rather rounded racemes; calyx teeth equal. Fruit with *thickened, prominent veins*. Local. MA, IT (incl. Sicily), MT.

Trigonella stellata A prostrate, sprawling annual. Leaflets narrowed at the base, toothed and notched at the tips. Flowers borne in dense clusters of 3–15, borne congested at the base of the plant and in the leaf axils along the stems. Fruit 5–8 mm long, cylindrical, divergently spreading. Bare, arid and desert fringe habitats in North Africa. DZ, MA, TN.

B. Flowers borne solitary or in pairs in the leaf axils; whitish.

Trigonella foenum-graecum CLASSICAL FENUGREEK A short to medium more or less hairy annual to 50 cm. Leaves trifoliate, oblong, toothed near the tip. Flowers cream, flushed with purple at the base, 12–18 mm long, solitary or paired in the axils of upper leaves. Fruit linear and held erect, 50 mm–10 cm long plus a long beak (10–35 mm), hairless, with longitudinal veins. Cultivated and sometimes naturalised. Throughout.

Medicago MEDICK

Annual or perennial herbs typically with trifoliate leaves. Flowers yellow (sometimes purple) borne in small lateral clusters; stamens 10, 9 in a tube and 1 free. Fruit highly variable (often curved, spiralled or spiny), 1–many-seeded; important for identification. Only common and widespread species are described.

M. lupulina

A. Fruit sickle- or bean-shaped.

Medicago lupulina BLACK MEDICK A spreading, hairy annual to 80 cm with leaves with rounded to diamond-shaped trifoliate leaflets, often notched at the tips. Stipules toothed or entire. Flowers small, 2–3 mm, yellow, and borne in ball-like, dense clusters of 20(50). Fruit 1.5–3 mm across, coiled, black when ripe and net-veined. Common in waste places. Thoughout.

Medicago monspeliaca (Syn. *Trigonella monspeliaca*) STAR-FRUITED FENUGREEK A prostrate or spreading, finely hairy annual to 20(40) cm with trifoliate leaves with oval, toothed or untoothed leaflets. Flowers small and yellow, borne 4–12(15) in stalkless clusters. Fruits 8–13 mm, borne in *spreading, star-like clusters* (similar to *Trigonella* with which it was previously grouped). Throughout.

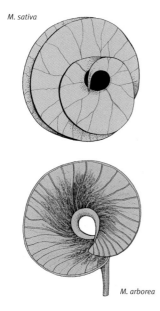

M. sativa

M. arborea

B. Fruit spineless, with a spiral of 1 or more turns.

Medicago sativa LUCERNE A variable (with many forms described), hairy perennial to 90 cm. Leaflets oblong to linear, toothed at the apex. Stipules toothed at the base. Flowers 5–12 mm, violet or blue, borne in rounded, dense racemes of 10–30(50) flowers. Fruit 5–9 mm across, spiralled with 2–3(4) turns and a hole through the centre, not spiny. Common on waste places and on roadsides. Thoughout.

Medicago arborea TREE MEDICK A silvery-grey leafy *shrub to 2 m* with silky-white younger branches. Leaflets narrowed at the base, slightly toothed at the apex. Stipules untoothed. Flowers yellow, 9–13 mm, borne in dense racemes of 8–20. Fruit 9–15 mm across, slender and coiled with 1 turn, hairless and net-veined. In disturbed habitats and gardens. Centre of region; naturalised elsewhere. ES, FR, IT, PT.

1. *Medicago lupulina*
2. *Medicago sativa*
3. *Medicago marina* (fruiting plant)
4. *Medicago marina*
5. *Medicago arborea*

M. orbicularis

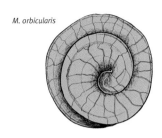

1. *Medicago orbicularis* (fruits)
2. *Medicago orbicularis*
3. *Medicago arabica*
4. *Medicago laciniata* (fruit)

Medicago orbicularis LARGE DISK MEDICK A low to short, hairless or slightly hairy annual 30–60 cm. Leaflets oval-wedge-shaped, toothed at the apex. Stipules deeply toothed. Flowers 5 mm, yellow, in racemes of 2–4(5). Fruit *large and disk-like*, 10–20 mm across, smooth and spiralled anticlockwise in 4–7 turns, without a central hole. Frequent in waste places. Thoughout.

Medicago blancheana A vigorous, spreading, slightly hairy annual 20–30 cm with deeply toothed leaflets. Flowers yellow, 6.5 mm long. Fruit 10 mm across, rounded, with 4–6 anticlockwise spirals, almost always without spines, the *spirals alternatively clockwise and anticlockwise*. Cultivated land. Thoughout.

C. Fruit normally with spines or projections, with a spiral of 1 or more turns (many very similar closely related and hybridising species; close attention to the fruits is required for accurate identification).

M. arabica

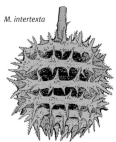

M. intertexta

M. marina

M. littoralis

M. rigidula

Medicago arabica SPOTTED MEDICK A low, prostrate and spreading more or less hairless annual to 60 cm with trifoliate leaves. Leaflets notched and with a *conspicuous dark spot*. Flowers yellow, 4–6 mm borne in racemes of 1–5. Fruit cylindrical, 4.5–5 mm across, hairless, and with 3–5(6) anticlockwise turns with stout *spreading spines* 2–3.5 mm and a margin with 3 grooves. Common on fallow land and in woods and pine forests. Throughout.

Medicago intertexta (Syn. *M. ciliaris*) A low to short annual 15–70 cm with spreading stems. Leaflets sometimes with a dark spot and finely toothed. Stipules toothed. Flowers dull yellow, 5–8 mm borne in clusters of 1–4(7). Fruit large, 10–17 mm across, *glandular-hairy to hairless pods, spiny*, with 5–8(10) anticlockwise spirals. Damp waste places. (*M. ciliaris* which has hairy fruits is now believed to be indistinct). Throughout.

Medicago marina SEA MEDICK A creeping, *prostrate, white-downy* perennial to 50 cm. Leaflets oval, toothed at the apex; stipules toothed or not. Flowers bright yellow, 7–9 mm in short clusters of 5–12(15). Fruit 4–6(7) mm across with 2–4 anticlockwise spirals and a hole in the centre, white-downy with 2 rows of spines to 1.5 mm (sometimes reduced to warts). Common on coastal sands and dunes. Throughout.

Medicago littoralis A low to short, spreading, hairy annual with purplish stems to 80 cm. Leaflets oval to heart-shaped, toothed at the apex and *hairy on both surfaces*. Stipules toothed. Flowers yellow to orange, 5–7 mm long in clusters of 2–5(8). Fruit 3.5–5(6) mm long, cylindrical with 3–7 turns, shiny, spiny or not, and hairless, with a groove along the margin. Coastal habitats. Throughout. *Medicago rigidula* has larger, spiny fruits 7–8(8) mm long with 4–7 tight anticlockwise spirals, spines 1.5–3 mm, with a netted glandular-hairy surface throughout. Fallow land. Throughout. *Medicago doliata* (Syn. *M. aculeata*) has solitary, spiny (rarely spineless), tightly spiralled fruits 7–12 mm across with *hairs along the margins*, hairless otherwise. Throughout.

M. polymorpha

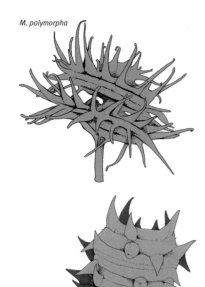

M. truncatula

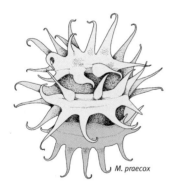

M. praecox

M. minima

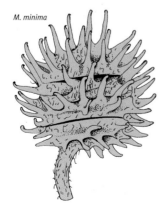

Medicago polymorpha HAIRY MEDICK A variable prostrate to spreading, hairy or hairless annual with purplish stems to 50 cm and trifoliate leaves, *toothed at the apex.* Stipules *deeply toothed with slender teeth.* Flowers yellow, 6–6.5 mm long, borne in clusters of 2–9. Fruit 6–7 mm across, flattened with 4–6 anticlockwise turns, with 2 rows of *gooved spines* 0.5–0.8 mm, the surface netted (*not shiny*) and the *apical spiral larger than the rest.* Common on roadsides, grassy places and cultivated fallow land. Throughout. *Medicago italica* (Syn. *M. tornata*) is similar to *M. polymorpha* but with fruits 5–8 mm with 2 (1–6) spirals and *spines 2 mm, not grooved*; margins of the spirals sharply defined. Throughout. *Medicago truncatula* has fruits 7–10 mm, with *much-thickened spines, swollen at the base,* 1.3–5 mm, and margins of the spirals thickened. Fields and grassy habitats. Throughout. **Medicago laciniata** is similar to *M. polymorpha* but with leaflets *deeply toothed or lobed,* and flower stalks rather longer than their leaves' petioles. Fruit 6–8 mm, spiny, hairless, with 4–6 turns. Native to North Africa and naturalised further north. DZ, ES, FR, IT, MA, TN. *Medicago praecox* has less clearly toothed leaflets and flower stalks much shorter than their leaves' petioles. Fruit 4–5 mm, spiny, with 3 turns. Rare and local, but widespread. Probably throughout.

Medicago minima SMALL MEDICK A prostrate to spreading, densely hairy annual, similar to the *M. polymorpha* group, but with *stipules entire or slightly toothed, the teeth if present short.* Flower stalks as long or a little longer than their leaves' petioles; flowers 2.5–4.5 mm. Fruits borne in clusters of 4–6, ball-like, spiny, 2.5–3.5(5) mm across (rather large relative to the plant) with 3–5 lax, sharp-edged spirals and spines to 1 mm that are weakly hooked; not distinctly flattened, and often rather shiny (less netted) on the surface. Similar habitats. Throughout, except for some islands. *Medicago coronata* is similar but with *flower stalks greatly exceeding their leaves' petioles* and 5–13 fruits with *broad-margined spirals with spines in 2 rows clearly pointing in opposing directions*; borne in dense clusters. Throughout.

1. *Medicago littoralis*
2. *Medicago polymorpha* (fruits)
3. *Medicago minima*
4. *Medicago polymorpha*

M. turbinata

M. murex

D. Fruit with spines or projections, with a spiral of 1 or more, *greatly thickened* turns, appearing turban-like.

Medicago turbinata A robust, prostrate or ascending, *densely hairy* annual to 50 cm with reddish stems and toothed, trifoliate leaves. Flowers yellow-orange, 6–6.5 mm long, borne in clusters of 2–9. Fruit barrel-shaped, 6–7.5 mm with 4–6 clockwise or anticlockwise turns, a single groove along the margin, and *short, broad spines* just <1 mm long. Various, often disturbed habitats. Throughout. *Medicago murex* is rather similar but hairless (except sometimes leaf undersides), with fruits with 5–9 anticlockwise turns, with 3 grooves along the margin, and *spines thick, thorn-like and pointed, 1–8 mm*. Similar habitats. Throughout.

Lotus BIRD'S-FOOT TREFOIL

Herbs with leaves with 5 leaflets (the lower 2 resembling stipules). Flowers often yellow, borne in flat-topped clusters; stamens 10, 1 free. Fruit several-seeded, splitting, smooth, or keeled and square in cross-section.

L. corniculatus (fruit)

L. corniculatus (flowers)

A. Plant a mid-green or blue-green, sometimes woody-based perennial. Fruit elongated, not winged or keeled.

Lotus corniculatus COMMON BIRD'S-FOOT TREFOIL A very variable (with many forms described), sprawling or creeping, almost hairless perennial to 50 cm with a woody stock and solid stems, and pinnately divided leaves that appear trifoliate because the lowermost pair is at the base of the stalk; leaflets oval, lanceolate or round. Flower-heads with 2–8 yellow-orange flowers 10–16 mm (red in bud) on long stalks to 80 mm long. Fruit 15–30 mm long. Grassy places. Throughout. *Lotus pedunculatus* (Syn. *L. uliginosus*) is similar but a vigorous, leafy perennial to 1 m, with *hollow and creeping, rooting stems*. Flowers 10–18 mm. Fruit 15–35 mm. Wet, marshy habitats. Throughout.

Lotus tetraphyllus A small perennial to 20(30) cm with *leaves with 3–4 leaflets* (rarely 5); leaflets blue-green and triangular. Flowers bright yellow, 6.5–9.5 mm lined with red, and *solitary*, held on erect stalks above the leaves. Fruit 15–25 mm. Rare, in limestone crevices and on rocks; an island endemic. ES (Balearic Islands).

B. Plant a *silvery-white-hairy* perennial. Fruit elongated, not winged or keeled.

Lotus creticus SOUTHERN BIRD'S-FOOT TREFOIL A variable (with many forms described), low, spreading, *grey-silver-hairy perennial* with stems to 1.5 m and leaves with oblong leaflets. Flowers 12–18 mm long yellow with a purple-tipped keel, borne in long-stalked racemes; calyx distinctly 2-lipped. Fruit straight or curved, many-seeded and 20–40 mm. Common in sandy maritime habitats; absent from some islands. ES, FR, IT, MA, PT. *Lotus cytisoides* is similar but less silvery and with small flowers 8–14 mm long, with a short *curved keel* and notched standard petal. Widespread. DZ, ES, FR, IT, MA.

C. Plant a herbaceous (non-woody) annual. Fruit elongated, not winged or keeled.

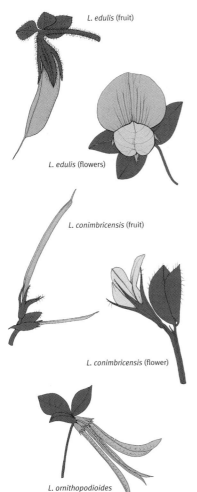

L. edulis (fruit)

L. edulis (flowers)

L. conimbricensis (fruit)

L. conimbricensis (flower)

L. ornithopodioides

Lotus edulis EDIBLE LOTUS A low to short, spreading, slightly hairy annual to 50 cm with oval leaflets. Flowers yellow with a purple-tipped keel, 10–16 mm long; calyx bell-shaped. Fruit oblong and curved, *very inflated*, with a groove along the back, 15–35 mm long. Garrigue and roadsides. Throughout.

Lotus conimbricensis A more or less hairless annual (or with sparse, long, white hairs) to 30 cm with oval to diamond-shaped leaflets. Flower stalks shorter than the leaves, with solitary flowers; corolla 4.5–7 mm, *white to pink* with a violet keel. Fruit slender, 25–40 mm and curved upwards. Dry grassy places and dunes. Absent from some islands and much of the east of the region. ES, FR, IT, MA, PT.

Lotus ornithopodioides A low to short, spreading, hairy annual to 50 cm with stalked leaves with oval to diamond-shaped leaflets, the lower 2 triangular and heart-shaped at the base. Flowers yellow, 6–8 mm long, borne in *stalked, terminal heads of 2–5 on stalks slightly longer than the leaves* (at least in fruit); calyx 2-lipped; *teeth very unequal*. Fruit 20–40 mm long, *constricted* between the seeds. Damp habitats near water. Throughout. *Lotus angustissimus* SLENDER BIRD'S FOOT TREFOIL is similar but with flowers in heads of 1–2(3) on stalks as long, or longer than the leaves (10–40 mm) and *calyx teeth nearly equal*. Fruit 15–25 mm, *much longer than calyx*. Throughout. *Lotus halophilus* is a low, whitish, prostrate annual similar to *L. ornithopodioides* but with *scarcely stalked or unstalked leaves*. Maritime sands. IT.

1. *Lotus corniculatus*
2. *Lotus corniculatus* (fruits)
3. *Lotus glinoides*
4. *Lotus tetraphyllus*
5. *Lotus creticus* (inset: fruits)

Lotus glinoides A prostrate, spreading annual with rather thick, blue-grey, trifoliate leaves with sparse to dense white hairs; leaflets broadest above the middle. Calyx with long, sparse, white hairs; flowers pink. Fruit narrowly cylindrical, reddish and shiny, often borne in divergently spreading pairs. Arid and bare ground, common in North African desert fringe habitats. DZ, MA.

D. Plant a herbaceous annual or perennial. Fruit *square* in cross-section, with a *keel or wing* along the angles.

1. *Lotus tetragonolobus*
2. *Ornithopus compressus*
3. *Tripodion tetraphyllum*
4. *Ornithopus sativus*
5. *Galega officinalis*

Lotus tetragonolobus (Syn. *Tetragonolobus purpureus*) ASPARAGUS PEA A low, sprawling to ascending, hairy annual to 60 cm. Leaflets 10–40 mm, oval to diamond-shaped and broadest above the middle. Flowers *bright dark red*, 18–23 mm long, solitary or paired on stalks shorter than or equalling the leaves. Fruit 35–75 mm, hairless, with broad wings along the angles. Roadsides, fallow land and garrigue. Absent from the far west and Corsica. DZ, ES, FR, MA, TN.

Lotus maritimus (Syn. *Tetragonolobus maritimus*) DRAGON'S TEETH A low, spreading, prostrate, glaucus perennial to 45 cm, often forming mats, variably hairy. Leaflets 10–28 mm oval and broadest above the middle; stipules oval. Flowers solitary, on long stalks (to 13 cm), *pale yellow*, 22–28 mm long, each with a leafy trifoliate bract. Fruit 30–60 mm, mostly hairless, narrowly winged on the angles. Damp fields, marshes and grassy habitats; local and uncommon. Widespread, virtually throughout. *Lotus biflorus* (Syn. *Tetragonolobus biflorus*) is a similar annual with flowers borne 1–4, *bright orange*, 17–25 mm long. Fruit 20–40 mm. IT (incl. Sicily).

Tripodion

Prostrate hairy annuals. *Calyx distinctly swollen* with sub-equal teeth; stamens 10, 1 free. Fruit non-splitting.

Tripodion tetraphyllum (Syn. *Anthyllis tetraphylla*) BLADDER VETCH A low, spreading, hairy annual to 40 cm. Leaves normally with 3–5 leaflets, the terminal lobe largest and oval. Flowers to 25 mm, yellow with darker wings and a red-tipped keel, borne in dense lateral clusters of 1–7; calyx with silvery hairs, becoming very inflated and bladder-like in fruit, pale green tipped with red. Fruit 8–10 mm. Very common in a range of dry and disturbed habitats. Throughout.

Ornithopus BIRD'S FOOT

Annual, erect or prostrate herbs with pinnately divided leaves with numerous leaflets. Flowers with 10 stamens, 9 in a tube, 1 free. Fruits *slender, often curved* (falcate).

Ornithopus pinnatus ORANGE BIRD'S FOOT A spreading, prostrate, slightly hairy annual. Leaves *pinnately lobed* with 6–14 of leaflets. Flowers *orange-yellow*, 4.5–5.5 mm, borne in heads of 2–5(7) without a bract below. Fruit linear, 16–33 mm long, flattened and markedly constricted between the seeds. Grassy and bare, stony habitats. Throughout. *Ornithopus perpusillus* has pinnately lobed leaves with 4–14 leaflets and a terminal leaflet. Flower-heads with a pinnately divided bract below, corolla 3.5–5.5 mm, *cream* to white with red veins. Fruit *curved and spreading*, in clusters of usually 2–3, and constricted between the seeds, 13–30 mm. Higher elevation habitats. ES, PT.

Ornithopus compressus is a *low, prostrate to ascending*, very hairy annual to 75 cm with pinnately divided leaves with 2–8, or *up to 18*, pairs of leaflets. Flowers 5–7 mm, yellow, borne in heads of 1–5, *subtended by a leafy bract*. Fruit slender, flattened and curved, slightly constricted between the seeds, 18–42 mm. Locally common in grassy places and on sand dunes. Throughout. *Ornithopus sativus* is similar, larger, ascending and with *pink-white* flowers 6.5–10 mm and fruits constricted between the seeds, 15–43 mm. Ruderal and grassy habitats. Absent from the east of the region. DZ, ES, FR, MA, PT.

Galega

Herbs with simple, pinnately divided leaves and prominent stipules. Flowers borne in erect clusters; stamens 10, 1 fused to the other 9. Fruit splitting, not inflated.

Galega officinalis GOAT'S RUE A vigorous, erect, hairless or virtually hairless perennial to 1.5 m with pinnately divided leaves with 9–17 narrowly oval leaflets. Flowers white, sometimes flushed blue-mauve, borne in erect, long-stalked, cylindrical racemes; calyx teeth bristle-like. Fruit cylindrical, not inflated, to 30 mm long. Ditches and damp habitats, rather local. ES, FR, IT. *Galega africana* is similar but has *hairy inflorescences*. Damp coastal habitats in southern Spain and northwest Africa. ES, MA.

Trifolium CLOVER

Annuals or herbaceous perennials with trifoliate leaves. Flowers numerous short-stalked, typically clustered into rounded, congested heads; stamens 10, 9 in a tube and 1 free. Fruit 1–9-seeded, often enclosed in calyx. Species numerous and only the most common are described here; calyx tube and vein number are important.

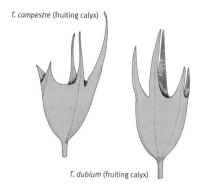

T. campestre (fruiting calyx)

T. dubium (fruiting calyx)

A. Flowers yellow.

Trifolium campestre HOP TREFOIL An erect or spreading *hairy* annual to 30 cm. Leaves alternate, the terminal leaflet longer-stalked than the laterals. Flowers, *yellow*, corolla 4–7 mm long, borne in *rounded heads of >20, 8–15 mm across*. Common in a range of habitats in the region. ES, FR, IT, MA, PT. ***Trifolium dubium*** LESSER HOP TREFOIL is similar but smaller, virtually hairless with flower-heads of 5–20, 5–9 mm across; corolla just 3–4 mm. ES, FR, IT, MA, PT.

B. Flowers typically whitish or pink (or red), borne in *elongated heads*.

T. angustifolium (fruiting calyx)

Trifolium angustifolium NARROW-LEAVED CRIMSON CLOVER A short to medium, somewhat hairy annual to 50 cm. Leaves with *linear-lanceolate*, pointed leaflets. Flowers pink, 10–13 mm, borne in stalked, *cylindrical* heads, opening all at about the same time corolla not (or scarcely) exceeding the calyx; calyx lobes unequal (lowest the longest). Common in the far west in a range of disturbed or sandy habitats. ES, FR, IT, MA, PT. ***Trifolium purpureum*** PURPLE CLOVER is similar but with *purple flowers opening from the base upwards,* the lowermost withered when the uppermost open. FR, IT.

Trifolium incarnatum CRIMSON CLOVER A robust, medium, hairy annual to 65 cm with branched, erect or ascending stems. Leaflets oval to rounded, toothed towards the apex; stipules oval. Flowers 9–16 mm, very conspicuous, *bright blood-red*, to 12 mm long, borne in dense cylindrical heads; calyx 10-veined. Grassy habitats, sometimes abundant though rather local. Throughout though absent from some islands.

T. arvense (fruiting calyx)

Trifolium arvense HARE'S-FOOT CLOVER A softly hairy annual to 40 cm with narrow leaflets, scarcely toothed. Flowers pale pink or white, *borne in dense, elongated ovoid or cylindrical, stalked heads* to 25 mm long, the flowers often exceeded by the calyx teeth; corolla just 3–6 mm; calyx lobes more or less equal. Common in sandy or disturbed areas, sometimes in large numbers. Throughout.

1. *Trifolium campestre*	4. *Trifolium purpureum*
2. *Trifolium dubium*	5. *Trifolium arvense*
3. *Trifolium angustifolium*	6. *Trifolium scabrum*

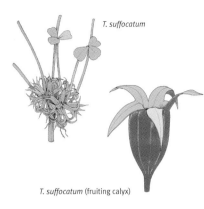

T. suffocatum

T. suffocatum (fruiting calyx)

C. Flowers typically whitish or pinkish, borne in *stalkless, rounded heads* in the leaf axils, often on spreading or prostrate stems.

Trifolium suffocatum SUFFOCATED CLOVER A distinctive *prostrate*, tufted, low annual <50 mm high. Leaves with oval leaflets over-topping the *unstalked* rounded flower-heads which are congested at the base of the plant. Flower-heads terminal and densely crowded; corolla white, 3–4 mm and shorter than the calyx; calyx tube nearly cylindrical. Local on coastal sands and bare ground. ES, FR, IT, MA, PT.

T. scabrum (fruiting calyx)

T. tomentosum (fruiting calyx)

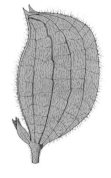

Trifolium scabrum ROUGH CLOVER A spreading, often *prostrate* and *rather downy* annual to 20 cm with *leaves with prominent paler veins* which are thickened at the leaf margins. Flowers 4–7 mm, rather inconspicuous, cream-white, turning pink with age, borne in small, unstalked heads to 10 mm across; calyx soon becomes stiff after flowering, with *starry, spreading, recurved, spiny teeth*. Various disturbed habitats, towns and gardens. ES, FR, IT, MA, PT.

Trifolium cherleri A spreading, *densely, long-hairy* annual. Flowers white or pink, the corolla *equalling or shorter* than the calyx; *flower-heads densely woolly-hairy*. ES, FR, IT, MA, PT. *Trifolium lappaceum* is similar (also similar to *T. hirtum* but see calyx), with rather spiky, greenish flower heads; *calyx virtually hairless* (never with long hairs), 20-veined; flower-heads short-stalked; corolla 4–8 mm. Throughout.

Trifolium tomentosum WOOLLY TREFOIL A slender, creeping annual to 15 cm, rather similar to *T. resupinatum,* but with leaflets 4–12 mm and smaller flowers 3–6 mm long and *inflated, spherical fruiting heads clothed in soft, white hairs*, short-stalked or stalkless (calyx teeth obscured). ES, FR, IT, MA, PT.

1. *Trifolium cherleri*
2. *Trifolium tomentosum*
3. *Trifolium repens*

1

2

3

D. Flowers typically whitish or pink (or yellowish), borne in *distinctly stalked, rounded heads* terminally or in the leaf axils.

T. isthmocarpum

T. repens (fruiting calyx)

Trifolium isthmocarpum A low to medium, spreading or ascending hairless annual to 75 cm. Leaves trifoliate with oval to elliptic leaflets; stipules membranous and pointed. Flowers pink or more rarely white, 8–12 mm long, borne in *large* cylindrical or hemispherical heads to 25 mm across, with stalks exceeding the leaves. Fruit oblong and compressed between the seeds. ES, FR, IT, MA, PT.

Trifolium repens WHITE CLOVER A variable creeping perennial to 50 cm with stems rooting at the nodes, often hairless. Leaves trifoliate with green, oval to elliptic leaflets that are paler along the veins. Flowers normally white, sometimes pink or reddish, 7–12 mm long, borne in dense, long-stalked globose heads; calyx tube longer than wide, the lobes triangular-lanceolate; flowers scented. Fruit linear and compressed between the seeds. Cultivated for fodder, and common in grassy places. Throughout. *Trifolium nigrescens* is superficially similar, low to short, usually hairless and with oval to heart-shaped leaflets, notched at the tips; stipules triangular. Flowers 5–9 mm, white, cream or pink, turning brownish, borne in rounded heads on stalks exceeding the leaves. ES, FR, IT, MA, PT.

Trifolium echinatum SPINY CLOVER A variable, low to short, erect or sprawling, hairy annual. Leaflets oval to oblong and broadest above the middle; the *uppermost leaves opposite*. Flowers whitish or cream, flushed pale pink towards the apex, to 12 mm long, borne in rounded heads on rather long, slender stalks. *Fruiting heads rather spiny, with spreading, pointed calyx lobes*; teasel-like. Grassy and damp habitats. IT (incl. Sicily).

T. stellatum (fruiting calyx)

T. stellatum

Trifolium stellatum STAR CLOVER A low to short, erect, hairy annual to 20(30) cm with stems simple or branched from the base. Leaflets oval and slightly toothed with oval, *toothed* stipules with bright green veins. Flowers pink or yellowish, 12–18 mm borne in solitary, large *spherical* heads to 25 mm across in fruit; calyx equalling the corolla, densely white-hairy with slender, reddish, *spreading lobes* (star-like). Common in disturbed and grassy habitats. ES, FR, IT, MA, PT.

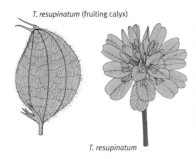

T. resupinatum (fruiting calyx)

T. resupinatum

Trifolium resupinatum REVERSED CLOVER A low to medium, spreading or sprawling more or less hairless annual to 30 cm. Leaves trifoliate with wedge-shaped leaflets 10–20(25) mm long. Flowers borne *upside down* in circular, flattened heads (superficially similar to *T. tomentosum*); pink to purple, 5–8 mm long, on rather short stalks; calyx conspicuously inflated and papery in fruit, rather hairy. Frequent in moist, sandy or grassy habitats, particularly in the west. ES, FR, IT, MA, PT.

Trifolium leucanthum A densely hairy annual to 45 cm distinguished by its wedge-shaped leaflets and white or pink flowers, 7–9 mm, in which the corolla equals (not exceeds) the calyx; calyx 10-veined with *equal*, lanceolate, 3-veined teeth. Flowers borne in more or less spherical heads, often in pairs, on stalks to 12 cm. Dry places and pine woods. ES, FR, IT, MA, PT.

Trifolium fragiferum STRAWBERRY CLOVER A creeping perennial with stems to 30 cm, rooting at the nodes. Leaflets oval, without whitish marks. Flowers pink or purplish, 5–7 mm long, borne in rounded heads to 15 mm across; calyx *greatly swollen in fruit to give the appearance of a grey-pink berry*. Damp and grassy habitats and marshes. ES, FR, IT, MA, PT. *Trifolium physodes* is similar but with stems not rooting at the nodes, and *larger flowers to 14 mm long*. Garrigue and pine forests. An island endemic. IT (Sicily).

Trifolium grandiflorum (Syn. *T. speciosum*) A rather hairy, short annual with oblong to elliptic leaflets, the terminal leaflet short-stalked; stipules oval. Flowers to 10 mm long, *pink-violet with darker veins and yellowish within*, borne in rather lax, egg-shaped heads; corolla peristing in fruit. IT (incl. Sicily).

E. Flowers typically whitish or pink (or yellowish), borne in terminal *short-stalked, rounded heads* with 2 leaves close below.

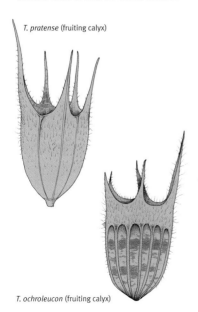

T. pratense (fruiting calyx)

T. ochroleucon (fruiting calyx)

Trifolium pratense RED CLOVER A perennial to 60 cm, initially rosette forming, later spreading; rather hairy. Leaflets circular to oval, more hairy below than above. Flower-heads large, *to 30 mm across*, spherical or ovoid, solitary or paired, and usually *stalkless*; calyx tube 10-veined with triangular teeth that are bristle-pointed at the tips; corolla 12–18 mm, usually pink, rarely cream or white. Throughout. **Trifolium ochroleucon** SULPHUR CLOVER is similar in most respects but has *yellow-white* flowers 15–18 mm long borne in virtually stalkless heads and unmarked leaves. Grassy habitats, usually above sea level; widespread but local. DZ, ES, FR, IT, MA.

1. *Trifolium stellatum*
2. *Trifolium squamosum*
3. *Trifolium resupinatum*
4. *Trifolium pratense*
5. *Trifolium ochroleucon*

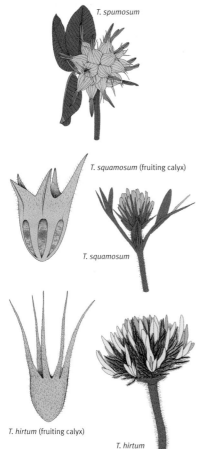

T. spumosum

T. squamosum (fruiting calyx)

T. squamosum

T. hirtum (fruiting calyx)

T. hirtum

Trifolium spumosum A spreading to erect annual to 30(70) cm, rather similar to *T. resupinatum*. Leaves trifoliate with wedge-shaped, toothed leaflets that are not strongly veined. Flower-heads more or less spherical, borne on short stalks; corolla 12–16 mm, pink, slightly exceeding the calyx; calyx *hairless* and inflated in fruit, forming somewhat spiky fruiting heads. ES, FR, IT, MA, PT.

Trifolium squamosum An erect or ascending more or less hairy perennial to 40 cm with elliptic leaflets; *stipules long and spreading*. Terminal flower-heads becoming ovoid, short-stalked and subtended by a pair of leaves; flowers 7–9 mm, pink-white; calyx teeth unequal and each with 3 veins >$1/2$ their length, exceeded by the corolla. *Fruiting heads resemble miniature teasels* due to spreading calyx teeth. Widespread though local, in coastal grassland; frequent in southwest Iberian Peninsula. ES, FR, IT, PT.

Trifolium hirtum HAIRY TREFOIL A short, spreading, *hairy*, branched annual to 35 cm. Leaflets wedge-shaped and finely toothed towards the tip. Stipules *long and straight* with a hairy tip. Flowers pink to purple, 12–17 mm long, borne in large, *densely hairy* heads to 20 mm across with a pair of leaves immediately below; corolla greatly *exceeding* the calyx; calyx 20-veined and with long hairs. Dry, stony habitats. Throughout.

F. Flowers borne solitary or in groups of 2–3, *not in dense heads*.

Trifolium uniflorum A low, prostrate or *mat-forming* perennial. Leaflets rounded to diamond-shaped, strongly veined and hairy beneath; stipules broadly triangular. Flowers white, cream or pink, to 22 mm long, borne *singly or in groups of 2–3, in the leaf axils*, rather large relative to the plant; standard petal strongly recurved. Rocky habitats and tracks. IT (incl. Sicily).

Dorycnium

Herbs with 5-foliate leaves. Flowers white or pink; stamens 10, 9 fused and 1 free. Fruits swollen, exceeding the calyces; 1–many-seeded, splitting.

A. Herbaceous perennials or subshrubs; flowers >10 mm.

Dorycnium hirsutum DORYCNIUM A sprawling perennial to 50 cm (1.5 m), *densely woolly hairy*. Leaves pinnately divided with 5 leaflets, the lowermost pair stipule-like, the true stipules minute. Flowers with a corolla *11–18 mm*, pale pink or white, with a dark red or blackish keel, borne in a compact, stalked raceme of 5–11. Fruit oblong, 6–10(13) mm. Common on sandy garrigue and dunes. Throughout.

1. *Dorycnium hirsutum* (flowers)   4. *Dorycnium fulgurans*
2. *Dorycnium hirsutum*   5. *Anthyllis hystrix*
3. *Dorycnium pentaphyllum*

B. Herbaceous perennials or subshrubs; flowers small, just 3.5–6 mm long.

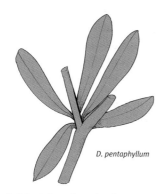

D. pentaphyllum

Dorycnium pentaphyllum is similar to *D. hirsutum* but taller, to 1.5 m, leaflets all arising from a single point (without a rachis). Flowers *small*, 4.5–6 mm, white with a red-black keel. Fruit egg-shaped, 3–4.7 mm. Similar habitats, common in the west; rare or absent in parts of the east. ES, FR, IT, MA, PT. *Dorycnium rectum* has pinnately divided leaves (the leaflets not arising from a single point), is slightly hairy, with *more numerous and smaller flowers*, the corolla 3.5–5 mm, white with a purplish keel. Damp and aquatic habitats. Throughout (rare in North Africa).

C. Intricately spiny shrubs.

Dorycnium fulgurans A *domed, spreading, densely and intricately spiny shrub* to 1 m (very different in appearance to the *Dorycnium* spp. described above) with rather bare, zig-zagging stems and leaves with 2–3 leaflets; leaves sparse. Flowers small, 3.2–5.5 mm, white and rather inconspicuous. Fruit 3–4 mm. Garrigue and bare rocks; an island endemic. ES (Balearic Islands).

Anthyllis

Perennials and shrubs with simple, trifoliate or pinnately divided leaves and stipules small or absent. Flowers numerous, borne in dense heads or interrupted racemes; stamens 10, 1 variably fused to the other 9. Fruit non-splitting.

A. *Spiny* shrubs with leaves with 1–3(5) leaflets. Main inflorescence axis forming a terminal spine.

Anthyllis hystrix A distinctive, dense, cushion-like, *spiny shrub* to 60 cm, superficially similar to gorses and *Stauracanthus*. Flowers 4.5–5 mm, *solitary or twinned* (unlike other *Anthyllis* spp.). *Leaves simple* (rarely trifoliate), and bluish and more or less hairless. Fruit 3–3.5 mm. A subdominant component of the coastal garrigue of Menorca; an island endemic. ES (Balearic Islands). *Anthyllis hermanniae* is similar but less intricately spiny, leafier, generally *trifoliate,* and flowers usually in *groups of 2–5*. FR (Corsica), IT (incl. Sardinia).

B. *Spineless* shrubs with trifoliate leaves.

Anthyllis cytisoides A dense, spineless shrub to 1.2 m with *white-downy stems and shoots*. Lower leaves simple, the upper leaves trifoliate with a disproportionately large terminal leaflet. Flowers 8–12 mm, pale lemon yellow, borne in long narrow racemes; calyx woolly with 5 equal teeth. Fruit 3–4 mm. Coastal pine forests, slopes and rocky scrub. Very common in southern Spain, local elsewhere. ES (incl. Balearic Islands), FR.

C. Spineless shrubs with 13–19 pairs of leaflets.

Anthyllis barba-jovis JUPITER'S BEARD A straggly, *silvery-white shrub* to 1 m with whitish, pinnately divided leaves with 13–19 pairs of rather crowded leaflets. *Flowers creamy-yellow, borne in compact spherical heads*, 7–9 mm, $1/2$ encircled by deeply cut, silvery bracts. Fruit 1-seeded, 4–5 mm. Coastal cliffs. Local and absent from many areas. ES, FR, IT, MA.

D. Plant herbaceous, or woody only at the base.

Anthyllis vulneraria Mediterranean kidney vetch A variable (with many forms described) low, tufted, hairy perennial. Lower leaves with a single leaflet, upper leaves with 1–9(15) elliptic leaflets. Flowers red or purple, 8–20 mm long borne in long-stalked heads with a pair of *leaf-like bracts* beneath; calyx shiny with silky hairs, purple-tipped. Fruit 3.5–7 mm. Common in a range of dry habitats, often coastal. Throughout.

E. *Hymenocarpos* group: annuals with multi-seeded pods, often treated as a separate genus but DNA-analysis indicates they are not distinct at the generic rank.

A. lotoides

A. cornicina

Anthyllis lotoides (Syn. *Hymenocarpos lotoides*) An ascending, hairy annual 12–30(40) cm. Lower leaves with a single terminal leaflet, the upper leaves with 5–7 more or less equal, narrowly lanceolate leaflets that are hairy on both sides, and especially on the margins. Flower-heads with *few flowers* (3–11), 20 mm, with bracts immediately beneath; *calyx straight*, tubular with 5 unequal teeth and covered in long hairs; corolla bright *yellow-orange. Fruit exserted from calyx and straight,* 13–18 mm. Local and sporadic on fixed dunes and cliff-tops in the region. ES, MA, PT. *Anthyllis hamosa* (Syn. *Hymenocarpos hamosus*) is very similar but with *flowers borne numerously in heads of up to 12–25*; calyx incurved. ES, MA, PT. *Anthyllis cornicina* (Syn. *Hymenocarpos cornicina*) is similar to the previous species but with an *ovoid (not tubular) calyx* and *fruit inserted within calyx*. ES, MA, PT.

Coronilla

Hairless shrubs with pinnately divided (rarely trifoliate) leaves. Flowers with 10 stamens, 9 in a tube, 1 free. Fruit a cylindrical, pod-like legume.

A. Plant a spreading shrub typically taller than 1 m.

Coronilla glauca (Syn. *C. valentina*) SCORPION SENNA A *blue-green dwarf shrub* to 1.5 m, with pinnately divided leaves with (normally) 2–3 pairs of notched, elliptic, not fleshy, *short-stalked leaflets*. Flowers 8.5–12(14) mm, borne in lateral clusters of 4–8(13), yellow, and *strongly scented*; calyx bell-shaped. Fruit 9–40(65) mm long, with 1–4(10) swollen segments. Common on scrubby cliff-tops. Throughout. *Coronilla valentina* (a name under which *C. glauca* is still often described) has been separated on account of having leaves with 4–6 pairs and heart-shaped stipules. DZ, FR (Corsica), IT (Sardinia and Sicily), MA, TN. *Coronilla minima* is similar to *C. glauca* but is smaller, and has 2–4 pairs of *unstalked leaflets* with papery margins. Subsp. *minima* has spreading branches. Subsp. *lotoides* is erect or ascending. ES, FR, IT, MA, PT.

Coronilla juncea RUSH-LIKE SCORPION VETCH A hairless shrub to 2 m with *rush-like stems* with long internodes and few leaves and leaves with 2–3 pairs of round to elliptic *fleshy leaflets* that are soon-falling. Flowers 8–12 mm, yellow, borne in very long-stalked clusters of 5–11(24). Maquis and hill slopes. Absent from some islands, otherwise Throughout.

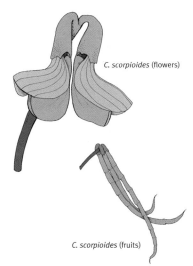

C. scorpioides (flowers)

C. scorpioides (fruits)

B. Plant largely herbaceous, often lax, and usually <50 cm.

Coronilla scorpioides ANNUAL SCORPION VETCH A prostate to sub-erect, short, hairless, blue-green, *lax annual* to 40(55) cm. Lower leaves simple or trifoliate, the terminal leaflets elliptic to rounded, *much larger than the other leaflets; upper leaves simple or trifoliate*. Flowers yellow, often with brownish veins, 3–8 mm long borne in stalked heads of 2–5. Fruit *long and cylindrical*, ridged, 18–75 mm, curved, with 2–11 jointed segments. Locally common in dry, open habitats. Throughout. *Coronilla repanda* is similar to *C. scorpioides* but with *upper leaves pinnately divided (2–4 pairs of leaflets) with a single terminal leaflet*, flowers becoming reddish on drying and a *curved*, segmented fruit. Coastal habitats. Throughout.

Coronilla vaginalis SMALL SCORPION VETCH A medium, straggling, woody-based blue-grey perennial or subshrub 10–25(50) cm with pinnately divided leaves with short-stalked, fleshy leaflets, and *conspicuous, eventually papery, oval stipules*. Flowers yellow, borne in long-stalked clusters. Fruits 20–30 mm, very slender, with 3–6(8) constrictions. FR, IT.

1. *Anthyllis lotoides*
2. *Anthyllis cytisoides*
3. *Anthyllis vulneraria*

1. *Coronilla scorpioides*
2. *Coronilla glauca*
3. *Coronilla juncea*
4. *Coronilla vaginalis* (fruits, inset: flowers)
5. *Securigera varia*
6. *Hippocrepis comosa*
7. *Hippocrepis multisiliquosa*
8. *Hippocrepis emerus*
9. *Hippocrepis biflora*

Securigera

Herbaceous perennials with furrowed or ridged stems. Flowers with 10 stamens, 9 in a tube, 1 free. Fruit straight or slightly curved, cylindrical with 3–8(12) segments, scarcely constricted or with constrictions absent.

Securigera varia (Syn. *Coronilla varia*) CROWN VETCH A straggling to ascending, often patch-forming, leafy perennial to 1.2 m. Leaves pinnately divided with 11–25 pairs of oblong to elliptic leaflets with narrow membranous margins. Flowers *white, and or, pink* (rarely purple), 8–15 mm long, borne in long-stalked, *globose heads.* Fruit 20–60(80) mm long with 3–8(12) segments. Grassy habitats. Throughout, except for most islands.

Hippocrepis

Herbaceous perennials with pinnately divided leaves. Flowers borne in axillary clusters; stamens 10, 9 fused and 1 free or partially so. Fruit divided into distinct, *horseshoe-like* 1-seeded sections, strongly compressed.

A. Plant an annual.

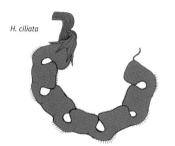

H. ciliata

Hippocrepis ciliata A slender, annual herb to 25(35) cm with pinnately divided leaves with 3–6 pairs of linear to oval leaflets. Flowers borne in clusters of 2–6 on stalks that *equal* the leaves; corolla yellow, 2–4.5 mm long. Fruit 15–30(40) mm, flattened and curved upwards (sometimes into an almost complete circle), with very pronounced segments, and hairs on 1 side; *swellings on the interior side of the curve.* Local on coastal sands, dry slopes and pine forests. Throughout.

Hippocrepis biflora (Syn. *H. unisiliquosa*) A spreading annual to 25 cm with pinnately divided leaves with 4–6 oblong, notched leaflets, and 1–2 small, almost stalkless flowers 5–8 mm, borne in the upper leaf axils on spreading (not erect) inflorescences. Fruit very distinctive: 10–35(40) mm long, slightly curved and with 7–10 horseshoe-shaped constrictions. Dry, stony habitats. *Hippocrepis salzmannii* is similar but with *long-stalked* inflorescences (especially in fruit, 50–80 mm). Maritime sands in southern Spain and northwest Africa. ES. MA. *Hippocrepis multisiliquosa* is similar to *H. biflora*, though variable, with larger flowers 4.5–7 mm long that are solitary or in clusters of 3–6, and the fruit *curved downwards* (sometimes into an almost complete circle); hairless, 30–35 mm; *swellings on the exterior side of the curve.* Similar habitats, probably throughout.

B. Plant a perennial or shrub.

Hippocrepis comosa HORSESHOE VETCH A spreading, almost hairless perennial to 30(50) cm with pinnately divided leaves with 7–25(31) leaflets to 8 mm long and yellow flowers 5–10(14) mm borne in dense, terminal, rounded, long-stalked clusters of 4–8(12); petal claw exceeds the hairy calyx. Fruit spreading, 10–30 mm long with 3–6(7) horseshoe-shaped constrictions, smooth or with minute swellings. Dry pastures and grassy habitats. ES, FR, IT. *Hippocrepis emerus* (Syn. *Coronilla emerus, Emerus major*) is a *woody shrub to 2 m* with pinnately divided leaves with 2–4 pairs of leaflets. Flowers yellow, borne in clusters of 2–4(5), 14–21 mm long. Fruit with 3–12 segments, 50 mm–11 cm long. Woods and thickets; rare or absent in most of the south. ES, FR, IT, TN.

Scorpiurus

Annual herbs with simple, alternate leaves. Flowers yellow; stamens 10, 9 fused and 1 free. Fruits elongated, strongly curved, with swellings or spines. Some authors recognise numerous taxa, most of which probably belong to the 2 species described below.

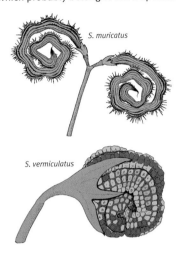

S. muricatus

S. vermiculatus

Scorpiurus muricatus A low, hairy or hairless sprawling annual to 60 cm with simple, elliptic, entire leaves with 3 prominent veins; the upper leaves short-stalked or unstalked. Flowers yellow, sometimes flushed with red, 9–13 mm long, borne in long-stalked clusters of 1–5; *calyx tube shorter than its teeth* (teeth 2–2.5mm). Fruit to 50 mm long, coiled and twisted (so appearing short), variably covered in short, robust hairs. Common in dry, rocky and sandy habitats. Throughout. Forms in the southwest with *calyx which has lower teeth shorter than their tube* (teeth 1.5–2.5 mm) are recognised by some authors as *S. sulcatus*. ***Scorpiurus vermiculatus*** is similar but with *solitary* flowers (rarely 2–3) 11–14 mm long, and fruits swollen, with *warts not hairs*, 5–7 mm across. Similar habitats. Throughout.

Hedysarum SAINFOIN

Annual herbs with pinnately divided leaves. Flowers borne in dense axillary clusters; stamens 10, 9 fused and 1 free. Fruit broad, flattened and constricted between the seeds.

H. coronarium

H. glomeratum

Hedysarum coronarium ITALIAN SAINFOIN A robust, rather hairy perennial to 1 m. Leaves pinnately divided with 3–5 pairs of elliptic to rounded leaflets that are hairy beneath, and untoothed. Flowers *magenta*, 15–18(20) mm long, borne in conspicuous, large, dense and long-stalked racemes of up to 30(50) flowers. Fruit with up to 4 spiny segments; hairless, 4–6 mm. Cultivated for fodder and naturalised on disturbed and fallow ground. Scattered throughout; more common further east.

Hedysarum spinosissimum SPINY SAINFOIN A low to short, more or less hairy annual to 20(35) cm with leaves with 5–8 pairs of elliptic leaflets. Flowers 7.5–11.5 mm, whitish, flushed and veined magenta, borne in dense, long-stalked clusters of 1–6(8). Fruit with 2–4 segments each 5.5–11 mm long, *flat and hairy with hooked, purplish spines*. ES, FR. *Hedysarum glomeratum* (Syn. *H. spinosissimum* subsp. *capitatum*) is also regarded as a subspecies of the former and has oval leaflets, bright pinkish-purple flowers 14–18 mm, borne in *small racemes* of 6–11, the corolla greatly exceeding the calyx. Fruit with *hooked*, purplish spines and 2–4 segments, each 5.5–10 mm. Dry, sandy places and pine forests. Locally common on sea cliffs. Throughout. *Hedysarum flexuosum* is similar but with leaves with *1–3(5) pairs of leaflets*, the terminal leaflet largest, and *smaller* rose-pink to purple flowers 9–11(12) mm long. Fruit with 2–4 segments, each 4–5 mm long. Maritime habitats. Iberian Peninsula and North Africa. DZ, ES, MA, PT.

Onobrychis SAINFOIN

Herbs with pinnately divided leaves. Flowers with 10 stamens, 9 in a tube and 1 free. Fruit hard, often rough-surface and jointed, non-splitting.

Onobrychis viciifolia PURPLE SAINFOIN An erect perennial 60(80) cm with pinnately divided leaves with 13–29 linear leaflets with adpressed hairs beneath. Flowers bright pink with darker veins, borne in dense, elongated heads of up to 50. Fruits 5–8 mm, pitted and with short spines. Widely cultivated and naturalised. Throughout.

---

1. *Scorpiurus muricatus* (fruits)   4. *Hedysarum glomeratum*
2. *Scorpiurus muricatus*   5. *Hedysarum coronarium*
3. *Scorpiurus vermiculatus* (fruit)

O. humilis

Onobrychis humilis (Syn. *O. peduncularis*) A hairy perennial to 60 cm with pinnately divided leaves with 3–10 pairs of elliptic leaflets. Flowers 9–12.5 mm, borne in racemes; corolla bright pale pink with *prominent purple, longitudinal veins*. Fruit 7–11.5 mm, not jointed, pale yellow-green with distinct reddish spines, and white woolly hairy. Local in sandy places. ES, MA, PT.

Onobrychis caput-galli COCKSCOMB SAINFOIN A slender, grey-hairy, sprawling annual to 60 cm with leaves with 4–9 pairs of leaflets. Flowers 4–5 mm, rather inconspicuous, pink, borne on stems as long or longer than the leaves, in clusters of 2–8; *calyx hairless*. Fruit 6.5–10 mm, *hard, rounded and compressed with deep pits and slender spines*. Various habitats. Throughout. *Onobrychis cristagalli* is similar but with a *hairy calyx tube*. Fruits spiny and with *spiralled lobes* (like in *Medicago*). North Africa only. DZ, MA, TN. *Onobrychis aequidentata* is similar to the previous species but with fruits with *cockerel-crest-like lobes* (not with many spines). FR, IT.

Wisteria

Deciduous, woody vines with pinnately divided leaves. Flowers borne in long racemes; stamens 10, 9 fused, 1 free. Fruit a flattened, pod-like legume.

Wisteria sinensis CHINESE WISTERIA A vigorous, deciduous climber to 15 m tall with twining stems. Leaves alternate and pinnately divided with 9–13, untoothed leaflets 20–60 mm long. Flowers pale lilac and pea-like, borne in long, drooping racemes 15–20(30) cm long. Fruit oblong and velvety. Native to China but widely planted. Throughout. *Wisteria floribunda* is similar but has leaves with up to 19 leaflets and *long racemes to >1 m in length*. Native to Japan but widely planted. Throughout.

1. *Onobrychis humilis*
2. *Onobrychis caput-galli*
3. *Wisteria sinensis*

POLYGALACEAE | MILKWORT FAMILY

Herbs and small shrubs with simple, opposite or alternate leaves. Flowers cosexual, zygomorphic, in slender terminal racemes or spikes; sepals 5, corolla with 3(5) fused petals, stamens 8(9–10 in non natives). Fruit a 2-seeded capsule.

Polygala

Annual herbs, perennials and shrubs. Flowers with 5 sepals and 3 (rarely 5) petals; stamens 8; ovary 2-parted.

P. monspeliaca

A. Plant an erect, slender annual with hairless stems.

Polygala monspeliaca A slender annual to 30 cm, *erect and simple or with few branches*. Leaves linear-lanceolate and broadest above the middle, blunt or finely pointed. Flowers borne in lax, terminal spikes; sepals rather large and white, 7.5–8 mm long with conspicuous veins, much-exceeding the petals; bracts shorter than the flower stalks. Fruit winged. Open woods and maquis; common and widespread. Throughout. *Polygala exilis* is similar but has purple-lilac flowers just 2.5–3.5 mm long. ES, FR, IT.

B. Plant a woody-based perennial with densely hairy stems.

Polygala rupestris ROCK MILKWORT A small, *spreading perennial with a woody stock* and minutely hairy stems to 20 cm. Leaves leathery and linear-lanceolate with inrolled margins. Flowers white and pink, 6–8 mm long, borne solitary or in groups of 1–4(8) in more or less terminal racemes; inner sepals greenish and exceeding the petals. Capsule narrowly winged. Rocky habitats. ES (incl. Balearic Islands), FR.

C. Plant a perennial with leaves in a basal rosette; leaves stalked.

Polygala calcarea CHALK MILKWORT A small, spreading perennial to 20 cm with *rosettes of blunt leaves slightly above ground level* (stem virtually leafless below); stem leaves smaller, alternate and widest above the middle. Flowers borne in racemes of 6–20, each 3–6(7) mm, usually bright dark blue (sometimes paler; rarely pink or white); veins on the inner sepals with 3–5 veins, scarcely netted near the margins; inner sepals slightly shorter than the corolla. ES, FR.

D. Plant a perennial *without* leaves in a basal rosette; leaves not distinctly stalked.

Polygala vulgaris COMMON MILKWORT A variable, short perennial with much-branched and erect or spreading stems to 30 cm with scattered, linear-lanceolate, all alternate leaves. Flowers borne in dense racemes, *small 4–7 mm*, and blue (frequently white or pink); outer sepals greenish, inner sepals slightly shorter than the corolla and with branched veins; bracts shorter than the flower buds. Various habitats; common. Throughout, though absent from many islands. *Polygala nicaeensis* NICE MILKWORT is similar but has *larger flowers 7.5–11 mm long*, borne in long, lax racemes of 10–40 flowers; sepals with netted veins. ES, FR, IT.

Polygala major LARGE MILKWORT A woody-based perennial to 30 cm with erect, dense, terminal spikes of *large*, rosy-purple (rarely blue or white) flowers which are *2 x the length of the bracts and the flower stalks*; inner sepals to 12 mm long with 3–7 veins that are netted at the margins, shorter than the petals. Fruit long-stalked. Grassy habitats above sea level; an eastern species that extends into Italy. IT.

Polygala preslii A scrambling or spreading, rather hairy perennial, with linear-lanceolate leaves that are minutely- oothed along the margins. Flowers generally white, flushed purple and with coral-coloured inner sepals to 9 mm; flower bracts equalling the flower stalks (both rather short). IT. *Polygala sardoa* is similar but smaller and hairless white *inner sepals to just 7 mm long*. IT (Sardinia).

E. Plant a shrub with woody branches.

Polygala myrtifolia An erect shrub to 2 m tall (often less) with oblong leaves that are blunt and broadest above the middle. *Flowers large and showy: bright pink-purple, to 20 mm long,* borne in short racemes; inner sepals slightly longer than the petals. A native of South Africa, widely planted (throughout), sometimes naturalised. FR (incl. Corsica), IT (Sicily), MT.

ULMACEAE | ELM FAMILY

Trees with simple, toothed, deciduous leaves, usually assymetrical at the base. Flowers borne in small axillary clusters, cosexual or male, regular and inconspicuous; perianth bell-shaped, 4–5-lobed; stamens 4–5; styles 2. Fruit a 2-winged achene, notched at the tip.

Ulmus ELM

Distinctive for having leaves assymetrical at the base and unmistakable flowers and fruits. A complicated genus in temperate regions.

Ulmus canescens MEDITERRANEAN ELM A deciduous tree to 20 m with greyish bark and slender, white-downy twigs. Leaves alternate, oval-elliptic, bluntly toothed and grey-hairy below. Flowers borne in clusters before the emergence of the leaves; stamens purplish. Fruit a brown, broadly winged nut, 15 mm, notched at the tip. Rocky gorges and ravines; uncommon. IT (south + Sicily), MT.

Celtis

Deciduous trees. Flowers male or cosexual; perianth 5-parted; stamens 5. Fruit a drupe.

Celtis australis EUROPEAN NETTLE TREE A deciduous tree to 25(30) m tall with smooth, grey bark and simple, alternate, rough, long-pointed, regularly sharp-toothed leaves 40 mm–15 cm long and small, green, solitary (sometimes 2–3) cosexual flowers that lack petals. Fruit a small, purple, berry-like drupe 8.5–12 mm, borne in clusters. Planted throughout. *Celtis tournefortii* is similar but smaller, to 6 m with pointed, oval to diamond-shaped, blunt-toothed leaves which are *hairy beneath*. IT (Sicily).

ROSACEAE | ROSE FAMILY

Annual to perennial herbs, shrubs and trees. Leaves alternate, simple or compound, and with stipules (often soon-falling). Flowers with 4(5)(–10) sepals and (0)4–16 petals; stamens usually 2–4 x as many; carpels 1–many. Fruit highly variable, for example an achene, drupe, follicle or capsule.

Rosa ROSE

Prickly shrubs, generally deciduous with pinnately divided, toothed leaves and stipules. Flowers terminal, usually with 5 sepals and petals; stamens numerous; styles separate or fused into a column. Fruit a nut, enclosed in a fleshy (often edible) structure, called a *hip*. Not all species are described; many absent from hot, dry areas.

A. Styles fused in a column; outer sepals entire.

Rosa sempervirens A vigorous, trailing or scrambling evergreen shrub to 5 m with stems with sparse, curved spines. Leaves leathery, hairless, dark green and shiny; leaflets 5–7, oval and sharply toothed (18–29 teeth) Flowers white, 25–60 mm across, with *entire*, soon-falling sepals; *styles hairless and united into a column*. Hip rounded to egg-shaped, to 10 mm long and red. Scrub and open woods. Absent from the far south. ES, FR, IT, PT.

---

1. *Polygala monspeliaca*   4. *Celtis australis*
2. *Polygala vulgaris*   5. *Rosa sempervirens*
3. *Polygala myrtifolia*   6. *Rosa canina*

B. Styles free; outer sepals with lobes.

Rosa pouzinii A lax shrub 50 cm –2(3) m with spines on the stems. Leaflets *hairless*; petiole with some glands. Flowers pale pink or occasionally white with 5 petals, *lobed sepals* and numerous yellow stamens and *hairless styles*; *flowering stem with many glands*. Hip red and ovoid, 12–19 mm long. Southwest only; frequent in the Algarve. ES, MA, PT. *Rosa micrantha* is similar, with arching stems, and leaves with hairy undersides, *hairless to sparsely hairy styles*, and *reflexed, glandular sepals* which are soon-falling. ES, FR, MA. *Rosa sicula* is similar to *R. micrantha*, with equal, narrow spines (bristles absent), leaves mostly hairless but glandular, at least along the veins, and with flowers with *densely hairy styles* and erect to spreading sepals with short lobes. Widespread but local. Throughout. *Rosa serafinii* is similar to *R. sicula* but with *hooked spines* and shiny, completely hairless leaves; sepals reflexed. FR (Corsica), IT (incl. Sardinia and Sicily). *Rosa spinosissima* (Syn. *R. pimpinellifolia*) Burnet rose A small, patch-forming shrub to just 50 cm (1 m), suckering freely. Stems with *numerous straight spines* along the main stems (shorter and less numerous on flowering branches). Leaves with 7–9(11) leaflets which are oval and up to 15 mm long. Flowers *solitary*, creamy-white, 30–40 mm across, with erect sepals and without a bract. Hip small and *purple-black* when ripe. Scrub and thickets. Probably throughout.

Rosa canina dog-rose A lax shrub to 3(4) m with arching stems with *stout, curved spines,* longer than the width of the base. Leaves with 5–7 leaflets to 40 mm long, hairless. Flowers borne in groups of up to 4, 30–50 mm across and pink or white, with *hairless stalks*; sepals with narrow, usually entire, projecting side lobes. Hip red and hairless, without sepals when ripe. Grassy places, generally above sea level. ES, FR, IT, MA.

Poterium BURNET

Herbs with pinnately divided leaves. Flowers inconspicuous, borne in dense, spherical, terminal heads, unisexual and cosexual mixed; petals absent; stamens 0–50. Fruit 1–3 achenes. DNA-based analyses support the distinction of this genus from *Sanguisorba*.

P. sanguisorba

Poterium sanguisorba (Syn. *Sanguisorba minor*) SALAD BURNET A variable (with many forms described), greyish, bushy perennial with densely hairy stems 20–70(80) cm. Leaves forming a basal rosette, pinnately divided with 9–12(25) pairs of elliptic, toothed leaflets. Flowers tiny, borne in egg-shaped heads (6–15 mm) with the upper flowers female with reddish styles, the lower male with yellow anthers; sepals bright green; petals absent. Fruit enclosed in a ridged receptacle. Very common in grassy places and fallow land in much of the region. ES, FR, IT, MA, PT. Subsp. *sanguisorba* (Syn. *Sanguisorba minor* subsp. *minor*) is the common form and has unwinged fruit receptacles. ES, FR, IT, MA, PT. Subsp. *balearicum* (Syn. *Sanguisorba minor* subsp. *balearica*) has winged fruit receptacles. Scattered throughout. ES, FR, IT, MA, PT. *Poterium verrucosum* (Syn. *Sanguisorba minor* subsp. *verrucosa*) is now generally recognised as a distinct species, differing in the fruit being enclosed by a sub-spherical receptacle with very prominent, lobed ridges. ES, FR, IT, MA, PT.

Sarcopoterium

Like *Poterium* but an intricately *spiny shrub*.

Sarcopoterium spinosum THORNY BURNET A low, *rounded, intricately spiny shrub with bare, grey, zig-zagging branches*. Leaves pinnately divided with 4–7 pairs of leaflets, soon-falling in summer. Flowers unisexual, borne in dense clusters. Fruit red and berry-like. IT (incl. Sardinia and Sicily).

Potentilla CINQEFOIL

Herbs or shrubs with lobed or pinnately divided leaves. Flowers with (4)5(6) petals; styles feathery in fruit; stamens (5)10. Fruit a head of achenes. A large genus but few species are common in the Mediterranean.

Potentilla reptans CREEPING CINQUEFOIL A *creeping perennial* a central rosette from which flowering stems to 1 m root at the nodes with long-stalked hairless leaves with 5(7) leaflets; stipules leafy and entire. Flowers solitary in the leaf axils, yellow, 15–25 mm across with 5 notched petals; petals 2 x the length of the calyx. Damp, grassy habitats above sea level more or less throughout but absent from hot, dry areas. Throughout.

Potentilla hirta A short to tall *hairy, bushy perennial* with stems 15–45 cm. Leaves 3–8, palmately lobed, linear and with blunt teeth or lobes; long-hairy. Flowers bright yellow, to 24 mm across borne in lax clusters (cymes) of 10–30, each with 5 petals and 5 sepals. Rocky and sandy habitats; absent from the far south and west. ES, FR, IT.

1. *Poterium verrucosum*
2. *Sarcopoterium spinosum*
3. *Potentilla reptans*
4. *Rubus ulmifolius*
5. *Rubus ulmifolius* (leaf undersides)

Rubus BRAMBLE

A complex genus, of typically woody, scrambling vines. Flowers with 5(–8) sepals and 5(–10) petals; stamens numerous. Fruit a head of 1–many 1-seeded drupes. Many closely related species occur which are difficult to identify in the field; only the most common species are described here.

Rubus ulmifolius A typically bramble-like spreading perennial with robust, thorny, arching or prostrate stems which are angled and grooved; hairy or hairless. Leaves often small with 3(5) leaflets, *leathery dark green above and shortly white-hairy beneath*, variously toothed and oval in shape. Inflorescence rather narrowly pyramidal, and leafy at the base, with hairy prickles; sepals deflexed after flowering, and white-hairy; petals 9–14 mm long, crumpled and fairly dark pink (rarely white) with stamens *equalling or scarcely exceeding* the styles. Fruit black and shiny (known as a blackberry). Thickets, and scrub. Throughout. *Rubus sanctus* is a closely related, more eastern species characterised by its abundance of star-like hairs on the mid-vein of the leaf, rounded terminal leaves with an abrupt end-point. IT.

Crataegus HAWTHORN

C. monogyna

Spiny trees and shrubs with lobed or toothed leaves. Flowers borne in flat-topped clusters stamens (5)10–20; carpels 1–5. Fruit a berry.

Crataegus monogyna HAWTHORN A thorny shrub or small tree to 10(15) m forming a dense crown above. Spines 10–25 mm, leaves oval to wedge-shaped, with (3)5–7 lobes, darker above, *with basal lobes with entire margins* (rarely with 1–2 teeth). Flowers white, 8–15 mm and borne in clusters; *1 style*. Fruit a bright red berry 8–10(13) mm. Common in thickets and hedges and on roadsides. Throughout. Forms with basal leaf lobes with 1–8 small teeth and flowers with 1 style are recognised by some authors as *C. granatensis*. DZ, ES (southeast), MA.

Crataegus azarolus MEDITERRANEAN MEDLAR A deciduous tree to 10 m forming a dense crown; young shoots white-cottony, later blackish and with few spines. Leaves hairless or virtually so, oval with 1–2 pairs of lateral lobes, the lowermost cut to 50–60% the depth of the leaf blade. Flowers white, to 20 mm across, borne in dense clusters on *hairy flower stalks*; styles 2–3. Fruit sub-globose, to 25 mm long and orange-red or yellow. Various scrubby and wooded habitats. Scattered from Spain eastwards. DZ, ES, FR, IT. *Crataegus laevigata* is similar to the previous species but has less stiff, less spiny twigs, hairless flower stalks and leaves normally with 3 rather shallow lobes and *many (to 13) small teeth* along the margins of the lower lobes. Styles 2–3. IT (Sicily).

C. azarolus

Pyracantha FIRE THORN

Spiny trees and shrubs. Flowers numerous, borne in flat-topped clusters; petals 5; stamens ~20; carpels 5. Fruit a berry.

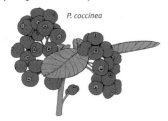

P. coccinea

Pyracantha coccinea FIRE THORN A small, very densely branched, spiny evergreen shrub to 2(6) m. Leaves 20–70 mm, elliptic, deep green and blunt-toothed, more or less hairless or hairy beneath when young, and with hairy stalks. Flowers cream-white, 8 mm across; stamens 20; styles 5. Berry orange or red with small persistent sepals. Cultivated and common as a hedge plant. Throughout.

Prunus CHERRY

Spineless or spiny deciduous or evergreen trees or shrubs with large, simple leaves. Flowers solitary or in racemes; sepals and petals 5; stamens 15–30(numerous); carpel 1. Fruit a drupe (often edible).

A. Flowers in clusters of 1–3; fruit characteristically plum-like, with a waxy bloom.

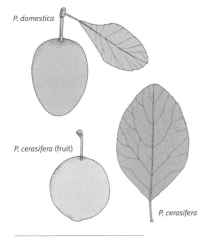

P. domestica

P. cerasifera (fruit)

P. cerasifera

1. *Crataegus monogyna*
2. *Crataegus azarolus*
3. *Prunus persica*

Prunus domestica PLUM A shrub or tree to 8(12) m with dull brown bark, and spiny branches in naturalised populations; *young twigs dull brown or grey.* Leaves 30–80 mm, *dull green and hairy, at least when young.* Flowers white, appearing with the leaves; petals 7–12 mm. Fruit a *hanging* round blue-black, red, green or yellow drupe 20–60 mm across (a plum). Widely planted. Throughout. *Prunus cerasifera* CHERRY PLUM is a similar shrub or tree to 8 m with *glossy, hairless* leaves markedly wavy-margined or toothed with forward-pointing lobes; *young twigs green, shiny and hairless.* Flowers white (or pink), borne mostly solitary, with or earlier than the leaves. Fruit a round, red or yellow drupe 20–30 mm across. Planted. Throughout. *Prunus prostrata* PROSTRATE CHERRY is a *low and spreading shrub* growing prostrate. Flowers bright pink, mostly solitary. Fruit a small, red drupe to 8 mm across. Absent from large areas. DZ, ES, IT, MA, TN.

1   2   3

P. persica

B. Flowers in clusters of 1–3; fruit characteristically peach-like and densely hairy.

Prunus persica PEACH A small tree to 6 m with familar *velvety* (hairless in var. *nucipersica*; the nectarine) yellow, orange or red, *large* rounded fruits with solitary flowers that remain *mid-dark pink*; petals 10–20 mm. Cultivated for peaches. Throughout. *Prunus armeniaca* has broader, almost round leaves on stalks to 40 mm long, flowers always pale *pink-white*; cultivated for apricots. Throughout.

Prunus dulcis ALMOND A deciduous shrub or small tree to 8 m, densely branched and spiny. Leaves oblong-lanceolate, hairless and toothed, 40 mm–12 cm. Flowers appear before the leaves, pink in bud, fading to white, borne in pairs; petals 15–20 mm. Fruit ovoid and laterally compressed, grey-green to yellow and velvety, becoming dry and splitting, 35–46(50) mm. Very common in cultivation and on maquis as a relic of planting. Throughout. *Prunus webbii* CRETAN WILD ALMOND is similar but has more strongly spreading branches, smaller leaves and flowers, and less hairy, *scarcely compressed fruits*. An eastern Mediterranean species. IT (incl. Sicily).

C. Flowers borne numerously in elongated clusters.

Prunus lusitanica PORTUGAL LAUREL A dense, evergreen shrub to 12 m with leathery, *shiny green* elliptic-oval leaves 60 mm–13 cm, with deep red young leaf stalks, and numerous whitish flowers borne in *long* erect racemes exceeding the leaves. Fruit ovoid, purplish black when ripe, 8–10 mm. Native to Iberian Peninsula, planted elsewhere. ES, PT. *Prunus laurocerasus* is similar but with pale green twigs and petioles, leaves lighter green, 50 mm–15 cm, and racemes scarcely (or not) exceeding the adjacent leaf. Fruits 10–12 mm. Planted widely in towns. Throughout.

1. *Prunus laurocerasus*
2. *Prunus dulcis*
3. *Prunus webbii*
   PHOTOGRAPH: GIANNIANTONIO DOMINA
4. *Cydonia oblonga*
5. *Eriobotrya japonica*

Cydonia QUINCE

Small, spineless, deciduous trees with simple leaves. Flowers solitary; stamens 15–25; carpels 5. Fruit hard and pear-like.

Cydonia oblonga QUINCE A shrub or small tree to 3(6) m with shoots hairy when young, later hairless. Leaves oval and entire, grey-hairy below. Flowers to 40–50 mm across, borne on short, hairy stalks; petals pink, exceeding the sepals. Fruit to 12 cm, many-seeded, yellow, pear-like and fragrant (a quince). Cultivated widely and persistent on derelict land. Throughout.

Pyrus PEAR

Small, spiny-branched deciduous trees with simple leaves. Flowers in clusters; stamens 20–30; carpels 2–5. Fruits large, fleshy and characteristically pear-like.

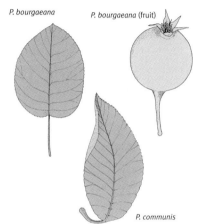

P. bourgaeana

P. bourgaeana (fruit)

P. communis

Pyrus bourgaeana WILD PEAR A tree to 10 m tall with an open crown and spiny lower branches. Branches grey and stout. Leaves 25–46 mm long, oval to heart-shaped, thick, *hairless, wavy-toothed-margined*, hairless, at least when mature, with long, slender stalks 13–40 mm long. Flowers white with rather long petals 8–12 mm, exceeding the sepals. Fruit 18–32 mm across, fig-shaped, and dull greenish yellow, later brown. Local, beside seasonal streams and rivers. ES, MA, PT. *Pyrus communis* is very similar, and distinguished by its *hairy-margined, untoothed leaves*, and larger flowers (petals 13–15 mm). Fruit the cultivated pear, 50–60 mm across (just 9–27 mm in wild populations). Cultivated widely in the region. Throughout.

Malus APPLE

Trees with simple, toothed, deciduous leaves. Flowers borne in corymbs; stamens 15–50; carpels 3–5. Fruit large, fleshy and characteristically apple-like.

Malus pumila APPLE A tree to 10(20) m with leaves to 15 cm, hairy on the undersurface, with petiole shorter than the blade. Flowers white (or pinkish), borne in clusters of 4–8; petals 13–29 mm long. Fruit to 12 cm across, variously coloured. Cultivated throughout; occasionally naturalised.

Eriobotrya LOQUAT

Small evergreen trees with large, leathery, entire leaves. Flowers borne in branched terminal clusters petals 5; stamens 15–25; carpels 5. Fruits plum-like.

E. japonica

Eriobotrya japonica JAPANESE LOQUAT A dense, robust, small tree to 10 m. Young stems covered in red-brown felted hairs. Leaves elliptic, large, 10–30 cm long, dark green and shiny above, felted below; strongly veined. Flowers white, borne in terminal panicles 70 mm–17 cm long; petals 6–11 mm. Fruit ovoid, 30–60 mm, yellow when ripe with 2–3(7) large seeds, 10–20 mm. Widely cultivated as an ornamental and for its edible fruits in gardens. Throughout.

RHAMNACEAE | BUCKTHORN FAMILY

Trees or shrubs with simple leaves with stipules. Flowers with 4–5 sepals and 4–5 petals small or absent, often hooded over the stamens; stamens 4–5. Fruit a fleshy black berry.

Rhamnus

Evergreen or deciduous shrubs with alternate or almost opposite toothed leaves and winter buds with scales.

A. Leaves not, or slightly leathery; deciduous.

Rhamnus pumila A *spineless, spreading shrub to just 20(40) cm tall* with hairy younger branches. Leaves not distinctly leathery, alternate, oblong to elliptic, 10–35(70) mm long, with sparse, small teeth, hairless but with downy stalks 5–28(40) mm, *deciduous*. Flowers 3–5 mm across, borne in clusters; flower stalks to 4 mm long; calyx lobes 4, triangular and hairless. Fruits green and hairless, 3–6 mm. Rare; usually in mountains. ES, FR, IT. *Rhamnus saxatilis* ROCK BUCKTHORN is similar but *taller, to 1.5(2) m*, with prominent spines; leaf stalks short, 1.5–5(6) mm. Limestone rocks and garrigue; frequent in southern France. ES, FR.

Rhamnus cathartica BUCKTORN *A medium deciduous shrub or small tree to 6(8) m with scattered thorns* and buds with dark scales. Leaves oval to elliptic, 40–90 mm long, hairless and finely toothed with 2–4(5) conspicuous side veins. Flowers borne in small clusters; sepals and petals 4, male and female flowers separate. Fruit a blackish berry, 6–10 mm with 3–4 seeds. Local in calcareous woodland. DZ, ES, FR, IT, MA.

B. Leaves thick and leathery, and persistent.

R. alaternus

R. lycioides

Rhamnus alaternus MEDITERRANEAN BUCKTHORN An, erect, superficially holly-like evergreen shrub to 5 m, not spiny and more or less hairless. Leaves 10–60 x 10–35 mm, alternate, *leathery*, oval and shiny dark green, toothed or sparsely toothed, with 3–6 side veins. Flowers borne in dense, cylindrical yellowish racemes, borne in small clusters in the leaf axils, with 5 sepals and lacking petals. Fruit a berry, red ripening black, 4–6 mm with 2–3 seeds. Very common on maquis and garrigue. Throughout. Forms with a *spreading* (not erect) habit and leaves to just 5–15(25) mm long are described as *R. myrtifolia* by some authors. ES, FR, MA. *Rhamnus ludovici-salvatoris* is similar to *R. alaternus* but has broader leaves 4–25 x 2.5–16 mm with spiny margins. ES (incl. Balearic Islands). *Rhamnus lycioides* is similar to *R. alaternus* but generally with narrower leaves, 4–40 x 0.5–8 mm; *spines present* and generally *4-parted* (not 5-parted) flowers. Similar habitats. Throughout.

1. *Rhamnus saxatilis*	4. *Rhamnus lycioides*	7. *Ficus elastica*
2. *Rhamnus alaternus* (flowers)	5. *Rhamnus lycioides* (fruits)	8. *Ficus microcarpa*
3. *Rhamnus alaternus* (fruits)	6. *Ficus carica*	

MORACEAE | FIG FAMILY

A large, mainly tropical family of trees including the fig, mulberry and breadfruit. Flowers small and inconspicuous, crowded; male flowers with 4–5 stamens; styles 1–2. Fruit a mass of drupes surrounded by a succulent perianth.

Ficus FIG

Trees with (normally) simple leaves. Flowers unisexual; male flowers with 3-parted perianth; female flowers with 5-parted perianth. Fruit succulent (a fig).

Ficus carica FIG A deciduous tree 4–5(10) m with greyish branches and large, palmately lobed leaves to 35 x 28 cm. Flowers minute, borne within a syncarp. The aggregate fruit is a fig; pear-shaped and purplish when mature and edible, 50–80 mm. Widely planted and naturalised on rocky slopes and near streams in the region; an important constituent of the western Mediterranean maquis. Throughout.

Ficus elastica RUBBER PLANT A tall tree 30–35 m with large, thick, leathery green or yellow-green leaves 10–35 cm long. Native to the tropics; commonly planted as an exotic ornamental in gardens, parks and on roadsides in the region. Throughout. *Ficus microcarpa* INDIAN LAUREL Generally a short tree with a grey trunk, oval, pointed, dark green, leathery leaves, and small yellow-brown fruits. Native to the tropics; ocasionally planted as an ornamental, in the very warmest parts of the region. IT (probably elsewhere).

URTICACEAE | NETTLE FAMILY

Annual to perennial herbs with opposite or alternate, simple leaves. Flowers typically unisexual, greenish with perianth with 1 whorl of 4, no petals; male flowers with 4 stamens; female flowers with 1 style. Fruit an achene.

Urtica NETTLE

Annual or perennial herbs with stinging hairs. Monoecious or dioecious; flowers usually inconspicuous and borne in dense clusters in the axils, forming elongated inflorescences; perianth of free tepals; stamens 4.

A. Plant a *perennial* with markedly elongated inflorescences.

Urtica dioica STINGING NETTLE A familiar, robust rhizomatous, little-branched perennial to 1.5 m tall, with many leafy stems and with stinging bristles; plant rather *dull green*. Leaves 25–80 mm across, oval, heart-shaped at the base and toothed. Racemes to 10 cm, elongated, and exceeding the petioles. Damp, disturbed habitats, grassy places and woods; absent from hot, dry areas but very widespread. Throughout.

B. Plant an *annual* with markedly elongated inflorescences.

Urtica membranacea (Syn. *U. dubia*) Membranous nettle A short to medium, bright green annual to 1.5 m. Leaves 20 mm–12 cm x 15–80 mm, with petiole of similar length to the lamina. *Racemes long and conspicuous*, 15–90 mm long, often arching; lower racemes female, shorter than petioles, upper racemes male and longer than petioles, often coiling. Common in disturbed and grassy places with nutrient-rich soils, often coastal. Throughout.

C. Plant an annual or perennial with inflorescences spherical or in clusters *not markedly elongated*.

Urtica urens small nettle A short annual 10–60 cm, with abundant stinging hairs but otherwise virtually hairless, similar to *U. membranacea* but with *very short, spreading clusters of flowers to 20 mm or less*, not in elongated racemes, the flowers spreading upwards in fruit; male flowers few; female tepals hairy-margined and unequal (outer segments smallest). Waste and agricultural land. Throughout.

Urtica bianorii (Syn. *U. atrovirens*) A medium, hairy, spreading *perennial* to 1 m with numerous leafy stems. Leaves rather broad, 20–35 mm across, lanceolate or *almost rounded* with course, zig-zagging teeth, and often prominent raised bumps below bristles on the upper surface. Flowers borne short, spreading clusters in the leaf axils, with *both male and female (unisexual) flowers*; female tepals hairy and more or less equal. Waste ground. ES (Balearic Islands), FR (Corsica), IT (incl. Sardinia).

Urtica pilulifera roman nettle A medium to tall, hairy annual to 1 m with oval-lanceolate leaves, toothed or untoothed, 50–80 mm across. Male flowers borne in interrupted racemes 40–70 mm long, female flowers borne on the same plant in *spherical heads which hang from the upper leaf axils,* 5–10 mm. Waste places and near buildings. Widespread but local. Throughout.

Parietaria Pellitory-of-the-wall

Annuals or perennials with alternate leaves. Flowers borne in clusters in the leaf axils, mostly unisexual; perianth of equal tepals, eventually enclosing the fruit; stamens 3–4.

Parietaria judaica pellitory-of-the-wall A short, tufted and *spreading, densely hairy perennial* with much-branched, reddish stems to 80 cm. Leaves oval, pointed, 10–50(70) mm. Monoecious; *perianth tubular*, 3–3.5 mm. A familiar plant on walls and damp rocks. Common in towns. Throughout. *Parietaria officinalis* is similar, and much-confused with the previous species, but much taller and *erect* with scarcely branched stems and bell-shaped, red perianth. Absent from the far west. ES, FR, IT, TN. *Parietaria mauritanica* is similar to the previous species, but a sparsely hairy *annual* with oval, long-tapering pointed leaves, perianth 2–3 mm long when in fruit, and olive or reddish (not blackish) achenes. DZ, ES, MA, PT. *Parietaria lusitanica* is very similar to the previous species but with abruptly tapering leaves and a much smaller perianth to just 1.5–1.7 mm long (in fruit), scarcely exceeding its bracts. Throughout.

---

1. *Urtica dioica*
2. *Urtica membranacea*
3. *Parietaria judaica*
4. *Parietaria officinalis*
5. *Parietaria officinalis*

CANNABACEAE | HEMP FAMILY

A small family of annual or perennial herbs with lobed to palmate leaves, including *Cannabis* (hemp) and *Humulus* (hops). Flowers usually unisexual, the male with 5 stamens, the female with 2 styles. Fruit an achene.

Cannabis

Aromatic herbs with alternate upper leaves (sometimes opposite below), often dioecious. Male flowers borne in panicles; female flowers borne in compact racemes.

Cannabis sativa HEMP An erect, rather lax, strong-smelling annual to 2.5 m with alternate leaves palmately lobed to the base into 3–9 narrow, toothed segments to 15 cm long. Male flowers with perianth segments 3.5 mm, borne in branched clusters, female flowers borne 5–8 in compact racemes in the axils. Generally naturalised in sandy or disturbed habitats, or cultivated. A native of Asia, now a cosmopolitan weed. Throughout.

ELAEAGNACEAE

A family of trees and shrubs with scale-like hairs and alternate leaves. Flowers unisexual or cosexual; perianth segments and stamens 4. Fruit drupe-like.

Elaeagnus

A genus of about 70 species of shrubs and trees, characterised by a covering of minute silvery scales, and 4-parted calyces and no petals; stamens 4. Fruit fleshy and drupe-like.

Elaeagnus angustifolia OLEASTER A variably spiny small tree or shrub to 7(10) m with silvery young branches covered in minute scales (which scrape off easily). Leaves oblong-lanceolate, green above and silver-scaly below, 15 mm–10 mm. Flowers butter-yellow and very fragrant, about 8–10 mm across, borne in axillary clusters. Fruit olive-shaped, 9–16 mm. Native to Asia but commonly planted for ornament. Throughout.

SIMAROUBACEAE

A predominantly tropical family of plants with alternate, pinnately divided leaves. Flowers small, usually unisexual with (3)5(7) sepals and petals and 2 x as many stamens; ovary superior, surrounded by a disk with 2–5 fused or free carpels. Fruit an aggregate of winged achenes.

Ailanthus

A. altissima

Trees with flowers with 5–6 sepals fused to the middle, the male flowers with 10 stamens; ovary superior surrounded by a disk. Fruit winged.

Ailanthus altissima TREE OF HEAVEN A rapidly growing tree to 20(30) m with smooth bark and large, pinnately divided leaves with 5–12 pairs of oval-lanceolate leaflets 40 mm–17 cm long and terminal panicles of small, strong-smelling greenish-yellow, 5(6)-parted flowers, unisexual; petals 2.2–4.5 mm. Fruit in clusters of 3-winged carpels, 25–50 mm, reddish brown. Native to China but very commonly planted as an ornamental. Throughout.

ZYGOPHYLLACEAE

A mostly tropical family of plants with pinnately divided leaves with stipules. Flowers regular, with a 5-parted perianth and 8 or 10 stamens. Fruit fleshy or a capsule (sometimes splitting into 5 mericarps).

Tribulus

Prostrate, annual herbs with leaves with 5–8 pairs of leaflets. Flowers yellow. Fruits generally spiny.

Tribulus terrestris SMALL CALTROPS A prostrate, spreading, hairy annual with long, trailing stems to 50(80) cm with pinnately divided leaves with 5–8 pairs of elliptic leaflets 6–10 mm long, somewhat slivery-haired. Flowers yellow, with 5 petals 8–14 mm long, borne in the leaf axils borne on short stalks (4–5 mm). Fruits 8–10 mm, *with prominent and robust spines.* Dry, bare areas and disturbed sand dunes; frequent. Throughout.

1. *Cannabis sativa*
2. *Elaeagnus angustifolia*
3. *Ailanthus altissima*
4. *Tribulus terrestris* (inset: fruits)

Fagonia

Perennial herbs, often woody-based with opposite, often trifoliate leaves. Flowers borne solitary in the leaf axils with 5 free sepals and petals and 10 stamens. Fruit a 5-parted capsule.

Fagonia cretica A short, hairless to hairy spreading, prostrate perennial 60–70 cm, woody at the base. Leaves 6–25 mm, trifoliate, leathery, with spine tips. Flowers bright reddish-purple with 5 free petals 8–9.5 mm, borne solitary between pairs of spine-tipped stipules. Fruit a 5-angled, egg-shaped capsule 7–9 mm. Mainly arid and desert fringe habitats. ES (incl. Balearic Islands), MA.

Tetraena

Spineless herbs and shrubs. Leaves opposite and succulent. Flowers borne solitary in the leaf axils, with 4–5 sepals and petals. Fruit a 3- or 5-parted capsule.

Tetraena fontanesii (Syn. *Zygophyllum fontanesi*) A fleshy, densely branched shrub to 1 m. Leaves with 2 lobes, broadly cylindrical, *strongly succulent, often yellow or reddish* with a white waxy bloom. Flowers inconspicuous, with 5 greenish sepals and 5 pinkish-white petals. Fruit 5-parted, succulent, 5–7 mm across, reddish, and appearing similar to the leaves. Arid and maritime dunes on North African Atlantic coasts, extending locally into the Mediterranean climate region. MA.

1. *Fagonia cretica*
2. *Tetraena fontanesii*
3. *Euonymus europaeus*
4. *Gymnosporia senegalensis*
5. *Ecballium elaterium*

CELASTRACEAE | SPINDLE FAMILY

Shrubs or trees (or woody climbers) with simple leaves. Flowers cosexual or unisexual with a nectar disk, with 4–5 sepals, petals and stamens; style 1. Fruit variable, often a succulent, 3–5-angled capsule with seeds with a bright orange-red aril.

Euonymus SPINDLE

Shrubs or small trees with deciduous or persistent, opposite leaves. Flowers cosexual (rarely unisexual) with 4–5 fused sepals and 4–5 stamens. Seeds with a conspicuous aril.

Euonymus europaeus SPINDLE A bushy deciduous shrub or small tree to 2–3(8) m with green, 4-angled twigs. Leaves elliptic to narrowly lanceolate, 30–80 mm (11 cm) long, opposite, pointed and finely toothed. Flowers borne in branched inflorescences in the leaf axils, greenish-white, 4-parted with lobes 8–15 mm across. *Fruit 4-lobed and coral pink, opening to expose orange seeds.* Woods and scrub, often on higher ground. Absent from the far south, west and arid areas. ES, FR, IT.

Gymnosporia

Spiny shrubs or small trees with persistent leaves and, often, unisexual flowers. Fruit a splitting capsule; seeds with an aril.

Gymnosporia senegalensis (Syn. *Maytenus senegalensis*) A much-branched, *very spiny* evergreen, hairless shrub to 2 m with grey to purple, long, arching stems with long, spreading, prominent thorns 20–50 mm. Leaves 10–35 mm, alternate and oval to diamond-shaped and wedge-shaped at the base, untoothed to shallowly toothed. Flowers 4–6 mm across, borne in clusters, white and yellow and 5-parted. Fruit a red, shiny capsule 3.5–4.5 mm across; seed with a fleshy aril at 1 end. Rocky slopes and garrigue in hot, dry areas. DZ, ES, MA.

CUCURBITACEAE | CUCUMBER FAMILY

Herbs with often climbing or trailing stems and tendrils, and alternate leaves. Flowers unisexual, sepals and petals 5; stamens usually 3; stigmas 2–3 on 1 style. Fruit succulent and berry- or pod-like.

Ecballium

Herbaceous, bristly monoecious perennials without tendrils. Corolla yellowish; stamens 5. Fruit an explosive pod.

Ecballium elaterium SQUIRTING CUCUMBER A spreading, very bristly perennial to 1.5 m without tendrils, with a tuberous rootstock. Leaves rough with bristles, more or less triangular and long-stalked (13 cm). Flowers small, 20–50 mm across and pale yellow; male and female flowers borne separately. Fruit bristly and pod-like, 30–45 mm long borne on long stalks, *exploding violently* from the point of attachment when ripe. Sandy waste ground. Common throughout the Mediterranean region.

Bryonia BRYONY

Climbing dioecious perennials with long, spiralling, unbranched tendrils. Flowers greenish white; stamens 5. Fruit a red berry.

Bryonia dioica (Syn. *B. cretica* subsp. *dioica*) WHITE BRYONY A dioecious climbing, hairy perennial to 4 m with simple, *coiled tendrils*. Leaves deeply palmately 5-lobed and plain green. Flowers greenish-white with darker veins, 5–12 mm across, borne several together. Fruit a berry 5–9 mm, green with white markings then red when ripe. Woods, thickets and maquis. Throughout.

FAGACEAE | OAK FAMILY

Evergreen or deciduous trees or shrubs with alternate, simple leaves. Monecious; male flowers borne in catkins, stamens 4–20; female flowers 1–few with 3–9 styles. Fruit 1–3(6) nuts enclosed in cup-like cupule of fused scales.

Castanea SWEET CHESTNUT

C. sativa

Deciduous trees with erect catkins; male flowers with 10–20 stamens; female flowers 3 at the base of long catkins. Fruits 1–3 enclosed in a spiny, splitting cupule.

Castanea sativa SWEET CHESTNUT A large deciduous tree to 35 m; trunk with grey-brown bark, often with spiralled fissures. Leaves 10–30 cm long, oblong-lanceolate, with pronounced veins and sharply toothed. Male flowers yellowish in long catkins (to 18 cm); female flowers few, on lower branches. Fruit a chestnut in a spiny, splitting husk. Widely planted in woods on acid soils. Throughout, except for some islands and hot, arid areas.

Quercus OAK

Evergreen or deciduous trees or shrubs. Male flowers with 4–12 stamens; female flowers 1–3. Fruit a distinctive nut (acorn) within a cupule.

A. Tree evergreen or semi-evergreen.

Quercus coccifera KERMES OAK A dense, evergreen shrub to 2 m with scaly, grey bark. Leaves 15–50 mm, oval-oblong with spiny teeth, short-stalked (1–5 mm), leathery, and *hairless when mature*, with veins prominent above but not beneath; midrib straight. Fruit 10–20 mm, ripening in the second year with conspicuous scales on the cupule; borne solitary or in pairs. A red dye is derived from a scale insect that commonly infects this species. An important component of garrigue and maquis ecosystems across the region. Throughout.

Q. coccifera

1. *Bryonia dioica*
2. *Castanea sativa*
3. *Quercus coccifera* (flowers)
4. *Quercus coccifera* (fruits)
5. *Quercus ilex* subsp. *ballota* (inset: fruit)
6. *Quercus suber*

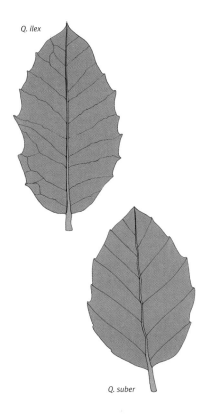

Q. ilex

Q. suber

Quercus ilex HOLME OAK A large evergreen tree to 27 m with downy young branches and grey bark. Leaves 20–90 mm, leathery, oblong-lanceolate, untoothed or sometimes spiny-toothed, *downy and with prominent veins beneath* when mature; stalks 3–10 mm. Fruit 15–35 mm, bitter-tasting with dense scales on the cupule. Common on dry hillsides, sometimes in great stands, and often planted. Throughout. Subsp. *ilex* has leaves with 7–14 pairs of secondary veins across the blade. Coastal zones throughout. Subsp. *ballota* (Syn. *Q. rotundifolia*) also often treated as a true species has broader leaves with 5–8 pairs of secondary veins across the blade; similar habitats, replacing the previous species in much of western Iberian Peninsula. Throughout.

Quercus suber CORK OAK A large evergreen tree 10–15(25) m with *thick, corky and deeply ridged bark,* red beneath, and downy young branches. Leaves 25 mm–10 cm, oblong, dark green above and grey and downy beneath, toothed; midrib wavy (sinuous). Fruit 10–20 mm, ripening in the first year, the cupule with long and spreading scales. Rocky hillsides and frequently cultivated for cork in the European region; common in southern Portugal. ES, FR, MA, PT.

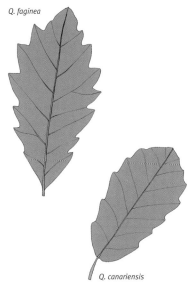

Q. faginea

Q. canariensis

Quercus faginea A semi-evergreen shrub or tree to 20 m. Leaves 30 mm–15 cm, leathery, oblong or elliptic with large, broad and forward-pointing teeth; shiny above and *densely downy beneath*. Fruit 15–35 mm, virtually stalkless. Iberian Peninsula and northwest Africa. ES, MA, PT. Subsp. *faginea* (Syn. *Q. valentina*) has winter leaves 60 x 40 mm in size. Absent from the far south. ES, PT. Subsp. *broteroi* has much larger winter leaves to 15 cm x 90 mm. Strongly southern in distribution. ES, MA, PT.

Quercus canariensis A very large semi-evergreen tree to 30 m with densely downy young branches, virtually hairless later. Leaves large, 50 mm–20 cm, oval-elliptic, with more or less acute teeth, *blue-grey beneath*. Fruit 15–30 mm with small and loosely adpressed scales. Rare, mainly in deciduous hill forests in the far southwest. ES, MA, PT.

1. *Quercus canariensis*
2. *Quercus pubescens*
3. *Quercus lusitanica*
4. *Juglans regia* (fruits)
5. *Juglans regia*

Quercus lusitanica A semi-evergreen small tree or shrub, generally to just 3 m. Leaves 25 mm–12 cm long, rather leathery, elliptic and wavy-toothed, hairless above and felted below and short-stalked. Fruits 10–16 mm, more or less stalkless, borne in downy cups with adpressed scales. Local on hill slopes in the southwest. ES, MA, PT.

B. Tree deciduous.

Quercus pubescens (Syn. *Q. humilis*) DOWNY OAK A deciduous shrub or tree to 25 m with densely downy young branches. Bark grey and fissured, finely cracking into scales. Leaves grey-green, oblong-lanceolate and bluntly lobed, to 12 cm long, *densely downy beneath*, particularly when young; short-stalked. Cup with narrow, closely pressed scales. Woods and rocky slopes; widespread but local, common in parts of southern France. ES, FR, IT.

Quercus pyrenaica PYRENEAN OAK A deciduous tree or large shrub to 25 m with woolly young twigs, and large, dark green, *distinctly narrowly pinnately lobed leaves*, 80 mm–16 cm, densely white-hairy beneath. Flower clusters stalkless or short-stalked. Fruit 15–45 mm; cup with narrow-lanceolate, blunt, loosely overlapping, grey-haired scales. Montane forests. Local throughout, particularly in the south. ES, FR, MA, PT.

JUGLANDACEAE

Trees with alternate, pinnately divided leaves. Flowers unisexual; male flowers in drooping catkins with 3–many stamens; female flowers in terminal spikes with 2 styles. Fruit a drupe or winged nut.

Juglans

Trees with pinnately divided leaves with 3–9 leaflets. Female flowers borne in racemes, erect in fruit. Fruit an unwinged nut.

Juglans regia WALNUT A large, deciduous tree to 24 m with pale, smooth bark which becomes fissured. Leaves alternate, pinnately divided with 7–9 elliptic, untoothed lobes to 15 cm long. Male flowers borne in drooping catkins to 15 cm long, female flowers in short, erect spikes. Fruit a large, edible nut >30 mm. Native to parts of southern Europe but widely planted throughout.

BETULACEAE

Deciduous trees or shrubs with simple, alternate leaves. Flowers unisexual, inconspicuous; stamens 2–14; styles 2. Fruit a nut, winged or not.

Alnus ALDER

Trees with entire leaves. Male flowers 3 per bract, with minute tepals and 4 stamens; female flowers borne in small, erect, stalked groups, 2 per bract. Fruit narrowly winged.

Alnus glutinosa ALDER A tree to 29 m with dark brown, fissured bark and hairless twigs. Buds purplish and short-stalked. Leaves oval with a cut-off or notched tip and *abrupt or wedge-shaped at the base,* often with double-toothed margins, and hairless. Male catkins cylindrical and pendent, to 50 mm long; female catkins oval, 8–28 mm long, with female flowers 3–8 per stalk, and woody when mature; all catkins appear before the leaves. Local in wet habitats. Throughout. *Alnus cordata* differs in having almost *round to heart-shaped leaves* with regular, rather rounded teeth and a blunt or short-pointed tip. Female flower group 1–3 per stalk on catkins 15–30 mm long. ES (planted), FR (Corsica), IT (Sardinia).

Carpinus HORNBEAM

Trees with entire leaves. Male flowers 1 per bract, without a perianth; stamens ~10; female flowers borne in pendent catkins, 2 per bract. Fruit winged or not, each nut with an enlarged, usually 3-lobed bract.

Carpinus orientalis EASTERN HORNBEAM A deciduous tree or shrub to 15 m with smooth, purplish-grey bark. Leaves 30–50 mm, oval to elliptic, pointed and finely toothed, slightly hairy. Male catkins pendent, female catkins borne on the same tree with large, leafy, *unlobed* bracts (unusual in the genus) to 20 mm. Fruit a small winged nut, 3 mm. IT (incl. Sicily).

RESEDACEAE | MIGNONETTE FAMILY

Annual or perennial herbs with alternate, simple or pinnately divided leaves. Flowers borne in long spikes; flowers with 4–6(8) free sepals and the same number of often yellow, distinctively shaped petals; stamens 7–numerous (25); style 0. Fruit a capsule.

Reseda

Herbs with flowers with 4–8 sepals and petals, 10–25 stamens and *bottle-shaped fruits*.

A. Flower spikes whitish; leaves clearly divided or lobed; capsule with (3)4 teeth.

Reseda alba WHITE MIGNONETTE An erect to ascending, hairless perennial to 75 cm with stems branched above. Leaves pinnately lobed with 10 or more narrow, untoothed lobes. *Flowers white*, with 5–6 sepals and petals, the petals all lobed to ¹/₃ or more of their length; *stamens 10–12*; flowers borne in long spikes. Fruit 6–15 mm, 4-angled (4 carpels) and elliptic, constricted apically and erect when ripe. Common in disturbed areas and sandy waste places. Throughout. *Reseda suffruticosa* has an erect, branching habit similar to *R. luteola* (see below) but with flowers similar to the previous species, with *13–20 stamens and oblong fruit capsules* 10–15 mm long. Very rare in southern Spain only. ES. *Reseda barrelieri* is very similar to *R. suffruticosa* but has erect, *cylindrical to 3-parted fruit capsules* 10–20 mm with protuberances on the surface. Southeast Spain above sea level. ES. *Reseda paui* is similar to *R. barrelieri* but with *smooth, cylindrical, spreading (not erect) fruit capsules* 4–14 mm. Rare in southern Spain above sea level only. ES.

B. Flower spikes whitish; leaves divided or not; capsule 3-parted.

R. phyteuma

R. media

Reseda phyteuma A rather lax, branched annual to 30(50) cm with leaves all linear-spatula shaped (sometimes with a single pair of lateral lobes). Flowers white with 6 sepals and petals with deeply cut lobes, and prominent stamens. Fruit 11–25 mm, 3-lobed and distinctly *nodding* in fruit, with conspicuous and persistent *sepals which are enlarged in fruit*. Common on cultivated ground and roadsides. Throughout, although absent from most islands. *Reseda media* is very similar to *R. phyteuma* but with many leaves *with up to 8 pairs of lobes*, (leaf segments 8 mm wide or less) and a small terminal lobe, the upper leaves rounded at the base; *sepals scarcely enlarged in fruit*; fruit capsule 14–19 mm long. Dry places. ES, MA, PT. *Reseda odorata* is similar to *R. media* but with *very fragrant flowers* and *fruit capsule smaller, just 5–8(11) mm long*. North Africa; possibly elsewhere. DZ, MA, TN.

C. Flower spikes *yellowish*; leaves divided or not; capsule 3-parted.

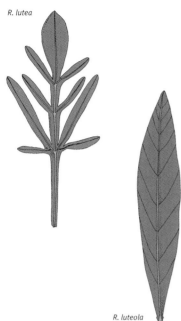

R. lutea

R. luteola

Reseda lutea WILD MIGNONETTE An ascending, leafy perennial to 75 cm, bushy and sometimes woody at the base. Leaves stalked and pinnately lobed with 1–4 pairs of leaflets. Flowers pale *greenish-yellow*, borne on short flower stalks (<6 mm), each with 6 petals and 5–6 sepals, borne in many long narrow spikes; stamens 15–20. Fruit 7–20 mm, oblong and erect, 3-parted. Very common in disturbed areas and sandy waste places. Throughout. *Reseda lanceolata* is similar but has flowers with *20–28 stamens and 6 or more sepals*; flower stalks long (exceeding 6 mm) and fruits noticeably larger, 20–30 mm long. Dry rocky slopes and cliffs in the far south. ES, MA.

Reseda luteola WELD A large, erect biennial to 1.3 m. All *leaves unlobed*, lanceolate and with wavy margins. Flowers greenish-yellow with 4 petals borne in long, slender spikes. Fruit 3–6 mm, rounded with 3 pointed lobes. Locally common in disturbed areas and sandy waste places. Throughout.

1. *Reseda alba*
2. *Reseda phyteuma*
3. *Reseda lutea* (inset: fruits)
4. *Reseda luteola* (inset: fruits)
5. *Reseda lanceolata*

Sesamoides

Similar to *Reseda* but smaller. Fruit consisting of 4–7 1-seeded carpels, spreading in a star; species similar and difficult to distinguish.

S. spathulifolia

Sesamoides interrupta (Syn. *S. pygmaea*) A slender, spreading perennial just 50 mm–15 cm with white flowers borne in long, terminal spikes, similar in form to *Reseda* but smaller. Leaves entire, the basal leaves in a rosette, *stem leaves oblong to linear-lanceolate*. Flowers with 5 sepals and petals, the petals cut into up to 7 lobes; stamens 7–12. ES, FR, IT. *Sesamoides spathulifolia* is similar but taller, to 35 cm, and has *spatula-shaped stem leaves* (oval below). Common on coastal sands in Portugal. ES, PT. *Sesamoides purpurascens* is similar to the previous species but rather *glaucous*, often with ascending or erect (rather than prostrate) stems that are rather woody at the base, and without a clear rosette. ES, FR, IT.

BRASSICACEAE | CABBAGE FAMILY

A large family of annual and perennial herbs (or shrubs). Leaves alternate and simple or pinnately divided. Flowers often in racemes, characteristically with 4 sepals and petals forming a cross; stamens usually 6 (2 shorter). Fruit dry, sometimes non-splitting; shape of the ripe fruit important for identification, classified as a *silicula* (broad and variously shaped), or a *siliqua* (long and thin).

Isatis WOAD

Erect annuals to perennials with simple leaves. Flowers small and yellow. Fruit winged, 1-seeded, pendent and non-splitting.

Isatis tinctoria WOAD A large, erect, much-branched biennial to 1.5 m with grey-green leaves; leaves arrow-shaped and clasping the stem above. Flowers yellow in dense and much-branched racemes. Fruit a *pendent*, oblong-elliptic silicula 10–25 mm. Used by the Romans as a medicinal plant, and cultivated in the Middle Ages as a source of the blue dye *indigotine*. Waste places and on disturbed ground, locally common, though rare and sporadic in the far east and west. Throughout, except for some islands, casual only in the far southwest.

Capsella SHEPHERD'S PURSE

Annuals to perennials with simple basal leaves. Petals normally white. Fruits triangular-heart-shaped.

Capsella bursa-pastoris SHEPHERD'S PURSE A distinctive, sparsely hairy annual to 40 cm with variable leaf shape, and scentless white flowers borne in a long raceme; petals 2–3 mm. Fruit 5–9 mm and *heart-shaped* with straight to slightly convex sides. A very common weed in dry waste places throughout.

1. *Sesamoides spathulifolia*	4. *Capsella bursa-pastoris*
2. *Sesamoides spathulifolia* (flowers)	5. *Cochlearia danica*
3. *Isatis tinctoria*	6. *Ionopsidium acaule*

Cochlearia SCURVYGRASS

Annuals or perennials with simple, long-stalked leaves. Petals white or mauve. Fruit an inflated or compressed siliqua. Usually coastal; mostly northern European (few species in the Mediterranean).

Cochlearia danica DANISH SCURVYGRASS A small annual, often with numerous short stems to 20 cm and fleshy, rounded basal leaves with heart-shaped bases, upper stem leaves stalked, *lobed and ivy-like*. Flowers white or pale mauve, usually <5 mm across, the petals 1.5 x length of the sepals. *Fruits ovoid and not compressed, 3–6 mm long* with persistent styles and <12 seeds. Coastal sands and shingle; rare, in the far west only (common in northern Europe). PT.

Ionopsidium (Syn. *Jonopsidium*) DIAMOND FLOWER

Hairless annuals with 4-parted, unequal flowers. All species rare and local. Some authors include some of the species below in the genus *Cochlearia* but most recent DNA-analysis supports their distinction.

A. Plants stemless or virtually so.

Ionopsidium acaule (Syn. *Cochlearia aucalis*) FALSE DIAMOND FLOWER A short-lived, small, tuft-like, *virtually stemless* annual to 60 mm with a rosette of rounded-oval, entire to 3-lobed leaves with long, slender stalks. *Flowers solitary, borne on long flower stalks in the axils of the leaves.* Petals white, usually *flushed lilac*, unequal and 2–4(8) mm (2–3 x as long as the calyx). Fruit 2–4.5(6) mm, ovoid. A rare Portuguese endemic of sand and woodland clearings (as a native; also cultivated elsewhere and possibly naturalised). PT.

B. Plants with distinct stems 50 cm–1.5 m tall, hollow; middle and upper stem leaves heart-shaped to clasping.

Ionopsidium megalospermum (Syn. *Cochlearia megalosperma*) A tall annual with few stems, to 1(1.5) m. Flowers white, borne on long, slender inflorescences becoming sparser below. Fruits bead-like, 2–3.5 mm long, borne on stalks 3–5 x as long as the fruits; seeds 6–8(10) per pod. Streams and springs on high ground; far south only; uncommon. ES (possibly MA). *Ionopsidium glastifolium* (Syn. *Cochlearia glastifolia*) is similar to the previous species but with fruits borne on stalks 1.5–2.5 x as long; seeds 15–30 per pod. Damp and aquatic habitats; uncommon. ES (possibly FR, IT).

C. Plants with distinct stems 50 mm–30 cm tall, solid; middle and upper stem leaves with wedge-shaped to lobed bases.

Ionopsidium prolongoi has a developed, usually solitary stem (to 40 cm) and *pure white* flowers; petals 3–7 mm. Fruits disk-like, 5–8.5 mm. Stony soils. ES (possibly DZ, MA). *Ionopsidium savianum* is very similar to *I. prolongoi* but sepals with translucent margins and fruits 4–7 mm *with conspicuous apical wings* to 0.5 mm. Very rare on high altitude limestone plains. IT (possibly ES). *Ionopsidium abulense* has sepals with white margins and apically wingless fruits 2–2.5 mm, exceeded by their stalks (x 1.5–2); fruits 30–40. ES, PT. *Ionopsidium heterospermum* has fruits 4 mm wide on stalks equal or shorter; fruits 18–26. DZ, MA.

Sisymbrium ROCKET

Annuals or perennials with simple, entire to deeply lobed leaves, the lowermost often withering in bloom. Petals yellow. Fruit a beakless siliqua, much longer than broad; seeds in 1 row under each valve.

A. Fruits adpressed to the stems.

Sisymbrium officinale HEDGE MUSTARD A bushy, more or less hairless annual or biennial to 1 m with grey-green dense, oval, deeply lobed leaves with a large terminal lobe. Flowers yellow, small, petals 3–4.2 mm, borne on slender, branched inflorescences *without bracts*. Fruits 10–20 mm, cylindrical, straight, *closely and densely adpressed to slender stems*. Common on roadsides and near buildings. Throughout.

B. Fruits erect to spreading.

Sisymbrium irio LONDON ROCKET An erect, much-branched annual to 60 cm with variously (some deeply) pinnately lobed leaves, the end lobe pointed. Flowers pale yellow, petals 2.5–3.5 mm, equal to or longer than the sepals. *Fruits 30–60 mm long, overtopping the open flowers* and erect to spreading. Throughout.

Sisymbrium altissimum TALL ROCKET An erect annual to 1 m, hairy below and hairless above. Leaves deeply cut into narrow, toothed, triangular lobes, *the stem leaves unstalked and with long, narrow, linear lobes*. Flowers yellow, petals 7–8 mm, 2 x the length of the sepals. Fruit 50–90 mm long. Native to southeast Europe but widely naturalised. Throughout. **Sisymbrium orientale** EASTERN ROCKET is similar, grey-hairy, and with short-stalked uppermost leaves with few (0–2) lateral lobes, the middle lobe linear-lanceolate. Fruits 40 mm–10(12) cm, slender, and erect to spreading. Throughout.

Bunias

Biennials to perennials with simple basal leaves. Petals typically yellow. Fruit an irregularly ovoid silicula with warty or wing-like crests, not splitting.

Bunias erucago SOUTHERN WARTY CABBAGE A bristly hairy annual or biennial 15–70 cm (1 m). Lower leaves pinnately lobed or wavy-margined, upper leaves unlobed, toothed or not. Flowers borne in racemes of 10–30, yellow, with notched petals 9–10(12) mm long. Fruit 8–12 mm, square in section, with *toothed wings on the angles*. Waste places and cultivated land. Throughout.

Erysimum WALLFLOWER

Annuals, herbaceous perennials or shrubs with erect, often slightly winged stems and leaves entire, covered in branched hairs. Petals yellow or red. Fruit a flattened to 4-angled siliqua. A large number of similar species occur in the region of which just a few are described here.

Erysimum × *cheiri* WALLFLOWER A perennial to 50(80) cm with woody, erect stems covered in branched hairs. Leaves narrowly elliptic and untoothed. Flowers usually dark *orange-yellow*, borne in racemes; petals 20–35 mm; sepals erect and 1/2 the length of the petals. Fruit flattened, 25–80 mm long and almost erect. Widely grown as an ornamental and naturalised in disturbed habitats. ES, FR, IT, PT.

1. *Sisymbrium officinale* (fruits)
2. *Sisymbrium orientale* (young plant)
3. *Sisymbrium orientale*
4. *Erysimum duriaei*
5. *Erysimum metlesicsii*
PHOTOGRAPH: GIANNIANTONIO DOMINA

Erysimum duriaei (Syn. *E. grandiflorum*) A perennial, rather similar to *E.* x *cheiri* in habit but with *bright yellow flowers*. Stems much-branched, with many linear-lanceolate leaves below. Flowers rather large, with petals 11–20 mm long, borne on short ascending stalks; stigma slightly grooved. Rather local in dry and bare habitats in Spain and France, possibly elsewhere. ES, FR. *Erysimum repandum* is similar but *an annual* to 35 cm, with flowers short-stalked (2–5 mm), *spreading more or less horizontally;* petals (6)9–10 mm. Fruit 45–80 mm (10 cm). Throughout. *Erysimum cheiranthoides* is also an annual, with *smaller flowers* with petals 3–5(6) mm, borne on stalks longer than the sepals. Fruit 10–25(30) mm. Throughout.

Erysimum metlesicsii A woody perennial with erect and ascending, densely leafy, striated branches; leaves linear to linear-lanceolate and short-stalked. Flowers *yellow upon opening, then fading to white*; petals to 23 mm long. Fruit slightly flattened and hairy. Rocks, cliffs and scree. IT (Sicily).

Moricandia

Hairless plants with simple, fleshy leaves. Petals *violet*. Fruit linear, 4-angled.

Moricandia arvensis VIOLET CABBAGE A hairless, blue-grey annual or perennial to 65 cm with slightly fleshy leaves which are shallowly lobed below and clasping the stem with heart-shaped bases above. *Flowers violet-purple* borne in racemes of 10–20; petals 21–29 mm. Fruit linear, 30–60 mm and 4-angled. Widespread through absent from Sardinia and the far west. DZ, ES, FR, IT, MA, TN.

Malcolmia

Annuals with branched hairs. Petals normally *pink or violet*. Styles absent, stigmas 2-lobed. Fruits linear with seeds in 1 row under each valve, not winged.

M. ramosissima

Malcolmia littorea A densely white-hairy perennial herb 10–40 cm tall, *woody at the base* and with numerous non-flowering shoots. Leaves 10–30 cm, lobed or not, *more or less stalkless*. Flowers 5–20 per cluster, pink-purple, with petals 15–18 mm. Fruit 30–60 mm and not beaded. Common in sandy and rocky maritime habitats in the west and centre of region. ES, FR, IT, MA, PT. *Malcolmia triloba* (Syn. *M. lacera*) is similar to *M. littorea* but *not woody at the base* and with *shortly stalked* basal leaves, smaller flowers, and a fruit 18–40 mm long and very narrow when mature. ES, MA, PT. *Malcolmia ramosissima* is similar to both species but with a *downy* fruit 15–35 mm long. Coastal dunes. Throughout.

Matthiola STOCK

Grey-leaved plants with simple basal leaves. Flowers with deeply 2-lobed stigmas; petals white or purple. Fruit linear with seeds in 1 row under each valve, broadly winged.

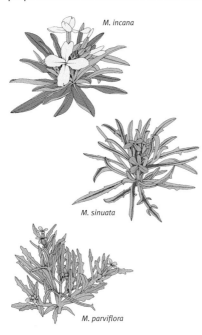

M. incana

M. sinuata

M. parviflora

Matthiola incana HOARY STOCK A stout, bushy perennial to 80 cm with a woody stock and numerous non-flowering shoots, stems and leaves grey-green and hairy. Leaves lanceolate, usually untoothed and unlobed. Flowers pink, purple or white, the petals 20–30 mm long. Fruit long and thin, 45 mm–15 cm, laterally compressed; hairy but not glandular. Local on sea cliffs and other coastal habitats. Throughout. *Matthiola sinuata* is similar to *M. incana* but with lower leaves deeply pinnately lobed and wavy-margined, the upper leaves entire; plant with conspicuous large glands on the upper parts. Flowers pale purple; petals 15–25 mm. *Fruit glandular and sticky*, 50 mm–15 cm. Throughout. *Matthiola parviflora* is similar, with deeply pinnately lobed lower leaves, but small flowers with *brownish purple petals, the limb just 2.5–4 mm,* and a *cylindrical* (not laterally compressed) fruit 50 mm–11 cm long that is *slightly constricted at intervals*. Southern and western in distribution. ES (incl. Balearic Islands), MA, PT. ***Matthiola tricuspidata*** has distinctive fruits with 3 apical horns. Throughout, except Portugal.

1. *Moricandia arvensis* PHOTOGRAPH: SERGE D. MULLER
2. *Moricandia arvensis* PHOTOGRAPH: SERGE D. MULLER
3. *Malcolmia littorea*
4. *Matthiola tricuspidata*
5. *Matthiola incana*

Lobularia SWEET ALISON

Annuals to perennials with narrow, untoothed leaves. Petals white or purplish. *Fruit a disk-like silicula.*

Lobularia maritima SWEET ALISON A greyish, downy, spreading woody-based *perennial* with stems to 30 cm, often prostrate. Leaves narrow, untoothed, linear-lanceolate and pointed. Flowers in dense rounded racemes that lengthen in fruit, white; sweetly scented; petals 2.5–4.5 mm long. Silicula flattened, oval or rounded, often hairy, 2–3.5 mm, usually with 2 seeds which are winged over at least some of their perimeter. Very common and widespread on cliffs, shingle and sand. Throughout. *Lobularia libyca* is a similar *annual* with smaller flowers (petals 1.5–2 mm long) and seeds winged along the entire perimeter. ES (incl. Balearic Islands), TN.

Fibigia

Perennials or subshrubs with entire leaves. Petals yellow. Fruit strongly compressed; seeds winged. Flowers yellow.

Fibigia clypeata An erect, densely grey-felted, tufted perennial 15–75 cm high with entire, elliptic leaves to 10 cm long. Flowers small and yellow, with petals 7–13 mm, borne in often branched racemes. Fruits distinctive: conspicuous flattened, bat-like and grey-felted, 21–28 mm long, borne on elongated stems 10–20 cm long. Rocks and walls from Italy eastwards (naturalised elsewhere). IT.

Iberis CANDYTUFT

Herbs with simple, entire to lobed leaves. Flowers borne in flat-topped clusters, the outermost pair of petals of each flower much longer than the inner. Fruit compressed with 1 seed under each valve. Many closely related species, which has led to confused distributions.

A. Plant a herbaceous annual.

I. umbellata

Iberis umbellata CANDYTUFT A robust, *hairless annual* to 70 cm with *entire, linear-lanceolate leaves*. Inflorescence a dense, rounded, flat-topped cluster (not elongating in fruit); petals pink to purple, unequal: 2 long and 2 short. Fruit 7–10 mm long, oval and broadly winged from the base; wings pointed at the tips. Cultivated and escaping; naturalised in Iberian Peninsula. ES, FR, IT, PT. *Iberis pinnata* ANNUAL CANDYTUFT is similar but with *pinnately lobed leaves*. Stony ground, and an escape from cultivation. ES, FR, IT.

Iberis pectinata (Syn. *I. sampaiana*) A short, procumbent, *bristly hairy* annual 10–30 cm high. Leaves spatula-shaped, *lobed* with 3–4 pairs of segments. Inflorescence a dense, rounded, flat-topped cluster; petals white, the outermost strongly radiate, to 9 mm long. Fruit oval-rectangular, not distinctly hairy, winged and to 5 mm long. Disturbed and degraded habitats. Iberian Peninsula only. ES, PT.

B. Plant a woody-based perennial or shrublet.

Iberis procumbens A short, compact and *bushy, short-hairy perennial* 15–30 cm high with a woody stock from which both non-flowering and flowering stems arise. Leaves broadly spatula-shaped, fleshy and entire or with 1–2 pairs of teeth near the apex. Inflorescence a dense, rounded, flat-topped cluster, *contracted in fruit*; *petals white or lilac-tinged,* 5–9 mm. Fruit oval, broadly winged from the base, 4–7 mm. On sea cliffs and dunes in Iberian Peninsula. ES, PT.

Iberis sempervirens EVERGREEN CANDYTUFT A small, branched, hairless, everygreen and woody-based shrublet to 25 cm with ascending flowers shoots and flat, narrowly spatula-shaped, thick leaves. Flowers *pure* white, borne in flat-topped clusters. Fruit rounded and broadly winged, especially at the tips, 4–7 mm. Mountains, and an escape from cultivation; absent from the islands. ES, FR, IT.

I. ciliata

Iberis ciliata An erect or spreading perennial 15–30 cm high with short central stems and oblong to spatula-shaped basal leaves which are entire or minutely toothed, and *hairy-margined*, the uppermost linear and v-shaped in cross-section. Inflorescence dense and flat, white pink or purple and very contracted in fruit; petals 7–10 mm. Fruit winged, 5–7 mm. ES, MA, PT. Subsp. *ciliata* has spatula-shaped leaves and white flowers. ES. Subsp. *contracta* (Syn. *I. linifolia*) has linear leaves and purple flowers. ES, MA, PT. *Iberis linifolia* is very similar, though entirely hairless. Previously considered to be more widespread (due to confusion with similar species). ES.

1. *Lobularia maritima*
2. *Fibigia clypeata*
3. *Iberis sempervirens*
4. *Iberis procumbens*

Biscutella BUCKLER MUSTARD

Perennials with simple leaves. Petals yellow. Fruit distinctive, with the appearance of 2 1-seeded disks fused edge to edge.

Biscutella laevigata BUCKLER MUSTARD A variable, tufted perennial with erect, branched stems 14–50 cm (with many subspecies described, considered to be true species by some authors). Leaves in a basal rosette; oval to spoon-shaped, entire, toothed or lobed. Flowers borne in branched racemes, yellow, petals 3–5 mm. Fruit distinctive, consisting of 2 disks fused together with membranous margins. Common in a wide range of habitats. Throughout. *Biscutella didyma* is a similar annual species to 40 cm tall with very small flowers with petals gradually narrowed at the base (not clawed). Dry habitats. IT (incl. Sardinia).

Biscutella sempervirens An erect perennial 20–50 cm high with a thick, woody rhizome. Basal leaves numerous, 2.5–15 mm long, and densely clustered; oblong, wavy at the margin, and minutely hairy. Inflorescence a dense raceme of yellow flowers, 3–5(9) mm, elongated in fruit. Subsp. *sempervirens* has fruits to 8 x 12 mm. ES. Subsp. *vicentina* (Syn. *B. vicentina*) is also treated as a true speces, and differs in that *the fruits are very large, to 10 x 17 mm, with membranous margins*. A rare endemic; restricted to sandy cliff-tops in the Algarve. PT.

Lepidium PEPPERWORT

Leaves simple to 2–3-pinnately divided. Flowers white or reddish. Fruits flattened and strongly keeled or winged.

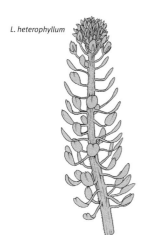

L. heterophyllum

L. draba

L. didymum

A. Inflorescence a dense, terminal raceme (many species occur besides the one described).

Lepidium heterophyllum SMITH'S CRESS A perennial herb with a woody stock and numerous ascending stems to 45 cm tall; grey-green with short, spreading hairs. Basal leaves oval to elliptic, and soon-falling; upper leaves narrowed into short stalks; uppermost leaves stalkless and triangular. Flowers small with petals 3–4 mm, *white with 6 violet anthers.* Fruit 4.5–6 mm, smooth and flat and shield-like with a notch at the tip, and projecting, persistent style. Disturbed grassy and sandy habitats. ES, FR, PT.

B. Inflorescence a corymb-shaped panicle.

Lepidium draba (Syn. *Cardaria draba*) HOARY CRESS An erect, hairless or slightly downy greyish perennial to 90 cm. Leaves oblong, pointed, toothed and long-stalked at the base, clasping the stem above. Flowers borne in dense, umbel-like clusters, white, petals 3–4 mm. Fruit 2.5–4 mm, heart-shaped with a protruding style (0.5–1.2 mm), inflated, not splitting. Common in waste places and on roadsides. Throughout.

C. Inflorescences mostly opposite the leaves (species previously described under *Coronopus*).

Lepidium didymum (Syn. *Coronopus didymus*) LESSER SWINECRESS A small, spreading or ascending biennial to 40 cm, strong-smelling when crushed. Leaves divided, feathery, at first in a rosette, later on spreading stems. Flowers inconspicuous, petals 0.5 mm, shorter than the sepals, or absent. Racemes elongated in fruit, flower stalks longer than the fruit. *Fruit dumb-bell-shaped,* 1.2–1.7 mm, veined and with a notch at the apex, style absent. Common in sandy waste places. Throughout. *Lepidium coronopus* (Syn. *Coronopus squamatus*) is similar, but with flowers with white petals to 2.5 mm across, and *ridged, kidney-* (not dumb-bell-) *shaped* fruits 2–3 mm. Throughout.

1. *Biscutella laevigata*
2. *Biscutella sempervirens* subsp. *vicentina*
3. *Lepidium draba*
4. *Lepidium didymum*
5. *Lepidium coronopus*

Diplotaxis WALL ROCKET

Annuals to perennials (normally) with deeply lobed leaves. Petals yellow. Fruits linear and flattened with seeds in 2 rows.

Diplotaxis siifolia An annual to 1 m with variably hairy, leafy to leafless stems and deeply divided leaves at the base; the segments unequal (the terminal larger than the laterals), with short hairs on both surfaces. Flowers pale sulphur-yellow; petals 9.5–12 mm. Fruit elongated and linear, 24–28 mm long with 2 rows of globose seeds. Subsp. *siifolia* has erect, leafless and scarcely hairy stems. DZ, ES, MA, PT. Subsp. *vicentina* (Syn. *D. vicentina*) is also treated as a true species, differing in having an ascending habit and hairy, leafless stems. Algarve only. PT. *Diplotaxis erucoides* is a similar annual 20–80 cm with *white flowers with violet veins*; petals 8–10(13) mm. Fruit 25–35(48) mm, erect to spreading. Local but widespread. Throughout (rare in Portugal). *Diplotaxis muralis* all rocket is similar to both the above species but a sparsely hairy annual to 60 cm with yellow flowers; petals 6–8(10) mm. Fruit 15–40 mm. Local but widespread. Throughout.

Brassica CABBAGE

Annuals to perennials with leaves entire or pinnately lobed. Petals yellow or white. Fruit a beaked siliqua with 1 row of seeds under each valve; valves rounded on the back with a single prominent vein.

A. Stem-leaves not distinctly clasping the base of the stem: either stalked or narrowed to the base.

Brassica balearica BALEARIC CABBAGE A hairy subshrub with woody stems to 35 cm and long-stalked, *fleshy, shiny green, lobed leaves resembling oak-like leaves*; those above stalkless. Flowers yellow, borne in short racemes; petals 12–14 mm. Fruits linear, 20–60 mm long with few constrictions. An island endemic of Mallorca. ES (Balearic Islands).

Brassica nigra BLACK MUSTARD An annual to 1.2 m with stems branched from the middle or from near the base. Lower leaves round to lyre-shaped, and pinnatisect (lobed to the midrib), with 1–3 pairs of lateral lobes and a much larger terminal lobe; *bristly on both surfaces*; upper leaves linear-oblong, usually entire and without bristles. All leaves *stalked*. Sepals erect-spreading; petals yellow, 7.5–12 mm. Fruit 8–25 mm long and slender - gradually more so towards the seedless beak, on short stalks *adpressed to the stem*. Local on maritime grassy and sandy places. Throughout.

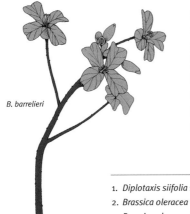

B. barrelieri

Brassica barrelieri An annual to 80 cm. Lower leaves numerous and clustered in a rosette, very shortly stalked, pinnatifid (lobed) with 7–10 pairs of lanceolate lobes; *bristly*, especially along the rachis; stem leaves few; *stalkless*, entire and hairless. Petals 9–12 mm, conspicuous, yellow with red veins, or whitish. Fruit 20–50 mm long and constricted at intervals. Local on sandy maritime soils. ES, MA, PT.

1. *Diplotaxis siifolia* subsp. *vicentina*   4. *Brassica macrocarpa* PHOTOGRAPH: GIANNIANTONIO DOMINA
2. *Brassica oleracea*   5. *Brassica rapa*
3. *Brassica oleracea* (flower)   6. *Hirschfeldia incana*

Brassica fruticulosa A perennial 30–90 cm, *woody at the base* with leaves stalked, lobed and with a large terminal lobe, very sparsely bristly, the upper leaves diminishing in size up the stem, entire above. Flowers overtopping the buds when open; *sepals erect*, petals pale yellow, 7–15 mm. Fruit 15–40 mm. ES, MA. *Brassica tournefortii* is an annual with *lax, flattish rosettes of leaves which are densely hairy beneath*, the upper leaves almost bract-like. Flowers with erect sepals and pale yellow petals 5–6 mm. Fruit 30–70 mm. ES, FR, IT, MA, PT.

B. Stem-leaves distinctly *clasping* the base of the stem.

Brassica oleracea WILD CABBAGE A robust, hairless, greyish biennial to perennial plant to 1.5(2) m with a thick woody trunk with leaf scars. Basal leaves large and fleshy, to 30 cm long, undulate, lobed and with winged stalks; upper leaves unlobed and *clasping at the base*. Flowers pale yellow, borne in *elongated racemes*; petals 12–30 mm; *sepals erect*. Fruit an ascending beaked siliqua with a single row of seeds, 50 mm–10 cm long. Tree forms (to 3 m) are cultivated in Spain and Portugal. Absent from most islands. ES, FR, IT, PT. Similar regional rarities include *Brassica macrocarpa*, which is restricted to 2 small islands in the Egadi archipelago of Sicily, and the white-flowered *Brassica insularis*, which is found in Corsica, Sardinia and Tunisia.

Brassica rapa TURNIP An annual or biennial to 1.5 m with a swollen taproot. Leaves bright green and often hairy below, those above with a grey-blue sheen, not fleshy and bristly hairy. Flowers pale yellow, *overtopping the buds*; petals 6.5–12 mm; *sepals erect-spreading*. *Fruits long and linear*, 50 mm–10 cm long, narrowed into a slender beak to 30 mm long. Cultivated and naturalised throughout.

Sinapis MUSTARD

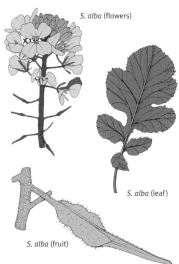

S. alba (flowers)

S. alba (leaf)

S. alba (fruit)

Annuals with crimped to lobed leaves. Flowers with spreading sepals and yellow sepals. Fruit a siliqua, splitting lengthways, with a distinct beak; valves with 3(7) veins.

Sinapis alba WHITE MUSTARD A tall, normally bristly annual (sometimes hairless) to 70 cm. Leaves *all stalked* and pinnately lobed. Flower pale yellow. Fruit 20–40 mm long, the beak flattened and sword-like, as long as, or exceeding the valves (10–30 mm). Common on disturbed, cultivated and waste areas. The source of white mustard. Throughout. *Sinapis arvensis* CHARLOCK is similar but with lanceolate *unstalked* upper leaves, and a larger fruit 25–45 mm long, with a *conical beak shorter than the 3–7 valves* (7–16 mm). Similar habitats. Throughout.

1

2

3

4

5

Hirschfeldia HOARY MUSTARD

Annuals or weak perennials with lobed to divided leaves. Petals yellow; sepals erect. Fruit a siliqua with a short, swollen, club-shaped beak; valves with 1–3 strong veins.

Hirschfeldia incana HOARY MUSTARD A tall, lax, erect annual to 1.2 m. Lower leaves stalked and pinnately divided with an oblong end-lobe and up to 9 pairs of lateral lobes; the uppermost unlobed. Flowers pale yellow, borne in crowded terminal racemes; petals 6–9 mm. Fruit 6–17 mm, closely adpressed to the stem, peg-shaped with a swollen, 1-seeded upper segment, and flattened, 2–6-seeded lower segment. Disturbed, sandy ground; frequent. Throughout.

Sisymbrella

Annuals or perennials with leafy stems, and entire or divided leaves. Petals yellow; flower stalks *thickened in fruit*. Fruit a siliqua, circular to elliptic in cross section.

Sisymbrella aspera A small annual 80 mm–50 cm, with basal *rosettes of pinnately divided leaves;* leaflets toothed. Flowers yellow, with 4 petals, 3–12 mm, exceeding the sepals which are visible between the petals; flowers borne in long, spreading inflorescences. Fruits 8–50 mm, long, smooth and slender, ascending, borne on thickened stalks. Aquatic habitats. Widespread. DZ, ES, FR, IT, MA, PT.

Cakile SEA ROCKET

C. maritima

Hairless annuals with succulent and glaucous leaves. Petals white, pink or violet. Fruit a 2-parted siliqua.

Cakile maritima SEA ROCKET A variable short, rather succulent, spreading, hairless annual to 50 cm. Leaves grey-green, irregularly pinnately lobed, the lobes narrow and untoothed, or undivided. Flowers lilac to white, the petals 4–10 mm long, borne in racemes that elongate significantly in fruit. Fruit 7–25 mm, brown and succulent; bipartite, *the lower segment with an arrow-shaped base*, the upper oval and 4-angled. Common on maritime sands. Throughout.

Rapistrum

R. rugosum

Annuals to perennials with toothed to lobes leaves. Petals yellow. Fruit a 2-parted siliqua: the lower slender with 0–1(3) seeds, the upper spherical, with 1 seed, wrinkled and with a persistent style.

Rapistrum rugosum BASTARD CABBAGE An annual to 80 cm, bristly hairy below and hairless above. Lower leaves pinnately divided, often toothed and stalked. Petals *lemon yellow*, 6–8 mm long. Fruit 4–10 mm, the upper segment ovoid-globose *and abruptly contracted into the beak*; the lower segment cylindrical or swollen. Common in ruderal maritime habitats. Throughout. *Rapistrum perenne* is similar but a perennial with *bright yellow flowers*. East of range only. FR, IT.

1. *Sisymbrella aspera*
   PHOTOGRAPH: SERGE D. MULLER
2. *Cakile maritima*
3. *Rapistrum rugosum*
4. *Raphanus raphanistrum*
5. *Cakile maritima*

Crambe SEAKALE

Large, bushy perennials with lobed leaves. Flowers with white petals and sepals erect to spreading. Fruit a 2-parted siliqua: the lower section stalk-like and seedless, the upper spherical and 1-seeded.

Crambe hispanica SEAKALE A slender, bristly hairy annual to 75 cm (1.2 m). Lower leaves rather shiny, round-lyre-shaped, divided and with a single large, kidney-shaped terminal lobe, and 0–2 pairs of inconspicuous lateral lobes. *Petals white*, 2–4 mm, sometimes purple at the base borne in a *much-branched panicle*. Fruit with a short lower part (1 mm) and a *globose* upper part (3–4.5 mm) *containing a single seed*. Arable land and garrigue. ES, MA, PT. *Crambe filiformis* is similar but an annual with leaves with up to 2 pairs of lateral segments much smaller than the terminal, and filaments with an appendage; lower sterile segment of fruit longer than the upper segment. DZ, ES, MA, TN.

Raphanus RADISH

R. raphanistrum

Annuals to perennials with a peppery smell; leaves shallowly lobed. Petals white, mauve or yellow; sepals erect. Fruit a cylindrical siliqua, elongated into a seedless beak, constricted between the seeds.

Raphanus raphanistrum WILD RADISH A variable short to tall, bristly annual to 80 cm, erect and branched. Flowers white to pale yellow or mauve, *with lilac or reddish veins*, borne in branched racemes; petals 10–25 mm. Fruit 20–60 mm long, *jointed and beaded*. Very common on arable land, sea cliffs, sandy and waste places. Throughout.

Eruca

Annuals with deeply lobed leaves. Flowers with erect sepals; petals white or yellow with purple veins. Fruits with seeds in 2 rows under each 1-veined valves.

Eruca vesicaria (Syn. *E. sativa*) A bristly annual to 1 m with stalked, lobed leaves with a large terminal lobe. Flowers with erect, purple sepals and flowers white or pale yellow and purple-veined; petals 15–20 mm. Fruit a small, *unbeaded siliqua 10–18 mm long with a flattened, sword-shaped beak*. Disturbed habitats. DZ, ES, MA, PT.

Erucastrum HAIRY ROCKET

Annuals to perennials with deeply lobed leaves. Petals yellow. Fruits linear, constricted between the seeds; seeds in 1 row under each 1-veined valve.

Erucastrum gallicum HAIRY ROCKET A rough-hairy annual to 60 cm with leaves deeply-cut into distant, parallel-sided toothed lobes. Flowers pale yellow with erect sepals; petals 7–8 mm; flower stalks with *pinnately lobed bracts*. Fruit curving upwards, 20–45 mm. Disturbed, sandy habitats. Centre of region, but also a casual introduction elsewhere. ES, FR.

1. *Eruca vesicaria*
2. *Teesdalia nudicaulis*
3. *Capparis spinosa*
4. *Capparis ovata*

Teesdalia SHEPHERD'S CRESS

Annuals with lobed or divided leaves. Petals white and unequal. Fruit a compressed siliqua, keeled to narrowly winged round the edges.

Teesdalia nudicaulis SHEPHERD'S CRESS An annual to 45 cm with few or no stem leaves and a basal rosette of pinnately lobed leaves. Flowers with white petals 0.8–2 mm, *2 of which are much shorter than the other pair; stamens 6*. Fruit heart-shaped and notched above, 3–5 mm. Widespread and locally common on rocky slopes on higher ground. ES, FR, IT, PT. *Teesdalia coronopifolia* is similar but with flowers with *more or less equal petals* and 4 stamens. Widespread on sandy soils. Throughout.

CAPPARACEAE | CAPER FAMILY

A small family of herbs and shrubs. Flowers with 4 petals, 4 sepals and 6–*numerous stamens*. Fruit a berry with numerous seeds.

Capparis CAPER

Shrubs with spiny stipules and simple leaves. Flowers with petals much shorter or longer than the sepals. Fruit a succulent berry (caper).

Capparis spinosa CAPER A low-growing (or hanging), spreading shrub, very often sprouting directly out of old walls. Leaves alternate, fleshy, grey-green and oval to circular, blunt or slightly notched at the tip with curved spines at the base of the petioles. Flowers conspicuous, with white petals 20–35 mm, and numerous fine, violet stamens. Fruit a large fleshy berry 20–30 mm. Common throughout most of the region, particularly in Italy. ES, FR, IT, MA, TN. *Capparis ovata* is often treated as a mere form of *C. spinosa* and has narrower leaves with a distinct spine at the very tip (as well as at the base of the petioles) and smaller flowers which are less regular. Walls, cliffs and coastal sands and shingle. Probably throughout the range of *C. spinosa*.

TROPAEOLACEAE | NASTURTIUM FAMILY

Herbaceous annuals and perennials, often with a rather succulent habit and hairless, alternate leaves. Flowers showy, with 5 sepals and petals and 8 stamens; stigmas 3. Fruit 3-parted, breaking into succulent segments.

Tropaeolum

Climbing or scrambling annuals with peltate leaves. Flowers zygomorphic with 5 free petals and sepals; stamens 8. Fruit 3-parted.

Tropaeolum majus NASTURTIUM A vigorous, creeping, hairless annual to 2 m with round, blue-green, shallowly 5–6-lobed, upward-facing, *peltate leaves* (stalk attached to the centre of the blade). Flowers large, solitary, long-spurred, yellow or orange-red 25–60 mm across borne in the leaf axils; spur 25–40 mm. Native to South America but often escaping, sometimes abundant. ES, FR, IT, PT.

MALVACEAE | MALLOW FAMILY

A large family of herbs or shrubs with star-shaped (stellate) hairs, and alternate, often palmately lobed leaves with stipules. Flowers often conspicuous, usually with both a *calyx and epicalyx* (an important character); sepals and petals 5; stamens numerous; styles 1 or 5–many. Fruit a nut with 1–3 seeds or capsule splitting into nutlets.

Malope

M. trifida

Similar to *Malva* (see below) but epicalyx with 3-heart-shaped lobes (wider than the sepals) and fruiting head globose.

Malope trifida A short to medium, virtually hairless annual to 1.5 m with a single, erect stem. Leaves 40 mm–10 cm, long-stalked, wavy-edged and rounded below, with 3–5 triangular lobes above. *Flowers deep purple-red*, the petals 30–60 mm long, not notched. Cultivated in gardens and sometimes naturalised. ES, MA, PT. *Malope malacoides* is similar but bristly hairy and with pale pink flowers. Throughout, except Portugal and some islands.

Malva MALLOW

Annuals to perennials with epicalyx of 3(–10) segments, free or fused, and white-pink petals; carpels numerous; fruit splitting into nutlets. The traditional distinction between *Lavatera* and *Malva* is based on fusion or non-fusion of the epicalyx but this character is now established to be artificial therefore the genera are now merged. A difficult genus; observation of mature fruits often neccessary.

1. *Tropaeolum majus*
2. *Malva moschata*
3. *Malva alcea*

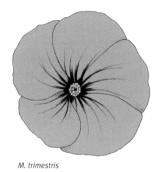

M. setigera

A. Plant an annual. *Epicalyx segments 6–10.*

Malva setigera (Syn. *Althaea hirsuta*) ROUGH MARSH MALLOW
A short to medium, slender annual to 60(70) cm with simple and star-like hairs with swollen bases. Lower leaves rounded and kidney-shaped, 10–40 mm across, long-stalked and toothed or shallowly lobed, more deeply palmately lobed further up the stem. Flowers pink, solitary, long-stalked and small, cup-shaped, the petals scarcely exceeding the sepals; petals 12–16 mm. Epicalyx segments 6–9(10), and nearly as long as the sepals. Fruit hairy and ridged. Dry fallow land and maritime waste places. Throughout.

B. Plant an annual with leaves often lobed but not deeply divided. Epicalyx segments 3; *petals large, 25–45 mm.*

Malva trimestris (Syn. *Lavatera trimestris*) ANNUAL LAVATERA
An erect, somewhat *bristly* annual to 1.2 m, with long-stalked, rounded to heart-shaped, toothed leaves; the upper shallowly 3–7-lobed, deep shiny green. Flowers large, the petals 25–45 mm long, pink (sometimes white), borne solitary. Epicalyx segments shorter than the sepals; sepals fused over most of their length, and *enlarging in fruit*. Fruits hairless, ridged, and *covered by a disk-like expansion* projecting from the central axis. On cultivated, fallow and waste ground. Throughout.

M. trimestris

1   2   3

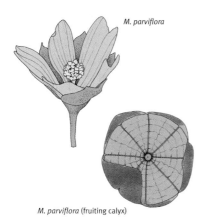

M. parviflora

M. parviflora (fruiting calyx)

M. moschata (fruiting calyx)

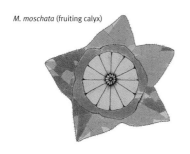

M. tournefortiana

C. Plant an annual with leaves often lobed but not deeply divided. Epicalyx segments 3; *petals small, <5 mm.*

Malva parviflora A short to medium, hairy or hairless annual to 50 cm. Leaves rounded to heart-shaped with 3–7, shallow, rounded and toothed lobes. Flowers borne in clusters of 2–4, pale mauve or lilac, *small* with the petals <5 mm long, *short-stalked* with linear-lanceolate epicalyx lobes; *sepals hairless at the margins. Fruit with enlarged, spreading, papery sepals,* fruit strongly netted and hairy or not; its stalk <10 mm. Throughout.

D. Plant a perennial usually <1 m with *upper leaves deeply divided*. Epicalyx segments 3; petals >16 mm.

Malva moschata MUSK-MALLOW An erect, branched perennial herb to 80 cm with scattered *simple hairs*. Basal leaves kidney-shaped, 3-lobed, to 80 mm long and long-stalked; stem leaves *deeply palmately cut* into narrow, further divided segments. Flowers pale to bright rose pink, borne singly or in pairs in the leaf axils in loose terminal clusters; petals >16 mm; epicalyx segments linear lanceolate and *>3 x as long as wide*. Nutlets smooth with *long hairs*. Grassy habitats, generally above sea level. ES, FR, IT, MA. ***Malva alcea*** GREATER MUSK-MALLOW is a similar perennial to 1.2 m with *star-like* (not simple) hairs on all vegetative parts. Epicalyx segments 7–8 mm, *<3 x as long as wide*. Nutlets hairless or hairy. Dry and waste habitats. ES, FR, IT. *Malva tournefortiana* is similar, with star-like (not simple) hairs on flower-stalks. Epicalyx segments 2–5 mm. Fruit angled and hairless. On cultivated land and dry, fertile places. ES, FR, PT.

1

2

3

E. Plant an annual to perennial, often to 1 m with leaves often lobed but not deeply divided. Epicalyx segments 3; *petals >9 mm*.

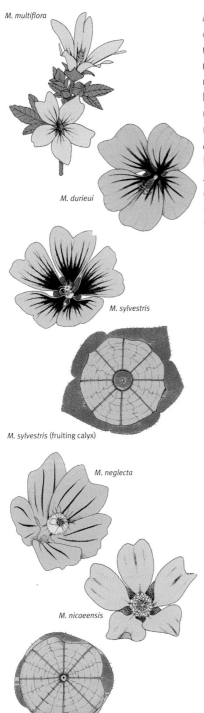

M. multiflora

M. durieui

M. sylvestris

M. sylvestris (fruiting calyx)

M. neglecta

M. nicaeensis

M. nicaeensis (fruiting calyx)

Malva multiflora (Syn. *M. pseudolavatera, Lavatera cretica*) SMALL TREE MALLOW An erect or ascending, *non-woody* annual or biennial to 1.5 m, with lower leaves rounded to heart-shaped, shallowly 3–7-lobed, the upper leaves more deeply 5-lobed. Flowers *lilac*, borne in the axils of the leaves in clusters of 2–8, on unequal stalks shorter than the leaves. Petals 10–20 mm long, epicalyx segments to 6 mm long, and free *almost* to the base (but fused there), shorter than the long-pointed sepals; *sepals just shorter or as long as sepals in fruit.* Fruit smooth or slightly ridged. Common on waste ground. Similar to *M. sylvestris* which has narrow epicalyx segments that are *free* completely to the base. *Malva durieui* (Syn. *Lavatera mauritanica*) is similar but with sepals *much enlarged* in fruit, and markedly ridged fruits. DZ, ES, MA, PT.

Malva sylvestris COMMON MALLOW A variable, erect to spreading annual to perennial herb to 1 m. Leaves roughly kidney-shaped or heart-shaped with 3–7 toothed lobes. Flowers borne in clusters of 2 or more, *bright pink or purple with darker veins*, the petals 12–30 mm long and bearded; petals about >2(4) x the length of the downy sepals. Epicalyx segments *free* to the base. Fruit sharply angled, hairy or not, and netted. Common on disturbed ground; frequently confused with *M. multiflora* which has epicalyx segments *fused* at the base, and less strongly marked flowers. Throughout.

Malva neglecta DWARF MALLOW An erect or ascending annual similar to *M. sylvestris* but smaller, to 60 cm with leaves less deeply lobed, petals *pale lilac* and 2–3 x the length of the sepals (9–13 mm); epicalyx segments linear-lanceolate. *Fruit smooth and hairless; fruit stalks remaining erect.* Waste places. Throughout. *Malva nicaeensis* is similar to *M. sylvestris* but with broader epicalyx segments, leaves not distinctly heart-shaped at the base, pale mauve petals 10–12 mm. Fruit *netted; fruit stalks recurved.* Similar habitats. Throughout.

1. *Malva arborea*
2. *Malva sylvestris*
3. *Malva multiflora*

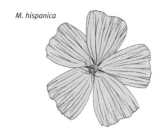

M. hispanica

M. arborea

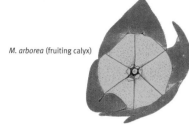

M. arborea (fruiting calyx)

M. subovata

Malva hispanica An erect annual to 70 cm (1 m) with long hairs. Leaves small, 5–60 mm wide, semi-circular and *scarcely lobed or unlobed*. Flowers normally solitary, borne in the leaf axils, epicalyx lobes linear to triangular, sepals diamond-shaped to oval, petals about 2 x the length of the sepals (15–25 mm), *very pale pink throughout*. Fruit hairless and without ridges. Sandy and dry habitats. ES, MA, PT.

F. ***Plant a robust, woody perennial* with leaves often lobed but not deeply divided. Epicalyx segments 3; petals normally >(10)15 mm.**

Malva arborea (Syn. *Lavatera arborea*) TREE MALLOW A robust, *woody* biennial or perennial to 3 m tall, downy above with star-shaped hairs. Leaves large, rounded and palmate with 5–7 lobes, up to 20(22) cm long and velvety. Flowers borne in the leaf axils in clusters of 2–7, forming a long, terminal inflorescence; petals 14–20 mm, *pink-purple with darker veins*. Epicalyx segments to 10 mm long, *exceeding the sepals* and greatly enlarged in fruit and fused at the base. Fruit hairy or not, and sharply angled. Common on maritime rocks. ES, FR, IT, MA, PT.

Malva subovata (Syn. *Lavatera maritima*) SEA MALLOW A shrub to 1.5 m with *white-felted* younger parts, later grey-woody, and with rounded leaves, greyish at least below. Flowers pale pink with a purple centre, borne on stalks as long or longer than their leaves, with gaps between the petal bases; petals 15–30 mm. Rocky coastlines; rare in the far west. ES, FR, IT.

Malva pallescens A sparsely glandular perennial 50–70 cm (1.5 m) with pale yellowish-green leaves up to 10 x 10 cm with 3–5(7) deep lobes and an undulate margin. Flowers pale white-pink; petals 20–25 mm, exceeding the calyx. Coastal limestone scree. An island endemic. IT (Sardinia). *Malva minoricensis* is a similar, compact perennial to 50 cm with pale green round, smaller leaves 35 x 35 mm with crimped margins, and only the upper leaves shallowly 3–5-lobed. Petals 10–15 mm, not exceeding the calyx. An island endemic of Menorca. ES (Balearic Islands).

---

1. *Althaea officinalis*
2. *Alcea rosea*

Althaea

A. officinalis (fruiting calyx)

Erect biennials or perennials with flowers with an epicalyx with 6–10 lobes fused at the base and shorter than their calyx. Fruit splitting into many disk-like, 1-seeded nutlets.

Althaea officinalis MARSH MALLOW A tall, velvety-downy perennial to 1.5 m with all hairs star-shaped. Leaves triangular-oval, entire or scarcely 3–5-lobed, often folded and fan-like, to 10 cm across. Flowers solitary or clustered in the leaf axils, pale lilac-pink, borne on stalks shorter than their adjacent leaves; petals 15–20 mm. Fruits hairy, sepals curved outwards in fruit. Damp, saline coastal habitats. Throughout.

Althaea cannabina A tall, erect, hairy perennial to 1.8 m with green, deeply 5-lobed leaves with oblong, double-toothed lobes. Flowers large and pale rose-pink, the petals >2 x as long as the calyx (12–30 mm). Widespread but absent from Corsica and the Balearic Islands. ES, FR, IT, MA, PT.

Alcea HOLLYHOCK

Tall, erect biennials or perennials with *very large* flowers borne in tall, wand-like inflorescences; epicalyx with 6–7 lobes fused at the base and shorter than their calyx. Fruit splitting into 1-seeded nutlets.

Alcea rosea HOLLYHOCK *A tall perennial to 3 m with stiff, erect, unbranched stems,* and large, flat palmate, bluntly toothed leaves. Flowers large and showy, usually pink but variable in colour, particularly in cultivated forms; petals 25–50 mm. Nutlets winged. Possibly of hybrid origin from the eastern Mediterranean but widely naturalised, often in rocky habitats and near buildings. DZ, ES, FR, IT, MA, TN.

1

2

Hibiscus HIBISCUS

Epicalyx with 3(10–13) segments; sepals 5, often fused and persisting in fruit; carpels 5, fused. Fruit a splitting capsule.

A. Woody-based shrubs.

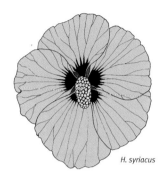

H. syriacus

Hibiscus syriacus COMMON HIBISCUS An erect shrub to 3 m with diamond-shaped, toothed leaves which are somewhat 3-lobed and short-stalked. Flowers variable in colour from pink to white with a purplish centre, borne solitary or in pairs in the upper leaf axils. Capsule yellow-hairy. Widely planted as an ornamental. Throughout. *Hibiscus rosa-sinensis* has shiny, oval, sometimes slightly irregularly toothed leaves, and conspicuous and showy bright red flowers. Very commonly planted in towns and gardens. Throughout.

B. Non-woody annuals or perennials (including *Abelmoschus*).

Hibiscus trionum BLADDER HIBISCUS A bristly, low to short, normally erect annual to 50 cm. Leaves deeply-divided into 3(5) pinnately lobed segments. Flowers solitary and long-stalked, cream-coloured with a *deep violet centre,* petals 15–25 mm long; epicalyx with 10–13 linear lobes; *calyx inflated, with purplish and bristly veins.* Fruit a capsule concealed within the inflated calyx. Native to the eastern Mediterranean but widely naturalised in the west. ES, FR, IT.

Abelmoschus esculentus (Syn. *Hibiscus esculentus*) OKRA A rather large perennial with broad, palmately lobed leaves to 20 cm across, held upright on long stems. Flowers cream with a dark purple centre, to 8 cm across with 5 petals. Fruit edible and pod-like, to 18 cm long. Occasionally cultivated in gardens. Throughout.

1. *Hibiscus rosa-sinensis*
2. *Ceiba insignis* (flower)
3. *Hibiscus trionum*
4. *Brachychiton populneus*
5. *Dombeya x cayeuxii*
6. *Ceiba insignis*

Dombeya

Trees or shrubs native to Africa and Madagasca. Flowers borne in pendent clusters; *petals persistent;* typically with 15–25 stamens + 5 conspicuous sterile stamens (staminodes); carpels 3–5. Fruit a capsule.

Dombeya × cayeuxii An exotic-looking tree, probably of hybrid origin (*D. burgessiae* x *D. wallichii*), to 5 m tall with large, drooping leaves and pendent inflorescences of pink flowers which have long-persisting petals; epicalyx not persistent. Sometimes planted in towns in Iberian Peninsula, possibly elsewhere. ES, PT.

Ceiba

Trees from the tropics with straight, swollen trunks, alternate, often palmately lobed leaves and regular flowers with 5 free petals and filaments fused in a column.

Ceiba insignis WHITE FLOSS SILK TREE A distinctive, deciduous tree with a grey, *markedly swollen, bottle-shaped trunk* when mature, sparsely *covered in stout spines*. Leaves dark green and palmately lobed to the base; lobes 5–7, lanceolate and pointed, often shallowly toothed. Flowers appear in late summer: large and showy with 5 free petals to 12 cm long, white, flushed with cream to golden yellow, often with reddish markings; stamens fused into a column. Fruit *large and pear-shaped* splitting to reveal *cotton-like seeds* which are conspicuous in the absence of leaves in winter. Commonly planted in towns and parks in Spain. ES.

Brachychiton

Trees and shrubs native to Australia. Monoecious; the unisexual flowers with a bell-shaped, lobed perianth; stamens typically 10–30 + the same number of sterile stamens (staminodes); carpels 5.

Brachychiton populneus A tree to 18 m (often much less) with a distended trunk, and pointed, simple or broad-lobed leaves. Flowers *bell-shaped*, variably greenish, yellowish or pinkish, with small red markings or entirely red within, and with usually 5 or 6 unequal, pointed lobes. Fruit a splitting pod. Occasionally planted in Spain and Mallorca on roadsides. ES (incl. Balearic Islands).

NEURADACEAE

A small, poorly understood family of hairy, prostrate annuals. Flowers solitary; sepals and petals 5; stamens and styles 10. Fruiting carpel dry, disk-like, convex.

Neurada

N. procumbens

Unmistakable for its white-hairy, prostrate habit and disk-like fruiting heads.

Neurada procumbens A prostrate, white-hairy annual to 14(32) cm, much-branched. Leaves 6–25 mm, oval, irregularly lobed. Flowers greenish, borne solitary in the leaf axils; petals 5, 2–4.3 mm long; stamens and styles 10. Fruiting carpel distinctly disk-like, 8–15 mm across, spiny above and 10-valved. Coastal sands and arid habitats in North Africa. DZ, MA, TN.

CYTINACEAE

A small family of obligate root-parasites of other shrubs. Flowers unisexual and ant-pollinated. Seeds numerous, in a pulp, dust-like when dry and windborne.

Cytinus

Parasitic plants without chlorophyll or obvious leaves or stems. Perianth 4-lobed; stamens 8–10. Fruit a capsule; seeds minute. Host-specific races are probably in the process of forming cryptic species.

A. Flowers bright yellow.

C. hypocistis subsp. hypocistis

C. hypocistis subsp. macranthus

Cytinus hypocistis CYTINUS An unmistakable, bright yellow parasite of various *Cistus* spp., with underground stems 30 mm–16 cm (stemless above ground), and yellow, orange or red oval-oblong scale leaves. *Flowers bright yellow,* in dense clusters of 4–14(20), subtended by 2 secondary bracts the same colour as the scale leaves. Garrigue, maquis and dunes. Throughout. The following subspecies are recognised but are difficult to distinguish in the field: Subsp. *macranthus* has orange-red bracts when young, *fading orange-yellow* when flowers mature, and a perianth that distinctly *exceeds* the secondary bracts at anthesis (female flowers 20–30 mm long). Parasitic on *Halimium* spp. Coastal dunes. MA, PT. Subsp. *hypocistis* is slightly later flowering, has *bright red bracts* contrasting the yellow flowers, *remaining red when flowers mature* and a *perianth scarcely exceeding the secondary bracts* (female flowers 12–15 mm long). Parasitic on *Cistus* spp. (especially *C. ladanifer* and *C. monspeliensis*). Throughout.

B. Flowers pale pink.

Cytinus ruber Similar to the above species, but with about 20 flowers with *crimson* (not red-orange) scale leaves and bracts contrasting a *white-pale pink* (not yellow) perianth 10–23 mm that slightly exceeds the secondary bracts. Parasitic on pink-flowered *Cistus* spp. (especially *C. albidus*). Garrigue and maquis, rather local. Throughout.

C. ruber

THYMELAEACEAE

Hairless shrubs with simple, untoothed, alternate leaves. Flowers in clusters or racemes, regular and cosexual; calyx a tube, petal-like; true petals absent; stamens 8, fused to the surface of the calyx-tubes; style solitary. Fruit a drupe, berry or nut.

Daphne

Shrubs with simple leaves. Flowers with 4 petal-like lobes; stamens 8. Fruit fleshy, enclosed in a persistent calyx.

D. gnidium

Daphne gnidium An erect, lax, virtually hairless shrub to 2 m with branches bare beneath; superficially rather *Euphorbia*-like when not in flower, but without a white latex when cut. Leaves 20–30 mm, pale green, leathery, linear and pointed. Flowers cream-white, 5–6.5 mm long, borne in dense panicles. Berry deep red then black, 7–8 mm. Common on garrigue and dry slopes. *Daphne oleoides* is rather similar to *D. gnidium* but shorter and more compact with ascending to almost prostrate stems to 50 cm bearing leaves which are completely hairless when mature and white to cream flowers borne in groups of 2–6 *with longer, pointed lobes;* 9–14 mm long. Fruit 10 mm. Rocky habitats at high altitude. Absent from most islands. ES, FR, IT. *Daphne rodriguezii* has leaves with hairy, down-turned margins and cream-white flowers 8–11 mm long with the corolla tube flushed with dull greenish-purple outside; fruit a red berry, 5 mm. Coastal scrub in Menorca; an island endemic. ES (Balearic Islands).

Daphne laureola SPURGE LAUREL An erect, hairless, evergreen shrub to 1.5 m tall with leathery, *broadly lanceolate*, alternate, *dark green and shiny* leaves 30 mm–12 cm. Flowers borne in short, tight clusters of 2–10; *yellowish-green*, 12–14 mm long, slightly nodding with 4 petal-like lobes; fragrant. Fruit globose and black when ripe, 8–10(13) mm. Scrub and deciduous woodland. Widespread, but rare or absent in the far east and west. DZ, ES, FR, IT, MA.

Daphne sericea A dense, rounded shrub to 70 cm with hairy young shoots. Leaves rather crowded, dark green and leathery, broadest above the middle and narrowed into short stalks; hairy beneath. Flowers pinkish (sometimes cream), borne in dense terminal heads, fragrant; 8–12 mm long. Fruit a red-brown berry. Maquis and open woodland, often coastal; also cultivated as a garden plant. IT (incl. Sicily).

Thymelaea

Dwarf evergreen shrubs or annual herbs with small, unstalked leaves. Flowers normally yellowish, borne singly or in clusters in the leaf axils. Fruit nut-like.

A. Plant a perennial shrub or subshrub.

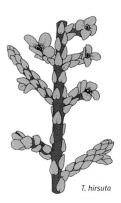

T. hirsuta

Thymelaea hirsuta A dwarf evergreen shrub to 2 m with white-downy stems; superficially similar to shrubby Amaranthaceae, with small, scale-like succulent leaves 2.5–6 mm overlapping along the stem, shiny green and *hairless outside, white-downy inside*. Flowers 3–5 mm, unisexual or cosexual; yellowish, borne in inconspicuous clusters of 6–12. Fruit 3–5 mm. Common in bare, dry and sandy places, often on sea cliffs. Throughout (possibly extinct in southern Portugal). *Thymelaea villosa* is similar in form to *T. hirsuta* but leaves 7–11 mm, less succulent, *markedly hairy* on both surfaces, and with clusters of 1–5 flowers. Local, in similar habitats. ES, MA, PT.

Thymelaea sanamunda A short to medium, hairless perennial to 40 cm with *erect annual stems* arising from a woody stock. Leaves narrowly elliptic and pointed, blue-grey green, 15–30 mm. Flowers 7–9 mm, yellow with a long slender tube with 4 rather long, pointed, deflexed lobes borne in the leaf axils in rather elongated inflorescences of 3–7. Fruit hairless, 3.5–3.8 mm. Garrigue. ES, FR.

Thymelaea tartonraira A low, *dense, almost prostrate and often cushion-like shrub* to 50 cm. Leaves broadest towards the tips and *greyish*, 5–15 mm. Flowers 3.5–6 mm, pale yellow, borne in clusters of 2–5 at the base of the leaves. Fruit 2.5–3.5 mm. Coastal garrigue and scrub. Absent from many islands. ES, FR, IT, MA. *Thymelaea velutina* (Syn. *T. myrtifolia*) is very similar but with *thickly felted leaves* 5–10 mm. ES (Balearic Islands).

Thymelaea tinctoria A short, spreading shrub with rather regularly arranged, greyish, *fleshy leaves that are rounded in cross-section*, 4–22 mm, almost spurge-like. Flowers 3.5–6 mm, yellow, unisexual, borne virtually stalkless in the leaf axils. Fruit 2.6–3.5 mm. Garrigue and stony slopes. ES, FR.

T. passerina

B. Plant an annual.

Thymelaea passerina A very slender, sparingly branched, more or less *hairless annual* with very erect stems to 50 (80) cm. Leaves linear and flax-like, alternate and pointed, 4–8 mm. Flowers 2–3 mm, greenish, borne in clusters of 3–7, in the leaf axils in elongated, lax inflorescences. Fruit 2–3 mm. Various dry habitats; widespread but rarely common. Throughout.

1. *Daphne laureola*   3. *Thymelaea hirsuta*
2. *Thymelaea villosa*   4. *Thymelaea tinctoria*

CISTACEAE | ROCK ROSE FAMILY

An important family of herbs and shrubs in the Mediterranean. Leaves normally opposite, stipules often present. Flowers often showy, cosexual, solitary or in lax, terminal clusters; sepals 3–5, petals 5, free, stamens numerous. Fruit a capsule with 3–5 valves.

Cistus ROCK ROSE

Shrubs with opposite leaves without stipules. Flowers solitary, often showy; stamens 50–150; carpels 5(6–12). Fruit a capsule.

A. Sepals 5; flowers pink.

Cistus crispus A short, compact, aromatic shrub 20–50 cm. Leaves oblong to elliptic, 12–35 mm long with a distinct *undulate margin*, 3-veined, grey-green, stalkless and hairy. Flowers short-stalked (1–5 mm), bright pink, borne in few-flowered clusters; petals 12–20 mm; sepals 5, hairy. Capsule 5–6 mm. Locally common in pine woods, garrigue and maquis. Throughout, except for many of the islands.

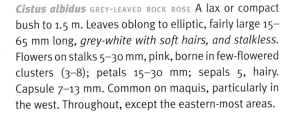

C. albidus

Cistus albidus GREY-LEAVED ROCK ROSE A lax or compact bush to 1.5 m. Leaves oblong to elliptic, fairly large 15–65 mm long, *grey-white with soft hairs, and stalkless*. Flowers on stalks 5–30 mm, pink, borne in few-flowered clusters (3–8); petals 15–30 mm; sepals 5, hairy. Capsule 7–13 mm. Common on maquis, particularly in the west. Throughout, except the eastern-most areas.

Cistus creticus A variable, rather compact, rounded bush to 1.4 m with sticky young branches. *Leaves short-stalked* (3–10 mm) oval to elliptic, 15–45 mm long, green to grey-green. Flowers pink-purple; petals 17–20 mm; sepals long-pointed. Capsules 7–10 mm. Stony pastures and field boundaries; common in the east, absent from western Iberian Peninsula. DZ, ES, FR, IT, TN. *Cistus heterophyllus* is similar but with smaller leaves 5–20 mm which are stalked below and *unstalked above,* and with sepals not long-pointed. Rare. ES (south). *Cistus parviflorus* is a more *compact* shrub than the previous species, often hemispherical, with 3-veined leaves, smaller, pale pink flowers and very short style (<0.5 mm, rather than 2.5–4 mm). IT (Sicily), MT.

B. Sepals 5; flowers white.

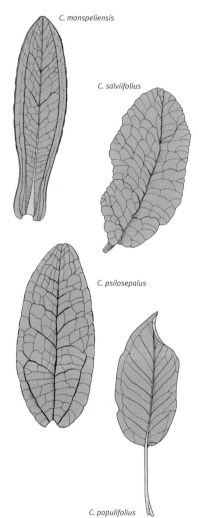

C. monspeliensis

C. salviifolius

C. psilosepalus

C. populifolius

Cistus monspeliensis NARROW-LEAVED ROCK ROSE A slightly sticky bush to 1.8 m, lax below, compact above. Leaves mid green, *narrow*, linear to lanceolate, scarcely tapered at the base and unstalked, 15–45(70) mm. Flowers small, white; petals 9–14 mm; sepals 5, the outer 2 wedge-shaped at the base. Capsule 4 mm. Common to abundant in pine woods, maquis and cliff-top garrigue. Throughout.

Cistus salviifolius SAGE-LEAVED ROCK ROSE A low, spreading, almost prostrate shrub 20–90 cm. Leaves *sage-like*; *short-stalked*, mid green, oval-elliptic with a rounded or wedge-shaped base, 8–18(45) mm long, 3-veined and hairy on both surfaces. Flowers white, borne solitary or in clusters of up to 4; petals 14–20 mm; sepals 5, the outer 2 with a heart-shaped base. Capsule 5–7 mm. Common sandy garrigue and dunes. Throughout. *Cistus psilosepalus* is similar to *C. salviifolius* and *C. monspeliensis*, but with *oblong, flat, stalkless leaves*, flowers borne in more or less symmetrical cymes of 1–5; petals 14–22 mm; sepals 5, the outer 2 *heart-shaped* at the base. Capsule 5–6 mm. ES, FR, PT.

Cistus populifolius A spreading, much-branched shrub to 2.5 m. *Leaves large, stalked, oval with a heart-shaped base,* 50–95 mm long, hairless and smooth with an undulate margin. Flowers white, borne in clusters of 2–6; petals 15–28 mm; sepals 5, the outer 2 with a heart-shaped base, becoming red when the petals fall. Capsule 5–7 mm. Woods and undisturbed habitats. ES, FR, MA, PT.

1. *Cistus crispus*
2. *Cistus albidus*
3. *Cistus creticus*
4. *Cistus monspeliensis*
5. *Cistus salviifolius*
6. *Cistus psilosepalus*
7. *Cistus populifolius*

C. Sepals 3; flowers white.

C. ladanifer

Cistus ladanifer GUM ROCK ROSE An aromatic, *extremely sticky*, lax, erect shrub to 2(4) m. Leaves linear-lanceolate, dark green (paler beneath), 40–80 mm long, 3-veined in the lower $^1/_3$, and scarcely stalked. Flowers solitary, *large*, (50–80 mm across), white, often with a crimson blotch at the base of each petal; petals 30–55 mm; *sepals 3*. Capsule 10–15 mm. Pine woods and maquis; dominant in parts of the Algarve. DZ, ES, FR, MA, PT. *Cistus palhinhae* (Syn. *C. ladanifer* subsp. *sulcatus*) is traditionally (perhaps doubtfully) described as distinct, and is *shorter*, to 50 cm with leaves oval to wedge-shaped and broadest above the middle, with clearly visible veins beneath. Flowers white and *unspotted*. Cliff-tops in the Algarve, Portugal. PT. *Cistus laurifolius* is similar to *C. ladanifer* but with 3-veined leaves 40–90 mm which are sticky above and *grey-woolly beneath* and with white flowers; petals 20–30 mm. Capsule 9–12 mm. Local. ES, FR, IT, MA, PT.

Cistus clusii A lax shrub to 1 m with *very narrow, linear leaves* 10–26 mm long, with *down-turned margins*, dark green above and white-hairy beneath. Flowers white, similar to those of to *C. monspeliensis* but with *3 hairy sepals*; petals 8–15 mm. Capsule 4–8 mm. Rocky scrub and maquis. ES, IT, MA. *Cistus libanotis* is similar but with *hairless flower stalks and sepals*; sepals sticky. Rare and local on coastal dunes and pine forests. ES, PT.

Halimium

Small, often greyish shrubs like *Cistus* but with 3 (not 5 or 10) valves to the capsule, sepals 3 or 5, the outer 2 much smaller, and flowers always yellow or white.

A. Flowers yellow.

H. halimifolium

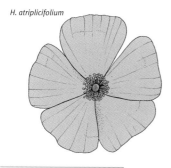

H. atriplicifolium

1. *Cistus ladanifer*
2. *Cistus ladanifer* (spotted form)
3. *Cistus palhinhae*
4. *Halimium halimifolium*
5. *Halimium ocymoides*

H. lasianthum

Halimium halimifolium A dense, silvery-grey shrub to 1.5(2) m. Leaves elliptic to oblong, broadest above the middle, 8–48 mm long, and downy when young, with silvery scales and star-like hairs when mature. Flowers yellow, often with a black spot at the base of the petals, borne in terminal clusters; petals 8–16 mm; *sepals 5*, the outer 2 smaller. Fixed sand dunes, common in southwest Iberian Peninsula. DZ, ES, IT, MA, PT, TN. Subsp. *halimifolium* has sepals without, or with few star-like hairs. Throughout the species' distribution. Subsp. *multiflorum* has denser panicles of flowers; sepals with numerous star-like hairs, and petals narrow and heart-shaped at the base, and broader above. ES, PT. **Halimium ocymoides** is similar to the previous species but with *small stem leaves 10–35 mm long*, grey-green with white hairs; unstalked. Flowers bright yellow; petals 10–18 mm; *sepals 3*. ES, PT.

Halimium atriplicifolium Similar in form to *H. halimifolium* but taller, to 1.75 m and whitish. Leaves of non-flowering shoots 5–15 mm, 3-veined and short-stalked, those of the flowering shoots 10–50 mm, pinnately veined with a heart-shaped base and unveined; all with silvery scales and star-like hairs. Flowers yellow, with petals 13–25 mm, spotted at the base, and with 3 hairy sepals, borne on *densely hairy, leafless branches*. ES, MA (records from Portugal are probably *Atriplex halimus*).

Halimium lasianthum A compact, grey, densely hairy shrub to 1.5(2) m with oblong leaves 5–40 mm long that are dark green above, woolly-white below, and *pointed*. Flower stalks and sepals with long, *silky hairs*; petals 10–20 mm, yellow and unspotted or with a brown spot at the base; sepals often with *purple bristles*. Local on sandy slopes and cork oak forests. ES, FR, MA, PT. Subsp. *lasianthum* (Syn. *H. lasianthum* subsp. *formosum*) has sepals 7.5–14 mm long, and gradually narrowing to a point. ES, MA, PT. Subsp. *alyssoides* has sepals 5.5–9 mm long, and abruptly narrowing to a point. ES, FR, MA, PT.

H. calycinum

H. umbellatum

Halimium calycinum YELLOW ROCK ROSE A low-growing, almost mat-forming shrub to 60 cm. Leaves small and *linear, with down-turned margins*, somewhat rosemary-like, 8–40 mm long and just 1–5 mm wide; shiny and hairless above, white-hairy below. Flowers borne solitary or in clusters of 2–5, bright yellow; petals 10–15 mm; sepals 3, hairless. Common to sub-dominant on coastal sands in the far west, also on garrigue at a range of altitudes. ES, MA, PT.

B. Flowers white.

Halimium umbellatum (Syn. *H. verticillatum*) A sticky, low-growing shrub to 70 cm with densely white-felted young stems and linear-elliptic, dark green leaves which are paler beneath, 10–35 mm. Flowers *white*, borne in clusters of 4–5 on very slender flower stalks; petals 7–15 mm; sepals hairy to woolly. In oak and pine woods; local. ES, FR, MA, PT.

Tuberaria (Syn. *Xolantha*)

Annuals or perennials with basal rosettes and erect flowering stems. Flowers yellow, sepals 5, the outer 2 smaller. Capsule 3-valved. Most floras classify under *Tuberaria* rather than *Xolantha*.

T. guttata

T. commutata

A. Plant an annual.

Tuberaria guttata (Syn. *Xolantha guttata*) ANNUAL ROCK ROSE A very variable, hairy, low annual to 42 cm with a basal leaf rosette that dies when mature, and a normally unbranched flowering stem. Leaves elliptic to oval, often with down-turned margins, 16–73 mm. Flowers yellow, with petals with or without a dark brown or purple spot at the base, 3–9 mm; flower stalks longer than the sepals at the point of flowering. The most common species in the region. Throughout. The following species are all similar: *Tuberaria commutata* (Syn. *T. bupleurifolia*) has narrow leaves (<5 mm) with strongly down-turned margins, and petals just 4–6 mm. Coastal sands in Iberian Peninsula. ES, PT. *Tuberaria praecox* is similar to *T. guttata* but rather *long-grey-hairy*. Flowers with small dark spots or none. ES (Balearic Islands), FR (Corsica), IT (incl. Sardinia and Sicily). *Tuberaria echioides* is also densely hairy, and distinguished by its rather *long and 1-sided, many-flowered, dense inflorescences*. Far south only. DZ, ES, MA.

B. Plant a perennial.

Tuberaria lignosa (Syn. *Xolantha tuberaria*) A medium, ascending to erect perennial with a branched to 57 cm woody stock and with persistent *plantain-like leaf-rosettes*. Leaves oval to elliptic, gradually tapered towards the base, 3-veined and white-hairy beneath, 36–65 mm (10 cm). Flowers yellow and unspotted; petals 10–15 mm. Local in scrubby and sandy habitats. ES, FR, IT, MA, PT. *Tuberaria major* similar to *T. lignosa* but with *distinctly stalked* leaves (stalk 20–40 mm) and flowers with *prominent brown-red dark spot at the base*; petals 20–24 mm. A rare endemic of the Algarve. PT. Some authors treat the species as a form of *T. globulariifolia* which is a temperate species found in the northern Iberian Peninsula.

T. lignosa

---

1. *Halimium lasianthum*
2. *Halimium calycinum*
3. *Tuberaria guttata*
4. *Tuberaria guttata* (spotted form)
5. *Tuberaria lignosa*
6. *Tuberaria lignosa*
7. *Tuberaria major*

Helianthemum

Dwarf shrubs or herbs with opposite leaves. Flowers borne in 1-sided clusters; sepals 5, the outer 2 smaller; stamens numerous, style long and S-shaped. Fruit a 3-valved capsule.

H. salcifolium

A. Plant an annual; flowers yellow.

Helianthemum salicifolium WILLOW-LEAVED ROCK ROSE A low, hairy, branched, erect or spreading *annual* to 30 cm. Leaves 5–25 mm, oval-lanceolate, flat and short-stalked. Flowers yellow, borne in lax clusters; petals 2–7 mm long and narrow, shorter to slightly longer than the sepals; bracts large and leafy; *flower stalks spreading in fruit,* upturned at the apex. Sandy and rocky slopes. Throughout. *Helianthemum ledifolium* is taller, to 45(60) cm, *hairier* and with *petals 6–8 mm, shorter than the sepals,* and *flower stalks erect* in fruit. Throughout, except many islands. *Helianthemum sanguineum* is very small, usually <10 cm, *sticky,* and with flower stalks *strongly bent downwards* in fruit. DZ, ES, MA, PT.

Helianthemum angustatum (Syn. *H. villosum*) A small, slender, erect *annual* to 35(40) cm with grey-green lanceolate leaves 10–20 mm with star-like hairs on both surfaces. Stipules linear and to almost $^1/_2$ as long as the leaves, and longer than the stalks. Flowers borne in *dense inflorescences,* more or less stalkless, yellow; the narrow petals 2–4.5 mm, inconspicuous and shorter than the woolly sepals. Dry scrub and clay soils. DZ, ES, MA, PT.

B. Plant a woody-based perennial, often (not always) with stipules; flowers yellow.

Helianthemum caput-felis A similar species to those described above, distinguished by its *dense white felt on the surface of the branches and both sides of the leaves;* leaves 6–15 mm. Flowers borne in compact, few-flowered inflorescences; petals 9–12 mm; sepals very hairy in bud. ES (incl. Balearic Islands).

H. nummularium

Helianthemum nummularium COMMON ROCK ROSE A variable, prostrate to ascending dwarf shrub 10–35 cm. Leaves oblong-lanceolate with *flat or almost flat margins,* hairy or hairless above, and grey, or white-hairy beneath, 10–25 mm. Stipules small and *leaf-like.* Flowers yellow, solitary or in clusters of up to 15(18); petals 8–12 mm; sepals hairless. Dry grassy habitats on higher ground. ES, FR, IT, PT.

---

1. *Helianthemum marifolium* subsp. *origanifolium*
2. *Helianthemum syriacum*
3. *Helianthemum canariense*

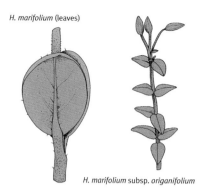

H. marifolium (leaves)

H. marifolium subsp. *origanifolium*

Helianthemum marifolium A low, spreading, prostrate or ascending dwarf shrub to 35 cm. Leaves *oval to rounded* and *heart-shaped at the base*, green above and *white-woolly beneath*, 3–20 mm. Flowers yellow, borne in branched inflorescences; petals 4–12 mm. Common in rocky and sandy calcareous habitats. DZ, ES, FR, PT. Subsp. *origanifolium* (also sometimes treated as a true species) is similar but with leaves with with sparse stellate hairs beneath (not markedly white-woolly). Stony and rocky habitats. Southern Iberian Peninsula. ES, PT.

Helianthemum syriacum (Syn. *H. lavandulifolium*) A distinctive, grey-felted, low shrub to 50(85) cm. Leaves lanceolate to linear-lanceolate, grey-green with recurved margins and white-downy beneath, 10–50 mm. Flowers bright yellow, numerous, borne in *distinctive forked inflorescences that are at first coiled*, later spreading, and bearing numerous drooping fruits; petals 5–10 mm. Pine woods and garrigue; a common species in the European Mediterranean, but absent from most islands. ES, FR, IT.

Helianthemum hirtum A small tufted shrub to 45 cm with erect or ascending branches. Lower leaves oval-rounded and narrower further up the stem, rather fleshy, green above and with starry hairs beneath, 3–15 mm. Flowers yellow (rarely white), borne in unbranched inflorescences of 6–20; petals 6–10 mm. Scrub and woodland, widespread in the southwest. DZ, ES, FR (incl. Corsica).

Helianthemum canariense A small, spreading subshrub to 25(50) cm with prostrate to ascending stems, much-branched below. Leaves softly white-grey-hairy, with downturned margins. Flowers yellow *with conspicuous yellowish sepals with dark reddish prominent ribs*, enlarging in fruit. Restricted to maritime cliffs on Atlantic North African coasts. MA.

C. Plant a subshrub or shrub. Flowers white.

Helianthemum apenninum WHITE ROCK ROSE A variable (many subspecies described), loose, spreading shrub to 40 cm with linear to oblong, grey-hairy leaves 4–20 mm. Flowers white (also pink on Balearic Islands) with a yellow claw to each petal, in clusters of 2–10(15); petals 8–13 mm. Grassy and rocky habitats. ES (incl. Balearic Islands), FR, IT, MA.

1 2 3

Helianthemum almeriense (Syn. *H. pilosum*) A low, ascending dwarf shrub to 45(60) cm with *white-hairy* branches. Leaves longer further up the stem, 3–14 mm; more or less *linear,* green above and grey beneath with down-turned margins. Flowers borne in lax clusters of 2–6; *white* with yellow stamens; petals 8–12 mm; sepals with few hairs or hairless. Dry clay soils in southern Spain. ES.

Fumana

Dwarf shrubs similar to *Helianthemum* with narrow, lanceolate to linear, usually *alternate* leaves. Outer stamens sterile.

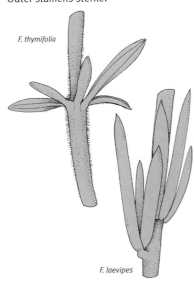

F. thymifolia

F. laevipes

A. Stipules present.

Fumana thymifolia THYME-LEAVED FUMANA A small, much-branched somewhat thyme-like, (not aromatic) dwarf shrub to 20 cm with erect or ascending branches. Leaves 4–12 mm, *opposite* at least below, unequal on the stem and reduced above; *oval to oval-lanceolate*, often *stickily hairy* and with *down-turned margins*; *stipules present*, as well as short shoots in the leaf axils. Flowers yellow, borne in 4–8-flowered clusters; the inner sepals much larger than the outer 2, membranous and green-veined; flower stalks 8–12 mm. Throughout. *Fumana laevipes* is similar to *F. thymifolia* but with alternate, linear, *bristle-like* leaves 8–10 mm. Flowers 5–10 per cluster; flower stalks 8–12 mm. Coastal habitats. Throughout.

Fumana arabica ARABIAN FUMANA A small, much-branched shrub to 25 cm with *leaves all alternate*, equally spaced along the stems and with stipules. Flowers yellow and rather large, 19–20 mm across, borne up to 7 in a raceme; the 2 outer sepals much smaller than the inner 3. Maquis and garrigue. East and south of region only. DZ, IT, MA, TN.

1. *Fumana thymifolia*
2. *Fumana ericoides*
3. *Aesculus hippocastanum*

1

2

3

B. Stipules absent.

Fumana procumbens PROCUMBENT FUMANA A small, finely-hairy stemmed shrub to 35 cm similar to *F. thymifolia* but more prostrate and with linear leaves 12–18 mm, *all alternate, with stipules absent. All flowers solitary* (or few), arising laterally; petals yellow with a dark golden spot at the base, *borne on short thick stalks* (0.8 mm thick), curving in fruit. North of the region. ES, FR, MA, IT. *Fumana scoparia* is similar but with many erect flowering stems 40–80 mm, curved in bud with 2–3 flowers; inflorescence very sticky-glandular-hairy. Rare and local but widespread. Throughout.

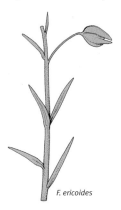

F. ericoides

Fumana ericoides ERICA-LEAVED FUMANA A variable (now with several distinct taxa recognised), erect (except in very exposed habitats) short dwarf shrub to 35(50) cm. Leaves 10–12 mm, *alternate, linear and fleshy, hairless, evenly spaced along the stem.* Flowers yellow, borne on arching, slender stalks *much longer* than the adjacent leaves (10–14 mm when mature) and spreading and down-turned at the apex in fruit. Dry slopes and garrigue. ES, MA. *Fumana ericifolia* is similar but less erect, with long, narrow, less fleshy leaves 8–15(20) mm long. Flower stalks 8–12 mm. Frequent in Iberian Peninsula from west Portugal eastwards. ES, PT.

SAPINDACEAE

A large family of perennials, lianas and trees (genera very distinct from each other). Leaves pinnately or palmately lobed or divided, with a petiole; stipules absent. Flowers unisexual or cosexual with 4–5 sepals and (0)4–5 petals; stamens 5–9; styles 1–2. Fruit variably dry or fleshy.

Aesculus HORSECHESTNUT

Deciduous trees opposite leaves. Flowers with 5 sepals forming a tubular calyx; stamens 5–9; style 1. Fruit a large capsule with 3 valves and 1–3 large seeds.

Aesculus hippocastanum HORSECHESTNUT A large, deciduous tree to 39 m with opposite, palmately divided leaves with 5–7 leaflets 10–25 cm long, the whole leaf to 60 cm across. Flowers white marked with pink-red, borne in erect panicles 15–30 cm in spring. Fruit a horsechestnut: a brown, shiny seed (conker) borne in a spiky green capsule 50–80 mm in the autumn. Native to the Balkans but planted in temperate areas at the northern boundary of the western Mediterranean. ES, FR, IT.

Acer MAPLE

Deciduous trees with opposite leaves. Flowers with 4–5 sepals and 0 or 4–5 petals; stamens 8; styles 2. Fruits with 2 parts, each 1-seeded with a long wing.

Acer monspessulanum MONTPELIER MAPLE A shrub or small tree 8–12(15) m with greyish, finely cracked bark and leathery, 3-lobed leaves 15–45 mm with blunt, untoothed lobes, shiny above and greyish below. Flowers yellow-green, borne in lax, erect clusters to 50 mm which appear before the leaves. Fruit 20–30 mm. Widespread but absent from the Balearic Islands and much of the north and west. *Acer granatense* (Syn. *A. opalus* subsp. *granatense*) is similar, with small, green sycamore-like leaves 25–80 mm across with 5 long, parallel-sided, shallowly toothed lobes. Fruit 25–35 mm. Dry wooded slopes at high altitude. ES, MA.

Cardiospermum

Herbaceous annuals or perennials, sometimes woody below, often with lobed leaves and small, soon-falling stipules. Flowers unisexual and zygomorphic; sepals 4(5); petals 4; stamens 8; style 1. Fruit a capsule.

Cardiospermum halicacabum BALLOON-VINE A tall, rather hairy annual 1–3 m, climbing with tendrils. Leaves trifoliate with deeply lobed and toothed leaflets 20–55 mm long. Flowers small and greenish-white with 4 sepals and 4 petals 2.5–5.2 mm long; stamens 8, of 2 sizes. *Fruit inflated and bladder-like*, 13–35 mm long, pendent; seeds black. Widespread as a native in tropical regions; naturalised widely. Probably local throughout.

ANACARDIACEAE

A large, mostly tropical family of shrubs, trees and lianas. Leaves alternate, simple or pinnately divided. Flowers with 5 sepals and petals and 5–10 stamens; styles 3. Fruit a small, 1-seeded drupe.

Pistacia MASTIC

Shrubs and trees (mostly) with alternate leaves. Dioecious; flowers unisexual; stamens 3–5. Fruit a drupe.

1. *Acer granatense* PHOTOGRAPH: SERGE D. MULLER   4. *Pistacia terebinthus*
2. *Pistacia lentiscus* (flowers)   5. *Schinus molle*
3. *Pistacia lentiscus* (fruits)

A. Leaves pinnately divided, without a terminal leaflet; stalks *winged*.

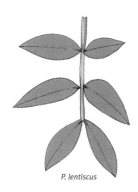

P. lentiscus

Pistacia lentiscus MASTIC A small, evergreen, dark green tree or shrub 6–8 m. Leaves dark green, pinnately divided *without* an end leaflet; leaflets 10–50 mm, borne 4–14, oval, leathery and untoothed, borne on winged stalks. Individual flowers rather inconspicuous, borne in dense *spike-like* clusters, the male with dark red anthers, the female greenish. Fruit 3.5–5 mm, spherical, red then black and shiny. One of the most abundant plants in the region. Throughout. Mastic (the resin) was formerly used as a chewing gum. *Pistacia × saportae* is (reportedly) a hybrid of *P. lentiscus* and *P. terebinthus*, with vegetative parts similar to *P. lentiscus* but with female flowers like *P. terebinthus;* fruit whitish. Widely reported but distribution improperly known. ES, FR, IT, MA.

B. Leaves pinnately divided, *with* a terminal leaflet; stalks *unwinged or narrowly winged*.

P. terebinthus

Pistacia terebinthus TURPENTINE TREE A hairless, aromatic *small, deciduous tree* 8–10 m with unwinged, pinnately divided leaves; leaflets 3–11 mm, and *with an end leaflet present*. Flowers brownish-red borne in lax, long-branched panicles. Fruit 5–9(12) mm. Maquis, dry slopes, rocks and pine woods; common, sometimes abundant, though more local than *P. lentiscus*. Throughout. *Pistacia atlantica* ATLAS MASTIC is a similar tree to 10 m with a broad, prominent trunk. Leaves with hairy stalks, the main axis *narrowly winged*; leaflets lanceolate. Open woods and maquis; abundant in parts of Algeria. DZ, MA, TN.

Pistacia vera PISTACIO NUT A tree to 10 m with pinnately divided leaves with 3–5 rather broad, pointed leaflets 50 mm–12 cm, which are minutely hairy when young. Fruit conspicuous: a pale brown nut 15–30 mm long, borne numerously in lax panicles. Cultivated for its edible fruit and sometimes naturalised. ES, FR, IT.

Schinus

Resinous trees native to South America with pinnately divided leaves with stalkless leaflets. Flowers with 4–5 sepals and petals and 8–10 stamens. Fruit berry-like.

S. molle

Schinus molle CALIFORNIAN PEPPER TREE A small evergreen tree to 15(25) m with slender, *hanging branches*. Leaves pinnately divided with 11–47 linear-lanceolate, toothed leaflets 15–60 mm that are hairy when young. Flowers 3–5 mm, yellow-white, borne in small, much-branched, hanging inflorescences; sepals and petals 5. Fruit 6–8 mm, pink, spherical. Widely planted as an ornamental; rarely naturalised. ES, IT (incl. Sardinia and Sicily), PT. *Schinus terebinthifolia* has leaves with 5–15 broader leaflets (30–60 mm) and numerous red fruits (4–5 mm); similar in general appearance to *Pistacia terebinthus* but with greenish-yellow (not red-purple) flowers with 10 stamens (not 3–5). Very occasionally planted.

Rhus SUMACH

Shrubs or small trees with pinnately divided leaves and thick twigs. Flowers often tiny and densely clustered, with 5 sepals and petals and stamens.

Rhus coriaria SUMACH A softly hairy shrub or tree 1–4(5) m with more or less evergreen leaves and densely downy shoots. Leaves pinnately divided with 7–21(25), toothed, green leaflets, *the stalk between the leaflets slightly winged*, at least towards the end; leaf stalk 20–30 mm. Flowers small and whitish, borne in very dense hairy panicles to 17–25 cm long. Fruit 4–6 mm, globular and woolly, ageing brown-purple. Widespread throughout but absent from most islands. *Rhus typhina* (Syn. *R. hirta*) STAGHORN SUMACH is very similar but with *leaf stalks 60 mm–10 cm, unwinged throughout*, 9–31 leaflets, and brighter crimson fruits 3–4.5 mm. Native to North America, widely naturalised. Throughout.

Searsia

Shrubs or small trees with alternate often 3–5-foliate leaves, sometimes with thorns. Flowers with (4)5(6)-parted perianth, the petals typically exceeding the sepals; styles 3(4). Fruit a drupe.

Searsia pentaphylla (Syn. *Rhus pentaphylla*) A *spiny shrub* 1–2 m, superficially similar to *Rhamnus*, with contorted, greyish-white branches. Leaves vaguely oak-like, *palmately divided* with 3–5 oblong to linear leaflets that are lobed or entire, often with 3 teeth at the apex. Flowers small and inconspicuous, green-yellow, borne in lateral racemes exceeded by the leaves. Fruit small and berry-like and dark purplish-red. Rocky habitats. Sicily and North Africa; common in Morocco. DZ, IT (Sicily), MA, TN. *Searsia tripartita* (Syn. *Rhus tripartita*) is similar but has *leaves with just 3 leaflets*. DZ, IT, MA, TN.

Cotinus

Deciduous shrubs with simple leaves slender twigs. Flowers borne in diffuse, plume-like panicles. Fruit 1-seeded.

Cotinus coggygria SMOKE TREE A rounded, deciduous, hairless shrub to 5 m. Leaves 30–80 mm, purple, simple, oval, untoothed and long-stalked. Flowers borne in large, lax panicles 15–20 cm long with many sterile branches bearing spreading hairs giving the impression of a *haze of smoke*. Fruit kidney-shaped, 3 mm long. Locally common in southern France and parts of Italy; absent from most islands. FR, IT.

MELIACEAE

A family of mostly trees and shrubs characterised by alternate, usually pinnately divided leaves without stipules. Flowers cosexual (or cryptically unisexual) borne in panicles, cymes, spikes, or clusters; stamens 3–10-numerous. Fruit a berry.

Melia

Trees with 2-pinnately divided leaves, usually toothed. Flowers with 5 sepals and petals; stamens 10–12, in 1–2 whorls. Fruits berry-like.

Melia azedarach INDIAN BEAD TREE A deciduous tree to 15 m tall with furrowed bark. Leaves 20–40(60) cm, alternate and 2-pinnately divided; leaflets elliptic and toothed or lobed, 50–70 mm. Flowers lilac and scented, borne in panicles; petals 5, each 8–12 mm. Fruit a berry 8–15(25) mm across, yellow and *long-persisting*, even when the tree sheds its leaves. Native to India and China but widely planted along roadsides, especially in Iberian Peninsula and southern France. Throughout.

NITRARIACEAE

A small family of annuals, perennials and shrubs with alternate, often fleshy leaves. Flowers with 5 persistent sepals and 5 petals; stamens 10–15; style 1. Fruit a drupe or capsule.

Peganum

Perennials with alternate, irregularly lobed leaves. Flowers with persistent sepals; sepals and petals 5; stamens 12–15. Fruit 3(4)-parted. Previously classified in the family Zygophyllaceae.

Peganum hamala An erect, hairless perennial with 5–13 branches, 40–45 cm high. Leaves rather fleshy, palmately, irregularly divided into 3–5 linear lobes 30–60 mm long and 1.5–3 mm wide. Flowers greenish- or yellowish-white, borne on stout stalks opposite the leaves; calyx lobes entire or slightly toothed. Capsule 9–13 mm, 3(4)-lobed, with 35–47 dark brown seeds. Derelict land and dry, semi-arid scrub. Local throughout, except Portugal and some islands.

1. *Rhus coriaria*   3. *Cotinus coggygria*
2. *Rhus coriaria*   4. *Melia azedarach*

RUTACEAE

A large and widely distributed family of aromatic herbs shrubs and trees. Leaves opposite or alternate with translucent glands. Flowers with 4–5 sepals and petals, free; stamens 2 x as many; style 1. Fruit a berry, capsule or drupe.

Ruta RUE

Strong-smelling perennials. Flowers yellowish with 4–5 sepals and petals, often lobed. Fruit a capsule.

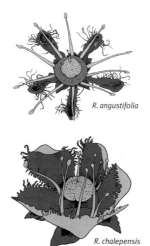

R. angustifolia

R. chalepensis

R. graveolens

Ruta angustifolia RUE A hairless, light grey-green woody-based and densely branched subshrub to 45(80) cm. Leaves 26 mm–13 cm alternate, 2–3-pinnately lobed with lobes to 7 mm, blue-green and very aromatic when crushed, short-stalked below, stalkless above. Lower bracts *scarcely wider* than their branches. Flowers yellow, borne in a *lax, glandular-hairy inflorescence*; petals spreading, 4–6 mm, widely separated and *with finely toothed margins*, the fringe as long as the width of the petal. Fruit persistent, 5–9 mm. Common in dry, rocky habitats. Throughout. *Ruta chalepensis* FRINGED RUE is similar to *R. angustifolia* but with lower bracts *much wider* than their branches, a *hairless inflorescence*, and petal fringes shorter than the width of the petal; petals 4–9 mm. Fruit 5–7 mm, ending abruptly in a spine-tip. Similar habitats. Throughout. *Ruta graveolens* COMMON RUE is similar but distinctly grey-blue, with denser branches, broad, oval-lanceolate leaf segments, and *toothed* petals. Fruits 3.5–9.1 mm. Similar habitats; cultivated and frequently naturalised. ES, FR, IT, PT. *Ruta montana* has flowers with *erect, unfringed petals* 3–6.5 mm, exceeded by the narrow, linear lobed leaves giving the inflorescence a leafy appearance; leaf segments narrow (just 0.5–1.4 mm wide). Fruit 2.1–4.3 mm. Throughout, though absent from islands except for the Balearic Islands.

Citrus

Small evergreen trees with glossy leaves. Flowers solitary to few, often fragrant; petals (4)5(–8); stamens numerous (16–20 to 100). Fruit large and edible. Relationships are complicated by a long history of cultivation and inter-breeding.

Citrus medica CITRON An evergreen shrub or small tree 2–5 m with reddish, spiny young branches. Leaves slightly toothed and scented when crushed, with scarcely winged stalks. Fruits yellow, very large, often mishapen to 25 cm. Locally cultivated throughout.

1. *Ruta angustifolia*
2. *Ruta chalepensis*
3. *Ruta graveolens*
4. *Citrus medica*
5. *Cltrus x sinensis*
6. *Citrus x limon*

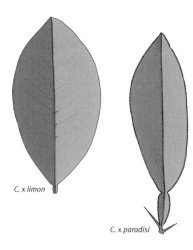

C. x limon

C. x paradisi

The following cultivars are frequently cultivated throughout the region: ***Citrus* × *limon*** LEMON A familiar evergreen, rounded tree to 4 m, flowering and fruiting throughout the year. Leaves elliptic-lanceolate and shallowly toothed. Flowers male or cosexual, with >4 x as many stamens as petals, petals white, often streaked purple. Fruit a lemon, yellow and warty when ripe. *Citrus* × *aurantium* SEVILLE ORANGE A small tree to 10 m with flexible spines. Leaves with a rounded base and *broadly winged stalks*. Fruit an orange. ***Citrus* × *sinensis*** SWEET ORANGE is similar but has narrowly *winged leaf stalks*. *Citrus* × *paradisi* GRAPEFRUIT is similar to *C. sinensis* but with a distinct, broadly winged petiole, and very large, yellow fruits.

Dictamnus BURNING BUSH

Perennials with pinnately divided leaves with a *winged axis*. Flowers slightly zygomorphic, borne in spike-like inflorescences; stamens 10; style 1. Fruit 5-lobed.

Dictamnus albus BURNING BUSH A strong-smelling, erect perennial to 90 cm with glandular-hairy, pinnately divided leaves with 7–9 oval, finely toothed, leathery leaflets 25–75 mm. Flowers large, zygomorphic, white or pink, veined violet borne in long, lax, leafless spikes; upper petals erect, 24–29 mm, the lower reflexed; stamens long-projecting, 25–31 mm. Fruits deeply 5-lobed. Open woods and maquis; absent from large areas and most islands. ES, IT. Some authors distinguish forms with *hairless leaves* with 13–17 segments 10–24 mm and shorter stamens 15–19 mm long as *D. hispanicus*. ES (southeast).

Cneorum

A small genus of plants with alternate, entire, leathery leaves and regular, cosexual flowers; sepals, petals, stamens and carpels all 3(4). Fruit made up of 3(4) mericarps.

Cneorum tricoccon A virtually hairless, small, evergreen shrub to 1.8 m. Leaves 23–40 mm, alternate, oblong, leathery, blunt and unstalked. Flowers yellow, solitary or in groups of up to 3, borne in the upper leaf axils, small, with petals to 5.5–7.4 mm long. Fruit 9.4–12.6 mm, conspicuous: bright red and strongly 3(4)-lobed, eventually turning black and splitting. Fairly common in woods and maquis or cliff-tops in the European western Mediterranean (absent from Corsica and Sicily). ES, FR, IT.

PRIMULACEAE | PRIMULA FAMILY

Herbaceous annuals or perennials. Flowers regular with free to fused sepals; petals 5(–9), fused (sometimes only just therefore seemingly free); stamens usually 5; ovary with a single style. Fruit a capsule.

Anagallis PIMPERNEL

Hairless annuals or perennials with opposite or alternate, simple leaves. Flower with 5 corolla lobes exceeding their tube. Fruit a splitting capsule.

A. arvensis

A. crassifolia

A. monelli

A. Plant a small, herbaceous annual.

Anagallis arvensis SCARLET PIMPERNEL A low, weedy annual with prostrate or ascending 4-angled stems to 40 cm. Leaves often >10 mm, oval, opposite and gland-dotted; unstalked. Flowers *blue, red or orange*, 4–7(10) mm, long-stalked (35 mm) and becoming curved in fruit; *petals with minutely hairy margins*, sometimes toothed at the tip. Very common in towns, waste places and near the coast. Throughout. Subsp. *foemina* (also treated as a true species *A. foemina*) has flowers always blue, with narrower petals, *without hairy margins*; corolla 5–8 mm. Throughout. *Anagallis crassifolia* is a similar annual or perennial with *stalkless leaves,* stems rooting at the nodes, and *flowers white or pink* with darker veins; corolla 3–5.5 mm. Absent from the east and most islands. DZ, ES, FR, MA, PT, TN.

B. Plant a woody-based, shrubby perennial.

Anagallis monelli SHRUBBY PIMPERNEL A low to short perennial with erect or spreading stems to 40(50) cm, sometimes shrubby and mat-forming in exposed areas. Leaves 6–15 mm, opposite or whorled, linear-lanceolate to elliptic, unstalked, often with small lateral shoots. Flowers bright blue (rarely pink or white), 15–25 mm across. Stamens with a tuft of hairs at the base. Locally common on stabilised dunes and other coastal sands; absent from much of the north of the region. DZ, ES, IT, MA, PT, TN.

1. *Cneorum tricoccon*
2. *Anagallis arvensis* (red-flowered form)
3. *Anagallis arvensis* (blue-flowered form)
4. *Anagallis monelli*
5. *Coris monspeliensis*

Samolus

S. valerandi

Hairless perennials with leaves in a basal rosette. Flowers white with 5 sepals and corolla lobes. Capsule with 5 teeth.

Samolus valerandi BROOKWEED A hairless perennial herb with a rosette of leaves 10–50 mm long, and erect flowering stems 50 mm–70 cm. Leaves rather shiny, spoon-shaped and scarcely stalked below; stalkless above. Flowers small and white, cup-shaped, to just 3 mm across and with 5 petals. Fruit 3 mm. Damp open habitats, often coastal; local. Throughout.

Coris

Flowers reddish-violet, corolla distinctly 2-lipped with 3 longer upper lobes and 2 shorter lower lobes; calyx with 10 or more lobes.

Coris monspeliensis A distinctive plant not instantly recognisible as a member of the Primulaceae. A short annual or biennial with ascending to erect stems 10–25(40) cm that are woody at the base. Leaves 4–9 mm, very narrowly linear (almost needle-like), densely arranged, unstalked and alternate up the stem. Flowers borne in dense terminal heads; corolla 9–16 mm, pinkish blue and 2-lipped and divided into unequal lobes; stamens 5. Fruit 1–2.5 mm. Dry, rocky habitats; very common around the coasts of Spain. DZ, ES, FR, IT, MA, TN. Subsp. *hispanica* (also treated as a true species *C. hispanica*) has down-turned leaf margins, white to pale pink flowers and 4 (not 6–14) outer calyx teeth. Coastal rocks and slopes in southeast Spain only. ES.

Cyclamen CYCLAMEN

A distinctive genus of perennials with simple leaves and solitary flowers with *strongly reflexed petal lobes;* lobes 5. Capsule 5-valved. Previously classified in the family Myrsinaceae, but most recent DNA analysis confirms the genus is in the Primulaceae.

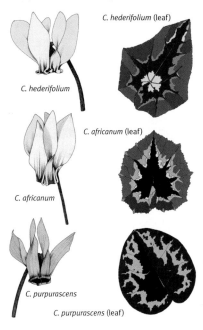

C. hederifolium (leaf)

C. hederifolium

C. africanum (leaf)

C. africanum

C. purpurascens

C. purpurascens (leaf)

A. Autumn-flowering.

Cyclamen hederifolium IVY-LEAVED CYCLAMEN An *autumn-flowering* tuberous perennial herb to 30 cm producing ivy-like, normally *lobed, slightly toothed, pointed, grey or cream-marbled leaves* 25–80 mm long in the autumn, during or shortly after flowering; leaves and flowers extend outwards in bud. Flowers pale pink or white with purple in the throat; corolla lobes (12)14–25 mm long, with crimson-purple at the base; style scarcely protruding. Fruit borne on coiled stalks. Deciduous woods and stony garrigue. FR, IT. *Cyclamen africanum* is very similar but with *larger flowers with lobes 18–35 mm long* which, like the leaves, *arise erect from the tuber in bud*. Rocky habitats in North Africa only. DZ, TN. *Cyclamen purpurascens* is similar to the above species but with indistinctly toothed leaves with converging basal lobes and *darker, reddish-pink flowers*; style protruding by 1 mm. Southern France and northern Italy only; local. FR, IT.

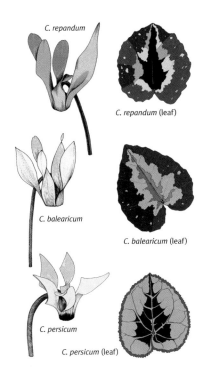

C. repandum

C. repandum (leaf)

C. balearicum

C. balearicum (leaf)

C. persicum

C. persicum (leaf)

B. Spring-flowering.

Cyclamen repandum A *spring-flowering* tuberous perennial to 15 cm similar in form to *C. hederifolium* with heart-shaped, slightly angled, bright green *leaves produced before the flowers*. Flowers typically *evenly purple-pink;* lobes to 30 mm long. Fruits borne on coiled stalks. Woods and shady maquis. FR (incl. Corsica), IT (incl. Sardinia). *Cyclamen balearicum* BALEARIC CYCLAMEN has leaves with obtuse tips and often slightly revolute margins. Flowers small, *white, veined pink* (without darker blotches); lobes 9–16 mm; style not protruding. ES (Balearic Islands), FR (local).

Cyclamen persicum PERSIAN CYCLAMEN A winter-to spring-flowering tuberous perennial with heart-shaped, finely toothed (not angled) green or marbled leaves to 14 cm across, purplish beneath. Flowers scented, white, pink or mauve, with lobes 20–37 mm long, *markedly twisted*; flowers appear when leaves are mature. Fruit *stalks not coiling*. Rocky garrigue in North Africa (commonly cultivated for ornament). DZ, TN.

CORNACEAE | CORNUS FAMILY

Trees or shrubs with simple, opposite leaves. Flowers borne in axillary or terminal inflorescences, cosexual; sepals, petals and stamens 4; ovary inferior, formed of 2 carpels. Fruit a berry or drupe.

Cornus mas CORNELIAN CHERRY A deciduous shrub or tree to 8 m with oval to elliptic, pointed, untoothed leaves 40 mm–10 cm. Flowers bright *yellow*, small, 4–5 mm across, borne in congested clusters of 10–25 *before the emergence of the leaves*. Fruit ovoid, 12–15 mm long, shiny red when ripe. Woods and thickets and cultivated as an ornamental. FR, IT.

Cornus sanguinea DOGWOOD A deciduous shrub 1.5–5 m with erect, *dark red twigs* and elliptic to oval, pale green, hairy, untoothed or toothed leaves. Flowers dull *white*, 7–11 mm across, borne in dense, umbel-like clusters. Fruit rounded, 5–8 mm across, black when ripe. Woods, scrub and ditches. Absent from the Balearic Islands. ES, FR, IT.

1. *Cyclamen balearicum*
2. *Cyclamen hederifolium*
3. *Impatiens balfourii*

BALSAMINACEAE | BALSAM FAMILY

Herbaceous hairless annuals with simple leaves and translucent, fleshy stems. Flowers strongly zygomorphic; sepals 3, petal-like; stamens 5; style 1. Fruit a capsule that violently expels seeds when ripe.

Impatiens

Herbaceous annuals with succulent stems swollen at the nodes and opposite or alternate leaves. Flowers zygomorphic. Fruit an explosive capsule.

Impatiens balfourii An annual herb to 1.2 m with hairless, reddish-translucent, fleshy stems. Leaves alternate, oval-lanceolate, toothed, stalked, to 40 mm long. Flowers borne in racemes, to 20 mm long, pink and white, the lower white sepal forming a long, slender spur >10 mm long. Native to the Himalayas but naturalised in towns in France and Italy. FR, IT.

ERICACEAE | HEATHER FAMILY

Evergreen trees and shrubs (often characteristically heather-like) with alternate, opposite or whorled leaves. Flowers often borne in clusters (rarely solitary); sepals and petals (3)4–5, petals fused; stamens 2 x as many as petals; style 1. Fruit a capsule, drupe or berry.

Arbutus

Evergreen shrubs with alternate leaves. Flowers with 5 petals, borne in terminal clusters; stamens 10. Fruit a warty berry.

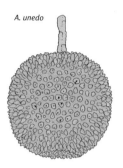

A. unedo

Arbutus unedo STRAWBERRY TREE An evergreen shrub or small tree 4–5(11) m. Bark dull brown and fissured. Leaves 80 mm long, oblong-lanceolate, short-stalked and somewhat toothed, more or less hairless. Flowers scented; white, tinged green or pink, bell-shaped with recurved petal lobes, 7–8(11) mm borne in drooping panicles. Fruit a globose berry 7–15(20) mm, ripening deep crimson; rather like a wild strawberry. Very common on undisturbed maquis and open woodland in the west; rarer further east and in North Africa. Throughout.

Erica HEATHER

Dwarf to medium shrubs with whorled, narrow to linear leaves. Flowers bell-shaped to globular, borne in spikes or panicles, generally 4-lobed; stamens 8. Fruit a dry capsule.

A. Stamens included in the corolla; shrub >1 m tall.

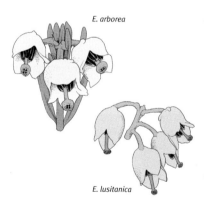

E. arborea

E. lusitanica

Erica arborea TREE HEATH A tall shrub or small tree to 4(7) m with densely hairy younger twigs. Leaves in groups of 3–4, to 4–9 mm long. Flowers pure *white*, broadly bell-shaped, 2–4 mm long with erect lobes, borne in dense terminal panicles; sepals 1.2–2 mm. Fruit capsule 2 mm. Common to abundant in a range of habitats, often above sea level. Throughout. *Erica lusitanica* is similar but with larger, paler green leaves 2–7.5 mm long and larger flowers 3.5–5.5 mm, white *tinged with pink*, particularly in bud, with red stigmas; sepals 1–1.2 mm. Capsule 1.8–2.3 mm. Similar habitats; especially damp places. ES, PT.

Erica scoparia GREEN HEATHER A tall shrub to 2.5(4) m with narrow, hairless leaves 4–7(10) mm, with inrolled margins borne in whorls of 3–4. Flowers borne in slender racemes, small, just 2.1–2.3 mm long, and *greenish-white* or red-tinged with anthers not protruding; stigma reddish and club-shaped; sepals 1–1.3 mm. Capsule 1.3–4.8 mm. Acidic rocky habitats and pine woods, often at low altitude. ES, FR, IT, PT.

---

1. *Arbutus unedo*
2. *Erica arborea*
3. *Erica scoparia*
4. *Erica sicula* PHOTOGRAPH: GIANNIANTONIO DOMINA
5. *Erica cinerea*

B. Stamens included in the corolla; shrub normally <1 m (except *E. terminalis*).

Erica cinerea BELL HEATHER A low, spreading, hairless undershrub to 60 cm. Leaves 4–7 mm, borne in groups of 3, dark green and hairless. *Flowers reddish-purple*, 4–6 mm long, borne in scattered groups (not just at the tops of the stems); stamens hidden within the corolla; sepals 2–3.5 mm. Capsule 2 mm. Damp, acidic habitats and at higher altitudes; rare in the Mediterranean. ES, PT.

Erica sicula A dense shrub to 50 cm (1.5 m) with *velvety* dark green leaves 5–6 mm with down-turned margins. Flowers pale pink-white with 5 sepals and petal lobes (4 in other species), and with 10 contrasting brown-red stamens with anthers not protruding from the corolla tube, borne on long and pendulous pinkish stalks in apical clusters. An island endemic of limestone cliffs. IT (Sicily), MT.

Erica ciliaris DORSET HEATH A medium shrub to 60 cm, often clumped, with erect branches from a spreading base. Leaves borne in groups of 3, 2–3 mm long with distinctive clusters of smaller leaves in their axils. Flowers large, 8–12 mm long and deep pink, borne in *distinct terminal, elongated racemes;* sepals 2.5–4 mm. Capsule 2–2.3 mm. Various damp or dry acidic habitats in the far southwest. ES, MA, PT. *Erica terminalis* CORSICAN HEATH is similar but taller, to 2.5 m with leaves in whorls of 4. Flowers 6–7 mm, bright pink, borne in terminal umbel-like clusters of 3–8; sepals 1.5–2.5 mm. Capsule 2–2.5 mm. Shady woods and ravines. ES, FR (Corsica), IT (incl. Sardinia).

E. australis

E. umbellata

C. Stamens longer than the corolla with anthers projecting.

Erica australis SPANISH HEATH A slender shrub to 2.5 m with suberect branches. Leaves linear, 3–7 mm in whorls of 4, with down-turned margins that completely conceal the lower surface. Flowers 6–9(10) mm, deep reddish pink, tubular-bell-shaped with reflexed petal lobes, borne in clusters of 4–8 forming lax, terminal panicles; flower stalks equalling the calyx; sepals 2–3.8 mm. Capsule 2.5–3 mm. Common on scrub and in open woods. ES, MA, PT. *Erica umbellata* is similar to *E. australis* but *shorter*, to 80 cm (1 m) with *smaller* flowers 3.5–5.5(7) mm long, borne in terminal umbels of 3–6; corolla rather globose with erect lobes; sepals 1.5–2 mm. Capsule 2–2.5 mm. Heaths, scrub and open woods; locally frequent in the far southwest. ES, MA, PT.

Erica vagans CORNISH HEATH A bushy, hairless, medium shrub to 80 cm. Leaves 5–10 mm, linear, and in groups of 4–5 with down-turned margins. Flowers small, to 2.5–3.5 mm long, borne on long stalks on long, leafy racemes, pink or lilac in colour with brown, protruding anthers; corolla lobes divergent; sepals short, just 1–1.3 mm. Capsule 1.8 mm. ES, FR. *Erica erigena* IRISH HEATH is similar in form; a rather tall, hairless shrub to 1.2(2) m with a well-developed main stem. Leaves borne in groups of 4, linear, 5–8 mm long. Flowers to 5–7 mm long, pale pink-purple, borne in long, leafy racemes; anthers reddish and $^1/_2$-protruding; sepals 2.5–3.5 mm. Local and strongly western in distribution. ES, FR, PT.

Erica multiflora An erect, branched shrub to 1(3) m with leaves in whorls of 4–5, linear, 6–11 mm long, and dense, rounded, terminal clusters of pale pink flowers 3.5–5.5(7) mm *borne on long stalks, up to 3 x as long as the calyx* (sepals 1.5–2 mm); anthers completely projecting. Capsule 2–2.5 mm. Garrigue and maquis; locally common, particularly in the east. ES, FR, IT.

Calluna LING

Heather-like, woody shrubs with opposite (not whorled), stalkless leaves. Flowers with 4 petals; stamens 8. Fruit a capsule.

C. vulgaris

Calluna vulgaris LING A short, compact scrub to 80 cm (1.5 m); superficially conifer-like when in leaf, typically heather-like when in flower. Leaves scale-like 2–3.5 mm, closely adpressed to the stem and with strongly downturned margins, making cross-sections triangular. Flowers 3–4.5 mm, abundant, pink, borne on narrow, crowded racemes; corolla lobed nearly to the base. Frequent in sand dune plant communities, heaths and moors in the north and west. ES, FR, IT, MA, PT.

Rhododendron

Evergreen shrubs with alternate leaves. Flowers showy, with 5 petals fused to form a bell-shaped corolla; stamens 5 or 10. Fruit a capsule.

Rhododendron ponticum RHODODENDRON An erect, evergreen shrub to 5 m with spreading branches. Leaves large, 60 mm–20 cm long, entire, leathery, dark green above and paler below; hairless. Flowers 40–60 mm across, showy, borne in lax racemes of up to 15 flowers; corolla pale purple, with 10 stamens. Local in rocky, wooded valleys and also grown for ornament and naturalised. ES, PT.

1. *Erica australis*
2. *Erica multiflora*
3. *Calluna vulgaris*

Corema

C. album

Densely branched, woody subshrubs with crowded, alternate leaves with down-turned margins. Flowers with 3 sepals and petals; stamens 3. Fruit a fleshy drupe.

Corema album A heather-like, erect, dull, brownish-green shrub to 1 m. Leaves to 10 mm, linear with a deep groove running along the centre. Flowers borne in clusters of 5–9, virtually stalkless. Male flowers pinkish with red anthers, female flowers with reduced petals or petals absent, and a reddish 3-lobed stigma. Fruit white (or pinkish) and berry-like. An Atlantic species extending to southwest Iberian Peninsula and naturalised in France. ES, FR, PT.

STYRACACEAE | STORAX FAMILY

Mainly shrubs and trees with spirally arranged simple leaves without stipules. Flowers regular, perianth with 4–5(7) lobes; stamens (5)8–10(20); styles and stigmas 1. Fruit a dry capsule, sometimes winged, or drupe.

Styrax officinalis STORAX A deciduous shrub or tree 2–5(7) m with white-hairy twigs. Leaves 50 mm–10 cm, alternate, oval to oblong, untoothed and short-stalked. Flowers *white*, bell-shaped and pendent, 18–22 mm, borne in clusters of 3–6, fragrant; petals 5–6, fused at the base; stamens yellow and prominent. Fruit 1-seeded, ovoid, white-woolly with a persistent calyx. Woods and near water. IT.

1. *Rhododendron ponticum*
2. *Corema album*
3. *Corema album* (fruits)
4. *Rubia balearica*
5. *Rubia peregrina*
6. *Plocama calabrica* PHOTOGRAPH: GIANNIANTONIO DOMINA

RUBIACEAE

A very large family of herbs with opposite or distinctly whorled leaves; stipules present between each pair of leaves, often leaf-like. Flowers typically small, funnel-shaped and tubular, borne in dense heads, branched cymes or panicles; sepals o or minute, petals 4–5; stamens 4–5; styles 1–2 (if 1, branched); ovary inferior. Fruit fleshy or dry and berry-like, 1–2-seeded.

Rubia

Scrambling perennials or shrubs with leaves in whorls of 4, equal and stalkless. Flowers with 4-lobed, yellow corolla; the terminal cosexual, the laterals male. Fruit a pair of smooth nutlets.

A. Plant scrambling or climbing.

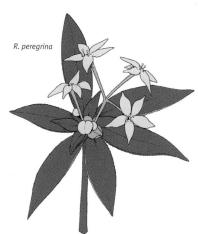

R. peregrina

Rubia peregrina WILD MADDER A medium to tall, trailing or scrambling hairless, evergreen perennial to 7 m with a creeping rootstock. Stems square and rough with down-turned bristles. Leaves in whorls of 4–6(8); oval to elliptic, tough, leathery and dark shiny green, 1-veined, *4–27 mm wide*. Flowers pale yellow-green, 3.5–8 mm, 5-lobed, forming dense, leafy panicles. Fruit 3–7(9) mm, black and fleshy when ripe. Common on scrub. Throughout. *Rubia agostinhoi* is similar to *R. peregrina* but with *narrower leaves just 2–5.5 mm across in whorls of 8(10)*. Flowers 4–7 mm. Fruit 4–7 mm. ES (south), MA. *Rubia tinctorum* MADDER is similar to *R. peregrina* but with leaves in whorls of 4–6 which are *paler green, not notably leathery, and with prominent lateral veins*. Flowers pale yellow, 3.5–6 mm. Fruit red-brown or blackish, 2.5–6.5 mm. IT (naturalised elsewhere).

B. Plant tufted and bushy.

Rubia balearica (Syn. *R. angustifolia*) BALEARIC MADDER A similar species to *R. peregrina*, but a *tufted, bushy* perennial to 1 m. Stems square, erect and persistent, much-branched, bearing leaves 0.7–3.5 mm wide in whorls of 4–6, which are linear with down-turned margins, and covered in minute spines. Flowers 3.5–6 mm, pale yellow-green, 5-parted, and borne in clusters in the axils. Fruit 3–4.5 mm. Rocky scrub and cliff-tops; an island endemic. ES (Balearic Islands).

Plocama

Herbs or shrubs, strong-smelling when crushed, with opposite leaves (sometimes appearing whorled). Corolla funnel-shaped with 4 spreading lobes; stamens 4, inserted to projecting. Fruit a drupe or splitting into 1-seeded mericarps.

Plocama calabrica (Syn. *Putoria calabrica*) PUTORIA A *strong-smelling,* dwarf, much-branched, prostrate to spreading, almost thyme-like shrublet to 1 m with opposite, elliptic to oval, leathery leaves 10–28 mm with down-turned, minutely hairy margins; stipules small and inconspicuous. Flowers 13–19 mm, borne in dense terminal heads; pink and funnel-shaped with a long tube and 4 spreading, linear-lanceolate lobes 2.8–4.7 mm; stamens projecting. Fruit 4–5 mm, 2-lobed and red or blackish when ripe. Garrigue and rocky ground, often by the sea. Absent from the far west and most islands. DZ, ES, FR, IT, MA, TN.

Crucianella

C. maritima

Annual or perennial herbs with whorled leaves and yellowish, stalkless or virtually stalkless flowers borne in 2- or 4-ranked *compact spikes*; corolla fused below into a tube.

A. Succulent perennials with *whitish stems*.

Crucianella maritima CRUCIANELLA A prostrate to spreading low shrub to 50 cm with whitish, hairless stems. Leaves 3–11 mm, borne in groups of 4, forming regular, symmetrical, close-set whorls; grey, leathery, spine-tipped with a whitish margin. Flowers yellow, 10–12.6 mm long, 5-lobed. Frequent on coastal sand dunes. Throughout.

B. Annual herbs; greenish or reddish.

Crucianella latifolia A rather lax and inconspicuous annual to 40 cm with leaves 3–28 mm borne in whorls of 6–8, linear-lanceolate and broader below, not spine-tipped. Flowers with corolla 4.5–6.2 mm, borne in rather flattened, *cylindrical inflorescences*, cream or purplish; corolla 4-parted, stamens 4; bracts fused. Bare places. Throughout. *Crucianella angustifolia* is similar but with slender, roughly *4-dimensional inflorescences* with flowers in whorls and bracts free; corolla 4–4.7 mm. Bare and dry habitats. Widespread but rare or absent on many islands. *Crucianella patula* is rather similar in form to the previous species but with flowers with *5-parted corolla 1.7–2.4 mm, exceeded by their bracts;* flowers borne on short stalks. ES (incl. Balearic Islands), MA.

Cruciata

Herbs with 3-veined leaves borne in whorls of 4–10. Flowers yellow, cosexual or male. Fruit 2 hairless nutlets.

Cruciata laevipes CROSSWORT A softly hairy, weak-stemmed, light green perennial 15–75 cm with weakly 3-veined leaves 10–27 mm, borne in whorls of 4 and almost stalkless clusters of 5–11 yellow flowers forming interrupted spikes; flower clusters exceeded by the leaves. Fruit 2.5–2.8 mm, smooth and hairless, borne on recurved stalks. Grassy habitats; absent from hot, arid areas. ES, FR, IT, PT. *Cruciata glabra* is similar but more lax and slender, with smooth, hairless, 4-angled stems, and shortly hairy, and later hairless leaves. Flowers 3–5(8) per cluster. Similar habitats; uncommon. ES, FR, IT, PT.

Asperula

Annual or perennial herbs or subshrubs with square stems and linear leaves in whorls of 4 or more. Flowers with 4-lobed corolla. Fruit a pair of nutlets.

A. Plant an annual; flowers blue.

Asperula arvensis BLUE WOODRUFF A short, slender, hairless annual to 50 cm with linear to linear-lanceolate, equal leaves in whorls of 6–8. Flowers blue or blue-violet, with slender tubes 5–6.5 mm long, much longer than the lobes, borne in clusters surrounded by leafy bracts. Fruit smooth and brown. Disturbed ground. Throughout.

B. Plant a perennial or subshrub; flowers white, pink or reddish.

Asperula laevigata A delicate perennial with long stems to 84 cm with *broadly elliptic to oval leaves* 2.4–8 x 1.7–5 mm, borne in whorls of 4 with a prominent mid-vein and netted veins either side. *Flowers white*, 1.3–1.9 mm long, narrowly funnel-shaped with 4 lobes. Absent from the far west and south. ES, FR, IT.

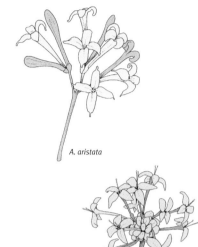

A. aristata

Asperula aristata A short to medium subshrub, woody at the base with green or greyish, slightly hairy, 4-angled stems to 95 cm. Leaves 3–8.5 x 0.3–1.1 mm, in *whorls only of 4*, lanceolate to linear and colourless at the tip, the margins often down-turned. Flowers 5–8(9) mm, *pale pink* (sometimes greenish or yellowish), funnel-shaped and with 4 petal lobes. Garrigue and maquis. *Asperula crassifolia* is similar in form but with *blue-grey leaves and stems*, leaves fleshy, and flowers yellowish and hairy. IT (incl. Sicily).

A. hirsuta

Asperula hirsuta A short to medium, scrambling or climbing subshrub, woody at the base with 4-angled, erect stems 15–70 cm. Leaves in *whorls of 4 or more*, linear and thin, 1.2–8 x 0.4–1 mm, dark green, *even when dry. Flowers pale pink to brownish red*. Garrigue and maquis. **Asperula paui** is similar to the previous species but a small and very slender, greyish, hairless plant to 48 cm with with *just 2 leaves per node* (rarely more); leaves narrow, 1.5–8 x 1.1–3.5 mm, with paired stipules. Flowers small and very pale pink, 3.1–7.7 mm. Pine forests and scrub. ES (east + Balearic Islands).

1. *Crucianella maritima*
2. *Crucianella angustifolia*
3. *Crucianella angustifolia* (flowers)
4. *Cruciata laevipes*
5. *Asperula aristata*
6. *Asperula paui*
7. *Asperula paui* (leaves)

Galium BEDSTRAW

Similar to *Asperula* but with rounded stems, and leaves in whorls of 4–12. Flowers with white to yellow 4-lobed corolla. Fruit a pair of bristly nutlets with hooked bristles.

A. Leaves 3-veined, in whorls of 4.

Galium rotundifolium ROUND-LEAVED BEDSTRAW A low to short, leafy, sparsely hairy perennial with *broad, almost circular, fine-pointed leaves* in whorls of 4, each with *3 parallel veins*. Flowers white or greenish, borne in lax clusters. Fruit with hooked bristles. Throughout, though absent from the Balearic Islands. *Galium scabrum* is similar but *densely hairy* and with long, lax, spreading inflorescences. ES, FR (Corsica), IT (incl. Sardinia and Sicily).

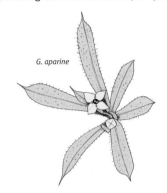

G. aparine

B. Leaves 1-veined in whorls of 4 or more, stems with bristles on the angles.

Galium aparine GOOSE-GRASS A low to short, familiar annual with spreading stems to 1(3) m, often stouter and more hairy at the nodes, with strongly recurved prickles on the stems. Leaves 3–5 mm, narrowly oblong, in whorls of 6–8, rough-margined. *Corolla short-tubed, 1.5–3 mm across, whitish.* Nutlets 1.1–4.1 mm with hooked bristles. A common weed on disturbed ground. Throughout.

C. Leaves 1-veined in whorls of >4, stems with smooth angles or no angles; flowers white and conspicuous.

Galium album (Syn. *G. mollugo*) HEDGE-BEDSTRAW A variable, scrambling perennial to 1.5(2) m, with 4-angled, smooth and solid stems. Leaves borne 5–8 per whorl, 10–25 mm long, and *wide, to 2.5 mm;* oblong to elliptic, 1-veined, pale green and bristle-tipped with forward-pointing bristles along the margins. Flowers short-tubed, white, 2.6–5.3 mm across with 4 strongly pointed lobes, borne in spreading, loose inflorescences. Nutlets hairless and wrinkled, 1–2 mm across. Common in a range of habitats including thickets and garrigue. ES, FR, IT, MA. *Galium lucidum* is rather similar, but with narrow leaves to just 1.5 mm wide, with downturned margins. Flowers white or cream. Absent from Corsica, otherwise throughout.

1   2   3   4

D. Leaves 1-veined in whorls of 4 or more, stems with smooth angles or no angles; flowers small and inconspicuous.

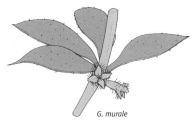

G. murale

Galium murale A low to short, typically goose-grass like, sprawling annual with stems to 25(30) cm, much-branched from the base and sparsely hairy. Leaves 1.7–6 mm long, narrowly elliptic with a *short spine-tip*, in whorls of 4–6. Flowers inconspicuous, yellowish, minute, just 0.4–0.65 mm, borne in lax, few-flowered inflorescences; corolla lobes pointed, erect. Nutlets 1.1–1.7 mm, rather cylindrical with spreading lobes and unevenly hairy. Common on waste and fallow land. Throughout. *Galium verticillatum* is similar but has whitish flowers 0.5–1.75 mm across and *markedly deflexed leaves* 1.7–5 mm long. Nutlets 0.7–1.2 mm, evenly hairy. ES, FR, IT.

E. Leaves 1-veined in whorls of 4 or more, stems with smooth angles or no angles; flowers *yellow* and conspicuous.

Galium verum LADY'S BEDSTRAW A creeping or scrambling perennial to 1 m with faintly 4-angled stems and erect flowering stems. Leaves linear and spine-tipped, 5–21(30) mm long, and *narrow*, to 0.8(1.4) mm wide, dark green above and downy below with inrolled margins, 8–12 in a whorl. Flowers *bright yellow* (most species in the genus with white flowers), 2–3.5 mm across, borne in erect, terminal inflorescences. Nutlets 0.8–1.8 mm, smooth and black. Grassy habitats, woods and woodland clearings, generally above sea level. ES, FR, IT, MA. *Galium crespianum* is a bushy perennial with long, narrow leaves 12–38 mm, and *broader, 0.9–2.7 mm wide,* in whorls of 4–8 and numerous *dull cream-yellow* flowers 3.5–5 mm across. Rocky slopes on Mallorca only; rare. ES (Balearic Islands).

Sherardia

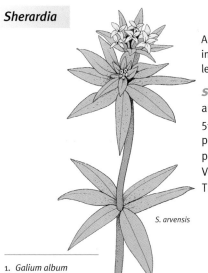

S. arvensis

Annuals with leaves in whorls of 4–6. Flowers lilac, borne in dense terminal and lateral clusters surrounded by 8–10 leafy bracts. Fruit a pair of nutlets with a persistent calyx.

Sherardia arvensis A small, slender, hairy or hairless annual with spreading stems to 40 cm or less. Leaves 5–18 mm, in whorls of 4–6, soon withering below, pale green. Flowers borne in clusters of 4–10; corolla pale lilac, 4–5 mm, with tube longer than the lobes. Very common on fallow land and in grassy places. Throughout.

1. *Galium album*
2. *Galium album* (flowers)
3. *Galium verum*
4. *Sherardia arvensis*

Valantia

Small annuals with leaves in whorls of 4. Flowers with 3–4 lobes, whitish or yellowish, borne in clusters of 3 in the leaf axils. Fruit 2 hairless nutlets.

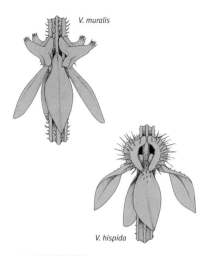

V. muralis

V. hispida

Valantia muralis A low, prostrate or ascending annual with fleshy, softly *hairy* stems to 23 cm. Leaves in whorls of 4, oval and broadest above the middle, 1–5(8) mm long, downward-pointing. Flowers greenish yellow, 1–1.5 mm, borne in clusters at the base of the leaves. Fruits borne on stalks with *conspicuous horn-like appendages*; nutlets usually solitary, and smooth. On rocks, cliffs and old walls; frequent. Throughout. *Valantia hispida* is very similar in form, but has stems that are very bristly towards the apex (not soft-hairy), and fruits without horn-like appendages, *but densely bristly;* nutlets usually 2, brown or blackish and rough. Throughout. *Valantia lainzii* is virtually identical but has *hairless stems*. Nutlets generally solitary, smooth and black. Dunes in southern Spain only. ES.

Theligonum

Virtually hairless annuals with opposite to alternate leaves. Monoecious; male flowers borne in 2s or 3s with 2–5 lobes; female flowers solitary or in 2s or 3s with 2–4 poorly defined lobes. Fruit nut-like.

Theligonum cynocrambe A small, prostrate or ascending annual 80 mm–40(50) cm, unpleasant smelling. Leaves opposite below, alternate above, 10–20(40) mm, untoothed and slightly succulent. *Flowers inconspicuous*, 2–3 mm solitary or in small lateral clusters of 2–3; stamens much-exserted. Fruit 1.6–2.3 mm, slightly fleshy. Rocky, damp and shady habitats. Probably throughout.

1

2

3

4

GENTIANACEAE | GENTIAN FAMILY

Hairless annuals and perennials with opposite, untoothed leaves. Flowers with 4–5(8) petals fused into a tube; calyx deeply lobed; stamens as many as petals; styles 1–2 (if 1, branched). Fruit a 2-parted, splitting capsule.

Centaurium CENTAURY

Hairless annuals to perennials, often small. Flowers pink (rarely white) with 4(5) calyx and corolla lobes; stamens 5; style 1, divided.

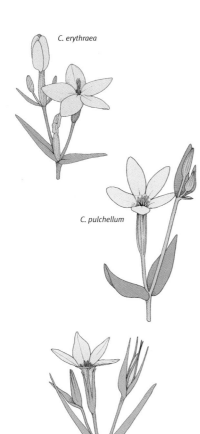

C. erythraea

C. pulchellum

C. maritimum

A. Flowers pink.

Centaurium erythraea COMMON CENTAURY A short biennial to 50 cm with a solitary stem, branched above. Leaves 20–60 mm, elliptic to oval, 3–7-veined, the lower in a *distinct rosette*, the stem leaves smaller. Flowers pink, purplish or sometimes white, virtually stalkless, borne in *flat-topped* clusters; corolla lobes 4.5–6 mm. Common in dry grassy places and maritime environments. Throughout. ***Centaurium pulchellum*** LESSER CENTAURY is similar but smaller, to 20 cm (often < 60 mm) *lacking a distinct basal leaf rosette*, with the stem usually forked in the lower part with *wide-spreading branches*, and with fewer flowers borne in *lax, open clusters*; corolla lobes 2–4 mm. Damp, grassy places. Throughout. *Centaurium tenuiflorum* SLENDER CENTAURY is similar to *C. pulchellum,* also lacking a basal rosette, and the *stems erectly branched above*, and the flowers are borne in *dense narrow* clusters; corolla lobes 2–4 mm. Damp, marshy, coastal places. Throughout.

B. Flowers yellow.

Centaurium maritimum YELLOW CENTAURY A short annual to biennial to 30 cm with a solitary stem that is branched above. Leaves 3–10 mm, fleshy, elliptic to oval, the lower 2 small, and *not* forming a distinct rosette, the upper leaves much longer. *Flowers pale yellow*, borne in loosely branched clusters, sometimes solitary; petals lobes 5, elliptic, 4–7 mm. Sandy and grassy maritime habitats. Throughout.

1. *Valantia muralis*
2. *Centaurium erythraea*
3. *Centaurium maritimum*
4. *Centaurium pulchellum*

Schenkia

Small annuals with straight, erect stems. Similar to, and still widely included in, the genus *Centaurium*.

Schenkia spicata (Syn. *Centaurium spicatum*) SPIKED CENTAURY An erect annual to 50 cm, similar to *Centaurium* but easily distinguished by its *spike-like inflorescences 60 mm–25 cm with erect-pointed, pink flowers;* flowers 10–16 mm, short-stalked (1.5 mm); corolla lobes 3.5–5 mm. Damp coastal habitats and salt-marshes. Throughout. *Schenkia elegans* is similar but has stalkless flowers 7–11 mm; corolla lobes 2–3 mm. Local, in Iberian Peninsula only. ES, PT.

Blackstonia YELLOW-WORT

Herbs with pairs of stem leaves fused at the base. Flowers yellow, short-tubed, with 6–8-numerous spreading lobes; style 2-lobed.

Blackstonia perfoliata YELLOW-WORT An erect, hairless, bluish plant to 40(50) cm tall with stems branching from the base or from the middle, from a basal rosette of broadly oval leaves which soon wither; stem leaves opposite, oval-triangular, pointed and narrowed at the base and joined there, *almost encircling the stem*. Flowers yellow, with a short corolla tube and 6–8 spreading lobes 4.2–10 mm; calyx divided into 12 *narrowly linear segments to 1 mm wide, free almost to the base*. Widespread and common. Throughout. *Blackstonia grandiflora* is very similar, but with 8–12 corolla lobes 11–22 mm. DZ, ES (incl. Balearic Islands), FR (Corsica), IT (Sardinia and Sicily), MA, TN. *Blackstonia acuminata* is similar too, and easily confused with the previous species but with stem leaves *scarcely or not fused at the base*, and with *broader, linear to almost lanceolate calyx segments to 3 mm wide*, fused at the base; corolla lobes 6–9 and 5–7.5(10) mm. Throughout.

APOCYNACEAE | PERIWINKLE FAMILY

Trees, shrubs and climbers with opposite, untoothed leaves, often with a milky latex when cut. Flowers with corolla with a distinct tube and 5 lobes; stamens 5, inserted on corolla tube; style 1. Fruit a pair of follicles; seeds often with hairy tufts. (Includes species formerly classified in the family Asclepiadaceae).

Nerium

Shrubs with opposite leaves. Large pink or white flowers in terminal clusters; stamens with woolly filaments. Fruit a pair of follicles.

Nerium oleander OLEANDER A robust, evergreen, upright shrub to 4(6) m with long, erect branches. Leaves opposite, leathery, linear-lanceolate, 16 mm–19 cm. Flowers pink, red or white, sweetly scented, borne in showy terminal clusters; corolla lobes 13–26 mm. Follicles large, 40 mm–16 cm long; seeds with brown hairs. Extremely poisonous. Very commonly planted on roadsides, in gardens for hedging and in damp places such as seasonally drying river banks and gullies. Throughout.

---

1. *Blackstonia perfoliata*
2. *Nerium oleander*
3. *Nerium oleander* (fruits)
4. *Vinca difformis*
5. *Vinca major*
6. *Vinca minor* (naturalised cultivar)

Vinca PERIWINKLE

Perennials with trailing stems and pairs of leathery leaves. Flowers characteristically periwinkle-like; solitary in the upper leaf axils; corolla propella-shaped. Fruit a pair of follicles.

V. major

V. minor

V. difformis

Vinca major GREATER PERIWINKLE A short to medium, spreading evergreen shrub with long trailing stems to 1.5 m, often rooting down. Leaves *large*, 25–90 mm, shiny dark green, oval with *minutely hairy margins*. Flowers *bluish-violet*; corolla tube 12–15 mm; calyx lobes 7–17 mm, hairy on the margin. Follicles to 50 mm; often not produced. Cultivated and widely naturalised. ES, FR, IT, PT. ***Vinca minor*** is similar to the above species but smaller in all parts (to 1 m), with leaves 15–45 mm, *hairless* along the margins, and flowers with triangular, hairless calyx lobes 3–4 mm; corolla tube 9–11 mm. Mauve-flowered varieties are commonly cultivated and naturalised. ES, FR, IT, PT.

Vinca difformis INTERMEDIATE PERIWINKLE Superficially similar to *V. minor* but larger, with stems reaching 2 m, and leaves 25–70 mm without hairy margins. Flowers *pale blue or whitish*, with very narrow, hairless or minutely hairy calyx lobes 5–14 mm; corolla tube long, 12–18 mm. Common as a wild plant in ditches, roadsides, embankments and shady maquis. ES, FR, IT, MA, PT.

1

2

3

4

5

6

Araujia

Climbing, hairless perennial vines with opposite leaves. Sepals large and leaf-like; corolla erect; stamens 5. Fruit pendent.

Araujia sericifera CRUEL VINE A creeping vine to 7(10) m with a milky latex. Leaves 45–70 mm, opposite, dark green, shiny and leathery, oval-triangular. Flowers 21–28 mm across, 5-lobed and with a tube 11–16 mm, white tinged with yellow or violet and generally moth-pollinated. Fruit 85 mm–13 cm, grey-green, pear-like with a smooth but clefted surface, containing numerous seeds, each with a silky parachute attached for wind-dispersal. Native to south America but widely naturalised in northeast Spain, possibly elsewhere, often on wasteground climbing up wire fences. ES.

Asclepias

Erect perennials. Corolla with 5 reflexed lobes and a 5-parted corona *with horned projections*; stamens 5. Fruit 1(2) spreading (not pendent) follicles.

Asclepias syriaca SILKWEED An erect, unbranched, stout, leafy perennial to 1 m with large, flat, oblong leaves and dense terminal umbels of sweet-scented white-purple flowers. Fruits appear in the autumn: erect, oblong-pointed, *inflated and soft-spiny*, containing numerous seeds with silky-white parachutes for wind-dispersal. A native of North America, locally naturalised in grassy and waste habitats; rare. ES, FR, IT.

Gomphocarpus SILKWEED

Erect perennials. Flowers with a corona of 5 horns *without smaller projections*. Fruit 1(2) inflated, bristly, erect-spreading follicles.

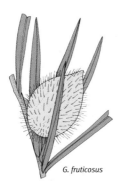

G. fruticosus

Gomphocarpus fruticosus BRISTLE-FRUITED SILKWEED An erect shrub to 1.2(2) m with linear-lanceolate, hairless leaves 39 mm–12 cm long. Flowers white, borne in hairy, stalked umbels; corolla lobes 6.2–8.3 mm. Fruit very large and conspicuous; pod-like, *inflated, pointed, bristly, 25–45(60) mm across* with 120 cottony seeds. Native to South Africa but planted as an ornamental in developed areas, and locally naturalised in the Iberian Peninsula. ES, PT. *Gomphocarpus physocarpus* is very similar but with larger, *rounded (not pointed) pods* 45–60 mm across with 100 seeds. Native to Asia. Naturalised. ES (south).

Vincetoxicum

Short, erect herbaceous perennials with oval to heart-shaped leaves. Flowers with 5-lobed corolla and a corona of rounded scales. Fruit 1–2 follicles.

Vincetoxicum hirundinaria SALLOWWORT A variable, erect (rarely climbing), hairless, short to tall perennial with stalked, pointed, heart-shaped or oval, opposite leaves. *Flowers greenish-white or yellowish*, with lobes 1.8–5 mm, borne in clusters of 5–12 at the bases of the upper-most leaves. Pods to 60 mm long, often paired, with silky seeds. Rocky garrigue, woods and pastures; local. Throughout. *Vincetoxicum nigrum* DARK SALLOWWORT is similar but generally with a more climbing habit with flexible stems to 2 m and *dark purple-brown flowers* with lobes 2.8–4.3 mm. Scattered in the European Mediterranean. ES, FR, IT, PT.

Cynanchum

Climbing perennials. Flowers with corona of 5 long appendages. Fruit 1–2 pendent follicles.

Cynanchum acutum STRANGLEWORT A hairless, twining, blue-green, woody climbing perennial to 4 m with slender stems. Leaves 22–78 mm, *arrowhead-shaped*, deeply heart-shaped at the base and stalked. Flowers tubular, scented, borne in lateral or terminal umbels, with a corolla of 5 triangular, projecting lobes 4.9–7.2 mm. *Fruits long and spindle-like*, pointed, 17 mm–18 cm long. Scrub and field margins. Local, but scattered throughout.

Periploca

Climbing perennials with oval-lanceolate leaves and greenish-brown flowers; corona with 5 slender appendages. Fruit 1–2 follicles.

Periploca graeca SILK-VINE A vigorous *twining deciduous woody perennial* to 12 m with dark green, stalked, shiny leaves 30 mm–12 cm. Flowers purple-brown, star-like with hairy, spreading-deflexed lobes 9.5–10 mm. Fruit long-pointed and pod-like 10–15 cm. Italy; naturalised in Spain. ES, IT. *Periploca laevigata* is a *branched shrub* with smaller flowers with *yellow-green* lobes 6–7 mm, with a brown-purple centre, with whitish appendages. Rocky scrub. DZ, ES, IT (Sicily), MA, TN.

1. *Araujia sericifera*
2. *Asclepias syriaca* (inset:fruiting plant)
3. *Gomphocarpus physocarpus*
4. *Vincetoxicum hirundinaria*
5. *Periploca laevigata*
6. *Periploca laevigata* (fruit)

Caralluma

Succulents with erect stems and soon-falling leaves (leaves often absent). Fruit 2 erect follicles.

Caralluma europaea CARALLUMA A short *succulent* to 20 cm with purplish or greyish, square stems to 30 mm wide; leaves soon-falling and inconspicuous. Flowers distinctive: starfish-like, 10–20 mm across, yellow with maroon stripes and a darker centre, fringed with short hairs; lobes 4.7–5 mm wide. Follicle 80 mm–13 cm. Dry, often salt-laden habitats. ES (southeast), IT (Sicily), MA. *Caralluma munbyana* is similar but has *grooved stems* and maroon *flowers 10–16 mm with long, narrow, spreading lobes just 0.8–2 mm wide.* DZ, ES (southeast), MA.

Orbea

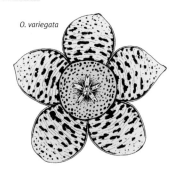

O. variegata

Succulents with often spiky stems and star-like, fly-pollinated flowers.

Orbea variegata STARFISH FLOWER A short, spreading *succulent* with erect and spreading stems to 15 cm high, without conspicuous leaves. Flowers very distinctive; *star-shaped,* warty, pale yellow with purple spots, unpleasant-smelling to attract flies, borne at the base of the stems, at the perimeter of the plant. Native to South Africa; planted in resorts as an ornamental (Iberian Peninsula, possibly elsewhere). ES, PT.

BORAGINACEAE | BORAGE FAMILY

Annual or perennial herbs or shrubs often with bristly stems and leaves; bristles often with swollen bases. Leaves simple and alternate. Flowers in spiralled clusters, short-stalked, blue in many species; stamens 5, borne on the corolla tube; ovary deeply 4-lobed; style 1 (sometimes split). Fruit 4 nutlets, often concealed in a persistent calyx.

Heliotropium

Small, bristly annuals to perennials. Flowers white or violet, borne on the upper side of outwardly coiled branches; stamens included in the corolla (or virtually so); style terminal (not basal).

Heliotropium europaeum HELIOTROPE A low to short, greyish, erect or spreading, much-branched, softly hairy annual to 50(60) cm. Leaves 30–70 mm, oval to elliptic, grey and stalked. Flowers white with a yellow throat, small 2–4 mm across, borne in distinctly 1-sided, spiralled spikes; *calyx divided almost to the base*. Fruit splitting into 4 nutlets. Bare, cultivated and waste ground; widespread and common. Throughout. *Heliotropium supinum* is similar but with spreading, *almost prostrate* stems and leaves green above, and a *flask-like calyx with short lobes, divided >1/3 its length*, which encircles the fruit. Throughout. *Heliotropium dolosum* has *scented flowers* and a *hairy stigma*. IT. *Heliotropium curassavicum* is a greyish, *fleshy perennial*, with narrow leaves, and *hairless throughout*. Fruit 4-lobed. Native to Americas, now naturalised throughout.

Neatostema

Small, bristly annuals. Flowers *yellow*, borne in branched or spike-like inflorescences; stamens included in the corolla.

Neatostema apulum YELLOW GROMWELL A short annual to 25(30) cm with erect, bristly stems that are branched above. Leaves 45(70) mm long, narrowly oblong, and bristly along the margins, those on the stems *erect*. *Flowers yellow*, borne in dense inflorescences to 90 mm long; corolla 2–3.5 mm across. Nutlets pale brown and warted. Fairly common in bare and stony habitats. Throughout.

Lithodora

Dwarf shrubs with entire leaves. Corolla tubular-bell-shaped; stamens included in the corolla. Fruiting heads dense.

Lithodora fruticosa SHRUBBY GROMWELL A dwarf shrub 20–80 cm (1 m) with erect and interwoven branches. Leaves 24 mm, oblong-linear and whitish with flattened bristly hairs and inrolled margins. Flowers blue, the corolla hairless outside and sparsely bristly hairy on the outside of the lobes; calyx 5–7 mm. Stony pastures and slopes. ES, FR, PT.

---

1. *Orbea variegata*
2. *Heliotropium europaeum*
3. *Heliotropium europaeum*
4. *Neatostema apulum*
5. *Glandora diffusa*

Glandora GROMWELL

Dwarf shrubs or tufted perennials, usually in rocky habitats. Corolla with glandular hairs. Fruiting heads contracted, with the calyx closely apressed to the nutlets. Similar to, and traditionally placed in, the genus *Lithodora* but established as genetically distinct.

G. diffusa

Glandora diffusa (Syn. *Lithodora diffusa*) SCRAMBLING GROMWELL A short to medium, bristly, spreading, shrubby perennial to 30(60) cm. Leaves 7–38 mm, elliptic, scarcely stalked. Flowers bright blue and funnel-shaped, the corolla with a hairy ring in the throat; calyx 5-lobed, 6–8 mm. Nutlets greyish brown and smooth. Acidic, sandy habitats. ES, FR, PT. *Glandora rosmarinifolia* (Syn. *Lithodora rosmarinifolia*) has linear leaves 10–60 mm long with inrolled margins, and corolla hairy outside. Rocky habitats. IT (incl. Sicily).

Cerinthe HONEYWORT

Virtually hairless, blue-grey annuals to perennials. Flowers with *cylindrical tubular corolla;* stamens included.

Cerinthe major HONEYWORT A rather fleshy, grey-green almost hairless annual 30–50(70) cm. Lower leaves 85 mm long, oblong, clasping the stem with a heart-shaped base, leaves covered with conspicuous *white swellings*. Bracts oval to heart-shaped, 12.5–20 mm. *Flowers dark purple throughout*, 17–22 mm long, the short petal lobes 3–4 mm, recurved; flowers borne in a *drooping* inflorescence. Common on coastal sands and garrigue. Throughout. ***Cerinthe gymnandra*** (also treated as a form of *C. major*) has *yellowish-white flowers* 17–20 mm, often with a purple-red band; lobes 3.5–5 mm. Similar habitats. ES, PT. *Cerinthe minor* LESSER HONEYWORT is an annual with *small yellow flowers, 11–12 mm long*, sometimes spotted violet at the base, with *straight lobes 3.5–5 mm*. Throughout, except for some islands. *Cerinthe glabra* is similar to *C. minor*, but a *perennial*, with small yellow flowers 11–12 mm long with deflexed lobes 1–2.8 mm. Mountain regions only. ES, FR, IT.

Nonea

Bristly herbs. Flowers borne in clusters with many bracts; corolla yellow, brown, purple or white with a tuft of hairs in the throat; stamens included or slightly protruding.

N. vesicaria

Nonea vesicaria A short to medium, greyish, bristly annual to 40(60) cm with lanceolate, sometimes toothed leaves 15–22 cm, clasping the stem above. Flowers with a *claret-coloured* corolla 3–5 mm across with spreading petal lobes 1.5–3 mm, borne in leafy, 1-sided clusters with conspicuous and enlarged, reddish sepals appearing inflated when in fruit; *calyx divided to $^{1}/_{2}$ its length*. Nutlets ovoid, ribbed, and constricted above the basal ring. Local in dry, grassy places. ES, IT (Sicily), PT. *Nonea echioides* is similar but has white to yellowish flowers. Local. ES, FR, IT, PT.

Echium BUGLOSS

Bristly herbs or shrubs. Flowers with zygomorphic, funnel-shaped corolla, borne in spiralled cymes making up dense or lax panicles; stamens unequal and normally at least some exserted (important trait).

A. Stamens not projecting from the corolla tube; corolla short, 7.5–13 mm long.

Echium parviflorum SMALL-FLOWERED BUGLOSS A rough, bristly haired annual with ascending stems to 40(65) cm, with oblong leaves, stalked below and stalkless above, with adpressed hairs. *Flowers small* with corolla 7.5–13 mm long, pale to dark blue with white throats, borne in lax, leafy clusters, with *stamens not protruding*. Coastal sandy and rocky habitats. Throughout. *Echium arenarium* is similar but often prostrate, with violet flowers with corolla just 6–10 mm long. Throughout; commonest on the islands and absent from many areas inland.

B. Stamens projecting from the corolla tube; corolla 9–16(18) mm long.

Echium italicum PALE BUGLOSS A large, erect biennial to 1 m with dense, spreading, white or yellowish bristles. Basal leaves lanceolate, to 20(25) cm long and 20 mm wide, narrowed into a basal stalk; stem leaves stalkless. Inflorescence spike-like or branched at the base, therefore pyramidal (depending on size of plant). *Corolla pale in colour*: whitish, pinkish or pale blue, 9–12 mm long and narrowly funnel-shaped; finely hairy outside; stamens 4–5 and markedly protruding. Pastures and fallow land. Widespread though local; frequent in eastern Italy. DZ, ES, FR, IT, MA, TN. *Echium asperrimum* is similar but generally more compact-spreading, with *flesh-coloured flowers with contrasting red stamens;* corolla 13–16(18) mm. Most of the European part of the region except the far west but local. ES, FR, IT.

1. *Cerinthe gymnandra*
2. *Cerinthe major*
3. *Nonea vesicaria*

1. *Echium plantagineum*
2. *Echium candicans*
3. *Echium italicum*
4. *Echium asperrimum*
5. *Echium vulgare*
6. *Echium sabulicola*

C. Stamens projecting from the corolla tube; corolla (10)14–35(40) mm.

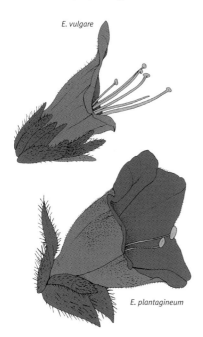

E. vulgare

E. plantagineum

Echium vulgare VIPER'S BUGLOSS A medium to tall bristly hairy biennial to 1 m with 1–several stems. Leaves covered in bristle-like hairs, elliptic and stalked at the base, narrowly lanceolate and stalkless along the stem. Inflorescence spike-like; flowers *blue* to blue-purple *with 4–5 long-protruding stamens*; corolla hairy all over, 10–21 mm. Nutlets ridged. Dry open habitats and fallow land; very common. Throughout. ***Echium plantagineum*** is similar to *E. vulgare* but *softly hairy* (rather than coursely bristly) with reddish bristles. Leaves with promiment midribs and lateral veins. Flowers dark blue-purple ageing red, with 2(4) protruding, reddish stamens, borne in branched, panicle-like inflorescences; corolla 18–30 mm *hairy on the veins and margins only*. Very common in sandy maritime habitats. Throughout. *Echium gaditanum* is distinguished by its grey-white leaves, and lax inflorescences of dark blue (ageing pink) flowers in which the corolla tube is 12–16(20) mm, narrowly funnel-shaped and abruptly inflated above the middle, with 3–4 stamens protruding. Coastal sands. PT.

E. creticum

E. tuberculatum

Echium creticum An erect, *bristly* biennial 25–90 cm high with few stems, covered with *2 types of bristle: long spreading hairs with a tubercle at the base, as well as abundant short hairs*. Basal leaves narrowly oblong, to 18 cm with sparse bristles with swollen bases; stem leaves narrowly elliptic. Flowers large, pinkish-red, often becoming purple with age; corolla 15–35(40) mm long, broadly funnel-shaped, with 1–2 stamens protruding. Common on roadsides and disturbed grassy places in the west. ES, FR, PT. ***Echium sabulicola*** is a similar, spreading biennial to 60 cm with 2 types of bristle including long spreading hairs with a tubercle at the base, as well as short hairs *recurved at the base*; long, 1-sided inflorescences coiled at the apex; flowers pink, ageing blue; corolla 14–20(25) mm. Dunes and fallow land. ES (incl. Balearic Islands). *Echium tuberculatum* is similar to *E. creticum* but with narrowly funnel-shaped flowers with corolla 18–28(30) mm, *dark-purple blue*, and with *2–4 stamens protruding*. Similar habitats. ES, MA, PT.

Echium candicans PRIDE OF MADEIRA A bristly hairy, *large, shubby perennial* to 2 m tall, with many ascending flowering stems, woody at the bases. Leaves dark green, rough, elliptic in shape and pointed at the tips. Flowers bright blue, borne in dense inflorescences on stems leafy below. Native to Madeira but commonly planted.

Borago BORAGE

Bristly annuals or perennials. Flowers blue (rarely white), short-tubed with widely spreading lobes, and stamens equal, forward-pointing in a cone (included to exserted).

Borago officinalis BORAGE A medium, bristly annual with robust, generally branched stems to 60 cm. Basal leaves oval and light green, stalked; stem leaves smaller and unstalked. Flowers borne in loosely branched cymes, bright blue and star-like with a white centre, reflexed lobes 7–10 mm, and an exposed *cone of blackish anthers*; calyx 7–15 mm (20 mm in fruit). Very common on cultivated, sandy and waste ground. Throughout.

Anchusa ALKANET

Bristly annual or perennial herbs. Flowers with blue, purple, white, funnel-shaped corolla tube; stamens equal and not exserted.

A. calcarea

A. undulata

A. azurea

A. Flowers blue.

Anchusa calcarea A short to medium hairy perennial 40 cm–1 m with elliptic leaves to 20 cm forming a bushy basal rosette; upper leaves unstalked and clasping the stem; leaves and stems covered in bristles *with white, pimple-like bases.* Flowers dark blue, funnel-shaped, bracts oval-lanceolate and *shorter* than, or scarcely exceeding, the calyx; corolla 6–10 mm across. Coastal sands. ES, PT. *Anchusa undulata* is very similar but with *leaves with undulate margins,* and hairs that lack white, pimple-like bases. Corolla 5–10 mm across. Dry, sandy places. Throughout.

Anchusa azurea LARGE BLUE ALKANET A robust species to 1.8 m with dense bristles, often with white, pimple-like bases. Leaves to 25(40) cm long and 50 mm wide; lanceolate. Flowers large, the corolla 8–15(20) mm across, deep blue or purple with a whitish centre (sometimes all white or cream); flowers borne in a large, loose and much-branched inflorescence. Frequent on bare ground. Throughout. *Anchusa humilis* is similar but much smaller, to 30 cm, with bright blue flowers *to just 6 mm across,* borne in compact, white-bristly clusters. DZ, IT (Sicily), MA, TN.

B. Flowers pale yellow or white.

Anchusa aegyptiaca EASTERN ALKANET A low, very bristly annual with prostrate stems 15–40 cm and oval leaves with undulate margins to 10(15) cm. *Flowers pale whitish-yellow,* the corolla 4.5–6.5 mm across, borne in lax, unbranched inflorescences. Various disturbed habitats. An eastern species, naturalised in the west. ES, TN.

1. *Borago officinalis*
2. *Anchusa calcarea*
3. *Anchusa calcareu*
4. *Anchusa azurea*
5. *Omphalodes linifolia*
6. *Myosotis ramosissima*

Omphalodes

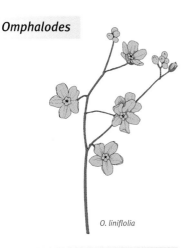

O. liniflolia

Small annuals. Flowers with corolla very short with a throat closed by 5, blunt, forward-pointing lobes; stamens included.

Omphalodes linifolia ANNUAL OMPHALODES A low to short, erect annual with simple or sparsely branched stems, to 50 cm. Leaves to 75–85 x 12–15 mm, spatula-shaped and stalked below, linear and unstalked above, all with sparsely hairy margins. Flowers white or blue-tinted, borne in long racemes; bracts absent; corolla 5.5–7 mm across. Garrigue and maquis; locally common. ES, FR, MA, PT. *Omphalodes brassicifolia* is a similar annual or biennial to 1 m with broader, oval leaves 13 cm x 35 mm which clasp the stem. Rare, in mountains. ES, MA.

Cynoglossum HOUND'S TONGUE

Hairy biennial to perennial herbs. Flowers borne in branched cymes without bracts; corolla with a short tube and 5 spreading petal lobes with scales closing the throat; stamens not protruding. Nutlets egg-shaped and with barbed spines.

A. Flowers bluish or purplish.

Cynoglossum creticum BLUE HOUND'S TONGUE A short, robust, softly hairy biennial 30–70(90) cm high with erect, angular stems branched above. Leaves to 30 cm, lanceolate, untoothed and densely hairy to felted on *both* surfaces, sometimes clasping the stem. Flowers purplish in bud, opening blue-violet with fine, *darker blue venation*, the corolla 5–9 mm across, borne in branched cymes that elongate in fruit. Nutlets without a distinct border, and with dense hooked spines. A range of habitats, often on disturbed ground. Throughout. *Cynoglossum columnae* has deep, dull blue or purple flowers without conspicuous veins, to just 6 mm long. IT (incl. Sicily).

B. Flowers whitish-pink to violet.

Cynoglossum clandestinum A grey-white biennial 30–50(65) cm high. Leaves to 12(20) cm. Flowers small, with corolla just 2.5–3.5 mm across and *pale whitish-pink or violet* and with *woolly* petal lobes. Nutlets without a distinct border. DZ, ES, FR (Corsica), IT (Sicily), MA, PT.

Pardoglossum

P. cheirifolium

Hairy biennial to perennial herbs closely related and morphologically similar to *Cynoglossum* but established as a genetically distinct group.

Pardoglossum cheirifolium (Syn. *Cynoglossum cheirifolium*) A biennial 25–40(65) cm high with leaves to 18 cm, conspicuously *white-felted* on both surfaces; somewhat undulate. Flowers with corolla 3–6.5 mm across, *reddish-purple* or bluish, borne in crowded racemes that elongate in fruit. Nutlets with a distinctly *thickened border*, and with dense hooked spines, or smooth. Dry, stony places and waste ground. ES, FR, IT, PT, TN.

Myosotis FORGET-ME-NOT

Hairy annuals to perennials. Flowers with corolla tube short and the throat enclosed by 5 short scales; stamens equal and included. Numerous species occur in northern Europe and in cooler parts of the region.

M. laxa

Myosotis laxa TUFTED FORGET-ME-NOT An erect to ascending annual to 40(70) cm, not conspicuously hairy, but with adpressed hairs on the stems and leaves; leaves to 55 mm. Flowers small, 4–5(6) mm across, pale blue with rounded petal lobes; flowers borne sparsely along the stem, densely at the apex; calyx lobed $^1/_2$ to the base. Local, in wet habitats. Throughout.

Myosotis ramosissima EARLY FORGET-ME-NOT A low, *bristly*, erect to spreading, short-lived annual to 25 cm. Flowers pale blue, just 0.8–2(3) mm across with corolla tube shorter than the calyx; calyx teeth spreading, flower stalks equalling calyx, and inflorescence longer than the leafy part of the stem when in fruit. Wet, sandy and grassy habitats. Throughout.

1. *Cynoglossum clandestinum*
2. *Cynoglossum creticum*
3. *Cynoglossum creticum*
4. *Atropa belladonna*
5. *Nicandra physalodes*
6. *Lycium chinense*

SOLANACEAE | POTATO FAMILY

Herbs or shrubs with simple, entire or pinnately divided, alternate leaves. Flowers with a star- or bell-shaped corolla, the 5 petals fused below; stamens 5, attached to the corolla tube; style 1; ovary superior with 2 compartments. Fruit a berry or 2–4-valved capsule.

Atropa

Herbaceous, almost hairless perennials with solitary or paired, entire leaves. Corolla bell-shaped and shortly 5-lobed, stamens not projecting. Fruit a berry.

Atropa belladonna DEADLY NIGHTSHADE A faintly unpleasant-smelling and poisonous, stout, erect, branched perennial to 1(2) m, glandular but not distinctly hairy. Leaves oval-pointed, rather large 80 mm–20 cm long. Flowers *bell-shaped and liver-coloured* (rarely green-yellow), stalked; corolla 24–30 mm long. *Fruit a black, shiny, fleshy berry* 15–20 mm across framed by the 5-lobed persistent calyx. Woods and thickets on limestone. Scattered more or less throughout on higher ground, but rather local. DZ, ES, FR, IT, MA. *Atropa baetica* is similar in form, but with *yellow flowers with some purple markings*, and the corolla tube markedly open. ES, MA.

Nicandra

Herbaceous hairless annuals with toothed to lobed leaves. Flowers with corolla broadly bell-shaped and shallowly-lobed; calyx much-inflated. Fruit a dry berry.

Nicandra physalodes APPLE OF PERU A hairless, unpleasant-smelling, vigorous, much-branched annual to 80 cm with stalked, oval and toothed to lobed, large leaves. Flowers borne singly in the leaf axils; corolla 12–20 mm, blue to violet with a white centre, soon closing; calyx much enlarging in fruit (25–35 mm). A native of Peru, widely naturalised in southern Europe. ES, FR, IT.

Lycium

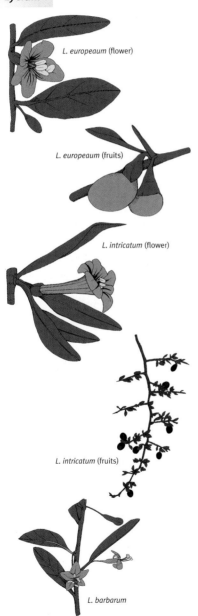

L. europeaum (flower)

L. europeaum (fruits)

L. intricatum (flower)

L. intricatum (fruits)

L. barbarum

Woody, almost hairless shrubs with simple, deciduous leaves. Flowers dull purple, rather deeply lobed, often with stamens protruding. Fruit a berry.

Lycium europaeum TEA TREE A deciduous, *conspicuously and robustly spiny* shrub to 3(4) m. Leaves elliptic, broadest above the middle, 20–73 mm long. Flowers borne in clusters of 2–5, with a pink or white corolla, narrowly funnel-shaped, 10–17 mm long; lobes 3–5 mm; stamens usually protruding; calyx 2–3.5 mm. Fruit a rounded, reddish berry, 5–6 mm. Hedges and thickets. Throughout. *Lycium intricatum* is similar to *L. europaeum* but *succulent,* with *smaller leaves* 2–26 mm long. Corolla tube 11–18 mm long; lilac, pink or white with stamens *not* protruding; calyx 1.7–3 mm. Fruit an orange-red or blackish berry 3–7(9) mm. Local on coastal cliffs and in semi-arid scrub. DZ, ES, MA, PT. *Lycium ferocissimum* is distinguished by its *glandular-hairy flower stalks, long calyx* (4.5–8 mm) and more or less spherical, *larger red fruits 6–13 mm*. Native to South Africa; naturalised. ES, MA.

Lycium barbarum DUKE OF ARGYLL'S TEAPLANT A deciduous shrub to 2.5 m with arched, grey-white branches with few, slender spines. Leaves alternate or in clusters, narrowly elliptic and broadest at the middle; untoothed, 20 mm–10 cm long. Flowers in small clusters of 1–3(7); corolla red-purple, turning brown with age, trumpet-shaped, *small,* 7–12 mm; lobes 4–5 mm; stamens long-protruding; calyx 3.5–5.5 mm. Fruit an orange-red oval (slightly elongated) berry, 10–20 mm. Native to China but widely naturalised. Throughout. *Lycium chinense* is very similar but with a *shorter calyx* just 1.8–3.3 mm; corolla 10–15 mm; lobes 5–8 mm. *Fruit smaller*, more rounded, 5–10 mm long. ES, FR, IT.

1. *Hyoscyamus albus*
2. *Hyoscyamus niger*
3. *Lycium intricatum* (inset: fruits)

Hyoscyamus HENBANE

Glandular-hairy annuals to biennials with simple, toothed to lobed leaves. Flowers borne numerously in rows along outwardly-coiled, leafy stems; corolla funnel-shaped. Fruit a splitting capsule.

H. albus

Hyoscyamus albus WHITE HENBANE A sticky annual or short-lived perennial with long, sparsely-branched stems to 80 cm, often woody below. Leaves 60 mm–20 cm long, broadly oblong and wedge-shaped or heart-shaped at the base, with wide teeth along the margin. *Flowers stalkless*, at least above, borne in long, dense, 1-sided spikes; calyx densely hairy and swollen below, ending in short, triangular teeth; corolla 20–30 mm, greenish or *yellowish-white* with a green or purple throat; *stamens not, or scarcely protruding*. Capsule 10 mm. Common on waste ground and near buildings; often casual only. Throughout.

Hyoscyamus niger HENBANE A sticky, unpleasant-smelling annual or biennial with long, sparsely branched stems to 80 cm, often woody below. Leaves 60 mm–20 cm long, oval and coarsely toothed, forming a basal rosette; stalked below but stalkless above and somewhat *clasping the stem*. Flowers virtually stalkless, borne in long, dense, 1-sided spikes; calyx densely hairy and swollen below (especially in fruit), ending in short, triangular teeth; corolla to 20–30 mm, slightly uneven (zygomorphic), *dull pale yellow with purple veins*; stamens slightly protruding. Capsule 10 mm. Disturbed habitats and walls; common. Throughout, except for smaller islands.

Physalis

Annuals to perennials with simple, often toothed leaves. Flowers with broadly funnel-shaped corolla with 5 spreading lobes. Fruit a berry, often concealed by a markedly inflated calyx.

P. ixocarpa

Physalis ixocarpa A *sparsely hairy, branched, erect annual* 40–60 cm. Leaves lanceolate, with course teeth or entire, and rounded bases; stalked. Flowers borne solitary in the leaf axils; calyx oval, green and purple-veined, with 5 teeth; corolla bright yellow with brownish markings in the centre; filaments purple. Fruit borne in a much inflated, greenish, purple-veined calyx 30–50 mm enclosing a berry 13–40 mm. Native of Mexico; naturalised in Iberian Peninsula. ES, PT. *Physalis peruviana* is similar, but a *densely hairy perennial* to 1 m with leaves with heart-shaped bases and paler, whitish-yellow, purple-spotted flowers. ES, IT. *Physalis alkekengi* is similar but a spreading perennial to 60 cm (1 m), with *white flowers*, and *red, inflated calyx* in fruit 25–50 mm; fruit 12–17 mm. Naturalised. ES, FR, IT.

Solanum NIGHTSHADE

Herbs or shrubs with simple leaves. Flowers borne in 1-sided cymes or umbels; leaf-opposed; corolla star-shaped with spreading petal lobes; stamens protruding, forming a cone around the stigma. Fruit a berry.

A. Plant a spiny shrub.

S. linnaeanum

Solanum linnaeanum APPLE OF SODOM A small to large, very *spiny shrub* to 2 m with much-branched, stout stems covered in yellow spines; slightly hairy. Leaves 50 mm–13 cm, oval and pinnately, bluntly lobed; prickly, and stalked. Flowers with a violet, 5-lobed corolla 20–30 mm across, borne solitary or in clusters. *Berry large and spherical, 20–40(50) mm across,* marbled green and white, later yellow or brown and shiny; rather tomato-like. Waste places, often near buildings. Native to Africa. Throughout. *Solanum carolinense* is similar but has short anthers (1.5–2.5 mm) and *smaller fruits,* to just 10–15 mm across. Native to America. ES.

B. Plant not spiny and flowers purple.

Solanum dulcamara WOODY NIGHTSHADE A scrambling, woody-based, shrubby perennial to 3(7) m with oval-lanceolate leaves to 80 mm long, simple, or the lower often lobed at the base. *Flowers with purple petals with contrasting yellow anthers,* borne in lax clusters. Fruit *a red berry 8–12 mm long.* Woods and damp habitats, often on higher ground; common. Throughout.

C. Plant not spiny and flowers yellow.

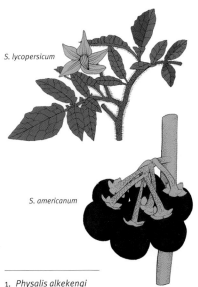

S. lycopersicum

S. americanum

Solanum lycopersicum TOMATO A glandular-hairy, aromatic annual. Leaves to 20 cm with oval leaflets. Cymes with up to 20 yellow flowers. Fruit an edible red or yellow berry (tomato). Cultivated and often naturalised as a casual weed. Throughout.

D. Plant not spiny and flowers white.

Solanum americanum (Syn. *S. nigrum*) BLACK NIGHTSHADE A variable, low to medium, hairless to hairy *annual* to 70 cm with stems spreading and blackish. Leaves oval-lanceolate, toothed, lobed or entire; stalked. Flowers with white petals, star-like with a yellow cone of anthers borne in clusters of 5–10. Berry 6–10 mm, round and green ripening matte-black, borne on erect-spreading stalks. A common weed on cultivated land. Throughout. *Solanum villosum* is very similar but *densely long-hairy* with more deeply lobed leaves *and fruit reddish or yellowish* (not black), 6–10 mm. Throughout. *Solanum chenopodioides* is similar to *S. nigrum* but *softy short-hairy* and with purplish-blackish fruits 6–8 mm borne on *reflexed stalks.* Originally from South America. ES, FR, IT, PT.

1. *Physalis alkekengi*
2. *Solanum linnaeanum*
3. *Solanum linnaeanum*
4. *Solanum dulcamara*
5. *Solanum laxum*

S. tuberosum

S. laxum

Solanum tuberosum POTATO A sparsely hairy, bushy perennial to 1 m with underground stolons bearing terminal tubers (potatoes). Leaves pinnately divided with up to 7 pairs of variably sized oval leaflets. Flowers borne numerously in cymes; petals purple or white with a yellow cones of anthers. Fruit green to purplish, 20–40 mm across. Cultivated widely and a casual weed. Throughout.

Solanum laxum (Syn. *S. jasminoides*) POTATO VINE *An ornamental evergreen climber* to 5 m. Leaves 25–75 mm, bright green, leathery and shiny; pointed and lanceolate with heart-shaped base; stalked. Flowers with a white corolla 15–18 mm across, star-like, with a yellow cones of anthers, borne in lax, many-flowered, showy cymes. Fruit 4–5 mm. Widely planted in developed areas. Throughout.

Mandragora MANDRAKE

Leaves basal. Flowers short-stalked and arising directly from the root stock. Fruit a berry. The roots were formerly used in the Mediterranean to relieve pain and induce sleep.

Mandragora officinarum MANDRAKE A ground-hugging, stemless, hairless or hairy, winter- or spring-flowering perennial with a robust rootstock. Leaves to 30(45) cm, forming a large, flat *rosette*, oval, bright green and shiny, stalked and distinctly wavy-margined. Flowers with a greenish-white to blue-violet corolla, 25–50 mm across with 5 triangular lobes 20–34 mm, are borne in the centre of the rosette. Fruit an orange or yellow, egg-shaped berry 16–25(40) mm across, held in a large, persistent calyx that equals or exceeds the fruit. Fallow land and bare places. Throughout, except for France and the Balearic Islands. Autumn-flowering forms referred to as *M. autumnalis* are no longer widely considered to be distinct.

Datura (including *Dutra*)

Erect annuals with simple leaves. Flowers regular, showy, trumpet-shaped, upward-pointing; calyx with 5 teeth. Fruit a large, spiny capsule.

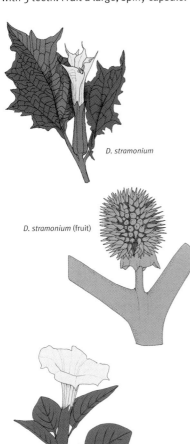

D. stramonium

D. stramonium (fruit)

D. inoxia

Datura stramonium A stout and vigorous, unpleasant-smelling, normally *hairless* annual to 1.5 m with stout, spreading stems. Leaves oval to elliptic, lobed with jagged teeth, 50 mm–18 cm. Flowers with a white corolla sometimes flushed with purple, trumpet-like, 50 mm–10 cm long, borne solitary in the leaf axils of the upper leaves; calyx large, to $^1/_2$ the length of the corolla (30–50 mm), sharply angled. Fruit an erect, large, spiny capsule 35–70 mm long. Frequent on bare waste ground. Throughout. *Datura ferox* is very similar but with smaller flowers; calyx 25–40 mm and corolla 40–60 mm. *Fruit with sparse, stout, long spines 10–30 mm.* Scattered throughout; often confused with the previous species.

Datura inoxia (Syn. *Dutra inoxia*) A perennial similar to the previous species (now classified by some authors under the separate genus *Dutra*) but *softly hairy,* often smaller, to 50 cm (2 m), with *larger flowers* with corolla 14–16.5 cm long, hairless outside, and capsules 30–50 mm, *nodding* when mature. Similar habitats. Casual throughout.

1. *Solanum americanum*
2. *Solanum villosum*
3. *Solanum villosum* (fruit)
4. *Datura stramonium*
5. *Datura stramonium* (fruit)
6. *Datura inoxia*

Brugmansia

Evergreen shrubs and trees with entire leaves and large, showy, drooping, fragrant flowers. Native to South America but very commonly planted. Fruit large, smooth and pod-like. A highly complex genus in which numerous hybrids and cultivars exist which are difficult to distinguish; below is a simplified and approximate guide only.

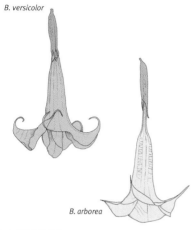

B. versicolor

B. arborea

Brugmansia versicolor ANGEL'S TRUMPET An ornamental shrub to 3(5) m with thick, stiff stems. Leaves oval, stalked and untoothed. Flowers pendent, white, flushed with yellow or pink with age, trumpet-like, *very large, 30–45(50) cm long* with a slender tube, and fragrant; borne solitary in the upper leaf axils; calyx inflated, slit along 1 side. Fruit egg-shaped. Widely cultivated. *Brugmansia arborea* is a similar small tree to 7 m with white to pale green, nodding flowers; *corolla smaller, 12–17 cm long.* Fruit soft-walled when ripe; seeds angular. Commonly planted. **Brugmansia sanguinea** is an ornamental tree to 10 m with *tubular, red, orange or yellow flowers;* calyx ribbed. Fruit smooth. Planted in resorts and gardens.

Cestrum

C. nocturnum

Spineless, hairless shrubs. Flowers with tubular corolla, borne in clusters. Fruit a berry. Native to tropical America.

Cestrum nocturnum NIGHT JASMINE An aromatic shrub to 4 m with alternate, simple, lanceolate, pointed leaves. *Flowers white*, fragrant, abundant and showy, 20–25 mm long with 5 stamens, not protruding. Fruit a 2-parted berry to 10 mm. Commonly planted in the region, possibly naturalised. *Cestrum parqui* is a similar shrub to 4 m with *yellow flowers* 19–27 mm. Commonly planted and naturalised. ES, IT.

Nicotiana TOBACCO

Shrubs or perennials, sticky, with simple, entire leaves. Flowers with corolla elongated and funnel-shaped, borne in branched, leafless clusters. Fruit a 2-valved capsule. Native to South America but widely planted and naturalised.

N. glauca

A. Plant a shrub or small tree.

Nicotiana glauca SHRUB TOBACCO A hairless, soft-wooded *shrub or small tree to 6 m* with lax, grey branches and *grey-green*, stalked, elliptic-lanceolate, pointed leaves 21 mm–12 cm. Flowers greenish-yellow, borne in lax panicles; corolla tubular 27–45 mm long. Fruit an egg-shaped capsule 8.5–15 mm long, exceeding the persistent, papery calyx. Widely naturalised in waste places, derelict sites and on walls. Throughout.

1. *Brugmansia sanguinea*
2. *Brugmansia* (cultivar)
3. *Nicotiana glauca*
4. *Salpichroa origanifolia*
5. *Solandra maxima*

B. Plant a herbaceous annual or perennial.

Nicotiana rustica SMALL TOBACCO A strong-smelling, glandular-hairy erect annual to 1 m with oval-heart-shaped, stalked shiny green leaves with unwinged stalks. Flowers borne in many-flowered terminal clusters; corolla *green-yellow*, 12–17 mm long, 2–3 x the length of the calyx; calyx lobes blunt. Widely cultivated and naturalised. Throughout. *Nicotiana tabacum* LARGE TOBACCO is similar but with broadly winged leaf stalks, clasping the stem at the base. Flowers with corolla *white flushed pink or purple*, 30–55 mm. Cultivated and naturalised more or less throughout.

Petunia

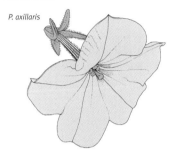

P. axillaris

A South American genus of herbaceous, sticky perennials with simple leaves. Flowers showy, trumpet-shaped and 5-lobed. Fruit a capsule.

Petunia axillaris A short, bushy, sticky-hairy annual to 60 cm with oval-elliptic, stalkless leaves. Flowers with corolla 50 mm–12 cm, large, white and trumpet-like and scented at night. Commonly planted as an ornamental in parks and gardens, sometimes naturalised as a transient casual in waste places. Throughout.

Solandra

Vigorous shrubby vines native to Central and South America with shiny foliage. Flowers very large, showy. Fruit a large capsule.

Solandra maxima HAWAIIAN LILY A shrubby climber with oval, dark green leaves and *large, showy, trumpet-shaped flowers to 20 cm long*; yellow-cream with purple veins and prominent stamens. Fruit a berry. A striking exotic ornamental widely planted in towns, gardens and resorts in the region. Naturalised in the Algarve. PT (possibly elsewhere).

Salpichroa

Fleshy, hairy, spineless perennial herbs native to South America with simple leaves. Flowers solitary, with regular, tubular corolla. Fruit a berry.

Salpichroa origanifolia A much-branched, hairy perennial to 1 m with woody stems below, usually scrambling among surrounding vegetation. Leaves oval, short-stalked and entire, 50 x 37 mm. Flowers with corolla white and bell-shaped, 6.5–11 mm long. Fruit a small cream-white berry 10–15 mm. Naturalised on roadsides and in coastal habitats. ES, FR, IT, PT.

CONVOLVULACEAE | CONVOLVULUS FAMILY

Typically climbing annuals or perennials (sometimes shrubs) with alternate leaves (or leafless parasites). Flowers 1–few in leaf axils, large and showy in some genera; sepals and petals 4–5, fused at the base or to form a funnel-shaped corolla tube; stamens 4–5; style(s) 1–2. Fruit a capsule.

Cuscuta DODDER

Leafless, parasitic twining herbs with no green pigment. Flowers small with 4–5-lobed corolla; styles 2. Fruit a 2–4-valved capsule. A number of species have become globally naturalised, but loss of characters useful for identification from herbarium specimens has led to conflicting information in the literature.

A. Flowers normally 4-parted, stems reddish, yellowish or whitish.

Cuscuta europaea A spreading, tangled, parasitic annual with twining, *purplish or reddish stems* 1.6 mm wide and *4-parted* flowers 2–3.5 mm borne in clusters of 5–20; *filaments very thin; styles + stigma shorter than the ovary.* Parasitic on various herbs, particularly *Urtica* spp. High altitude habitats in woods and fields. Throughout. *Cuscuta nivea* has pink, yellowish or whitish stems 0.4 mm wide; white flowers, which are normally 4-parted (rarely 3- or 5-parted), borne in *lax heads of 3–7(10); flowers with many glandular hairs*; flowers 1.4–2.2 mm. Parasitic on small herbs. Eastern Iberian Peninsula and North Africa only. DZ, ES, MA.

B. Flowers normally 5-parted, stems reddish.

C. epithymum

Cuscuta epithymum COMMON DODDER A spreading annual with *reddish* and thread-like stems 0.8 mm wide. Flowers 1.5–4.9 mm, pale pink with a reddish calyx, scented and 5-parted with spreading petals borne in ball-like clusters of 5–30; *styles + stigma longer than the ovary.* Parasitic on various herbs and shrubs; one of the commonest species. Throughout. *Cuscuta suaveolens* has grey-purple stems, 5-parted, and stalked white flowers with greenish sepals, borne in lax clusters. A South American weed that appears only sporadically, and soon disappears. DZ, ES, FR, IT, PT.

C. Flowers normally 5-parted, stems often yellowish.

C. campestris

Cuscuta campestris FIELD DODDER A twining annual parasite with *yellow-orange* stems 1.2 mm wide and small 5-parted flowers 1.5–3 mm across borne in lax clusters of just 2–4; *stigma capitate (ending in a knob).* Native to North America but very widely naturalised and parasitic on various herbs. Fields and garrigue; common. Throughout. *Cuscuta australis* is very similar to *C. campestris,* also with yellowish stems 1.4 mm wide and 5- (or 4-) parted flowers 2.9 mm *with stamens very short and adpressed to the corolla tube* (not large and covering the ovary). Parasitic on *Polygonum* and *Calystegia* in fields. Probably native to Australia, locally naturalised in Iberian Peninsula, possibly elsewhere. ES. *Cuscuta planiflora* also has yellowish stems (often reddish in places and in juveniles) 0.5 mm wide and 5–20(30) 5-parted (sometimes 4-parted) *whitish flowers 2.5–3.5 mm borne in compact, stalkless clusters.* Parasitic on various herbs and shrubs (especially legumes). Locally throughout. *Cuscuta approximata* is similar to *C. planiflora* but with calyx lobes with smooth (not crystalline) edges and bright golden yellow tips. Parasitic on various herbs and shrubs. Most common in hot, arid areas. Probably local throughout.

1. *Cuscuta epithymum*
2. *Cuscuta approximata*
3. *Cuscuta planiflora*
4. *Calystegia soldanella*

Calystegia BINDWEED

Perennial herbs with twining or prostrate stems with a white latex and triangular leaves. Flowers typically large and solitary, with scarcely divided, broadly funnel-shaped corolla; style 1. Fruit not lobed.

C. soldanella

A. Prostrate perennials.

Calystegia soldanella SEA BINDWEED A low, hairless, prostrate, spreading perennial to 50 cm. Leaves 40 mm–12 cm, kidney-shaped, deep green and veined, rather fleshy and long-stalked (30–90 mm). *Flowers pale pink* with white stripes, 40–45 mm; solitary. Very common on coastal dunes. Throughout.

B. Climbing perennials.

Calystegia sepium HEDGE BINDWEED A vigorous climbing perennial with twisted stems to 4 m which clamber on other plants for support. Leaves light green, matte (not glossy), arrow-shaped, 40 mm–13 cm long. *Flowers white*, 40–55 mm along with 2 epicalyx bracts which exceed the sepals and surround them, *not, or scarcely, overlapping at the base*. Capsule 10–16 mm. Various waste and damp habitats. Throughout. *Calystegia silvatica* GREAT BINDWEED is very similar, differing only in having larger flowers 47–65(75) mm across and the 2 epicalyx bracts strongly inflated and markedly overlapping at the base, concealing the sepals. Capsule 10–12 mm. Waste habitats and thickets. Throughout.

1

2

3

4

5

6

Convolvulus BINDWEED

Shrubby, trailing or climbing herbs with entire or variously divided and lobed leaves. Flowers 1–few; calyx 5-parted, corolla funnel-shaped, opening fully in daylight; style 1. Fruit unlobed.

A. Flowers normally at least partly blue.

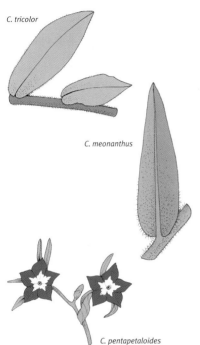

C. tricolor

C. meonanthus

C. pentapetaloides

Convolvulus tricolor DWARF CONVOLVULUS A hairy, short, spreading annual to 40 cm. Leaves 13–40 mm, oval to elliptic, untoothed and unstalked. Flowers 20–30 mm, conspicuously *with 3 bands of colour*: yellow in the throat, then white, and bright blue on the perimeter, solitary or paired on short stalks. Capsule hairy, 5–6 mm. Frequent on bare and sandy ground and waste places near the sea. Throughout, except for some islands. *Convolvulus meonanthus* is similar to *C. tricolor* but with narrower, almost linear leaves (5–8 mm wide), smaller flowers only 15–20 mm across, and a hairless capsule 4.5–5.5 mm. Cultivated ground. DZ, ES, IT, MA, PT. *Convolvulus pentapetaloides* is similar but smaller still; the flower just 8–9 mm and distinctly 5-lobed. Throughout. *Convolvulus siculus* is similar to *C. pentapetaloides* but with leaves 12–60 mm, lanceolate to oval, untoothed and *stalked* (0.5–35 mm). Flowers blue with a yellowish centre, 7–12 mm across, distinctly 5-lobed and solitary or paired on short stalks. Capsule 4–6 mm. Dry, bare places. *Convolvulus humilis* has *unstalked flowers* and *hairy fruits*. DZ, ES, IT, MA, TN.

Convolvulus sabatius A vigorous trailing perennial with a branched, woody stock. Leaves lanceolate to elliptic and pointed, deep green, not toothed and hairy. *Flowers light blue* (rarely pink), 25–40 mm across, bone solitary or in groups of 2–3. Dry, rocky habitats and old walls. IT (incl. Sicily), MA (possibly naturalised elsewhere). *Convolvulus valentinus* is similar but with *very narrow leaves* (4–9 mm) and variously blue, pink, white or yellow flowers 22–26 mm. Dry, rocky habitats. ES, MA.

B. Flowers pink, white or yellowish. Leaves linear or lanceolate. Plant not climbing.

Convolvulus cantabrica PINK CONVOLVULUS A spreading to ascending or prostrate, tufted perennial to 75 cm with a woody stock. Leaves 12–93 mm long, linear and often broadest above the middle, *covered in long, silky, spreading hairs. Flowers pale pink*, 17–23 mm across, borne in lax clusters on stems exceeding their adjacent leaves; ovary hairy. Capsule 6–8 mm. Various dry and bare habitats; common and widespread. DZ, ES, FR, IT, MA, TN. *Convolvulus lanuginosus* is similar in form but is usually *densely white-woolly* (but with a hairless ovary). ES, FR, MA.

---

1. *Calystegia sepium*
2. *Calystegia sepium* (calyx)
3. *Calystegia silvatica*
4. *Convolvulus tricolor*
5. *Convolvulus tricolor*
6. *Convolvulus sabatius*

C. Flowers pink, white or yellowish. Leaves heart-shaped. Plant generally climbing (or trailing).

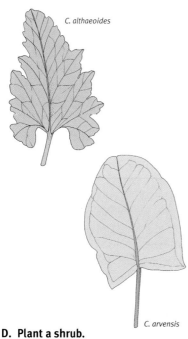

C. althaeoides

C. arvensis

Convolvulus althaeoides MALLOW-LEAVED BINDWEED A trailing or twining, hairy perennial to 2 m with a slender, creeping rootstock. Leaves 30 mm–12 cm, mallow-like; greyish, heart-shaped below and stalked, deeply-lobed above with linear divisions; all hairy with somewhat wavy margins. Flowers pale pink with a deep pink throat, 27–37(40) mm across, solitary or paired, and long-stalked. Capsule 8–10 mm. Garrigue and exposed maquis; common. Throughout.

Convolvulus arvensis BINDWEED A creeping or twining, more or less hairless perennial to 2 m with arrow to oblong-shaped, stalked leaves 10–75 mm. Flowers pale pink to white with paler stripes, 15–20 mm across, solitary or paired on stalks shorter than their leaves; scarcely scented; *ovary hairless*. Capsule 6–8 mm. Frequent on fallow land and waste ground. Throughout. *Convolvulus betonicifolius* is similar but more *densely hairy* with larger, darker pink flowers 21–30 mm in clusters 1–2(3); ovary *hairy*. ES, FR, IT (incl. Sicily).

D. Plant a shrub.

Convolvulus cneorum SILVERY CONVOLVULUS An erect to spreading *silvery subshrub* to 50 cm tall, woody below. Leaves linear to elliptic, widest above the middle and untoothed. Flowers white or pinkish, to 35 mm across, borne in dense terminal heads. Fruit hairy. Rocky habitats by the sea. IT (incl. Sicily). *Convolvulus oleifolius* is similar but less robust and with *very narrow, linear leaves*. Flowers pink, to 30 mm across, borne in lax clusters. Garrigue and rocky habitats. Predominantly eastern Mediterranean. MT. *Convolvulus dorycnium* is similar to *C. cneorum* but a woodier shrub that is densely hairy but not silvery. Flowers pink, to 25 mm across, borne in branched inflorescences. Dry garrigue and scrub. North Africa only. TN.

Ipomoea

Climbing annuals or tuberous perennials with oval, cordate leaves. Flowers with funnel-shaped, unlobed, often showy corolla; style 1. Fruit not lobed.

A. Flowers white or yellowish.

Ipomoea imperati (Syn. *I. stolonifera*) FIDDLE-LEAF MORNING GLORY A fleshy, *creeping* perennial to 1 m with alternate, entire to deeply 3-lobed leaves 40–60 mm, heart-shaped at the base. *Flowers large and white* or pale yellow, sometimes with a purple centre, 35–50 mm across. Coastal sands. Native to the Americas but locally naturalised. DZ, ES, IT, MA.

---

1. *Convolvulus arvensis*   4. *Ipomoea purpurea*
2. *Convolvulus althaeoides*   5. *Jasminum fruticans*
3. *Convolvulus cantabrica*   6. *Fraxinus excelsior*

B. Flowers blue or purplish.

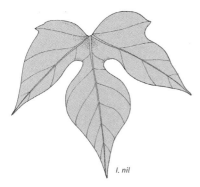

I. nil

Ipomoea indica (Syn. *I. acuminata*) MORNING GLORY A perennial herb with long trailing stems to 15 m. Leaves to 18 cm, oval and *entire* to deeply 3-lobed, heart-shaped at the base. Flowers large and conspicuous: sky-blue or purplish (rarely white) but fading to pink, 60–86 mm. Naturalised in thickets, usually near towns. ES, FR, PT. ***Ipomoea purpurea*** is similar but with always entire, oval leaves 80 mm–16 cm and *deep purple flowers* 40–50(60) mm. A native of tropical America, commonly naturalised in Spain, France and Italy. ES, FR, IT. *Ipomoea nil* (Syn. *I. hederacea*) is an exotic ornamental native to tropical America with *3-lobed* leaves and light purple flowers 20–40 mm. Planted in developed areas and naturalised. ES, PT.

Ipomoea batatas SWEET POTATO A herbaceous perennial vine with 3–7-lobed leaves and typically *Convolvulus*-like purplish, violet or white trumpet-shaped flowers 30–50 mm. The large reddish tubers are cultivated as a food source. Cultivated occasionally in gardens in the region.

OLEACEAE | OLIVE FAMILY

Trees and shrubs, generally with opposite, simple leaves. Flowers in cymes or panicles; calyx with 4 small teeth; corolla with 4(–6) free or fused petals; stamens usually 2; style 1. Fruit a 2-valved capsule, 2–4-seeded-berry or winged nut or achene.

Jasminum JASMINE

Woody climbers and shrubs with alternate compound leaves (rarely simply and opposite). Flowers yellow or white with 4–6 petals united into a tube. Fruit (usually) a 2-lobed black berry.

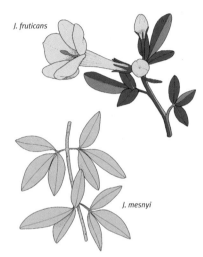

J. fruticans

J. mesnyi

A. Flowers yellow.

Jasminum fruticans WILD JASMINE An evergreen shrub to 2 m with slender, 4-angled stems. Leaves alternate, usually trifoliate, shiny, with oval-lanceolate leaflets 7–25 mm. *Flowers yellow*, 9–18.5 mm long, unscented, borne in clusters of 1–5; lobes 4–10.5 mm. Fruit a black, shiny berry 7.8–13.2 mm. Common on garrigue and maquis. Throughout, except for most islands. *Jasminum mesnyi* is an arching, evergreen shrub with angular stems and opposite, deep green, trifoliate leaves with elliptic leaflets. Flowers 15–22 mm long, borne singly and laterally; bright yellow, often with an orange-tinged centre, to 40 mm with 6–8 overlapping, oval lobes 6–8 mm. Native to China but widely cultivated.

B. Flowers white.

Jasminum officinale WHITE JASMINE A vigorous twining deciduous climber with pinnately divided, *opposite leaves* with 5–7 leaflets. *Flowers white,* 22 mm long, and strongly fragrant, borne in umbels; corolla lobes 15–20 mm. Fruit a blackish berry. Commonly cultivated in gardens and hedges.

Fraxinus

Deciduous trees with compound leaves. Flowers with 4 petals or petals absent; calyx 4-lobed or absent; stamens 2. Fruit a 1-seeded winged samara.

F. angustifolia

Fraxinus excelsior ASH A tall, erect tree to 30(45)m with smooth silvery bark and large compound leaves with 9–13 pointed, oval, *stalkless lateral leaflets* 22 mm–11 cm. Flowers unisexual or cosexual, brownish-purple, borne on the previous year's twigs before the leaves in *paniculate axillary clusters*; calyx and corolla absent. Fruit a winged samara 28–48 mm long. Widespread in open woods and river banks; rare or absent from the far south and west and all arid zones. ES, FR, IT, MA. *Fraxinus angustifolia* NARROW-LEAVED ASH is similar to *F. excelsior* but with *pale brown (not blackish) winter buds and flowers borne in racemes*. Mixed woods and riversides. Often the most common *Fraxinus* species in the far southwest. ES, FR, IT, MA, PT. *Fraxinus pennsylvanica* has leaves with a relatively long-stalked end leaflet (to 32 mm) and unisexual flowers. Native to North America but planted and naturalised, at least in Spain. ES. *Fraxinus ornus* FLOWERING ASH has at least some short-stalked leaflets (stalks 24–88 mm) and conspicuous whitish cosexual or male flowers borne in sweetly scented, *terminal pyramidal clusters at the same time as the leaves*. Rare in the west. ES, FR, IT.

Olea OLIVE

Small evergreen trees with opposite, entire leaves and flowers with 4 corolla lobes and sepals borne in axillary clusters. Fruit a berry (the olive).

Olea europaea OLIVE A familiar, much-branched, often small tree to 15 m with a grey trunk. Leaves 7–60 mm, opposite; grey-green, silvery beneath, minutely scaly, lanceolate, untoothed and short-stalked. Flowers small, 6–8.5 mm, whitish, borne in erect clusters. Fruit an olive, 6–15(20) mm. An important species that is extensively cultivated, and very commonly naturalised on maquis and thickets. Throughout. The wild variety (var. *sylvestris*) differs from the cultivar (var. *europaea*) in having spiny lower branches. **Subsp.** *maroccana* (Syn. *O. maroccana*) is a much larger tree, to 20 m, which is believed to be a relic of an ancient tropical African lineage which grows in the High Atlas. MA.

1. *Olea europaea*
2. *Olea europaea* (flowers)
3. *Olea europaea* (fruits)
4. *Olea europaea* subsp. *maroccana*
PHOTOGRAPH: SERGE D. MULLER

Phillyrea

Evergreen shrubs with simple leaves. Flowers greenish or yellowish with 4 corolla lobes and projecting stamens. Fruit a berry.

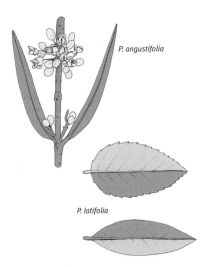

P. angustifolia

P. latifolia

Phillyrea angustifolia An olive tree-like evergreen shrub to 2(4) m with upright branches and grey bark. Leaves 31–78 mm long, and *narrow, 3–12 mm wide*; opposite, linear-lanceolate, untoothed to finely toothed, with 4–6 pairs of *obscure* veins. Flowers small, 2.7–4.2 mm, greenish, 4-lobed and borne in lateral clusters to 10 mm across; *fragrant*. Fruit small, 4–6.6 mm, fleshy with a point, blue-black when ripe. Common on garrigue, maquis and rocky, open woods. Throughout. *Phillyrea latifolia* is similar but with finely hairy twigs and leaves broader, 10–33 mm and of *2 types*: juvenile leaves oval and heart-shaped at the base, adult leaves elliptic; both leathery dark green, toothed or not, and with 7–10 pairs of *distinct* lateral veins. Flowers 3–7.8 mm, whitish. Fruit similar to *P. angustifolia*. Similar habitats. Throughout.

Ligustrum PRIVET

Deciduous shrubs with simple leaves. Flowers white, densely clustered; petals 4, united into a tube. Fruit a black berry.

Ligustrum lucidum CHINESE PRIVET A small, *evergreen* tree to 10(15) m tall with elliptic, opposite, glossy, dark green leaves 55 mm–15 cm long, and *hairless young branches*. Flowers typically privet-like: white, small and borne in dense panicles to 20 cm long; corolla 3.8–5.2 mm across, tubular with 4 spreading lobes. Berry black when ripe, 4.5–10 mm long. Native to China and widely planted. *Ligustrum vulgare* COMMON PRIVET is similar but a *deciduous* shrub with hairy young branches and *smaller leaves*, 12–64 mm long. Damp habitats and woods; local, and rare in the far south. ES, FR, IT, MA, PT. *Ligustrum ovalifolium* is similar to *L. vulgare* but with *oval leaves* 17–59 mm, hairless young branches, and leaves remaining for most of the winter. Native to Japan but widely planted.

1   2   3   4

SCROPHULARIACEAE | FIGWORT FAMILY

Herbs (rarely shrubs or trees) with opposite or alternate leaves. Flowers zygomorphic, usually in spikes or racemes; calyx 4–5-lobed or 2-lipped; corolla 5-lobed, often 2-lipped; stamens 2 or 4. Fruit usually a 2-parted capsule. A family much-reduced following recent DNA-analysis (many genera transferred to the families Plantaginaceae).

Buddleja

Shrubs with simple opposite leaves. Flowers with a 4-lobed corolla fused below into a tube; stamens 4.

Buddleja davidii A lax shrub 1–5 m tall with long, arching branches with grey-green, opposite, short-stalked leaves 10–22 cm long. Flowers lilac (sometimes white) with an orange centre, 4.4–5.5 mm across and fused into a tube with 4 lobes, borne in dense, many-flowered inflorescences; fragrant. Native to China but widely naturalised in disturbed habitats across Europe. ES, FR, IT.

Scrophularia FIGWORT

Annuals or perennials, often with square stems and opposite leaves. Flowers yellow or greenish; corolla with 5, small, spreading lobes; calyx with 5 lobes; fertile stamens 4 (and 1 sterile).

A. Leaves divided into narrow, toothed or lobed segments.

S. canina

Scrophularia canina FRENCH FIGWORT A medium, hairless, much-branched perennial to 1.25 m. *Leaves below pinnately, narrowly lobed with teeth*; upper leaves elliptic and toothed, usually *alternate. Bracts small and not leaf-like*. Flowers small, 3.5–6 mm long, borne on *stalks exceeding the calyx*; purple and white, with *stamens clearly projecting*, and borne numerously in lax terminal, cylindrical clusters; calyx lobes with a membranous margin. Dry rocky places and garrigue. Throughout. *Scrophularia lucida* is similar but normally with solitary stems and with *flower stalks much shorter than the calyx* and with stalkless, glandular hairs and *stamens scarcely projecting*. Rocky cliffs. FR, IT. *Scrophularia ramosissima* is also similar to *S. canina* but with flowers borne in small clusters of 1–3 (not 1–23), and the *flower-stalks persistent and spine-like*. Maritime sands. FR (Corsica), IT (Sardinia).

Scrophularia sambucifolia ELDER-LEAVED FIGWORT A robust, virtually hairless perennial to 1.9 m with 1–2-pinnately divided lower leaves to 15 cm long, and smaller, entire upper leaves. Flowers rather large, 12–19.5 mm green with a *large* brownish-pink or rust-coloured upper lip, borne in short-stalked whorls in the leaf axils. Wet places in southern Iberian Peninsula. ES, PT.

---

1. *Phillyrea angustifolia*
2. *Buddleja davidii*
3. *Scrophularia canina*
4. *Scrophularia lucida*

B. Leaves undivided (toothed or untoothed); calyx lobes *without a papery margin*.

Scrophularia peregrina NETTLE-LEAVED FIGWORT A short to tall annual to 75 cm (1 m) with hollow, 4-angled, often reddish stems. *Leaves entire*, 12 mm–12 cm long, light green, paired and nettle-like, oval and pointed, with unevenly toothed margins. Flowers borne in clusters in the leaf axils in groups of 1–3(6); corolla 4.5–7 mm long and dark brown-purple; *sepals green without a papery margin*. Cultivated land, old walls and pastures; widespread and quite common. DZ, ES, FR, IT, MA, TN.

C. Leaves undivided (toothed or untoothed); calyx lobes *with a papery margin*.

Scrophularia nodosa COMMON FIGWORT An erect perennial to 1.4 m with square, unwinged stems which are hairless below the flowers. Leaves 50 mm–20 cm, oval, pointed and coarsely toothed and short-stalked. Flowers 5.2–8 mm long with a greenish tube and purplish upper lip; calyx 5-lobed with a *narrow white, papery border*. Damp, cool habitats; not at sea level. DZ, ES, FR, IT, MA, TN.

S. frutescens

Scrophularia frutescens A medium, hairless, sparingly branched perennial to 70 cm, woody below, rather similar to *S. canina* in appearance but with (mostly) undivided leaves; leaves 10–58 mm, lanceolate, entire or toothed, leathery, and mostly *opposite*. Bracts *leaf-like*, at least below. Flowers small, 3–5.5 mm long, dull purple, and borne numerously in lax terminal, cylindrical clusters. Locally common on maritime sands in southern Iberian Peninsula. ES, PT.

Scrophularia scorodonia BALM-LEAVED FIGWORT A medium to tall more or less *downy* perennial to 1.75 m with slightly wrinkled, oval, pointed leaves 55 mm–14 cm; *entire, with toothed margins*. Flowers purple, 6.5–10 mm long, borne in erect, sparse inflorescences; calyx lobes with *broad membranous margins*. Local in woods and in damp habitats on higher ground. Woods in west and southwest Iberian Peninsula only in the region covered. ES, PT.

Scrophularia scopolii A rather hairy medium to tall perennial to 1 m with *double-toothed leaves*; leaves oval-lanceolate and heart-shaped at the base; bracts generally not leaf-like. Flowers greenish with a purple-brown upper lip, to 12 mm long, borne in clusters of 4–7 with glandular-hairy stalks. Cool, damp woods on higher ground. IT.

Verbascum

Herbs with large basal rosettes, and often grey-hairy leaves. Flowers usually with yellow or white, or purple 5-lobed corolla; calyx with 5 equal lobes; stamens 5; style 1. Hybrids arise frequently.

A. *Flowers in clusters of 2 or more*; filaments with purple hairs.

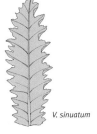

V. sinuatum

Verbascum sinuatum A stout, erect, medium, *grey or yellow-woolly* biennial to 1.5 m. Basal leaves 15–45 cm, dense, *forming distinctive rosettes*; *conspicuously wavy-pinnately lobed*. Flowers yellow, 13–25 mm across, borne in clusters on a *widely branching inflorescence*; stamens with violet-hairy filaments; bracts 1.5–4 mm. Rocky, grassy and coastal habitats; one of the commonest species. Throughout.

Verbascum creticum A mid-green perennial to 1.5 m with rosettes of *toothed or lobed leaves* 9–21 cm without an obvious terminal lobe. Flowers bright yellow and *large, 40–50 mm across* with violet-hairy filaments and *brownish inverted V-shaped markings* on the upper petals; bracts 15–19 mm long. ES (incl. Balearic Islands), IT (incl. Sardinia and Sicily).

Verbascum virgatum TWIGGY MULLEIN An erect biennial to 1.5 m similar to *V. blattaria* (p.422) but more glandular hairy and with flowers 28–35 mm *borne on short stalks* 2–4 mm (shorter than calyx) in clusters of *2–5 per bract* (not solitary); bracts 6–10 mm. Europe only. ES, FR, IT, PT.

Verbascum boerhavii An erect, grey-hairy biennial to 2 m similar to *V. thapsus* (p.423) with leaves 10–30 cm with *leaf bases not running down the main stem,* and with bright yellow flowers 22–30 mm *borne in groups of 3–5 per bract, with filaments with purple hairs*; bracts 10–25 mm. FR (incl. Corsica), IT.

---

1. *Verbascum sinuatum* (inset: rosette leaves)
2. *Scrophularia nodosa*
3. *Scrophularia frutescens*

B. *Flowers solitary*; filaments with purple hairs.

Verbascum blattaria A slender biennial, hairless below with erect stems to 1.2 m, stickily hairy above. Leaves 9–19 cm with paler mid-veins, rather upward-pointing, and coarsely and irregularly toothed along the margins, shiny, hairless and wrinkled. Flowers *long-stalked,* yellow, 20–35 mm across, borne in long, slender, lax, *hairless spikes*; filaments with reddish or violet hairs; the upper petals with small reddish blotches at the base; bracts 5–11 mm. Throughout.

V. barnadesii

Verbascum barnadesii A stout, erect, medium, glandular, white-woolly biennial to 1.8 m. Basal leaves 50 mm–12 cm, narrowly lanceolate and *lobed*, with toothed margins; short-stalked. Inflorescence solitary, *lax* and spike-like. Flowers bright yellow, 25–35 mm across, often with purple-brown markings; calyx rather *small* (5–6 mm); bracts 4.5–5 mm. Dry, rocky places. Iberian Peninsula. ES, PT.

Verbascum chaixii NETTLE-LEAVED MULLEIN An erect biennial to 1.2 m, similar to *V. lychnitis* but with leaves 20–35 cm slightly lobed at the base. Flowers yellow or white, 18–22 mm across, borne on long, slender inflorescences with *upward-arching branches; filaments with violet hairs*; bracts 5.5–7 mm. ES (northeast), FR, IT (incl. Sicily).

Verbascum rotundifolium ROUND-LEAVED MULLEIN A *thickly white-downy*, tall biennial to 1.5 m tall, woolly above. *Basal leaves broadly oval to rounded*, blunt-toothed or untoothed. Flowers yellow, 20–35 mm across, borne in simple racemes (or with few branches); stamens with violet-hairy filaments; bracts 10–20 mm. Rocky habitats. Mainly southern and eastern, and on the islands except the Balearic Islands. DZ, ES, IT, MA, TN.

1. *Verbascum blattaria*
2. *Verbascum lychnitis* (habit)
3. *Verbascum lychnitis*
4. *Verbascum thapsus*

C. Filaments with white or yellow hairs; *anthers elongated down the filament.*

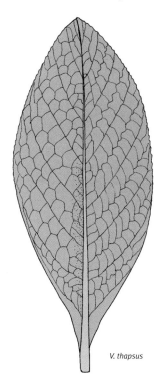

V. thapsus

Verbascum thapsus (Syn. *V. phlomoides*) GREAT MULLEIN A tall, white-woolly biennial to 2 m. Basal leaves 10–50 cm, elliptic, rather wavy-margined and blunt, toothed or not, with a narrow, winged stalk. Stem leaves with *winged bases running down the stem (decurrent) almost to the leaf beneath.* Flowers yellow, 18–23 mm across, borne in an often solitary, woolly, *dense* terminal spike-like inflorescence; petals with 5 more or less equal lobes; stamens 5, the upper *filaments 3 with yellow-white-hairy filaments, the lower 2 almost hairless.* Commonest in centre of range, casual only in parts of south and west. Throughout. *Verbascum giganteum* (Syn. *V. thapsiforme*) is similar to *V. thapsus* but with coarsely toothed leaves and flat, *larger flowers 10–35(50) mm across,* with anthers attached to filaments over their entire length. Centre-west of region; absent from most islands. ES, FR, IT. *Verbascum longifolium* is similar to *V. thapsus,* rather yellowish-white-woolly, the upper leaves clasping the stem. Flowers flat, to 30 mm across, the longest filaments hairless; flowers borne in dense spikes with flower stalks of unequal length (either longer or shorter than calyx). East only. IT.

D. Filaments with white or yellow hairs; *anthers attached to centre of filament.*

Verbascum lychnitis WHITE MULLEIN An erect biennial to 1.5 m with angled, *white-mealy stems.* Leaves 16–35 cm, *dark shiny green and hairless above, powdery-white below.* Inflorescences branched, the *branches erect, parallel to the main stem*, giving a slender appearance; flowers 14–18 mm, white or yellow; *all stamens clothed with white hairs*, borne 2–7 per bract; bracts 10–15 mm. Widespread but absent from far west and many hot, dry areas. DZ, ES, FR, IT, MA. *Verbascum pulverulentum* is similar but the whole plant thickly covered in mealy white hairs which rub off. Flowers 18–25 mm. Absent from far south. ES, FR, IT.

Verbascum speciosum A densely grey-hairy, erect biennial with entire, lanceolate leaves. Flowers borne in clusters, yellow with white or yellow hairs, on larger specimens borne in *pyramidal, top-heavy-seeming inflorescences* with many long, ascending branches. IT.

E. Flowers *purple.*

Verbascum phoeniceum PURPLE MULLEIN A medium to tall perennial with erect stems, glandular-hairy above. Basal leaves oval and toothed or untoothed, upper leaves stalked. *Flowers violet,* to 30 mm across, borne in lax racemes; filaments with purple hairs. Widely cultivated as an ornamental. East only. IT.

Myoporum

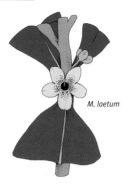

M. laetum

A genus of trees and shrubs native to Australasia, formerly classified in the Myoporaceae. Flowers with 5 corolla lobes; stamens 4; ovary 2–4-parted. Fruit a berry.

Myoporum laetum An evergreen shrub or tree to 13 m with sticky shoots. Leaves 45 mm–17 cm, alternate, narrowly lanceolate, slightly toothed towards the tip, hairless and dotted with oil glands. Flowers 10–15 mm across, borne in lateral clusters of 5–6(10), white with purple markings; stamens 4. Fruit a purple berry 7–10 mm. Native to New Zealand, frequently planted in Iberian Peninsula, and probably elsewhere. ES, PT.

PAULOWNIACEAE

Deciduous trees native to China, Laos and Vietnam, with large, opposite, heart-shaped leaves. Flowers zygomorphic, borne in large panicles; calyx and corolla 5-lobed; stamens 4; style 1. Fruit a 2-parted capsule.

Paulownia

P. tomentosa

Deciduous trees with heart-shaped leaves. Flowers with tubular, foxglove-shaped corolla, borne in erect panicles; stamens 4. Fruit a capsule.

Paulownia tomentosa FOXGLOVE TREE A deciduous tree with a rounded but open crown, to 16 m. Leaves opposite, 3–5-lobed and *large* 12–25(50) cm across, long-stalked and downy, heart-shaped at the base. Flowers 35–60 mm, blue-purple with a paler throat, 2-lipped and funnel-shaped, rather foxglove-like, borne in terminal panicles. Fruit 2-parted, pointed, egg-shaped and woody, to 50 mm long. Commonly planted in France and locally elsewhere. FR.

ACANTHACEAE

Herbaceous perennials with simple, often lobed leaves and erect, (usually) unbranched stems; bracts conspicuous and spiny. Flowers borne in dense spikes; calyx 4-lobed and 2-lipped; corolla zygomorphic, 1–2-lipped, the lower lip 3-lobed; stamens 4, not protruding; style 1. Fruit a capsule.

Acanthus BEAR'S BREECH

Leafy perennials with flowers borne in long, dense, terminal spikes. Easily distinguished by the robust habit, large pinnately lobed leaves, spiny bracts and zygomorphic flowers.

1. *Myoporum laetum*
2. *Paulownia tomentosa*
3. *Acanthus spinosus*
4. *Acanthus mollis* (habit)
5. *Acanthus mollis*

A. mollis

Acanthus mollis BEAR'S BREECH A robust, bushy perennial 75 cm (1 m). Leaves large, 20 cm–1 m, shiny dark green, pinnately lobed, and soft; hairless or nearly so, and long-stalked; stem leaves small and few. Bracts purple-tinged and spiny. Flowers white, 35–50 mm long, borne in dense spikes; corolla 3-lobed; calyx purple and hairless. Various habitats, often in woods or in damp, shady places. Throughout. ***Acanthus spinosus*** SPINY BEAR'S BREECH is similar but with leaves with *markedly spiny lobes*, and the whole plant often hairy. An eastern species extending to southern Italy. IT.

Thunbergia

Tropical vigorous climbing vines native to southern Asia, Madagasca and Africa. Flowers conspicuous. Fruit a splitting capsule.

Thunbergia grandiflora BENGAL TRUMPET A vigorous tropical climbing vine to 20 m with stalked, opposite, rough, triangular to oval leaves to 20 cm, variably lobed to entire. Flowers pale blue or mauve with a yellow throat, 70–80 mm across, borne in hanging racemes. Frequently planted in the Côte d'Azur, scattered elsewhere. FR.

VERBENACEAE

Herbaceous annuals perennials or shrubs with opposite leaves and square stems. Flowers borne in clusters or heads; corolla a slender tube and flat limb, often 2-lipped; stamens 4, not protruding; style 1. Fruit berry like or 4 1-seeded nutlets.

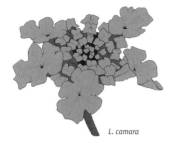

L. camara

Lantana camara LANTANA A small prickly shrub to 1.5(4) m with square and prickly branches that are hairy when young. Leaves 40 mm–13 cm, opposite, oval, toothed and short-stalked. Flowers 10–11 mm, bright yellow or orange, ageing red, congested in tight, often paired heads carried on long stalks; corolla slightly 2-lipped. Fruit a small black berry 4–7 mm. Cultivated in gardens and frequently naturalised. ES, IT (Sicily); probably elsewhere.

1. *Thunbergia grandiflora*
2. *Lantana camara*
3. *Verbena officinalis*
4. *Jacaranda mimosifolia*
5. *Catalpa bignonioides*
6. *Podranea ricasoliana*

Verbena

Herbaceous annuals or perennials. Flowers borne borne in elongated or flat-topped clusters; calyx and corolla 5-lobed. Fruit 4 nutlets.

V. officinalis

Verbena officinalis VERVAIN A medium, rough-hairy perennial to 1.8 m with slender, stiffly erect and square stems; superficially mint-like. Leaves 40–80 mm, opposite, lanceolate to diamond-shaped in outline but deeply pinnately lobed; stalked below, stalkless and often unlobed above. Flowers stalkless, pink, 4.5–6.5 mm, slightly 2-lipped, borne in long, slender, leafless spikes to 30(55) cm long. Fruit separating into 4 nutlets. Bare waste ground. Throughout. *Verbena litoralis* BRAZILIAN VERVAIN is similar but with *toothed* (not deeply lobed) leaves, and flowers 2.5–3 mm borne in *very crowded* spikes at the ends of the branches. Native to South America, widely naturalised. ES, FR, IT.

Phyla

Woody-based perennial herbs with trailing stems, simple, opposite leaves. Flowers small, borne in dense clusters. Fruit 2 nutlets.

Phyla nodiflora A *prostrate* perennial herb with trailing stems, rooting at the nodes, and ascending flowering stems to 75 cm with oval to elliptic, toothed leaves 20–40 mm. Flowers 2.5–3 mm, purple in bud, white when open, borne in dense, long-stalked, *cylindrical spikes*. Damp and disturbed, often coastal habitats. ES, IT, MA. *Phyla canescens* (Syn. *P. filiformis*) is very similar but with *rounded flower spikes with pale pink or lilac flowers* 4–5 mm. Native to south-central America but naturalised on the islands and some mainland coasts. ES (Balearic Islands), FR (Corsica), IT (Sardinia), PT.

BIGNONIACEAE

A large tropical and subtropical family of trees and climbers with opposite, sometimes pinnately divided leaves. Flowers with *large, tubular* corolla with *4 stamens in 2 pairs*. Fruit a pod-like capsule splitting lengthways into 2, containing numerous, often winged seeds.

Jacaranda

Trees native to tropical America, normally with pinnately divided leaves. Flowers conspicuous, borne in panicles. Fruits pendent, flattened.

J. mimosifolia

Jacaranda mimosifolia JACARANDA A small tree to 8(15) m. Leaves to 45 cm, 2-pinnately divided with numerous, opposite, oval to diamond-shaped, downy leaflets. Flowers blue, trumpet-shaped, 50–60 mm long, 2-lipped, borne in showy, terminal panicles, before the leaves have developed. Fruit oval, brown and papery when ripe, 50–80 mm across; flattened. Frequently planted along roadsides in towns in the Iberian Peninsula and sometimes elsewhere. ES, PT.

Catalpa

Deciduous trees native to warm-temperate North America with heart-shaped (sometimes 3-lobed) leaves. Flowers with bell-shaped corolla with 2 stamens. Fruits slender.

Catalpa bignonioides INDIAN BEAN TREE A deciduous tree to 15(18) m with a spreading crown. Leaves large and heart-shaped, long-stalked, 20–30 cm long. Flowers bell-shaped, white with purple spots and yellow markings within, 25–35(50) mm long, borne in large, showy panicles. Fruit slender and bean-like, green and later brown, (15)20–40 cm long, persisting after the leaves. Native to North America, widely planted in France and Italy. FR, IT.

Podranea

Evergreen shrubs and vines native to Africa with compound leaves. Flowers showy, with bell-shaped corolla, borne in panicles. Fruit linear, flattened.

Podranea ricasoliana PORT ST. JOHN'S CREEPER A vigorous climber with long, twining stems and large, pinnately divided leaves. Flowers pale pink with darker linear markings in the throat, tubular-bell-shaped and 5-lobed, borne in hanging bunches. Native of South Africa; planted. ES.

Campsis TRUMPET VINE

Deciduous, woody climbers with aerial roots native to North America and China. Flowers with showy, trumpet-shaped corolla. Fruit elongated, bean-like.

Campsis radicans TRUMPET VINE A vigorous, deciduous climber to 10(12) m with robust, woody stems clinging with aerial roots. *Leaves hairy*, pinnately divided, with 9–11(13) oval, toothed leaflets 20–50 mm. *Flowers bright orange to scarlet, trumpet-shaped* and weakly 2-lipped, 65–95 mm long. Fruit spindle-like, 60 mm–24 cm; seeds linear, hairless, 20–30 mm. Commonly planted in parks and gardens across the region. *Campsis grandiflora* is similar but has twining stems, *hairless leaves* and larger flowers borne in pendulous clusters. Fruit capsule obtuse. Commonly planted, as are hybrids of the above 2 species.

LAMIACEAE | MINT FAMILY

A large and important family of herbs and shrubs; often glandular and aromatic. Leaves opposite and simple, often toothed or lobed. Flowers zygomorphic, often borne in whorls around the stem; calyx 5-lobed, often a tube with teeth; corolla 2-lipped with 3–5 lobes, the upper lip sometimes reduced; stamens 4, 2 long and 2 short; style 1. Fruit comprising 4 1-seeded nutlets concealed within the persistent calyx.

Ajuga BUGLE

Annual or perennial herbs with entire to deeply divided leaves. Flowers with 5-lobed calyx; corolla pink, white, blue or yellow with a very short upper lip and conspicuous 3-lobed lower lip and with a ring of hairs within; stamens 4, shorter than lower lip.

A. Inflorescence with *many-flowered whorls*; flowers blue.

Ajuga reptans BUGLE A variable, short perennial herb *with leafy runners above ground* and erect flowering stems 15–40 cm, usually *hairy on 2 faces only*; leaves 25–50 mm, spoon-shaped and toothed or not, not especially hairy. Flowers blue-violet, 15–18 mm long. Restricted to mountains in the region, and rare or absent from many islands and the far south. Scattered throughout. *Ajuga genevensis* is similar but with *aerial runners absent*, stems hairy on all faces (sometimes just on opposite sides) and rather hairy leaf blades. Local, and absent from hot, dry areas and the islands. FR, IT. *Ajuga orientalis* EASTERN BUGLE has *distinctively symmetrical, dense pyramidal, very hairy* spikes of flowers which are exceeded by their leaves. IT (incl. Sardinia and Sicily).

B. Inflorescence normally with *2-flowered whorls*, and leaves undivided.

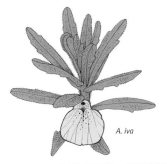

A. iva

Ajuga iva SOUTHERN BUGLE A tufted or sprawling, low to short, softly hairy perennial to 15 cm. Leaves 15–40 mm, broadly linear, normally with short lobes, sometimes unlobed. Bracts similar to the leaves and exceeding the flowers. Flowers white or cream to pink-purple, often with small darker spots; corolla 13–24 mm long, the *upper lip entire and highly reduced*. Common on stony garrigue and cliff-tops. Throughout.

1. *Ajuga iva*
2. *Ajuga chamaepitys*
3. *Campsis radicans*
4. *Teucrium pseudochamaepitys*

C. Inflorescence normally with 2-flowered whorls, and *leaves deeply 3-lobed*.

Ajuga chamaepitys GROUND PINE A short, hairy, ascending annual to 19 cm. Basal leaves 23–52 mm, soon withering, stem leaves divided into *3 long, linear lobes,* and smelling of pine when crushed. Flowers rather sparse, 10–16 mm long, yellow with purplish markings, borne 2–4 at each node, and greatly exceeded by their bracts. Throughout.

Teucrium GERMANDER

Herbs or shrubs with toothed to lobes leaves. Flowers with a 2-lipped calyx and corolla with a *single, 5-lobed lower lip*; stamens 4, shorter than the lower lip.

A. Plant a shrubby perennial; leaves deeply divided into narrow lobes.

T. pseudochamaepitys

Teucrium pseudochamaepitys GROUND-PINE GERMANDER A sparsely branched, often sparsely woolly-hairy perennial 25–30(40) cm, woody at the base. Lower leaves 1–2-divided, the upper 3-parted, all with *linear, untoothed lobes with downturned margins.* Inflorescence unbranched, the bracts equalling or exceeding the flower stalks; calyx 8.5–10 mm, bell-shaped, usually glandular-hairy with triangular teeth exceeding their tube; *corolla 2 x as long as the calyx,* 13–14 mm; whitish and marked, with long-protruding stamens. Common on garrigue in far south and west. DZ, ES, FR, MA, PT, TN.

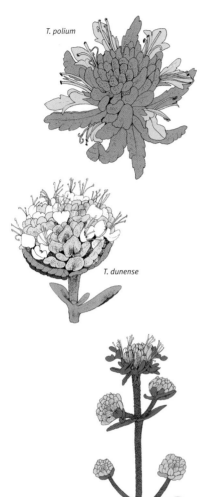

T. polium

T. dunense

T. capitatum

B. Plant a shrubby perennial; leaves entire or toothed. *Flowers in dense rounded heads* (many similar species).

Teucrium polium FELTY GERMANDER A *dense, dwarf shrub to 20(30) cm tall, densely white, yellow or grey-felted.* Leaves 10–13 mm long, *oblong*, often with down-turned margins, bluntly toothed; bracts leaf-like but untoothed. Flowers borne in felty, rounded *heads >10 mm across of up to 20 flowers*; corolla 7–9 mm, *white or pink-tinted*; calyx 5–6 mm. Low garrigue. Throughout (in its broader description). The following are a few of a large number of very closely related species which are treated by some as mere variants of *T. polium*: ***Teucrium vincentinum*** (Syn. *T. polium* subsp. *vincentinum*) has *long* leaves 15–18(35) mm, densely yellow or green-grey-felted stems, and cream or white flowers with hairless corolla lobes. Calyx 4.4–5 mm. Rocky cliffs and dunes. PT. ***Teucrium dunense*** (Syn. *T. polium* subsp. *dunense*) is virtually indistinguishable from *T. vicentinum* but is grey-hairy, and has markedly swollen, dense flower-heads >10 mm across; corolla pink cream or white; calyx 5–6 mm. ES, FR, PT. ***Teucrium cossonii*** (Syn. *T. polium* var. *pulverulentum*) has *purplish flowers*. ES (Balearic Islands). *Teucrium haenseleri* has an ascending habit (45 cm) and *unfelted,* but hairy stems, narrower, *linear* leaves 12–15 mm, often deflexed, and heads of 3–10 *cream* flowers with triangular corolla lobes; calyx 4–4.5 mm. ES, PT. *Teucrium capitatum* (Syn. *T. polium* subsp. *capitatum*) has *compound, branching inflorescences* with small heads, <10 mm of cream, white or pink flowers; calyx 3–3.5 mm. Throughout.

C. Plant a *large shrub to 2 m*; leaves entire or toothed. Flowers pink, purple or blue and *not* in dense rounded heads.

T. fruticans

Teucrium fruticans TREE GERMANDER A spreading, white-felted evergreen *shrub to 2 m* with square stems. Leaves 25–35 mm, lanceolate, flat, short-stalked, entire and white-felted; bracts leaf-like. Flowers paired; corolla blue or lilac, 20–24 mm with long-protruding and conspicuous stamens; calyx bell-shaped. Maquis and planted in gardens. Throughout.

1. *Teucrium polium*
2. *Teucrium vicentinum*
3. *Teucrium dunense*
4. *Teucrium cossonii*
5. *Teucrium fruticans*
6. *Teucrium scordium*

D. Plant a shrubby perennial; leaves entire or toothed. Flowers pink, purple or blue and *not* in dense rounded heads.

Teucrium scordium WATER GERMANDER A softly white-hairy perennial to 40(60) cm with creeping rhizomes and leafy runners from which arise erect or ascending stems. Leaves 35 mm long, oblong and coarsely toothed, rounded at the base, grey-green and more or less unstalked; garlic-scented when crushed. Flowers with a corolla pink-purple, 9–10 mm long, borne in loose terminal inflorescences. Damp or aquatic habitats; widespread. Throughout.

Teucrium chamaedrys WALL GERMANDER A tufted, spreading, green perennial to 20(35) cm, woody at the base with erect or ascending stems. Leaves small, 10–15(20) mm long, oval-oblong, 2 x as long as wide, blunt and short-stalked, shiny green and usually with distinct, rounded teeth. Flowers pink-purple; corolla 2–17 mm long; calyx reddish, hairy and bell-shaped. Common in various habitats. Throughout. Numerous subspecies are recognised, though some are continuously variable: Subsp. *pinnatifidum* is characterised by lobed leaves, glandular on the underside, and many-flowered inflorescences. ES (Balearic Islands), FR. Subsp. *albarracini* has white stems and lower leaf surface. ES, FR.

Teucrium marum CAT THYME A leafy, spreading, *pale grey undershrub to 50 cm with numerous dense, terminal, spike-like inforescences.* Leaves small, to 10 mm, oval-lanceolate, entire with inrolled margins, greenish above and whitish below. Flowers pink, to 10 mm, borne in whorls; calyx densely hairy. FR (Corsica), IT (north + Sardinia).

E. Plant a *very spiny subshrub*; leaves entire or toothed. Flowers pink, purple or blue and *not* in dense rounded heads.

Teucrium subspinosum A perennial or subshrub similar to *T. marum* but shorter, denser, grey-green, to 40 cm *with prominent spines*, leafy below but leaves sparser above. *Flowers sparse*, borne on short stems; corolla pale pink, 10–13 mm. Rocky and stony habitats and short garrigue. ES (Balearic Islands), IT (Sardinia). *Teucrium balearicum* BALEARIC GERMANDER is very similar to *T. subspinosum* but a *more intricately spiny, dense, cushion-like shrub* with broader, more bell-shaped calyx; corolla 10–12 mm. Sea cliffs and scrub. ES (Balearic Islands).

F. Leaves entire or toothed. Flowers yellow, greenish or white and *not* in dense rounded heads.

Teucrium flavum YELLOW GERMANDER A shrubby perennial to 35(65) cm with few ascending stems, stalked, leathery, toothed leaves 18–25 mm, and rather dense, leafless, terminal, rather 1-sided clusters of dull, pale sulphur- or green-yellow flowers; corolla 12–20 mm-up to 2.5 x the length of the densely hairy calyx. Rocky places and woods; common, especially in the east. Throughout.

Teucrium scorodonia WOOD SAGE An erect, shrubby perennial to 50 cm (1 m) with stalked, oval-heart-shaped, *wrinkled, sage-like leaves* 40(80) mm long with rounded teeth, and green-yellow flowers forming loose, leafless spikes; flowers borne numerously in pairs in terminal, leafless spikes; corolla pale yellow-green, the tube 6–10 mm long and much exceeding the calyx. Mixed woods and scrub, often above sea level. DZ, ES, FR, IT.

---

1. *Teucrium chamaedrys*	3. *Teucrium marum*	6. *Teucrium flavum*
2. *Teucrium chamaedrys*	4. *Teucrium subspinosum*	7. *Teucrium scorodonia*
subsp. *pinnatifidum*	5. *Teucrium balearicum*	8. *Prasium majus*

T. spinosum

Teucrium spinosum SPINY GERMANDER A *glandular*-hairy and rather 'spiky', bushy annual to 45 cm. Stems much-branched and more or less spiny and leafless when in flower. Lower leaves 15–30 mm, oblong, narrowed at the base, sharply toothed, the uppermost small and untoothed. Flowers white, 8 mm long and held *upside down* (resupinate) in small whorls; calyx with spiny teeth. Fallow land or sandy places. ES, MA, PT.

Prasium

Small shrubs with flowers with 2-lipped corolla, the upper lip arched over the stamens, the lower 3-lobed; stamens 4.

Prasium majus PRASIUM A subshrub to 1 m, hairless or slightly hairy. Leaves 16–43 mm, *shiny dark green*; oval, pointed and toothed, heart-shaped at the base; all stalked; bracts similar. Flowers with a white corolla with tube 10 mm long, upper lip 6–8 mm, lower lip 8–11 mm, borne in terminal racemes; calyx 2-lipped with bristle-tipped teeth; corolla with the middle lobe the largest. Nutlets shiny black when ripe. Widespread and common on rocky maquis. Throughout.

Marrubium HOREHOUND

Perennial herbs with toothed leaves. Flowers with calyx with 10 teeth, spreading when in fruit; corolla white (cream or purple) with a flat upper lip; stamens 4, none protruding.

Marrubium vulgare WHITE HOREHOUND An erect, medium white-downy, aromatic, rather nettle-like perennial to 85 cm with erect, square and *branched, white-cottony stems*. Leaves 17–65 mm, oval, heart-shaped at the base and wrinkled on the surface, with stalks shorter than their blades. Flowers small, and rather inconspicuous, borne in dense whorls in the leaf axils up the stem; corolla 2-lipped and white, the tube 3.5–5 mm, upper lip 2–3.5 mm, lower lip 1.8–3.5 mm; *calyx with 10 or more equally short, hooked teeth*. Common on rocky and stony ground. Throughout.

Marrubium incanum A perennial with ridged, *unbranched flowering stems* to 60 cm tall, woody at the base, often also with short, erect non-flowering stems. Leaves oblong and wedge-shaped at the base with wavy margins, the leaf stalks shorter than their blades and *densely felted*. Flowers white, borne in many-flowered, interrupted false whorls along the stem with awl-shaped secondary bracts curving upwards; calyx tube star-like and with a ring of hairs in the thoat. Maquis, garrigue and pastures. IT (incl. Sicily). *Marrubium peregrinum* BRANCHED HOREHOUND is similar but has narrow, toothed leaves, white flowers that are smaller, particularly relative to the leaves, borne in rather few-flowered, well-spaced false whorls. FR, IT. *Marrubium alysson* is similar to *M. peregrinum* but has unstalked, wedge-shaped leaves and *purple flowers*; *calyx with 5 teeth, rigid and star-like*. ES, IT (incl. Sardinia). *Marrubium supinum* is similar, with cream or purple flowers; calyx *with 5 teeth, curved* and not rigid or star-like. Rare and local. DZ, ES, MA, TN.

Galeopsis HEMP-NETTLES

Annuals usually with toothed leaves. Calyx bell-shaped or tubular, 10-veined, with 5 spine-pointed lobes, exceeded by the 2-lipped corolla which has a flattened upper lip and 3-lobed lower lip with projections at the base; stamens 4, exceeded by upper corolla lip.

Galeopsis ladanum NARROW-LEAVED HEMP-NETTLE An erect, divergently branched, hairy annual to 76 cm with narrow leaves 13–59 mm with forward-pointing teeth and whorls of few (up to 14) pale purple to bright rosy-purple flowers with yellowish markings; corolla 12–24 mm, hairy and calyx densely hairy, the tube much longer than its teeth. Disturbed habitats, shingle and sandy river banks. Rare or absent in most of the far south. ES, FR, IT. **Subsp.** *ladanum* has oval-lanceolate leaves 4–35 mm wide. ES, FR, IT. **Subsp.** *angustifolia* (Syn. *G. angustifolia*) has narrower, more linear leaves 3–15 mm wide. ES, FR, IT.

1. *Marrubium incanum*	4. *Sideritis romana*
2. *Marrubium vulgare*	5. *Sideritis romana*
3. *Galeopsis ladanum* subsp. *angustifolia*	6. *Sideritis hirsuta*

Sideritis

Erect, aromatic herbs, perennials and shrubs. Flowers with calyx bell-shaped, 5-toothed and 10-veined, corolla usually yellow and 2-lipped with a flat upper lip; stamens 4, not protruding. A difficult genus for which floras give conflicting information. Hybrids exist.

S. romana

A. Plant a herbaceous annual; bracts and leaves similar.

Sideritis romana A very variable, shaggy-hairy, generally *short annual* to 36(50) cm. Leaves 10–40 mm, oblong, toothed, and mostly unstalked; *bracts similar to leaves*. Flowers with yellow, sometimes purple or white corolla 8–9 mm long, borne in distant whorls; corolla upper lip flat and undivided; calyx *strongly veined* and with spreading teeth in fruit; *upper calyx tooth solitary much wider than the lower teeth*. Common in various dry, scrubby habitats. Throughout. *Sideritis montana* is similar but a calyx with *3 upper teeth, all of similar size*. Throughout, except for most islands.

B. Plant a woody perennial or lax shrub; bracts different from leaves.

S. hirsuta

Sideritis hirsuta A tall, hairy perennial 10–69 cm with *stems hairy on all faces, and to the base.* Leaves hairy, oval, 11–28 mm long with a sharply toothed margin; green or greyish. Flowers with a whitish corolla, often with a yellow lower lip, 11–14 mm long, borne in *densely long-hairy inflorescences* with distant whorls; calyx regular and *not* 2-lipped; bracts 9–15 mm wide. Garrigue. ES, PT. *Sideritis hyssopifolia* is similar to *S. hirsuta* but with sparsely hairy leaves and pale yellow, often purple-spotted flowers *borne in dense, cylindrical clusters*; bracts 7–13 mm wide. Various habitats and altitudes. ES, FR, IT. *Sideritis incana* is similar to *S. hirsuta* but with white-hairy leaves, and *stems hairless along the angles*, *untoothed bracts* 6–8 mm wide and flowers with corolla *8–11 mm, dull yellow throughout*. ES. *Sideritis lacaitae* is similar to *S. incana* but with lanceolate, green (not white-hairy) leaves. Flowers with yellow corolla 11 mm; bracts 7–11 mm wide. ES.

Sideritis pungens (Syn. *S. linearifolia*) A more or less hairless, rather tall, erect perennial 15–53 cm with greenish, 4-angled stems with forwardly curved hairs on 2 opposite sides. Leaves 14–45 mm long, narrowly lanceolate and entire with spine-tipped teeth. Flowers borne in congested clusters of about 6; calyx with a ring of hairs inside; corolla 7–9 mm, yellow, often purple-tinged; bracts 9–15 mm, the lowermost with a long terminal tooth. ES, PT. *Sideritis arborescens* is similar but taller, with distant whorls of *6–10 flowers* that with a *white-yellow* corolla 7.5–10 mm; bracts 6–10 x 8–15 mm. ES, MA, PT (records of *S. linearifolia* from the Algarve may correspond with this species). *Sideritis grandiflora* is similar to *S. arborescens* but has *20(>10) flowers* in each whorl, and inferior bracts 12–40 x 17–23 mm. ES, MA.

Phlomis

Herbs or shrubs with entire leaves. Flowers with calyx with 5 equal teeth; corolla 2-lipped, the upper lip notably hooded and notched at the tip, the lower lip 3-lobed; stamens 4, protruding or not.

A. Flowers yellow.

Phlomis fruticosa JERUSALEM SAGE A large, sage-like, grey-felted, erect or spreading, lax evergreen shrub to 1.5 m. Leaves elliptic and usually untoothed, thick, stalked, greyish above and white-felted below. *Flowers large and yellow*, to 35 mm, borne in dense whorls. Rocky ground, maquis and field boundaries. Common in southern Italy and Sardinia; planted and naturalised elsewhere. ES, IT (incl. Sardinia and Sicily), PT.

P. lychnitis

Phlomis lychnitis A small shrub to 50(70) cm with white-felted stems and leaves; leaves 30 mm–14 cm, 14 mm wide, mostly basal and *narrow*, linear to spatula-shaped and mostly untoothed, wrinkled above and densely white-felted below; *scarcely stalked*. Flowers yellow, 25 mm, borne on whorls of 6, each with a pair of broadly oval leaves immediately beneath. Garrigue and maquis. ES, FR, PT. *Phlomis crinita* is similar but with broader, *oval leaves 18 mm wide*; flowers 22–25 mm, dark yellow, often brownish above. ES, MA.

B. Flowers pink or purple.

Phlomis purpurea A sage-like, grey-felted, erect and lax evergreen shrub 80 cm (1.5 m); hairy but not glandular. Leaves 46 mm–11 cm, thick, leathery and wrinkled; greyish above and *white-felted beneath*; all leaves stalked. *Flowers pale purple* (sometimes white), borne in congested clusters of 10–12, to 26 mm long, borne in distant whorls; *calyx grey-felted with lobes lacking spine-teeth*. Common on maquis in southern Iberian Peninsula. ES, PT. *Phlomis italica* is similar but with *leaves white-felted on both surfaces*, and larger flowers to 20 mm long; *calyx teeth <2 mm*. ES (Balearic Islands).

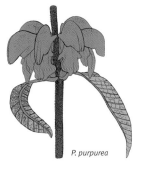

P. purpurea

1. *Sideritis hyssopifolia*
2. *Phlomis fruticosa*
3. *Phlomis lychnitis*
4. *Phlomis purpurea*

P. herba-venti

Phlomis herba-venti A robust, hairy, branched perennial 25–75 cm with lanceolate, toothed, leathery, shiny green leaves 42 mm–22 cm which are paler beneath. Flowers purple, borne in 2–6 dense whorls of 8–16, to 20 mm long; *calyx with spine-tipped teeth >4 mm*. Dry slopes and maquis; absent from most islands. ES, FR, IT (incl. Sicily), MA, PT.

Nepeta CAT-MINT

Perennial herbs with toothed leaves, plants often male-sterile. Flowers with 5-toothed calyx; corolla white, blue or purple, with flat to slightly hooded upper lip; stamens 4, shorter than upper lip of corolla.

Nepeta cataria CAT MINT An erect, hairy rhizomatous perennial with branched stems 34 cm–2 m and grey, toothed leaves 27–84 mm long, grey-woolly beneath. Inflorescence whitish with numerous false whorls; flowers cosexual; corolla 5.7–9 mm, white with *small purple spots*; secondary bracts just 0.2–0.5 mm wide; calyx teeth straight. Fields, cultivated land and woods, often above sea level. Probably throughout, except for hot, dry areas. *Nepeta italica* has creamy white flowers; corolla 11–13 mm. IT (east-central).

Nepeta tuberosa An erect, hairy, tuberous-rhizomatous perennial herb 30 cm–1.6 m high with oval-lanceolate leaves with heart-shaped bases 27–98 mm long. Inflorescence generally unbranched and dense, *purple*, spike-like, with false whorls slightly separated below; flowers cosexual; corolla 10.2–14 mm; secondary bracts 3-veined, 3.5–4.3(5.7) mm wide. Maquis, open pine woods and dry fields. ES, IT (Sicily), MA, PT.

Prunella SELF-HEAL

Perennial herbs with entire to divided leaves. Flowers with 2-lipped calyx, the upper broad with 3 short teeth, the lower with 2 narrow lobes; corolla yellow, blue, pink or white with strongly hooded upper lip; stamens 4, shorter than upper corolla lip.

A. Flower-heads with a pair of leaves immediately below; flowers purple.

Prunella vulgaris COMMON SELF-HEAL A *hairy*, tufted perennial with stems ascending from a creeping base, 50 mm–60 cm high. Leaves 17–96 mm, *broadly lanceolate*, toothed or not, stalked. Flowers blue-purple (rarely white), 11–12 mm long, emerging from very dense, small, cylindrical heads of conspicuous purple, spiky calyces, with *both bracts and a pair of leaves immediately below*. Damp and wooded places above sea level. Throughout. *Prunella hyssopifolia* HYSSOP-LEAVED SELF-HEAL is similar but with *hairless stems* and *narrow linear-lanceolate leaves* 30–70 mm. ES, FR, IT.

B. Flower-heads with a pair of leaves immediately below; flowers white.

Prunella laciniata CUT-LEAVED SELF-HEAL A short, densely hairy, patch-forming perennial with erect flowering stems 10–37 cm. Leaves 35 mm–11 cm, *pinnately lobed*, stalked, with narrow segments. *Flowers pale yellow-white* (rarely pink or purplish), 14–16 mm long, borne in heads with a pair of leaves immediately below; calyx scarcely toothed. Dry scrub; widespread but local. DZ, ES, FR, IT, MA, TN.

---

1. *Nepeta tuberosa*   3. *Prunella grandiflora*   5. *Lamium amplexicaule*
2. *Prunella laciniata*   4. *Prunella vulgaris*   6. *Lamium maculatum*

C. Flower-heads interrupted from the next pair of leaves by a short stalk; flowers purple.

Prunella grandiflora LARGE SELF-HEAL A leafy perennial 10–48 cm, similar to *P. vulgaris* in general appearance but with *larger* flowers, with corolla 18–32(48) mm long; calyx large (10–16 mm) and purplish and inflorescence with bracts, but *not leaves immediately below*. Woods and shady places; absent from the islands and the far south. ES, FR, IT.

Lamium DEAD-NETTLE

Annual or perennial herbs with crowded whorls of flowers. Flowers with white pink or purple corolla with hooded upper lip and tubular or bell-shaped calyx with 5 fine-pointed lobes; stamens 4, exceeded by upper corolla lip.

A. Small annuals 10–25 cm, with flowers <15 mm long.

Lamium amplexicaule HENBIT DEAD-NETTLE A short, scarcely branched, hairy annual 10–40 cm. Leaves 9–20 mm, rounded or oval, blunt-toothed and stalked below; unstalked and clasping the stem above. Flowers held *aloft*; pink-purple, 13–20 mm long, with a slender, straight tube; calyx hairy. A common weed of cultivated ground. Throughout. *Lamium purpureum* RED DEAD-NETTLE is similar but with stalked, not clasping upper leaves and bracts, and densely leafy inflorescences, flushed purple above. Flowers 8–12 mm. Common on disturbed ground. Throughout, except for the far southeast and some islands.

B. Medium perennials 30–80 cm, with flowers >15 mm long.

Lamium maculatum SPOTTED DEAD-NETTLE A variable, hairy, aromatic, patch-forming perennial with erect stems to 18–50 cm. Leaves 30–65 mm, nettle-like, oval to triangular, pointed, toothed and stalked, often with a pale central blotch. Flowers whitish or purplish-pink with darker markings; corolla tube curved, 18–30 mm long. Woodlands and shaded habitats inland. Absent from most islands. ES, FR, IT (incl. Sardinia). *Lamium garganicum* LARGE RED DEAD-NETTLE is similar but with *large flowers 25–40 mm long, the corolla tube straight* and hairless within, greatly exceeding the calyx, and rose-purple, often with darker markings. Mainly in the mountains. FR, IT.

Ballota

Perennials with sterile leaf rosettes. Flowers with 2-lipped corolla with concave upper lip; calyx funnel-shaped, 10-veined and usually 5-lobed; stamens 4, exceeded by upper corolla lip.

Ballota hirsuta A soft-hairy perennial with spreading, woody stems 24–60 cm. Stem leaves 18–80 mm, simple, toothed and glandular, those below heart-shaped at the base. Inflorescence composed of false whorls of many flowers; corolla 13–18 mm, pink-purple with paler markings (sometimes all white), to 16 mm long. Roadsides, scrub and hillsides. ES (incl. Balearic Islands), IT, MA, PT.

Stachys

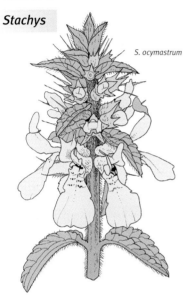

S. ocymastrum

Annual or perennial herbs, often with toothed leaves. Flowers borne in dense, spike-like inflorescences; calyx tubular or bell-shaped with 5 equal teeth; corolla yellow, pink or purple, 2-lipped with a flat or hooded upper lip and 3-lobed lower lip; stamens 4, exceeded by upper corolla lip.

A. Flowers whitish or yellowish.

Stachys ocymastrum An erect, pale green, hairy annual to 70 cm (1.1 m). Leaves 16–65 mm long, oblong and pointed, slightly heart-shaped at the base, toothed and wavy along the margin. Flowers borne in dense whorls of 4–6 along the stem, congested above, *laxer below*; calyx densely hairy with 2 upper teeth as long as the tube and 3 shorter lower teeth; *corolla white*, 10–16 mm (shorter than the calyx), the upper lip entire. Garrigue. Throughout.

Stachys recta PERENNIAL YELLOW WOUNDWORT A sweet-smelling, slightly hairy, tufted and woody-based perennial with many erect stems to 87 cm. Leaves 18–70 mm, hairy or hairless, oblong and *short-stalked below*, lanceolate to linear and stalkless above, all toothed. Flowers with corolla 12–13 mm long, *yellowish-white*, borne in long, slender, leafless spikes, in *whorls of 6–16, crowded above*. Grassy places, usually above sea level. Throughout.

Stachys maritima COASTAL WOUNDWORT A rather short, hairy, somewhat straggling perennial 13–43 cm with ascending stems leafy below; leaves oblong to elliptic and densely hairy and toothed, *with long petioles 10–60 mm*. Flowers pale yellow, *corolla just 10 mm*, borne in crowded whorls of 4–6; calyx densely hairy. Absent from most islands. DZ, ES, FR (incl. Corsica), IT, TN. *Stachys pubescens* is similar in form but with *untoothed leaves and calyx*, and generally white flowers. IT (south). *Stachys arenaria* is similar to *S. maritima* but has an almost prostrate habit; plant bristly with narrow leaves. Flowers whitish, tinged and veined with red-purple, borne in distant whorls. Far south and east of range only. DZ, IT, MA, TN.

---

1. *Ballota hirsuta*
2. *Stachys ocymastrum*
3. *Stachys recta*
4. *Stachys germanica*

5. *Stachys byzantina*
6. *Stachys heraclea*
7. *Stachys arvensis*

B. Perennials with pink or purplish flowers.

S. germanica

Stachys germanica DOWNY WOUNDWORT A *densely white-felted*, erect perennial to 1.3 m. Leaves 58 mm–17 cm, *oblong and heart-shaped at the base*, grey-woolly below and grey-green and less hairy above; calyx with unequal teeth, the upper 2 <$^1/_2$ as long as the tube, but longer than the lower 3. Flowers with corolla to 20 mm; cream-white or bright pink-purple, borne in congested terminal inflorescences. Dry grassy places, fallow land and roadsides; common. Throughout. *Stachys cretica* MEDITERRANEAN WOUNDWORT is similar to *S. germanica*, and also white-felted, with pink-purple flowers, but with *leaves with rounded or wedge-shaped bases* (not heart-shaped). Absent from Iberian Peninsula. FR, IT, MA, TN. **Stachys byzantina** is similar in form to *S. germanica* but *extremely densely white-felted on all surfaces*, such that most green parts are obscured; flowers borne in dense, cylindrical inflorescences; corolla 13 mm. Native to Turkey but naturalised in Sicily, perhaps elsewhere. IT (Sicily).

Stachys heraclea A *shaggy-long-hairy* perennial 17–56 cm, similar in form to the *S. germanica* group. Leaves 70 mm–15 cm, elliptic and with small teeth along the margins. Flowers borne in whorls of 4–10; calyx *very hairy*, with triangular, long-pointed teeth; corolla 15 mm, cream to pink-mauve with darker veins. Field margins and grassland inland; commonest in the east. ES, FR, IT.

C. Annuals with pink or purplish flowers.

S. arvensis

Stachys arvensis FIELD WOUNDWORT A small, erect, hairy annual to 20(45) cm. Leaves 15–50 mm, oval and heart-shaped at the base, hairy, toothed along the margin and wavy-edged. Flowers borne in whorls of 4–6, crowded above, distant below; calyx with teeth as long as the tube; corolla 6–8 mm, white or pale pink and scarcely exceeding the calyx; upper lip entire. Cultivated land and sandy places; common. Throughout.

Satureja

Shrubs or perennials with entire leaves. Flowers with bell-shaped, 10-veined calyx with 5 nearly equal lobes; corolla straight and 2-lipped; stamens 4, shorter than the upper corolla lip.

1. *Satureja montana*
2. *Clinopodium acinos*
3. *Clinopodium nepeta*
4. *Clinopodium vulgare*
5. *Micromeria graeca*
6. *Micromeria nervosa*

Satureja thymbra SATUREJA An aromatic *thyme-like dwarf shrub* to 35 cm tall, grey-hairy. Leaves oblong, broadest above the middle, pointed and rather bristly. Flowers pale pink, to 12 mm long, *borne in rather dense, rounded, distant whorls* (unlike most *Thymus* spp.); calyx reddish and bristly. Garrigue and roadsides. A predominantly eastern Mediterranean species. IT (Sardinia).

Satureja montana WINTER SAVORY A short to medium, aromatic, hairless to slightly hairy, rather densely spreading *dwarf shrub* 13–45 cm. Leaves 12–24 mm, linear to oblong and broadest above the middle, hairless but with *short-hairy margins*. Flowers white to purple, 7.5–12 mm long, borne in many-flowered, leafy terminal whorls; lower bracts longer than the flowers; calyx with teeth shorter than their tube. Stony pastures and rocky slopes. Absent from the south, far west and most islands. ES, FR, IT. *Satureja cuneifolia* is similar but with *distant whorls of flowers* (corolla 5–8 mm) and leaves with inrolled margins. ES, IT. *Satureja hortensis* SUMMER SAVORY is the cultivated culinary herb distinguished by its annual habit, toothed leaves and *small flowers 5 mm long with calyx teeth longer than their tube*. ES, FR, IT.

Clinopodium

Perennial or annual herbs with entire or toothed leaves. Flowers borne in stalked axillary clusters (sometimes reduced to solitary flowers); calyx 5-lobed, tubular with (11)13(15) veins; corolla tube 2-lipped and straight to curved; stamens 4, not protruding. The genus now includes species traditionally classified under *Acinos* and *Calamintha*.

A. Flowers 6–18 mm long.

Clinopodium acinos (Syn. *Acinos arvensis*) BASIL THYME A short, hairy annual or perennial to 28 cm with spreading stems, branched at the base. Leaves 4–17 mm, lanceolate to oval and blunt or pointed and net-veined, sometimes toothed along the margins. Flowers with corolla 6–8.5 mm, violet with white markings on the lower lip, borne in lax whorls in a leafy raceme with bracts similar to the leaves; calyx tube slightly curved. Various dry habitats. Throughout except Sicily and the far west. DZ, ES, FR, IT, MA.

C. nepeta

C. vulgare

Clinopodium nepeta (Syn. *Calamintha nepeta*) LESSER CALAMINT A mint-like, medium, greyish, hairy perennial 20–75 cm with creeping rhizomes; stems erect and branched. Leaves 17–70 mm, oval and shallowly toothed to untoothed; stalked. Flowers with corolla 6–17 mm, pale pink-purple with darker markings, borne in leafy whorls; calyx ribbed and purplish, with white *hairs protruding from the mouth*. Dry, fallow land and waste places. DZ, ES, FR, IT, MA, PT.

Clinopodium vulgare WILD BASIL A mint-like, short to medium, softly hairy, aromatic perennial 16–95 cm with erect, branched or unbranched stems. Leaves 14–50 mm, oval-lanceolate, slightly toothed and short-stalked. Flowers with corolla 9–18 mm pink-purple, borne in distant whorls along an interrupted spike with prominent calyces; calyx green with purple ribs, more or less 2-lipped and prominently toothed. Disturbed, dry, grassy and stony habitats; common almost throughout. ES, FR, IT, MA, PT.

1

2

3

B. Flowers 25–35 mm long.

Clinopodium grandiflorum (Syn. *Calamintha grandiflora*) LARGE-FLOWERED CALAMINT A spreading, sparsely hairy, medium perennial with weakly ascending stems 25–35 cm with oval, coarsely toothed leaves 30–51 mm. Flowers pink, protruding and conspicuous, borne in leafy inflorescences; corolla 25–35 mm long. Mountain woods; local and absent from most islands. ES, FR, IT.

Micromeria

Small, thyme-like subshrubs. Flowers with calyx with 13–15 veins and 5 pointed lobes.

Micromeria graeca A slender, lax, *dwarf shrub* 13–60 cm; woody at the base and variably hairy. Leaves small, 7–14 mm long, oval below and linear-lanceolate above, *with down-turned margins. Flowers stalked*, borne in distant whorls forming long, slender, terminal inflorescences; corolla 5–7 mm, pink-purple; calyx 4–6 mm, woolly-hairy in the throat with teeth almost as long as their tube and unequal. Dry, scrubby places; common. Throughout. *Micromeria microphylla* is similar but to just 22 cm, with the *upper leaves flat*, 35–55 mm. *Calyx small*, 2.5–4 mm. ES (Balearic Islands), IT (incl. Sicily), MT.

Micromeria juliana A dwarf shrub 12–40 cm tall with numerous stiffly erect, unbranched and very densely small-leaved stems; leaves 4–8 mm long, and *linear with down-turned margins*. Flowers tiny, borne in *dense, spike-like, continuous stalkless whorls*; corolla 2.5–3 purple and hairy; calyx 3.2–3.5 mm, *hairless in the throat*. DZ, ES, FR, IT, PT.

Micromeria nervosa An erect subshrub 15–40 cm, similar to the previous species but with oval to triangular leaves 5–15 mm long, flowers borne in laxer, interrupted inflorescences and *calyx covered in sparse, white, long fine hairs*. ES (Balearic Islands), IT (incl. Sicily).

Hyssopus HYSSOP

Aromatic subshrubs with entire leaves. Flowers with tubular-bell-shaped calyx with 15 veins and 5 sub-equal lobes; corolla blue (or white); stamens 4, protruding.

Hyssopus officinalis HYSSOP An aromatic, virtually hairless, much-branched subshrub 15–52 cm with narrow leaves 11–22 mm, clustered at the nodes. Flowers pale pink, blue or violet, borne rather compact, elongated spikes; corolla 7–8 mm, *open with stamens long-projecting*. Dry slopes and maquis, also planted. ES, FR, IT, MA.

Thymus THYME

Dwarf shrubs, woody at the base, and characteristically aromatic, with entire leaves. Flowers borne in heads; calyx 2-lipped, the upper lip with 3 short teeth, the lower with 2 long teeth; corolla 2-lipped; stamens 4, protruding (in cosexual flowers).

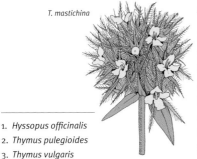

T. mastichina

1. *Hyssopus officinalis*
2. *Thymus pulegioides*
3. *Thymus vulgaris*

A. Leaves flat.

Thymus mastichina A rather irregular dwarf shrub to 50(80) cm, not notably cushion-like or dense in profile with erect, woody branches. Leaves 3.5–13 mm long, elliptic and flat (or wavy-margined), variably hairy. Inflorescence spherical (10–18 mm across); *calyx to 4–7 mm with long, narrow, all similar teeth longer than their tube and white flowers*; corolla scarcely exceeding the calyx. Sunny, exposed dry places. ES, PT.

Thymus pulegioides LARGE THYME An aromatic, tufted, spreading perennial to 30 cm with elliptic, flat leaves 3.5–16 mm, *with long hairs at the base, otherwise hairless.* Flower stems *strongly 4-angled with hairs only on the angles; flowers pink*, borne in rather *distant whorls;* calyx 3–4 mm. Local in grassy and often damp habitats; absent from hot, dry areas. ES, FR, IT.

B. Leaves with inrolled margins.

Thymus vulgaris THYME A variable, short, dark grey-green, densely branched subshrub to 40 cm, strongly aromatic, with spreading, woody stems. Leaves 3.5–6.5 mm, narrowly elliptic, with leaf clusters in the axils, *hairy and with down-turned margins* (without long hairs at the base). Flowers whitish, pink or purple, borne in rounded heads; calyx 3.5–5.5 mm, 10–13-veined and stiffly hairy. West-central. ES, FR, IT.

T. zygis

Thymus zygis A low to short dwarf shrub 10–30 cm with erect, woody stems. Stem leaves 4.5–9 mm, exceeding the axillary leaf clusters; linear, stalkless, hairy and with down-turned margins; fringed with hairs at the base. *Inflorescence forming long, spike-like, interrupted heads, to 10 cm;* calyx 2.5–5 mm, greyish-green and bell-shaped with broad teeth; *not ciliate; corolla white.* Dry habitats, maquis and fields. ES, MA, PT. *Thymus baeticus* is a similar, grey-shortly-hairy subshrub to 50 cm with small elliptic leaves 4–7 mm with down-turned margins and white flowers with corolla lobes with *ciliate teeth; calyx densely hairy*, 3–3.5 mm. ES (south). *Thymus willdenowii* is similar to *T. baeticus* and also has corolla lobes with ciliate teeth, but a *spreading, not erect habit*, finely hairy leaves and *pinkish flowers;* calyx 4.5–5 mm. Rare. ES (Gibraltar), MA.

Thymus lotocephalus (Syn. *T. cephalotus*) A dense dwarf shrub to just 15 cm with erect, woody branches with axillary leaf clusters. Stem leaves 3.5–8.5 mm long, linear and *densely hairy* with a few longer hairs, with down-turned margins. *Inflorescence large*, to 40 mm; oblong. Bracts purplish, oval and leathery; calyx 4–6mm, cylindrical with lanceolate teeth; corolla purple. Pine forests and garrigue; rare. PT.

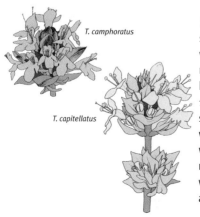

T. camphoratus

T. capitellatus

Thymus camphoratus A dense, aromatic (strawberry-scented) cushioned dwarf shrub 15–30 cm with erect, woody branches with axillary leaf clusters. Leaves 6–8 mm and *broadly oval-triangular, 2–4.5 mm;* stalked, hairy, and with down-turned margins. Inflorescence to 18 mm across; bracts flushed with purple, noticeably so when in bud; *corolla dark pink-purple;* calyx 4–6 mm with thread-like teeth. Fixed sand dunes in the Algarve, where common. PT. *Thymus capitellatus* is similar has narrow leaves just 1–2 mm wide; calyx teeth as long as wide (*not thread-like*), greenish (not purplish) bracts, and *white* flowers. Rare. PT.

Thymbra

Very similar to *Thymus* (and still widely described under this genus).

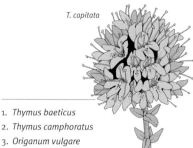

T. capitata

1. *Thymus baeticus*
2. *Thymus camphoratus*
3. *Origanum vulgare*
4. *Thymbra capitata*
5. *Thymbra capitata*

Thymbra capitata (Syn. *Thymus capitatus*) A mid-to late-summer-flowering, dense, often cushion-like dwarf shrub 10–40 cm with erect, woody branches *with axillary leaf clusters*. Leaves fleshy; *somewhat 3-angled* 5–10 mm long, stalkless, linear, more or less hairy and with inconspicuous lateral veins; gland-dotted with a *flat margin*. Flowers with corolla *mid-pink*, 6–10 mm long, borne in *dense, oblong terminal clusters* with red-tinged bracts that overlap, forming cone-like heads; calyx *many-veined* (20–22). Dry, rocky places. Probably throughout.

Cleonia

Small herbaceous annuals with 10-veined, bell-shaped calyx with 5 unequal teeth.

Cleonia lusitanica A short, rather hairy plant with erect, purplish, solitary stems often to just 10(44) cm tall. Leaves 19–45 mm, pinnately lobed below, grey-green and covered with short, white hairs. Flowers borne upward-facing in leafy, terminal clusters; corolla 15–22 mm, 2-lipped and tubular with rounded petal-lobes, violet (rarely yellow); calyx 5–8 mm. Low turf and scrubby habitats; local, in the southwest only. ES, MA, PT.

Origanum MARJORAM

Herbaceous perennials. Flowers clustered in terminal heads; calyx bell-shaped and white-hairy within, with 13 veins and 5 lobes; corolla purplish (or white); stamens 4, protruding in cosexual flowers.

Origanum vulgare MARJORAM A medium, rather thyme-like, lax, aromatic perennial with erect, purplish stems to 1.3 m. Leaves 15–42 mm, oval and scarcely toothed or untoothed; short stalked or unstalked and with leaf tufts in the axils. Flowers with corolla 4.5–10 mm, pink to reddish purple, darker in bud, borne in *broad, branched, panicle-like clusters*; *calyx bell-shaped with 13 veins and 5 equal lobes*. Grassy places; common. Throughout. *Origanum compactum* is very similar but with *purple, leathery (rather than greenish, papery) bracts*. Local. ES, MA.

Origanum majorana SWEET MARJORAM A subshrub similar to the previous species, but with very dense, pyramidal inflorescences with long branches, with purple (rarely white) flowers extending down the stems; *calyx split deeply along 1 side* only (not into 5 equal teeth). Garrigue; also cultivated and widely naturalised. DZ, ES, MA. *Origanum onites* POT MARJORAM is similar but with *flowers always white*, and stems irregularly covered with minute protuberances. Garrigue. IT (Sicily); also widely planted.

1. *Mentha suaveolens*
2. *Lycopus europaeus*

1

2

Mentha MINT

Aromatic perennial herbs with creeping rhizomes. Flowers cosexual or female in dense whorls; calyx regular or weakly 2-lipped, corolla weakly 2-lipped with 4 subequal lobes; stamens 4, protruding (in cosexual flowers). Frequently hybridising.

A. Inflorescence terminating in a cluster of leaves.

Mentha pulegium PENNY-ROYAL A strong-smelling spreading or ascending perennial 12–78 cm with erect flowering stems; leaves roughly oval, 8.5–30 mm and finely downy and blunt-toothed. Flowers mauve, borne in *rounded, well-separated heads, without a clear terminal head;* corolla 4–5 mm; calyx 2.5–3.5 mm. Pondsides; rare in much of the north of the region. Throughout.

M. aquatica

B. Inflorescence terminating in a cluster of flowers; terminal cluster rounded.

Mentha aquatica WATER MINT A short to tall, vigorous and leafy, hairy to hairless perennial to 1.5 m, strongly smelling of mint when crushed, with angled, purple stems. Leaves 20–50 mm, oval, pointed, toothed and stalked. Flowers purple-white, borne in very *dense, oblong heads* with 1 or 2 whorls of flowers below; corolla 5.5–7 mm; calyx 3.5–4.5 mm, veined and hairy. Always near water. Throughout.

C. Inflorescence terminating in a cluster of flowers; terminal cluster cylindrical.

Mentha spicata SPEAR MINT A variable, green, mint sauce-smelling perennial 43–84 cm with lanceolate, toothed, *scarcely hairy leaves* 17–88 mm. Flowers pink or white, borne in long, cylindrical, interrupted, *lax spikes;* corolla 2.5–4 mm; calyx 1.5–2.5 mm, the *tube hairless* (teeth sometimes hairy). Eastern, but widely naturalised elsewhere. IT. *Mentha longifolia* HORSE MINT is similar, though greyish and distinctly *white-downy below,* and leaves 15 mm–12 cm, *grey and densely-silkily-hairy.* Flowers white or lilac, borne in long, terminal, *dense spikes* which separate as they mature; calyx 2–3 mm; corolla 3–3.5 mm. Wet habitats. Throughout, except some islands.

Mentha suaveolens APPLE MINT A rather small perennial to 40(87) cm with a sickly sweet scent and stems variably hairy. Leaves 18–52 mm, *bright green,* stalkless or virtually so, round to oblong and broadest near the base, toothed, hairy above, and grey-hairy beneath. Flowers borne in many congested whorls forming long, dense inflorescences; often branched; corolla 3–3.8 mm, pale pink or white; calyx 1.2–2.5 mm, hairy with subequal teeth. Cultivated and in damp habitats; common. Throughout.

Lycopus GYPSYWORT

Herbaceous perennials. Flowers with 5-toothed calyx; corolla with 4 sub-equal lobes; stamens 2, protruding.

Lycopus europaeus GYPSYWORT An erect, slightly hairy, light, bright green perennial, often much-branched, with angled stems to 92 cm. Leaves 21 mm–11 cm, oval and very deeply cut into narrow triangular teeth except at the tip. Flowers small and inconspicuous, 3–4 mm across, white, with 4 roughly equal lobes, borne in interrupted clusters. Common on river banks and near ponds at all altitudes. Throughout.

Lavandula LAVENDER

Aromatic shrubs with narrow leaves and distinct bracts. Flowers borne in crowded, long-stalked terminal spikes; calyx with 5 small teeth, 13-veined; corolla purple, 2-lipped, weakly zygomorphic; stamens 4, not protruding.

A. Flower spikes topped by coloured bracts.

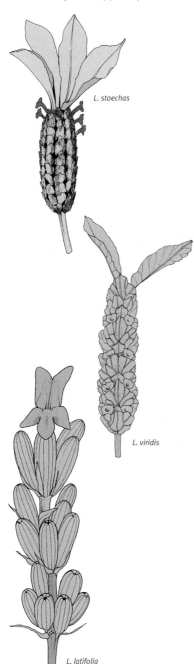

L. stoechas

L. viridis

L. latifolia

Lavandula stoechas FRENCH LAVENDER A greyish shrub to 1.5 m with erect, much-branched stems. Leaves 6–37 mm, linear and untoothed. Flowers with a deep mauve corolla 4.5–5 mm borne in short, dense spines, topped by *conspicuous purple flower-like bracts* 8–36 mm long; inflorescences longer than their spikes. Common on maquis, dry scrub, open woods and fixed dunes. ES, FR, IT, MA, PT. Subsp. *pedunculata* (described by some authors as a distinct species) has stalks longer than their short inflorescences, and lower bracts *exceeding* the calyx. *Lavandula dentata* is distinguished by its lanceolate leaves which have inrolled margins *deeply cut into teeth*. ES, IT, PT.

Lavandula viridis A small shrub 37–44 cm, similar to *L. stoechas* but with leaves and stems *shortly* hairy, stalks longer than their inflorescences, and *white flowers with the bracts above the spike green*; corolla 4–5 mm; terminal bracts 10–18 mm. Hill forests and maquis in southern Iberian Peninsula. ES, PT.

B. Flower spikes not topped by coloured bracts.

Lavandula latifolia A strongly aromatic (camphor-scented), grey-hairy shrub to 1.1 m with erect, much-branched stems. Leaves 17–62 mm, linear and untoothed; *densely* grey-hairy, and *broad, to 13 mm*. Flowers borne on long-stalked spikes, *not topped by flowery bracts; bracts linear-lanceolate, green and 1-veined;* flowers small, the corolla 8 mm, pale violet. Dry, rocky slopes; absent from most islands, rare, or naturalised, in Portugal. ES, FR, IT (incl. Sicily), PT. *Lavandula angustifolia* is similar but with *leaves narrow, to just 5 mm* and with broadly oval, purple, pointed, papery bracts with 7 veins. Flowers blue, borne in dense spikes. Grown as an ornamental and cultivated widely in southern France. Sometimes naturalised. ES, FR. *Lavandula angustifolia × L. latifolia* is intermediate in characteristics, and the commonly cultivated crop. ES, FR.

L. multifida

Lavandula multifida CUT-LEAVED LAVENDER A grey-hairy, faintly aromatic shrub to 78(88) cm. Leaves 10–48 mm, green, sparsely hairy and oval in outline, *2-pinnately lobed*. Flowers with corolla violet-blue, 9–14 mm, borne in spikes up to 70 mm long; calyx 15-veined. Dry, stony fallow land and low garrigue, patchy in distribution. ES (incl. Ibiza of Balearic Islands), IT (incl. Sicily), MA, PT.

1. *Lavandula stoechas*
2. *Lavandula dentata*
3. *Lavandula viridis*
4. *Lavandula latifolia*
5. *Lavandula x intermedia*
6. *Lavandula multifida*

Rosmarinus ROSEMARY

Evergreen, aromatic shrubs with linear leaves with inrolled margins. Flowers with 2-lipped calyx with a virtually entire upper lip; corolla blue or pink, strongly zygomorphic; stamens 2, protruding.

Rosmarinus officinalis ROSEMARY A familiar evergreen shrub to 1.8 m, characteristically aromatic; branches brown and woody, erect to spreading. Leaves needle-like, 10–41 mm, linear and leathery, mid to dark green, sharply pointed and with down-turned margins. Flowers white, flushed pale purple, 8.5–13.5 mm long, borne in small lateral clusters; corolla with 2 protruding stamens; calyx bell-shaped. Very common on maquis, garrigue, open woods and fixed dunes. Throughout. *Rosmarinus eriocalyx* is similar, though often prostrate or low-spreading and with *short leaves 7–20 mm long*; inflorescences both lateral and terminal. ES, MA, TN (rare). *Rosmarinus tomentosus* is very similar to *R. eriocalyx* but with *white-grey-woolly leaves* 5–13 mm long. Sea cliffs in southern Spain only. ES.

Salvia SAGE

Herbs and shrubs with distinct whorls of (often purple) flowers forming a lax inflorescence. Flowers with both calyx and corolla 2-lipped, the upper corolla lip hooded, the lower 3-lobed; stamens 2, hinged in the middle, joined beneath the corolla hood (shorter than upper corolla lip).

A. Plant a shrub or subshrub.

S. officinalis

1. *Rosmarinus officinalis*
2. *Salvia fruticosa*
3. *Salvia africana-lutea*
4. *Salvia sclarea*
5. *Salvia pratensis*
6. *Salvia nemorosa*
7. *Salvia verbenaca*

Salvia officinalis SAGE A strongly aromatic, greyish shrub to 60 cm with erect branches that are woody beneath. Leaves broadly elliptic, greenish above and white-felted below, with a finely toothed margin. Flowers pale violet, blue or white, 15–25 mm long; calyx 10–14 mm, flushed purple, the upper lip 3-toothed; bracts oval and hairy. Dry, stony pastures in the centre of the region, also cultivated and naturalised elsewhere. ES, FR, MA. Subsp. *lavandulifolia* (Syn. *S. lavandulifolia*) is also treated as a true species and has leaves *narrowly oval to elliptic*, stalked, the immature leaves whitish-grey; leaves crowded below. Flowers borne in well-separated false whorls on hairless stems; corolla pale purple-blue, 15–40 mm long; calyx 8–14 mm, tinged purple and divided into *5 teeth of equal size*, >$^1/_2$ its length. Rocky habitats above sea level. Throughout the range of *S. officinalis*.

Salvia fruticosa THREE-LOBED SAGE An aromatic subshrub to 1.5 m with felted stems. Leaves grey-green and wrinkled above, grey-felted below, and stalked; narrowly oval usually with 3(5) *lobes at the base*. Inflorescence spike-like with 2–6-flowered false whorls; corolla blue-purple or pink, 15–30 mm long; calyx 5–11 mm, bell-shaped and indistinctly 2-lipped, with 5 triangular teeth 1.5–3.5 mm long. Garrigue in Sicily and south Italy; naturalised in the west. ES, IT (south + Sicily), PT.

Salvia africana-lutea A *medium shrub* to 2 m with greyish, oval-elliptic leaves. Flowers bright yellow when young, later *rust-coloured;* calyx papery and persistent when in fruit. A native of South Africa, widely planted in Spain and Portugal in towns and gardens. ES, PT.

B. Plant a herbaceous annual, biennial or perennial; bracts beneath flower whorls conspicuous and partially hiding calyx.

Salvia sclarea CLARY A robust, sticky, strong-smelling biennial or perennial to 1.5 m tall. Leaves hairy and oval to heart-shaped. Flowers 4–5 per whorl, *lilac or pale blue*, rather large, the corolla 18–30 mm long, with a strongly curved hood, and with *prominent oval or heart-shaped, lilac or white bracts that exceed the flowers*; calyx 10–16 mm with spiny teeth. Locally common in eastern Spain, and southern France; rare elsewhere and absent from the south. ES, FR, IT. *Salvia aethiopis* is similar but with white-felted stems and smaller *white flowers 10–15 mm* and *greenish bracts equalling the calyx*. FR, IT. *Salvia argentea* SILVER SAGE is similar to *S. sclarea* but with *very silvery-white leaves* with shaggy, cobweb-like hairs, *6–10 pinkish-white* flowers *per whorl*; corolla 17–22 mm; *bracts green or white*. Scattered and absent from many areas. ES, IT, PT.

C. Plant a herbaceous annual, biennial or perennial; bracts beneath flower whorls *not* conspicuous or hiding calyx.

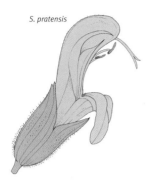

S. pratensis

Salvia pratensis FIELD SAGE A green perennial herb to 1 m, erect and branched; glandular above. Leaves long-stalked, oval and heart-shaped at the base, crinkled along the margin, with spreading hairs above and more or less hairless beneath. Inflorescence lax and shortly branched. Whorls of 4–6 cosexual or female flowers; *violet-blue*, and *large 20–30 mm*. Grassy places and open woods. ES, FR, IT, MA, PT. *Salvia nemorosa* is similar but with more numerous stem leaves, purple bracts and *smaller, dark blue-purple flowers* to 15 mm long, borne in dense spikes. Native to southeast Europe; naturalised elsewhere. IT.

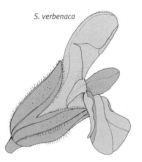

S. verbenaca

Salvia verbenaca WILD CLARY A short to tall perennial to 80 cm with erect stems that are glandular above. Basal leaves form a rosette, wrinkled, *deeply lobed*, long-stalked below, and short-stalked or stalkless above. Flowers pale blue or violet, 6–16 mm long, borne in lax whorls forming a spike; calyx 5–12 mm, bell-shaped and distinctly veined; enlarged in fruit. Fallow land, grassy places, roadsides and garrigue; common. Throughout.

Salvia viridis RED-TOPPED SAGE A small *annual* sage with stems simple or branched below, to 50 cm. Leaves hairy, pale green, blunt and long-stalked, with finely toothed margins. Inflorescence an elongated spike of small, pink flowers borne in false whorls of 4–6(8); corolla 10–14 mm; calyx 7–12 mm; bracts pointed, *the terminal bracts reddish or violet forming a terminal tuft*. Dry habitats on limestone, also planted; local. Throughout.

Salvia verticillata WHORLED CLARY An erect, hairy, rather unpleasant-smelling perennial 30–80 cm with purple-tinged stems and leaves. Leaves stalked and toothed, oval to heart-shaped with basal lobes. *Flowers lilac-blue or purple, borne in numerous tight whorls of 20–30*; corolla 8–15 mm long; calyx 5–7 mm. Dry habitats above sea level; absent from most islands. ES, FR, IT.

1

2

LENTIBULARIACEAE | BLADDERWORT FAMILY

Insectivorous perennials which trap small insects with small, bladder-like traps. Flowers zygomorphic; calyx 2–5-lobed, usually 2-lipped; corolla tubular, the upper lip 2-lobed, the lower 3-lobed; stamens 2; style 0 (or short). Fruit a capsule. Rare in the Mediterranean.

Utricularia BLADDERWORT

Submerged aquatic insectivorous perennials. Flowers with 2-lipped, yellow corolla borne on erect stems above water. Capsule opening irregularly.

A. Flowers 12–18 mm.

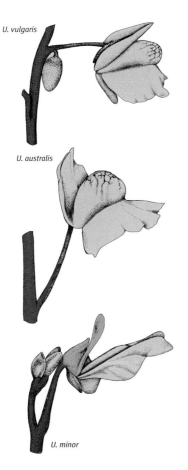

U. vulgaris

U. australis

U. minor

Utricularia vulgaris GREATER BLADDERWORT An aquatic perennial with submerged, free-floating stems which bear numerous tiny 'bladders' which trap tiny invertebrates. Flowers *deep yellow*, borne few on erect, leafless, reddish stems 10–25 cm; corolla with *lower lip 12–15 x 14, deeply turned backwards at the edges; upper lip 11 x 10 mm, equalling the palate*; spur *broadly* cone-shaped, 6–8 mm; flower stalks 6–12(15) mm, recurved but not elongated after flowering. Stagnant water in marshes and ditches. Recorded throughout (often doubtfully); always local. *Utricularia australis* is very similar to (and often confused with) *U. vulgaris* but has *lemon yellow flowers* with with *lower lip not strongly deflexed at the edges, upper lip exceeding the palate;* spur *narrowly* cone-shaped 6–8 mm; flower stalks 8–15 mm when flowering, elongating to 10–30 mm in fruit. Lakes and ponds. Distribution poorly known due to confusion with the previous species; possibly throughout.

B. Flowers very small: 6–8 mm.

Utricularia minor LESSER BLADDERWORT An aquatic perennial with submerged stems, distinguished by its very slender purple scapes to 15(25) cm carrying *very small, pale yellow flowers with a small corolla, just 6–8 mm long* (upper lip 4 x 3 mm, lower lip 7 x 6 mm); spur 1–2 mm; flower stalks 4–8 mm. Very local in pools in marshes on higher ground. Scattered throughout.

1. *Utricularia vulgaris* (habit)
2. *Utricularia vulgaris*

OROBANCHACEAE | BROOMRAPE FAMILY

A distinctive family of root parasitic (hemiparasitic or holoparasitic) herbs that attach to the roots of their host plants as seedlings. Flowers borne in spikes or racemes; corolla zygomorphic and more or less 2-lipped; stamens 4; style 1. Fruit a 2-parted capsule. Hemiparasitic species were previously classified with the Scrophulariaceae.

Odontites

O. vernus

1. *Odontites vernus*
2. *Odontites luteus*
3. *Odontites bocconei*
PHOTOGRAPH: GIANNIANTONIO DOMINA

Small hemiparasitic annuals to perennials with opposite entire to toothed leaves. Calyx tubular or bell-shaped with 4 entire lobes; corolla open-mouthed. Capsules with few seeds.

A. Flowers pink-purple.

Odontites vernus RED BARTSIA A variable, short, hairy hemiparasitic annual to 50 cm. Stems erect, dark reddish-purple and slightly squared. Leaves 10–50 mm, opposite, lanceolate and normally toothed; bracts similar, exceeding the flowers. Flowers red-pink, 2-lipped, the lower deflexed, the anthers protruding; corolla 7–12 mm. Grassy and sandy places, ruderal sites and roadsides. Throughout.

B. Flowers normally yellow (sometimes purple).

Odontites luteus An erect, much-branched, virtually hairless annual to 60 cm with linear-lanceolate leaves 10–32 mm with inrolled margins, shallowly toothed. Flowers *bright yellow*, borne in dense, 1-sided spikes with narrow bracts; corolla 6–8 mm long, *finely hairy and with ciliate margins* and anthers and style long-projecting; *filaments hairy*. Various habitats including woods, maquis and mountains; absent from the far south and west. ES, FR, IT. *Odontites viscosa* is similar but sticky, aromatic, with *long stem hairs at the base* (1.5–3 mm) and with *pale yellow or purple* flowers; corolla 5–6.5(8) mm, virtually hairless. ES, FR, IT, MA, PT. *Odontites bocconei* is similar to the previous species; a bushy, erect, woody perennial with *very leafy stems* with long, linear, rather backward-curling leaves, and numerous yellow flowers. IT (Sicily).

1

2

3

Odontitella

Virtually identical to *Odontites* (often still described under this name), differing chiefly in pollen characeristics.

Odontitella virgata (Syn. *Odontites tenuifolia*) A weak, densely hairy hemi-parasitic annual to 40(60) cm with linear leaves 7–22 mm and dense, yellow flowers with protruding anthers borne in 1-sided terminal clusters; corolla 8–15 mm. Heaths, pinewoods and maritime sandy places; rare and local. Iberian Peninsula. ES, PT.

Parentucellia

Annuals with opposite leaves. Flowers with tubular to bell-shaped, regularly 4-lobed calyx; corolla long-tubed with open mouth. Capsule with numerous seeds.

P. viscosa

P. latifolia

Parentucellia viscosa YELLOW BARTSIA A short glandular-hairy hemiparasitic annual with erect, normally unbranched stems to 35(60) cm. Leaves 17–35 mm, opposite, lanceolate, deeply toothed and sessile; bracts similar, and decreasing in size along the stem. *Flowers yellow* (rarely white); corolla 18–23 mm, 2-lipped, the upper lip hooded, the lower lip 3-lobed; calyx teeth linear-lanceolate and long, 4.5–6.5 mm. Damp grassy places; locally common. Throughout.

Parentucellia latifolia SOUTHERN RED BARTSIA A short annual to 20(40) cm with *triangular-lanceolate and deeply-lobed leaves* 6–15 mm. Flowers small, with a corolla 8–10(15) mm long, pale *reddish-purple* (rarely yellow or white); calyx teeth 1.4–1.6 mm long. Sandy and stony places, often coastal. Throughout.

Bartsia

Perennials with opposite, toothed leaves. Flowers with tubular to bell-shaped, regularly 4-lobed calyx and long-tubed corolla with an open mouth. Capsules with few, winged seeds.

Bartsia trixago (Syn. *Bellardia trixago*) A short glandular-hairy hemiparasitic annual with erect, simple stems to 60(70) cm. Leaves 20–40 mm, opposite, linear-lanceolate, toothed and sessile; bracts similar, and decreasing in size along the stem. Flowers with corolla white, normally flushed with pink, or bright yellow, 17–24 mm, borne in a dense, *4-sided spike*. Ruderal sites, grassy places and garrigue; locally common. Throughout.

Lathraea TOOTHWORT

Parasitic rhizomatous perennials that lack green pigment. Flowers with bell-shaped, 4-lobed calyx and 2-lipped corolla.

Lathraea clandestina PURPLE TOOTHWORT A highly distinctive parasite that lacks leaves, roots and green pigment; patch-forming. Flowers bright purple, borne in clusters of 4–12(18); corolla 40–60 mm long borne in clusters on short stalks from scaly underground rhizomes. Capsule 10–15 mm. Hill forests on the roots of various tree species; north of range only. ES, FR, IT.

Cistanche

Robust, obligate parasites with no true leaves or green pigment. Flowers borne in dense spikes; calyx tubular with 5 teeth; corolla with 5 almost equal teeth. Fruit an ovoid capsule.

Cistanche phelypaea CISTANCHE *A tall, robust, bright yellow parasite* to 1 m with hairless, unbranched, thick-stemmed inflorescences with oval scale leaves. *Flowers bright, shiny and yellow* borne in cone-shaped spikes; calyx 12–18 mm and 5-lobed; corolla 30–50 mm with 5 more or less equal lobes. Parasitic on shrubby Amaranthaceae on dunes and in salt marshes. Rare and local; frequent in the Algarve, southwest Spain and parts of North Africa. Plants with pale yellow flowers with deep yellow and violet markings in southeast Spain are sometimes

refered to as subsp. *lutea*. DZ, ES, MA, PT, TN. *Cistanche violacea* is similar in form, but has *whitish or pale lilac flowers with dark purple markings* at the tips, and yellow in the throat, 25–40 mm. Mainly a desert species, extending to coastal dunes. North Africa only; local (forms of *C. phelypaea* from Spain are sometimes incorrectly referred to as this species). DZ, MA, TN. *Cistanche mauritanica* is similar to *C. violacea* but much shorter, often under 30 cm, and with *densely woolly bracts and calyx; calyx lobes longer than broad*. Parasitic on *Atriplex* spp. Rare and local. DZ (west), ES (Chafarinas Islands), MA.

PHELIPANCHE | BROOMRAPE

Branched or unbranched parasites, often bluish. Flowers with *2 secondary bracts* (bracteoles) below the calyx; corolla zygomorphic, 2-lipped. Capsule with numerous minute seeds. Host identification is useful but difficult in mixed vegetation. Closely related to *Orobanche* (still included in that genus by some authors). Morphological traits are of limited identification value; reliable identification for closely related species may not be possible in the field.

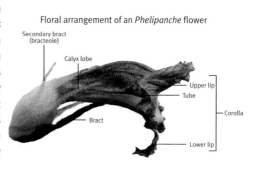

Floral arrangement of an *Phelipanche* flower

Secondary bract (bracteole)
Calyx lobe
Upper lip
Tube
Corolla
Bract
Lower lip

A. Species with hairless or almost hairless anthers.

P. mutelii

1. *Parentucellia viscosa*
2. *Bartsia trixago* (yellow form)
3. *Lathraea clandestina*
4. *Bartsia trixago*

Phelipanche ramosa BRANCHED BROOMRAPE A variable, blue-purple (sometimes yellow-white), glandular-pubescent annual with, usually, *branched stems*, *usually short* to 20(50) cm. *Flowers small*, 10–15(22) mm long, pale blue, violet or cream with a white patch at the base; stigmas white or pale blue; calyx teeth 5–15 mm (shorter than the corolla), lower corolla teeth *blunt*. On various hosts, *usually cultivated* Linaceae, Cannabaceae, Solanaceae and Leguminosae. Sandy and disturbed habitats. Throughout. *Phelipanche nana* (also regarded as a subspecies of *P. ramosa*) is *usually unbranched*, with *pointed* lower corolla lobes, and calyx teeth almost as long as the corolla. *Phelipanche mutelii* is much-confused with the previous species but has more erect flowers with distinctly open corolla mouth with markedly divergent lower teeth. Parasitises various herbs including *Hedypnois, Helichrysum, Hyoseris, Medicago* and *Scorpiurus*. ES, FR, IT, MA.

Phelipanche rosmarina has *unbranched and thick stems* to 15 cm. Corolla 14–16 mm, very *constricted* and *dark blue* with virtually hairless filaments and a white stigma. Always on *Rosmarinus*, rare and local; mainly Iberian Peninsula. ES, FR (incl. Corsica), IT, PT. *Phelipanche rumseiana* also grows on *Rosmarinus* but is typically shorter, more robust with erect-spreading flowers, calyx lobes with long-pointed, triangular teeth and *hairy* filaments. ES (Balearic Islands).

P. purpurea

Phelipanche purpurea A (normally) simple annual, robust, to 30(45) cm with short-hairy, not glandular, bluish stems. Inflorescence dense, with a blue-purplish or greyish appearance. Corolla long, 18–26(30) mm and distinctly constricted at the base; stigma white or pale blue; anthers normally hairless. Parasitic on various hosts in the Asteraceae (rarely Leguminosae). Absent in parts of North Africa. ES, FR, IT, MA, PT.

1. *Cistanche phelypaea*
2. *Cistanche violacea*
   PHOTOGRAPH: NICKLAS STRÖMBERG - IGOTERRA
3. *Phelipanche ramosa*

4. *Phelipanche purpurea*
5. *Orobanche minor*
6. *Orobanche calendulae*
7. *Orobanche picridis*

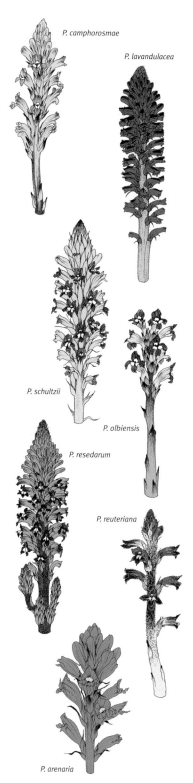

P. camphorosmae

P. lavandulacea

P. schultzii

P. olbiensis

P. resedarum

P. reuteriana

P. arenaria

Phelipanche camphorosmae Stems normally unbranched, cream with translucent hairs. Flowers crowded; corolla 18–21 mm, tubular, with upper and lower lips with hairs, the upper lip bilobed, the lower lip with 3 subequal *narrowly triangular* lobes; filaments inserted above the corolla base (4–8 mm) and scarcely hairy; stigma white; calyx lobes *acutely triangular*. Parasitic on *Camphorosma monspeliaca*. ES, FR. *Phelipanche lainzii* is very similar but has *very pale blue* flowers, less acute, awn-shaped calyx lobes, and is always parasitic on *Cleonia lusitanica*. ES (east).

B. Species with hairy anthers (many similar species difficult to distinguish, host plant important).

Phelipanche lavandulacea A robust annual to 45(59) cm with dense flowers and with a *dark purple or blackish appearance*. Corolla 19–22 mm with *lower lip obtuse*; stigma white, bluish or yellowish; filaments slightly hairy; calyx teeth *not exceeding* corolla tube. Lower portion of the stem with stalked flowers. Various hosts, particularly *Bituminaria bituminosa*. Common. DZ, ES, IT, MA, PT, TN. *Phelipanche schultzii* has few or no stalked flowers, *lower corolla lip acute* with divergent lobes, white stigma, and calyx teeth *greatly exceeding the corolla tube, and filiform at the tips*. Parasitic on Apiaceae. Local and rare. ES, IT. *Phelipanche cernua* (Syn. *P. inexpectata*) is *grey-blue* with an abruptly arched corolla; *stigma yellow; calyx teeth not as long as length of corolla tube*. Parasitic on *Lactuca*. DZ, ES, FR, MA. *Phelipanche arenaria* has large, *funnel-shaped*, mauve flowers 25–38 mm; stigma lobes distant and whitish; *filaments woolly*. Parasitic on *Artemisia* spp., especially *A. campestris*. Widespread but rare. DZ, ES, FR, IT, MA, TN. *Phelipanche portoilicitana* is tall, to 35 cm with flowers densely clothed in short glandular hairs; lips ciliate at the margins. Parasitic on *Centaurea* on coastal sands. ES (southeast). *Phelipanche olbiensis* has small flowers to 15 mm long, and is parasitic on *Helichrysum*. Coastal. ES (incl. Balearic Islands), FR. *Phelipanche resedarum* A robust species clothed in long white hairs; corolla 18–24 mm. Parasitic on *Reseda*. ES. *Phelipanche aedoi* is similar, with corolla 15–19 mm, but parasitic on *Launaea arborescens* and *Sonchus tenerrimus*. ES (southeast). *Phelipanche reuteriana* has a corolla markedly *woolly hairy* on the outer surface; bright blue within, with prominent white folds. Parasitic on *Plantago, Crambe, Malcolmia* and *Cleome* on coastal dunes in southwest Spain and North Africa. DZ, ES, MA, TN.

OROBANCHE | BROOMRAPE

Unbranched parasites, often dull reddish or yellowish, similar to *Phelipanche* but *lacking 2 secondary bracts* below the calyx; corolla zygomorphic, 2-lipped. Capsule with numerous minute seeds. Host identification is useful but difficult in mixed vegetation; morphological traits are of limited identification value; reliable identification for closely related species may not be possible in the field.

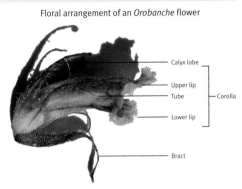

Floral arrangement of an *Orobanche* flower

Calyx lobe
Upper lip
Tube ⎱ Corolla
Lower lip

Bract

A. Spikes usually pinkish, purplish or yellowish, slender, often <30 cm; lax to dense; *corolla narrow, tubular, and small*, 10–20(25) mm. Many similar species difficult to distinguish.

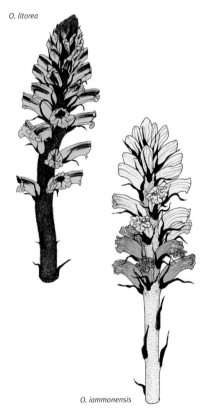

O. litorea

O. iammonensis

1. *Orobanche balsensis*
2. *Orobanche amethystea*
3. *Orobanche hederae*
4. *Orobanche foetida* (purple form)
5. *Orobanche pallidiflora*
6. *Orobanche caryophyllacea*
7. *Orobanche foetida* (yellow form)

Orobanche minor COMMON BROOMRAPE A variable annual with lax to dense spikes to 60 cm. Flowers *small with an evenly arched corolla* 10–18 mm, yellow tinged violet; lower lip evenly lobed without a hairy margin, stigma often pink (but variable); *filaments not densely hairy or woolly*, usually inserted close to corolla base; bracts equalling or exceeding the corolla; calyx lobes variably toothed. Parasitic on various hosts across many families, in disturbed habitats; the commonest species. Throughout. *Orobanche litorea* is similar but with filaments *densely hairy* in the lower $^1/_2$; stigma purple. Parasitic on *Anthemis maritima* in coastal habitats. FR, IT (incl. Sardinia and Sicily). *Orobanche iammonensis* is similar, also parasitic on *A. maritima* but *yellow* in all parts. Coasts of Menorca. ES (Balearic Islands). *Orobanche calendulae* CALENDULA BROOMRAPE has pale yellow flowers 15–18 mm tinged with violet or dull red, *stigma yellowish*. Parasitic on *Calendula suffruticosa* (possibly other Asteraceae) on sea cliffs. Rare. DZ, ES, MA, PT. *Orobanche ballotae* has *milky-white* flowers 10–14 mm with filaments glandular-hairy *to the apex*; stigma lobes whitish. Parasitic on *Ballota hirsuta*. ES (south + Balearic Islands). *Orobanche clausonis* is *yellowish* with pale flowers flushed red, 15–19(22) mm with *contrasting dark pink to red stigmas*. Parasitic on Rubiaceae; rare. ES, PT.

Orobanche picridis OXTONGUE BROOMRAPE An annual to 60 cm, similar to *O. minor* but paler. Corolla 15–22 mm, *whitish, straight along the back*; stigma lobes dark purple or orange (sometimes pink); *filaments inserted above the base of the corolla, and densely hairy* at the base;

O. artemisiae-campestris

bracts long and filiform. Parasitic on Asteraceae, often *Picris hieracioides*, *Crepis* and *Artemisia* (occasionally on *Daucus*). Throughout but rare. **Orobanche balsensis** is similar but has a markedly *crimped corolla lip*. Only recently redescribed. Rare, on *Carlina corymbosa*. ES, FR (incl. Corsica), IT (Sicily), PT. *Orobanche artemisiae-campestris* has red-orange stems and *dull yellow flowers* 16–22 mm, and pink-red stigma lobes. Parasitic on *Artemisia campestris*. Local. ES (north), FR, IT (incl. Sicily), MA. *Orobanche santolinae* differs from the above species in having *densely woolly stems and bracts,* more widely tubular flowers and is parasitic on *Santolina*. ES (incl. Balearic Islands), FR.

Orobanche amethystea Like *O. minor* but often more robust and dense, 15–50 cm. Corolla 15–20(25) *straight then abruptly arched (geniculate)*, pale yellow tinged and veined with violet (rarely all yellow or orange); bracts exceeding the corolla, often markedly. Parasitic on various hosts, most commonly *Eryngium campestre*. Common on the mainland Mediterranean coast (especially the west) and many islands. Throughout. *Orobanche hederae* IVY BROOMRAPE is similar, but often *strongly purple with long spikes*; corolla 10–22 mm, cream tinted maroon, inflated at the base, narrow and straight-backed, with notched lobes, the middle lobe square; stigma lobes always *yellow*; filaments scarcely hairy. Parasitic on *Hedera helix* (rarely other species). Common from Spain to Italy. DZ, ES (incl. Balearic Islands), FR, IT (incl. Sardinia), MA, (possibly PT).

O. canescens

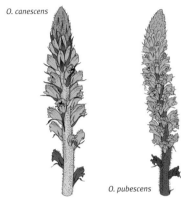

Orobanche canescens An annual to 30(50) cm with lax spikes of small flowers; corolla 10–19 mm, cream with purple veins and evenly curved, lobes of lower lip equal or the middle lobe slightly larger; filaments *densely hairy below* (inserted 2 mm above the base); *stigma lobes yellow to white*. Parasitic on Asteraceae including *Galactites*, *Carlina*, and *Glebionus*. IT (incl. Sardinia and Sicily). *Orobanche pubescens* is similar but *densely shaggy-white hairy throughout*, especially on the corolla which is 10–20 mm. Parasitic on Asteraceae and Apiaceae. Primarily eastern Mediterranean. FR, IT (incl. Sardinia and Sicily).

O. pubescens

B. Spikes uniformly deep red or pale yellow, robust, often exceeding 30 cm, *dense and many-flowered*; corolla tubular to slightly bell-shaped, (13)20–25 mm long.

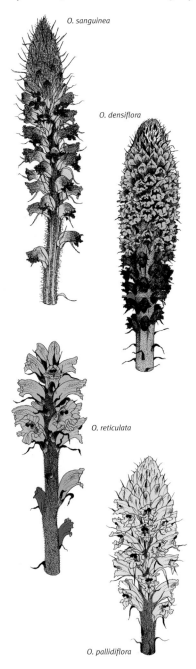

O. sanguinea

O. densiflora

O. reticulata

O. pallidiflora

Orobanche foetida A tall, robust plant with densely flowered red to *blackish-red spikes* to 70 cm (1 m). Flowers 20–25 mm, purplish-red with a curved dorsal line, and bilobed upper lip; stigma *yellow* (ageing red); filaments inserted above the corolla base; scarcely hairy; *calyx lobes long and slender*. Yellow forms recorded (in error) as *O. densiflora* in Portugal. DZ, ES, MA, PT, TN. *Orobanche sanguinea* (Syn. *O. crinita*) is less robust, to 27 cm tall with *smaller, narrower flowers* 13–14 mm long with a weakly bilobed upper lip and pointed lower lobes; stigma *red* (even when fresh; rarely orange, never yellow). Parasitic on *Lotus cytisoides*. ES (Menorca of Balearic Islands), FR, IT (Sardinia and Sicily).

Orobanche densiflora A robust annual to 30(50) cm with dense spikes of *pale lemon yellow flowers*; corolla 16–22 mm, *white-yellow* with a shallowly curved dorsal line, *exceeded by long, narrow bracts; bracts brown, dense and regular at the base of the stem*, filaments inserted at the base of the corolla, and hairy. Parasitic mainly on *Lotus creticus* (also *Medicago* and *Ononis*). Locally common on dunes in southern Spain. ES.

C. Corolla *not* narrowly tubular, but broader or more bell-shaped (18)20–33 mm long.

Orobanche reticulata THISTLE BROOMRAPE A robust species to 35 cm with stout, purplish or yellowish stems. Flowers dense, few to numerous; corolla 18–22 mm, yellowish with purple markings and *dark glands, abruptly curved along the back*, the lower lobes equal, upper lobes spreading and notched; stamens inserted above the base of the corolla, filaments hairy above; stigma lobes touching and *dark purple*. Parasitic on *Cirsium, Carduus, Scabiosa* and *Knautia*. Widespread but absent from many areas and the islands. ES, FR, IT. *Orobanche pallidiflora* is similar, grows exclusively on *Cirsium* and *Carduus* and has less reddish pigmentation, dark glands and less densely hairy filaments. Rare. FR, IT.

1. *Orobanche rapum-genistae*
2. *Orobanche rapum-genistae*
3. *Orobanche latisquama*

Orobanche caryophyllacea CLOVE-SCENTED BROOMRAPE A robust annual to 40 cm with pale, curved stems and darker bracts. Flowers few and *large* relative to plant; 20–32 mm long and *bell-shaped*, weakly *carnation-scented*; pale pink or yellow to violet-brown; *stigma lobes distant and purple*; filaments hairy below; bracts shorter than the corolla; lateral calyx lobes *oval*, toothed and short. Parasitic on *Galium* and *Asperula*. Rather rare and restricted to sub-alpine areas; absent from most islands. ES, FR, IT (incl. Sicily), MA.

Orobanche rapum-genistae GREATER BROOMRAPE *A robust*, tall, brownish perennial to 90 cm. *Flowers large*; 20–25 mm long, dull brown, yellow or red, slightly foetid; lower corolla lip with a hairy margin, the *upper lip scarcely 2-lobed*; bracts exceeding the corolla, often markedly; stigma yellow. Parasitic on shrubby legumes, especially *Cytisus*. Absent from the Balearic Islands. ES, FR (incl. Corsica), IT (incl. Sardinia and Sicily), PT.

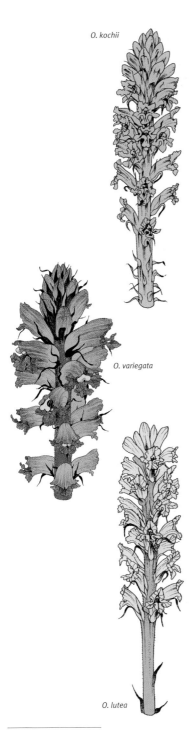

O. kochii

O. variegata

O. lutea

Orobanche elatior KNAPWEED BROOMRAPE A stout annual to 75 cm, with *numerous, very dense* flowers 18–25 mm; with corolla lobes toothed and crisped; *stamens inserted well above* (3–5 mm) *the base of the corolla*; stigma lobes bright yellow and diverging. Parasitic on *Centaurea scabiosa* and *C. nigra*. Rare in the Mediterranean and absent from the islands. FR, IT. The following species are similar to *O. elatior*: *Orobanche icterica* has sparsely, minutely hairy filaments and is parasitic on *Centaurea aspera* and *C. linaresii*. ES. *Orobanche haenseleri* has filaments inserted *just* above the base of the corolla; corolla 22–28 mm. Parasitic on *Helleborus foetidus* and possibly *Rubus ulmifolius*. ES (south). *Orobanche kochii* has pale pinkish or yellowish flowers with reflexed upper and lower lips (flowers with an open appearance) and bright yellow, more or less touching stigma lobes. Parasitic on *Centaurea aspera*. FR (south).

Orobanche gracilis SLENDER BROOMRAPE A short-medium parasite to 40 cm with reddish, lax spikes. Flowers large relative to the stem 18–24(29) mm, broadly campanulate, yellowish veined with red externally and *shiny red inside*; *bracts triangular, and shorter than the corolla*; filaments short-hairy. Parasitic on legumes (sometimes *Cistus*). Locally common throughout, above sea level. Var. *deludens* (Syn. *O. austrohispanica*), previously regarded as a distinct species has *yellow-brown or reddish* interior corolla. Parasitic on *Ulex*. ES, MA. *Orobanche variegata* is similar but with the lower middle lobe of the corolla *much larger* than the laterals; *flowers foetid*; filaments hairy; calyx lobes triangular-lanceolate. Parasitic on shrubby legumes. Often confused with *O. foetida* (see p.465). DZ, IT (Sardinia and Sicily), TN. *Orobanche tetuanensis* has bracts covered in long, woolly, rust-coloured hairs, shorter corolla, 14(16)–20 mm and virtually *hairless filaments*. Probably parasitic on legumes. Mountains near Tétouan. MA.

Orobanche lutea YELLOW BROOMRAPE A yellowish annual with few to many flowers. Corolla 20–33 mm, rather pale, yellow, pink or brown; corolla inflated at the base, then evenly arched, the lobes spreading; stamens inserted up to *7 mm* above the base and *woolly hairy below*; stigma lobes bright yellow and distant. Bracts dark and not exceeding the corolla. Parasitic on legumes. Rare. FR, IT (incl. Sardinia).

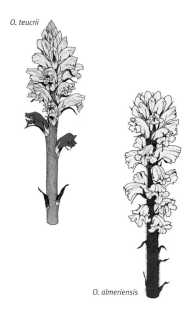

O. teucrii

O. almeriensis

Orobanche alba THYME BROOMRAPE A short, *reddish* (sometimes pale) annual to 30(50) cm. Flowers few, *fragrant;* corolla 15–202(25) mm, pink to dull red, bell-shaped with toothed, *hairy* lobes and *dark hairs* on the surface; upper lip spreading; stamens inserted at least 3 mm *above* the base of the corolla and *hairy; stigma lobes reddish and touching*; bracts short and triangular; calyx lobes *entire* (or weakly toothed). Parasitic on Lamiaceae. Frequent on maquis from Northern Spain eastwards but absent from most islands. ES, FR, IT (incl. Sicily). *Orobanche teucrii* has an irregularly curved corolla 23–28 mm, and *always toothed calyx segments*. Parasitic on *Teucrium*. Rare and local, absent from the islands. ES, FR, IT (incl. Sardinia). *Orobanche almeriensis* has a pale yellow corolla 17–20 mm with darker markings, and *pale hairs*; lobes *hairless* along the margins; stigma pale orange or reddish; calyx lobes *toothed* >$^1/_2$ their length. Parasitic on *Andryala rugusina*. ES (southeast).

Orobanche crenata BEAN BROOMRAPE A short to tall annual to 1 m with densely flowered spikes. Stems purplish, flowers large; corolla (15)20–30 mm, broadly campanulate and *white with violet veins; fragrant*; stigma variable in colour. Parasitic on various legumes, particularly on cultivated peas and beans, sometimes in large numbers. Locally common, particularly in the west, on cultivated and fallow ground; otherwise casual. Throughout.

D. Corolla large, 25–40 mm and *markedly and broadly bell-shaped*.

Orobanche latisquama A robust, often clumped species to 50 cm with many flowers on long spikes. *Corolla large, 25–40 mm, pink-mauve and white at the base* (and at the tips when fresh), abruptly arched near the mouth; lower lobes more or less equal and small; upper lip weakly 2-lobed; stigma white or yellow; filaments inserted *7–9(12) mm above the base*; bracts white at the base, brown at the tip; calyx not divided. Parasitic on *Rosmarinus officinalis* (rarely *Cistus*). Rare inland in Portugal, locally common in coastal Spain. ES, PT.

E. Corolla 12–18 mm, *markedly curved* to appear downward-pointing.

O. cernua

Orobanche cumana SUNFLOWER BROOMRAPE A very slender, lax, erect , bluish or brownish plant to 50 cm tall with *flowers from the base* to the apex of the spike. Corolla 15–18 mm, white or blue, narrow, and curved to appear *downward-pointing*; stigma lobes white; bracts and calyx lobes distinctly short and abrupt. A pest on sunflower crops throughout much of the Mediterranean (absent from Portugal and many islands). DZ, ES, FR (incl. Corsica), MA, TN. *Orobanche cernua* is similar but with short stems to 23 cm tall. Corolla 12–18 mm, white to yellow with a *blackish margin; downward-pointing*; stigma lobes yellow-white. Parasitic on *Artemisia*. Absent from most islands. DZ, ES, FR, IT (incl. Sicily), MA, TN.

PLANTAGINACEAE | PLANTAIN FAMILY

Annual to perennial herbs or shrubs with opposite or whorled, simple or compound leaves. Flowers variable, but usually zygomorphic and 2-lipped; sepals and petals 2–4; stamens 4; style 1. Fruit a capsule or 1-seeded nut. The family includes numerous genera traditionally in the Scrophulariaceae (though the revised classification and taxonomy are not universally accepted).

Plantago PLANTAIN

Small annual or perennial herbs with a basal rosette of leaves, opposite or alternate along the stem. Flowers small and inconspicuous, borne in dense heads or spikes; 4-parted; corolla papery; stamens protruding. Fruit a splitting capsule.

A. *Stems leafy and branched* (spikes borne in axils opposite the leaves).

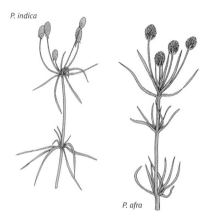

P. indica

P. afra

Plantago indica (Syn. *P. arenaria*) BRANCHED PLANTAIN A short to medium, hairy (but *not* markedly sticky) annual to 30(50) cm with *much-branched stems*, not typically plantain-like when in leaf. Leaves 40–80 mm, *linear* to linear-lanceolate, opposite or whorled, not fleshy and normally untoothed. Flowers brownish-white, to 4 mm, borne in round or conical spikes 5–15 mm, on spreading stalks; anthers pale yellow; *inner bracts larger than the outer.* Fairly common on coastal garrigue and in dry, sandy places. Throughout. *Plantago afra* is similar but usually (not always) extremely *sticky* and glandular-hairy above, *bracts all similar.* On waste and fallow ground. Throughout.

B. Plant a small, woody-based *shrub*.

Plantago sempervirens SHRUBBY PLANTAIN A much-branched *shrublet* 10–40(60) cm high, similar in form to *P. arenaria* but *woody at the base* and very tufted. Leaves linear, 10–60 mm long and *narrow*, just 0.5–2 mm wide. Flowers borne in ovoid heads 5–17 mm, the lowermost bracts with a green point, the upper lanceolate and pointed; calyx lobes unequal. ES, FR, IT, PT.

1. *Plantago afra*
2. *Plantago coronopus*
3. *Plantago maritima*

C. *Leaves borne in a rosette*, linear or narrowly lanceolate, stems not ribbed; spikes borne on leafless stems.

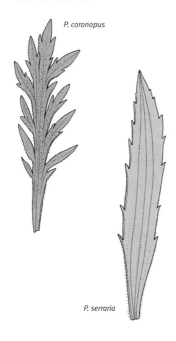

P. coronopus

P. serraria

Plantago coronopus BUCK'S HORN PLANTAIN A very variable, low annual to perennial to 20 cm with solitary or clustered leaf rosettes. Leaves 20 mm–20 cm, linear-lanceolate, usually *pinnately lobed*, though sometimes unlobed, *not* particularly fleshy; hairless or finely hairy. Flowers yellowish-brown, to 3 mm, borne in spikes 40–70 mm long, terminating from ungrooved, curved stems exceeding the leaves; anthers pale yellow. Very common in disturbed coastal habitats in the region. Throughout. *Plantago serraria* is similar to *P. coronopus* but with leaves 50 mm–27 cm, *regularly toothed, not lobed, with 2–12 pairs of teeth*, often flushed purple, and flowering stems equalling or exceeding the leaves, often with long spikes of flowers, 20 mm–14 cm. Common in maritime rocky and sandy habitats. Throughout. *Plantago macrorhiza* is similar to *P. coronopus* but *always a perennial* with a branched stock and often multiple leaf rosettes. Leaves often short, 20 mm–15 cm, *fleshy* and rigid, spatula-oblong-shaped. Flower stalks few, and exceeding the leaves, with spikes 25 mm–10 cm long. ES, PT.

Plantago maritima SEA PLANTAIN A small, glaucous and *fleshy* perennial to 30 cm with multiple rosettes of rigid, long, *linear, untoothed leaves* 50 mm–38 cm long and 1–15 mm wide with 3–5 veins. Flowers brownish, borne in spikes to 70 mm (10 cm) on unridged stalks usually exceeding the leaves; anthers yellow. Salt-marshes and other saline habitats; local. Throughout. *Plantago subulata* is similar but with a much-branched stock and *leaves thread-like*, just 0.5–2 mm across; flower-stalks shorter than their leaves; spikes 10–35 mm. Coastal habitats. Throughout (absent from some islands).

Plantago bellardii A low, *densely hairy* annual to 80 mm (16 cm) high with 1 or more leaf rosettes with linear-lanceolate leaves 15–60 mm, scarcely toothed or entire, 3-veined and white-hairy. Flowers brownish, borne in spreading spikes 8–20(48) mm, rather large relative to the leaves, *borne on stalks not longer than the leaves* (to 13 cm); bracts hairy. Dry, bare ground and garrigue. Common on Italian coasts, rarer further west. Throughout. *Plantago ovata* is similar but with flower stalks with *adpressed* (not spreading) hairs and flat, sub-equal sepals. Arid and bare habitats. Strongly southern; common in parts of North Africa. DZ, ES (southeast), MA, TN.

Plantago albicans SILVERY PLANTAIN A small, low to short, tufted, *silver-woolly* perennial to 28 cm with a woody stock. Leaves 30 mm–15 cm, linear and often slightly twisted, 3-veined (obscured by hairs), and untoothed. Flowers greenish, borne in small, oblong spikes to 5 mm–11 cm long on long, spreading or erect stems; stamens not markedly protruding. *Seeds 4–5 mm.* Locally frequent on dry, bare ground, often coastal. Throughout. *Plantago amplexicaulis* is similar, less hairy, and *less silvery* with elliptic leaves 20–50 mm broadest above the middle, tapered at the base, and faintly veined. Spikes 10–20(30) mm. *Seeds just 2.5 mm.* Various dry habitats; strongly southern in distribution. DZ, ES, IT, MA, TN.

D. Leaves borne in a rosette, linear or narrowly lanceolate, *stems grooved or ribbed;* spikes borne on leafless stems.

P. lanceolata

1. *Plantago serraria*
2. *Plantago ovata*
3. *Plantago bellardii*
4. *Plantago lanceolata*
5. *Plantago major*

Plantago lanceolata RIBWORT PLANTAIN A variable low to medium, hairy or hairless perennial to 50 cm with 1–*several* leaf rosettes. Leaves 15 mm–20 cm, linear-lanceolate or lanceolate, toothed or untoothed, 3–5-veined, *strongly ribbed* and stalked. *Bracts hairless.* Flowers brown, borne in short, blackish spikes 40(80) mm long on grooved stalks that markedly exceed the leaves; anthers pale yellow. A common weed on fallow land and grassy places. Throughout. *Plantago lagopus* is similar to *P. lanceolata* but smaller to 15(47) cm and more *white-hairy, especially the bracts.* Spikes 10–30 mm. Common in similar habitats to the previous species, especially near the coast. Throughout.

E. Leaves borne in a rosette, *broadly* oval or elliptic; spikes borne on leafless stems.

Plantago major GREATER PLANTAIN A low to short, hairy or hairless perennial with broadly oval to elliptic leaves 50 mm–37 cm in a *single* basal rosette, 3–9-veined, narrowing abruptly into a broad stalk at the base; stalk equalling the blade. Spikes *long, dense and slender, 30 mm–32 cm* borne on unfurrowed, hairy stalks to *shorter than the leaves*; corolla whitish, anthers yellowish. Very common in cultivated and grassy places. Throughout. *Plantago media* HOARY PLANTAIN is similar, with 1–few rosettes with elliptic (not oval) leaves 80 mm (28 cm) long which are grey-downy and *gradually narrow into short stalks at the base*. Inflorescence greatly exceeding the leaves; anthers purple and white, and *long and prominent;* spikes 15–60 mm (12 cm). Absent from most islands. ES, FR, IT. *Plantago cornuti* is similar to *P. major* but with leaves to 33 cm with narrow, cylindrical petioles, and flower-stalks *longer than the leaves*. Spikes to 12 cm. ES, FR, IT.

Antirrhinum SNAPDRAGON

Dwarf shrubs or woody-based herbs with entire leaves. Flowers zygomorphic and 2-lipped; stamens 4. Capsule opening by 3 apical pores. Some authors recognise numerous species in Iberian Peninsula but in the absence of a region-wide study based on DNA-sequence study, the number of true species remains unclear.

A. Leaves and stems hairless or shortly glandular-hairy.

Antirrhinum majus SNAPDRAGON A variable (many forms described, accepted by some as separate species), bushy perennial, to 65 cm (1 m) much-branched below, with stems woody at the base. Leaves lanceolate to linear and wedge-shaped at the base, opposite or alternate. Flowers with corolla 33–45 mm, bright pink-purple (pale yellow in cultivated forms); calyx 6–10 mm. Fruit capsule 12–15 mm. Rocky slopes and fixed dunes. Throughout (in its wider description). Subsp. *cirrhigerum* is often taller, 50 cm–1 m with *climbing* stems supported by other vegetation, and linear leaves widest at the middle. Corolla purple, 12–18 mm; calyx 6–10 mm. Fruit capsule 12–18 mm. Dunes in southern Portugal and Morocco. MA, PT. *Antirrhinum onubense* has a similar form with straggling stems but with *white-pink corolla* 20–25 mm with yellow markings. Fruit capsule 8–10 mm. ES, PT.

Antirrhinum controversum (Syn. *A. barrelieri*) An erect, sometimes scrambling or climbing perennial to 1.5 m, usually hairless below but glandular-hairy above with slender branches. Leaves narrow, to 6–40 mm long, linear-lanceolate. Inflorescence a raceme of 10–40 pale pink flowers with corolla 16–24 mm long; calyx 3–4.5 mm. Fruit capsule 6–9 mm, glandular-hairy. Rocky slopes and thickets in the southern Iberian Peninsula. ES, PT.

Antirrhinum siculum SICILIAN SNAPDRAGON A medium, erect or spreading perennial, more or less hairless below and glandular-hairy above; stems freely branched, with leaf tufts in the axils. Leaves 20–60 mm. Flowers borne in terminal racemes, with corolla 17–25 mm, pale yellow, often flushed with darker yellow and with violet veins. Fruit capsule 10–12 mm, glandular-hairy. Rocky habitats. ES (naturalised), FR (naturalised), IT, MT.

B. Leaves and stems hairy.

Antirrhinum graniticum A tall, erect or climbing *glandular-hairy* perennial to 1 m. Leaves 15–62 mm, oval-lanceolate, blunt, opposite below and alternate towards the apex of the stem. Flowers with corolla 25–35 mm, pink or white, borne stalks to 15 mm long (exceeding their bracts); calyx 3–10 mm. Fruit capsule 8–13 mm long and glandular-hairy. Rocky scrub. ES, PT. *Antirrhinum australe* is similar, with leaves opposite or in whorls of 3 and *larger flowers* with corolla 30–43 mm long, bright pink-purple; calyx 5–8 mm. Capsule 10–15 mm, glandular-hairy. Rocky slopes. ES (south, southeast).

Antirrhinum hispanicum An ascending or spreading shrub to 60 cm with numerous branched, *glandular to long-hairy* stems. Leaves 4–25 mm, oval-lanceolate, opposite below and alternate above. Inflorescence a spike-like cluster of 6–20 flowers with corolla 16–25 mm, pink or white flowers; calyx 3.5–8 mm. Fruit capsule 6–10 mm long and glandular-hairy. Rocky habitats. ES (south).

Antirrhinum latifolium LARGE SNAPDRAGON An erect perennial to 60(90) cm, glandular-hairy at the base. Leaves 15–60 mm, oval and slightly longer than broad, blunt-tipped, opposite below and alternate above. Flowers borne in terminal clusters; *corolla pale yellow,* 35–45 mm long and rather broad; calyx 5.5–8 mm, deeply cut into 5 teeth with equal, blunt lobes. Fruit capsule 10–18 mm long and glandular-hairy. Old walls and stony ground from northern Spain to Italy, absent from the islands. ES (north), FR, IT.

1. *Plantago lagopus*
2. *Plantago media*
3. *Antirrhinum controversum*
4. *Antirrhinum siculum*
5. *Antirrhinum majus* subsp. *cirrhigerum*

Misopates

M. orontium

M. calycinum

Annuals similar in form to *Antirrhinum* with distinctly unequal, rather long, linear calyx lobes; stamens 4.

Misopates orontium LESSER SNAPDRAGON A short, sparingly branched, more or less hairless annual to 30(70) cm. Leaves 10–55 mm, linear to elliptic, untoothed, opposite below and alternate above. Flowers with corolla 10–17 mm, pale pink or whitish, snapdragon-like; *calyx 12–20 mm with long lobes.* Fruit capsule 5–10 mm, glandular-hairy. Common in a range of habitats. Throughout. ***Misopates calycinum*** is similar but usually hairless with a raceme that elongates in fruit, and *larger, whitish* flowers with corolla 18–22 mm, *exceeding the calyx lobes*; calyx 14–20 mm. Common on waste ground near the sea and other sandy places. Throughout.

Linaria

Herbs with simple unstalked leaves, opposite or whorled, alternate above. Flowers in spikes or racemes, snapdragon-like but small; calyx unequally 5-lobed and short; stamens 4. Fruit a capsule opening by slits. A diverse and difficult genus, with numerous species in the Iberian Peninsula; seed characteristics are important. The more common species and a few endemics are described here.

A. Flowers mainly violet, red or purple; *inflorescence glandular-hairy*.

Linaria aeruginea A variable annual or perennial to 40(70) cm tall with spreading to ascending stems and *linear* leaves 3–18(35) mm long and *just 0.3–0.9 mm wide*. Corolla 13–27 mm long, variably brick-red and yellow, violet or yellowish tinged purple or brown; calyx 3.5–7.5 mm long with linear-lanceolate lobes; spur 5–11(14) mm. Capsule 3–7.5 mm, globose. ES (east + Balearic Islands), PT (south).

Linaria clementei An *erect*, sparsely leafy grey-green perennial to 90 cm (1.5 m) with fleshy, linear leaves 7–25 mm long. Corolla 12–22 mm, *bright pale violet and yellow*, borne in dense terminal clusters of 2–16; calyx 2.5–4 mm; stalks 2–2.5 mm long and glandular-sticky, just exceeding their bracts (longer in fruit); stigma deeply 2-lobed; spur 2.2–6.8 mm long. Capsule 2.8–4 mm. Dry, sandy habitats. ES (south).

Linaria elegans A slender, erect annual 15–45 cm high with linear-lanceolate leaves 7–70 mm long. Inflorescence long, lax and glandular-hairy. Corolla 16–26 mm, *bright violet to purple* with strongly divergent lips; calyx 2–3.5 mm; *spur long,* 8–14 mm; stigma club-shaped. Capsule 2.2–4.5 mm, seeds angled. Dry habitats. ES, PT.

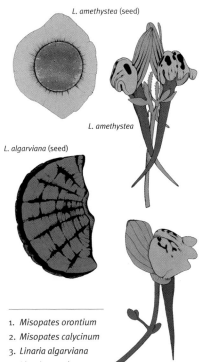

L. amethystea (seed)

L. amethystea

L. algarviana (seed)

1. *Misopates orontium*
2. *Misopates calycinum*
3. *Linaria algarviana*
4. *Linaria amethystea*
5. *Linaria repens*

L. algarviana

Linaria amethystea A short, usually unbranched, *erect annual* to 38 cm, hairless below, but hairy (with *violet hairs*) above. Leaves 2–11 mm, linear-lanceolate, at least 3 x as long as wide, whorled below, alternate above. Racemes few-flowered and lax with 1–8(15) flowers borne in short stalks 0.5–2 mm, not exceeding their bracts. Corolla 10–28 mm, *blue-violet and spotted*; calyx 2.5–5.5 mm; spur 6–8(12) mm and slender. Capsule 3–5 mm; seeds brown to black with a wing *engulfed by a disk* at point of attachment. Coastal, open habitats. ES, PT *Linaria algarviana* is virtually identical, with *spreading to ascending stems* to 25 cm with 1–8 flowers in which the corolla lips are more or less touching (*not divergent*); corolla 16–21 mm; calyx 2.5–4 mm. Capsule 2.1–3.5 mm; seeds black and crested. A rare endemic of coastal shales and sands in the Algarve. PT (south). *Linaria incarnata* is a slender *erect annual* to 45 cm with linear leaves 2.1–10 mm and short, lax inflorescences with *scented flowers*; corolla 15–21 mm long, violet, white and yellow, with gaping lips and *prominent, erect-diverging upper lobes*; calyx 3–5 mm; spur slender, 8.5–11 mm long. Capsule 2.8–4.2 mm. Dry grassland. ES, PT.

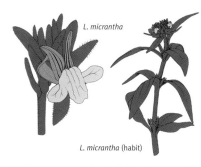

L. micrantha

L. micrantha (habit)

Linaria micrantha A leafy, slightly glaucous annual to 45(55) cm, hairless throughout except on the inflorescence. Leaves 5–40 mm, linear-lanceolate. Racemes dense in fruit; flowers *very small* with corolla 2.5–5 mm, lilac; calyx 2.5–5 mm; *spur minute,* 0.5–1 mm, straight or curved; borne in rather inconspicuous, small terminal clusters of 8–25. Capsule 3.5–6 mm. Widespread in cultivated and waste places. Throughout.

B. Flowers mainly violet, red or purple; *inflorescence hairless*.

Linaria repens PALE TOADFLAX A *hairless perennial* to 1.2 m with numerous linear-lanceolate leaves 10–50 mm, generally in whorls, 0.4–5 mm wide and slender, elongated inflorescences with many flowers (5–45); corolla 9–14 mm, pale white-violet with darker veins; calyx 3–4 mm; spur 1–4(5) mm long and conical. Capsule 4–4.5 mm. Grassy habitats and scree, generally on higher ground in eastern Spain and France. ES, FR. *Linaria purpurea* PURPLE TOADFLAX is similar in form with greyish, linear leaves and many stems, and with normally *dark purple flowers* (sometimes pale) with a *long, curved spur* (not conical). IT (also a casual of waste places throughout).

Linaria pelisseriana JERSEY TOADFLAX An erect, slender and hairless, grey-green annual with unbranched stems 15–50 cm with narrow, strap-like, pointed leaves 5–47 mm. Flowers many (2–35); corolla 15–20 mm, bright violet with a white throat-boss and slender spur 6–9 mm; calyx 4–6 mm, hairless with white-margined lobes. Capsule 3.5–5 mm; seeds flattened with hairy, *irregularly winged margins*. Widespread in sandy and dry habitats. Throughout.

Linaria maroccana MOROCCAN TOADFLAX An upright annual to 40 cm with broadly linear to elliptic leaves 20–40 mm long and terminal heads of small, violet-purple flowers with corolla 15–20 mm long with erect, divergent upper lips and whitish-yellow throat bosses; spur more or less straight. Seeds black and unwinged. Open ground and sandy habitats. Hybrids derived from this species are cultivated in gardens. MA (naturalised elsewhere).

1

2

L. hirta

L. hirta (habit)

C. Flowers mainly white, yellow or tri-coloured; inflorescence glandular-hairy.

Linaria hirta A *densely glandular-hairy*, rather robust, unbranched annual to 50(80) cm tall. Leaves 14–75 mm, narrowly oblong-lanceolate, opposite below but *mostly alternate*, semi-clasping the stem at the base, and 3–4 x as long as wide (2.5–35 mm). Corolla *large,* whitish or cream *25–40 mm*; calyx 6.5–13 mm; *spur long* and reddish, 10–16 mm long. Capsule 5–8.3 mm. Local on coastal sands and sandy waste ground. ES, PT. *Linaria viscosa* is similar to *L. spartea* but with a *densely glandular-hairy* inflorescence, at least above; corolla 16–25 mm, *bright yellow* (rarely violet), borne on erect stalks 3.5–8 mm; calyx lobes long-pointed, 3.5–6.2 mm; spur 5.3–14 mm. Capsule 2.5–5.8 mm. Similar habitats. ES, PT. *Linaria bipunctata* is similar to *L. hirta* but *low to short* to 15(40) cm tall. Leaves oblong-elliptic, 2.5–20 mm long, and crowded on the stem, alternate above; *narrow, 0.25–1 mm wide.* Corolla 10–19 mm, bright yellow and spotted with a reddish spur 4–11 mm long. Capsule 1.5–4 mm; *seeds minute, 0.4–0.8 mm*. Sandy habitats. ES, PT. **Linaria ficalhoana** (Syn. *L. bipunctata* subsp. *glutinosa*) (also treated as a subspecies) is a similar, short annual with *much broader* lanceolate leaves 0.8–3.5 mm wide. A very rare, legally protected endemic of coastal sands in the Algarve. PT.

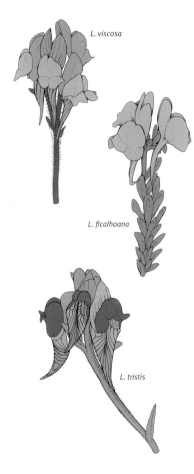

L. viscosa

L. ficalhoana

L. tristis

Linaria tristis A distinctive, more or less prostrate perennial 9–50(90) cm high with many crowded, oblong-elliptic leaves 6–40 mm long and 1–5 mm wide, and terminal clusters of 2–25, *maroon-coloured flowers* with corolla 17–28 mm *with a paler, broad spur* 8.5–12 mm with darker veins, curled at the tip. Capsule 3.5–8 mm. Limestone rocks. Southern Iberian Peninsula. ES, PT.

Linaria supina PROSTRATE TOADFLAX A low, grey-green annual or perennial with alternate linear-lanceolate leaves 5–20 mm long. Inflorescence few-flowered, *very glandular hairy* with pale yellow flowers tinged with violet; corolla tube 2–4 mm wide; sepals oblong-lanceolate shaped; spur 1–15 mm long. Seeds grey to black with a broad wing. ES, FR, IT. *Linaria caesia* is similar but with a broader corolla tube, 3.5–5 mm wide, narrower oblong-elliptic sepals and *shiny seeds*. ES.

1. *Linaria purpurea*
2. *Linaria ficalhoana*

D. Flowers mainly white, yellow or tri-coloured; *inflorescence hairless.*

L. pedunculata

Linaria pedunculata A rather glaucous, hairless annual to 30 cm, spreading to erect, stout, and branched above. Leaves 2–20(30) mm, oblong-lanceolate, fleshy, whorled below and alternate above. Racemes lax and leafy, flowers *long-stalked*; flower stalks 8–30 mm in flower and 13–43 mm in fruit, held erect-spreading; corolla 13–18 mm, cream with violet veins; calyx 3.5–5 mm; spur violet, 5.5–9 mm long. Capsule 3.4–6 mm; *seeds smooth.* Maritime sands. DZ, ES, MA, PT. *Linaria reflexa* is similar but less robust with *larger, white flowers* to 30 mm long, often flushed violet or cream, with orange throat bosses. FR (incl. Corsica), IT (incl. Sardinia).

Linaria spartea A more or less hairless, erect annual to 40(55) cm. Leaves *small*, 6–29 mm, linear, mostly alternate but whorled or opposite below; *distant.* Racemes short, *sparsely hairy and lax*; even more so in fruit; flower stalks erect-spreading, 2–11 mm long, and greatly *exceeding* their bracts (4–22 mm in fruit); corolla bright yellow, 12–24 mm long with a yellow spur 4–12 mm; stigma distinctly 2-parted. Capsule 2.5–5 mm. Common in dry, sandy habitats. ES, FR, PT.

L. spartea

Linaria vulgaris COMMON TOADFLAX A medium, hairless, grey-green perennial *with numerous erect stems* to 1.2 m with dense linear leaves 15–53 mm. Flowers rather numerous (5–85); corolla 19–28 mm, yellow with an orange spot on the lower lip, and slightly scented; spur long, 8–15 mm and curved. Capsule 5–9 mm. Widespread in grassy habitats above sea level; absent from hot, dry areas. ES, FR, IT, PT.

L. polygalifolia subsp. lamarckii

Linaria polygalifolia A *more or less hairless, glaucous perennial* to 45(60) cm, usually unbranched and leafless at least for 20 mm below the lowermost flower; spreading, *not erect.* Leaves 1–22 mm, oblong-spatula-shaped, flat and somewhat fleshy, mostly alternate, at least above. Racemes dense with 2–10 flowers, short-stalked, 1–3 mm (1–8 mm in fruit), but stalks always *shorter than their bracts*; corolla 16–31 mm, deep yellow, with a *long spur* 10–20 mm long, striped orange-brown. Capsule 6–8 mm. Subsp. *lamarckii* (Syn. *L. lamarckii*) has a long corolla, 25–31 mm long. ES, PT.

Linaria triphylla THREE-LEAVED TOADFLAX A robust, hairless annual with thick, erect stems 10–65 cm tall, branched at the base. Leaves 6–36 mm, *broadly oval, 8–25 mm wide*, 3-veined and whorled in groups of 3. Inflorescence cylindrical and rather dense; corolla 20–30 mm, pale yellow-white, often flushed violet, with an orange throat base; spur 4–7 mm; calyx 9–12 mm with oval lobes. Capsule 6–9 mm. Widespread on open ground. Throughout.

Linaria chalepensis WHITE TOADFLAX A slender, hairless, erect and usually unbranched, delicate annual to 40(50) cm with linear leaves 15–50 mm, the lowermost in pairs or in groups of 3. Inflorescence long and lax, with small, pure white flowers with corolla 12–22 mm with a long, slender spur 8–12.5 mm long; calyx 3–5.5 mm, *with lobes markedly unequal*. Capsule 4–5 mm. Common in grassy and stony habitats. Throughout.

Kickxia

Annuals with oval to elliptic or arrow-shaped leaves, entire to sparsely toothed, with pinnate veins. Corolla strongly zygomorphic; stamens 4. Capsule opening by 2 oblique lids.

A. Flower-stalks *hairless*. Flowers whitish or yellowish, with at least some purple markings.

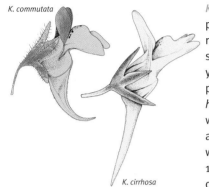

K. commutata

K. cirrhosa

Kickxia commutata A low, spreading, glandular-hairy perennial with slender stems to 80 cm, sometimes rooting at the nodes. Leaves 10–45 mm, oval to arrow-shaped, long-hairy. Corolla 7–17 mm, whitish or yellowish with a blue or violet upper lip and spotted palate; spur strongly curved; flower-stalks 5–25 mm, *hairless*. Capsule leathery, 2–4 mm. Cultivated and waste ground. Throughout. *Kickxia cirrhosa* is a similar annual to 90 cm with *smaller* corolla 4–6 mm long, whitish, *faintly* tinged and veined violet; flower-stalks 10–31 mm, hairless. Capsule papery, 2–2.5 mm. Damp, coastal habitats. ES, FR, IT (incl. Sicily).

1. *Linaria vulgaris*
2. *Linaria supina*
3. *Kickxia sagittata*

1

2

3

K. elatine

K. lanigera

B. Flower-stalks *hairy*. Flowers whitish or yellowish, with at least some violet markings.

Kickxia elatine An annual to 60 cm, similar to *K. commutata* with leaves 3–10 mm. Corolla *yellow* with a purple upper lip, 7–15 mm long; *spur straight*; flower-stalks 5–20 mm, *sparsely hairy* and *longer* than their bracts. Capsule 3–5 mm, hairy above. Throughout.

Kickxia lanigera A long-hairy annual to 1 m, similar to the above species but with *densely white-hairy,* ridged stems and *heart-shaped leaves.* Corolla 8–11 mm long, white or yellowish with a blue to violet upper lip and spotted palate; spur curved; flower-stalks hairy and *shorter* than their bracts. Capsule 2–3 mm, sparsely hairy. Cultivated and waste ground. ES (incl. Balearic Islands), MA, PT.

C. Flower-stalks *hairless*. Flowers typically bright yellow throughout.

Kickxia sagittata A delicate *prostrate or clambering* annual, much-branched. Leaves rather fleshy, narrow, pointed, with 2 divergent lobes at the base. Flowers solitary, bright yellow, sometimes with darker spots on the palate, with a long and conspicuous, *backward-pointing spur;* flower-stalks hairless. Arid and desert fringe habitats in northwest Africa. MA.

Cymbalaria IVY-LEAVED TOADFLAX

Trailing herbs with very slender stems and palmately-veined leaves. Flowers with zygomorphic, 2-lipped corolla with spur at the base. Species all similar.

A. Plants hairless or virtually hairless.

Cymbalaria muralis IVY-LEAVED TOADFLAX A trailing, purplish, tufted, hairless perennial to 60 cm. Leaves alternate, rounded with 5–9 lobes; long-stalked. Flowers small; corolla 9–15 mm, lilac with a yellowish palate, borne on long, slender stalks; *spur 1.5–3 mm long*. Capsule hairless. Local in damp shady places on rocks on higher ground. Native to the eastern Mediterranean, naturalised throughout. *Cymbalaria pallida* is similar, to 20 cm with mostly *opposite leaves* and *larger* lilac-blue flowers with corolla 15–25 mm long; *spur 6–9 mm long*. Rocky crevices. IT. *Cymbalaria muelleri* is similar but *tufted rather than spreading* and with small, whitish-violet flowers with corolla to just 10 mm long with a spur to just 2 mm. Rocky cliffs. IT (Sardinia).

B. Plants hairy.

Cymbalaria aequitriloba A tufted, *hairy* perennial to 40 cm with rounded, rather *Cyclamen*-like leaves with paler veins. Flowers with corolla 7–11 mm, pale violet; spur 2–3 mm long. Damp, shady habitats. ES (Balearic Islands), FR (Corsica), IT (incl. Sardinia). *Cymbalaria fragilis* is very similar but *woolly hairy* and with flowers with corolla 12–15 mm, *white* or very pale violet. An island endemic of Menorca. ES (Balearic Islands).

Digitalis FOXGLOVE

Tall biennial to perennial herbs with alternate leaves. Flowers showy with long, tubular-bell-shaped, 2-lipped corolla; stamens 4. Capsule opening by 2 valves.

A. East of range only; middle lobe of lower corolla lip large, almost as long as the tube.

Digitalis ferruginea RUSTY FOXGLOVE An erect, almost hairless perennial to 1.2 m with lanceolate leaves. Flowers borne in long, dense spikes; corolla 15 mm, dull brownish or reddish-yellow with darker veins; *middle lobe of lower corolla lip large, almost as long as the tube*. Woods at high altitude in the far east of region only. IT.

B. West of range only (*D. purpurea* throughout); middle lobe of lower corolla lip much shorter than its tube.

Digitalis purpurea FOXGLOVE A familiar biennial or perennial to 2 m with soft, oval-lanceolate basal leaves, rather hairy and long-stalked. Flowers showy; borne in long, robust racemes; many-flowered and unbranched; flower stalks 8–16 mm; corolla large and bell-shaped, 40–55 mm long, pink-purple and spotted internally; *lower lip slightly 3-lobed*. Locally common in damp, shady woods on higher ground. Throughout. *Digitalis minor* is similar though shorter, with fewer, pale pink flowers; *lower lip markedly 3-lobed*. Rocky cliffs. ES (Balearic Islands). *Digitalis thapsi* is similar to *D. purpurea* but *sticky-hairy*, and with flowers *long-stalked* (10–20 mm), rather outward-spreading. Mountain woods. ES, PT.

---

1. *Cymbalaria muralis*   4. *Digitalis ferruginea*
2. *Cymbalaria pallida*   5. *Digitalis parviflora*
3. *Digitalis purpurea*

Digitalis parviflora An erect perennial to 90 cm with many leathery, oblong-lanceolate leaves which are entire or slightly toothed. Inflorescence a dense, rather white-woolly spike of *small, rust-coloured flowers* 9–13 mm; calyx lobes oval, 4–5 mm. Among rocks, generally at high altitude. ES (centre, north). *Digitalis obscura* Spanish rusty foxglove has hairless stems to 1.2 m and laxer inflorescences of larger, orange-red flowers 21–31 mm; calyx lobes 5.5–12 mm. Dry mountain scrub. ES, MA.

Veronica SPEEDWELL

Annual or perennial herbs (sometimes shrubs) with opposite leaves. Flowers often blue, short-tubed and with 4 unequal lobes longer than their tube; stamens 2. Capsule opening by 2 valves. Numerous species exist in the area; the most common are described.

A. Flowers in terminal clusters, not borne in the leaf axils.

Veronica serpyllifolia THYME-LEAVED SPEEDWELL A more or less hairless *perennial herb* with creeping, rooting stems to 30 cm as well as *erect or ascending*, flowering stems. Leaves oval, 8–20 mm. Flowers 5–10 mm across, pale blue to white, borne in lax terminal spikes; flower stalks longer than calyx. Capsule wider than long, with equal style. Scrub and open woods; a cosmopolitan weed. Throughout. *Veronica acinifolia* FRENCH SPEEDWELL is a *glandular annual* to 15 cm with smaller leaves 5–14 mm. Bracts and leaves shallowly toothed. Flowers blue, 2–3 mm across. *Fruit longer than broad, and notched;* seeds flat. Throughout, except for most islands. *Veronica arvensis* WALL SPEEDWELL is an *erect annual* to 30 cm with oval leaves 2–35 mm; many hairs non-glandular. Flowers 2–3 mm across, blue. *Fruits hairy* and as long as *broad and heart-shaped*. Throughout. *Veronica triphyllos* FINGERED SPEEDWELL is a sub-erect annual to 20 cm with *palmately lobed leaves* 5–18 mm. Lower bracts and upper leaves deeply lobed at the base (3–7 lobes). Flowers 3–4 mm across. *Fruit as long as broad and shorter than the calyx* with spreading glandular hairs. Throughout.

B. Flowers in clusters borne in the lower leaf axils, with a leafy, non-flowering shoot at the tip of the plant.

V. anagallis-aquatica

Veronica anagallis-aquatica BLUE WATER-SPEEDWELL A medium, hairless or slightly hairy *erect perennial* to 50 cm with branched or unbranched stems. Leaves opposite, oval-lanceolate and scarcely toothed; stalked below, unstalked and semi-clasping the stem above. Flowers 4–9 mm across, pale blue with darker veins, borne in slender, paired racemes. Capsules 2.5–4 mm, rounded or elliptic; hairless. Frequent in damp, seasonally flooded habitats or by streams. Throughout. *Veronica catenata* PINK WATER-SPEEDWELL is similar but with all leaves narrow and stalkless. Flowers 3–8 mm across, *pinkish*. Capsule 2–3 mm, wider than long. Widespread but local, in wet habitats on dunes. Throughout.

1. *Veronica serpyllifolia*   4. *Globularia alypum*
2. *Veronica persica*   5. *Hydrocotyle vulgaris*
3. *Veronica polita*

C. Flowers solitary in the leaf axils.

Veronica persica COMMON FIELD-SPEEDWELL A prostrate, hairy annual to 50 cm long with triangular-oval leaves 15–17 mm, coarsely toothed (>7 teeth) and hairy below. Flowers 8–12 mm across, *bright blue with a paler or white lower lip*, borne solitary in the leaf axils; calyx lobes with rounded bases. Capsule with spreading hairs. Bare and cultivated land. Throughout. *Veronica polita* GREY FIELD-SPEEDWELL is a similar hairy annual with *dull green* leaves 6–17 mm and *flowers bright blue throughout*, 4–8 mm across. Capsule with many short arched hairs (some glandular). Throughout. *Veronica hederifolia* IVY-LEAVED SPEEDWELL has *kidney-shaped leaves 8–28 mm, with 3–7 large, shallow teeth near the base*. Flowers 4–9 mm across, pale lilac, the corolla shorter than the calyx; calyx lobes heart-shaped at the base. Capsule hairless. Throughout.

Globularia

Perennial herbs or small shrubs with alternate, undivided leaves, and flowers in dense rounded heads with an involucre of bracts; stamens 4. Fruit a 1-seeded nut. Most mountain-dwelling species in the region are not described here.

A. Plant a small shrub.

Globularia alypum SHRUBBY GLOBULARIA A low-growing, much-branched evergreen shrub, 40–60 cm (1 m) high with brittle twigs and alternate, leathery, spine-tipped, short-stalked, sometimes apically 3-toothed leaves 15–25 mm. Flowers lilac, borne in *dense, rounded heads*, 10–25 mm across; fragrant; corolla 7 mm with a single 3-lobed lip; bracts oval. Cliffs, scree and garrigue, common in southern and eastern Spain. DZ, ES, FR, IT, MA, TN. *Globularia cordifolia* MATTED GLOBULARIA is a *low, mat-forming perennial* with spreading rooting branches and short, erect, almost leafless stems bearing globular heads of blue flowers 6–15 mm across; corolla 6 mm. Mountains. ES.

B. Plant a herbaceous perennial.

Globularia vulgaris COMMON GLOBULARIA A hairless perennial herb to 40 cm with a rosette of stalked, oval to spoon-shaped and undulate leaves 10–45 mm, notched or 3-toothed at the tip with lateral veins scarcely visible. Stems erect and unbranched, with alternate, lanceolate leaves, bearing terminal dense rounded heads of bluish flowers 15–25 mm across; corolla 9–13 mm long. Rocky scree and scrub at higher altitude. ES, FR. *Globularia majoricensis* (Syn. *G. cambessedesii*) is similar but larger in all parts, with *flat-margined leaves* 20–85 mm, and larger flowers with *corolla 12–13 mm long* borne in heads 15–30 mm across. Stony ground. ES (Balearic Islands).

APIACEAE | CARROT FAMILY

A large and important family. Mostly herbs with alternate leaves, often pinnately divided with sheath-like bases. Flowers normally borne in very characteristic compound *umbels*; often green, yellow or white with 5 free petals and 0 or 5 sepals; stamens 5; styles 2, often arising from a swelling. Fruit a dry, 2-parted schizocarp (made of 2 mericarps), often with a central carpophore. Fruit characteristics are an important diagnostic.

Typical floral arrangement of Apiaceae flower-heads (umbels)

Hydrocotyle PENNYWORT

Aquatic, creeping perennials *with entire (rounded) leaves*. Flowers very small and inconspicuous. Fruit strongly laterally compressed. Unlike all other genera in the family in the region covered.

Hydrocotyle vulgaris MARSH PENNYWORT A low, spreading perennial with creeping stems to 30 cm rooting at the nodes, and with long- and hairy-stalked, roughly *circular leaves* 8–35(50) mm wide with blunt, curved teeth and 7–9 radiating veins. Inflorescence small and inconspicuous, stalkless, to 3 mm across with pinkish-green flowers obscured by the leaves. Fruit 1.8–2.3 mm, rounded. Damp, marshy habitats. Throughout. *Hydrocotyle verticillata* is similar but with *hairless leaf stalks* and leaves with 9–13 veins; flowers stalkless, borne on long, erect stalks. Native to the tropics and naturalised. ES. *Hydrocotyle bonariensis* is similar to the previous species but with *prominent, long-stalked umbels of greenish-white flowers*. Native to the Americas; naturalised on dunes. ES, FR, IT, PT.

Eryngium SEA HOLLY

Hairless, stiff, spiny herbs with simple or pinnate leaves. Flowers dense, borne in rounded, thistle-like heads, surrounded by spiny bracts. Fruit rounded and often scaly or bristly. A confused genus with conflicting accounts.

E. maritimum

A. Involucral bracts oval and toothed or deeply lobed.

Eryngium maritimum SEA HOLLY A short, rigid perennial to 50 cm with persistent, leathery, *blue-grey, white-veined* leaves, 3–5 lobed with an undulate, spiny margin. Flowers pale blue, borne in dense, rounded heads 10–30 mm framed by *broadly lanceolate, lobed, coarsely spiny bracts* with 1–3 pairs of spines; upper stems flushed bright blue. Fruit >6 mm, with hooked bristles. Very common on coastal sands and dunes. Throughout.

1. *Eryngium campestre*
2. *Eryngium maritimum*
3. *Eryngium amethystinum*
4. *Eryngium atlanticum* PHOTOGRAPH: SERGE D. MULLER

B. Involucral bracts linear or lanceolate and spiny-toothed or entire; *basal leaves 3-lobed or entire*.

Eryngium aquifolium A short, stiffly-spiny, blue-grey perennial 15–50 cm, similar in general appearance to *E. maritimum*, though sparingly branched above, and with spiny-toothed lower leaves, gradually narrowing at the base, with a *dense network of veins*. Flower-heads 10–20 mm across with *linear-lanceolate, blue-flushed, spiny bracts*. ES, MA.

Eryngium ilicifolium A low *annual with prostrate stems* to 30 cm and often a rosette-like arrangement of entire, spiny-margined lower leaves; stems branched from the base. Flower-heads 19–14 mm, borne on very short stems; bracts linear-lanceolate, with membranous extensions at the base (auricles) and with spiny margins. Arid habitats. ES, MA.

Eryngium corniculatum A short, spreading, sparse (with rather few, spreading branches), spiny perennial 10–60 cm with *undivided basal leaves* with *long* swollen, segmented stalks, and bluish flower-heads 7–10 mm, *framed by long-pointed, linear, entire bracts* 10–60 mm long. Seasonally flooded habitats; mainly western. ES, FR (Corsica), MA, PT.

Eryngium tricuspidatum A medium, often rather *intricately branched above, spiny white-grey perennial* 70 mm–90 cm with *soon-withering and falling tender, green, oval-rectangular leaves* with heart-shaped bases. Flower-heads 2–15(20) mm, blue, framed by whitish, linear, spiny-margined bracts. Widespread. DZ, ES, FR (Corsica), IT (Sicily), MA, TN.

E. campestre

C. Involucral bracts linear or lanceolate and spiny-toothed or entire; *basal leaves pinnately lobed*.

Eryngium campestre FIELD ERYNGO A short, rigid, *much-branched* perennial 20–60(75) cm high with green or yellowish, leathery, lobed basal leaves which are spine-toothed with unwinged stems; stem leaves unstalked and clasping the stem. Flowers pale green-white or yellowish borne in dense, rounded heads 8–15 mm; bracts 5–7, *linear-lanceolate and entire or spiny-toothed*, spreading. Very common in dry, sandy or grassy places both coastal and inland. Throughout.

Eryngium dilatatum A greenish or greyish perennial to 55 cm with 3-lobed to *pinnately lobed* lower leaves which are softly spine-toothed, and with *winged stalks*. Flower-heads 10–12(15) mm, framed by *blue, linear, entire bracts* with 3–6 pairs of spines, often strongly blue-tinted; *inflorescence a long, erect, narrow raceme*. Dry habitats. ES, MA, PT.

Eryngium amethystinum BLUE ERYNGO A short to medium, stiff, much-branched perennial with *pinnately-leaved basal leaves* with spine-toothed margins and broadly winged stalks. Flower-heads *strongly flushed blue;* bracts 7–8, blue, and *linear and entire* or sparsely spine-toothed, 2–5 x as long as the flower-heads. Garrigue and pastures. IT (incl. Sicily).

Eryngium atlanticum A branched, *strongly blue-tinted perennial* with spreading, almost horizontal branches. Flowers *few, borne in small clusters* (not dense, rounded heads), framed by few, blue-flushed, lanceolate, leathery, pointed bracts. A rare Moroccan endemic, threatened by habitat destruction. MA.

Lagoecia

L. cuminoides

Annual, hairless herbs with pinnately divided leaves. Umbels with 8 bracts and 4 secoary bracts; petals minute. Fruits with very fine ribs.

Lagoecia cuminoides LAGOECIA An annual to 55 cm with pinnately divided leaves 20–70 mm with toothed segments; upper leaves yarrow-like, deeply divided into 3–13 pairs of narrow lobes. *Umbels feathery, white, rounded and dense*, 7–15 mm; petals 1 mm. Fruit 1.5 x 1 mm with brittle hairs. Widespread but local in dry, fallow habitats. ES, IT, PT.

Echinophora

Stiff, spiny perennials. Flowers borne in umbels with 5–8(12) rays. Fruits angled.

Echinophora spinosa A very spiny, much-branched, robust perennial to 70 cm with fleshy, 2–3-pinnately divided, rigid leaves 40 mm–22 cm with thick, spine-tipped lobes. Flowers white, borne in flattish umbels to 30 mm across, short-stalked, with 5–8(12) rays; bracts linear and spiny. Fruit 8 x 4 mm, egg-shaped and angled. Coastal sands, widespread but patchy. DZ, ES, FR (Corsica), IT (Sardinia and Sicily). *Echinophora tenuifolia* is similar, though greyish, often very densely branched and with *numerous yellow umbels*. Fallow and waste ground. IT (incl. Sicily).

Smyrnium

Biennial herbs in which the 2–3 leaf lobes arise from a single point (ternate). Umbels yellow, the flowers lacking sepals. Fruit egg-shaped with slender ridges and oil glands, hairless.

S. olusatrum

Smyrnium olusatrum ALEXANDERS A tall, pungent, hairless biennial to 1.5(2.2) m. Stem robust, hollow when mature, bearing 1–2-pinnately *divided leaves* with triangular, toothed lobes, borne on short, somewhat inflated stalks; shiny green below, yellow above. Umbels terminal and borne in the leaf axils with 7–15(18) unequal rays; flowers yellow, to 3 mm across without sepals. Fruit ovoid, 5–7 mm and *black* when ripe. Common in damp, coastal habitats. Throughout.

1. *Echinophora spinosa* (flowers)
2. *Echinophora spinosa*
3. *Smyrnium olusatrum*

S. perfoliatum

Smyrnium perfoliatum Perfoliate alexanders A virtually hairless, tall biennial to 60 cm (1.5) m. Stem robust, angled with longitudinal narrow wings, solid and hairy at the nodes. Basal leaves with 1–2 lobes which are oval and toothed; *upper leaves yellowish, simple, oval and markedly clasping the stem* with heart-shaped bases. Umbels yellow, rounded with 4–8(12) rays and no bracts. Fruit 2–4 mm, broader than long, brown-black. Wooded and open grassy slopes; local. ES, FR, IT, PT. *Smyrnium rotundifolium* is similar but with the upper leaves entire (not toothed) and often yellow, and the stems ribbed but not winged; leaves greyish. FR (Corsica), IT.

Levisticum LOVAGE

Robust perennial herbs with 2–3 pinnately divided leaves with wide leaflets. Flowers with pale yellow petals. Fruits with 5 primary ribs.

Levisticum officinale LOVAGE An erect, robust perennial to 2.5 m with a basal rosette of shiny, hairless 2–3 pinnately divided leaves with broadly triangular segments, smelling of lime when crushed. Flowers pale greenish-yellow, to 3 mm across, borne in rounded umbels of 8–20 rays 7–40 mm long; petals 1 mm. Fruit 5–7 mm, yellowish. Generally above sea level. Not native (origin unknown) but sporadic throughout.

Bunium

Tuberous, hairless, erect, hairless perennial herbs with divided leaves. Flowers borne in umbels with 5–20(25) equal to unequal rays; petals white or pink. Fruits hairless, with 5 primary, rounded ribs.

Bunium pachypodum A short to medium perennial 15–50 cm tall with leaves 2–3-pinnately divided with *linear*, blunt-tipped lobes, withered at the base when mature. Umbels 50–80 mm across, white, lax, with 14–20(22) spreading, *rigid rays*, enlarged in fruit. Fruit 3.5–6 mm long, with thick ridges. Local, on cultivated land; distribution patchy. ES (Balearic Islands), FR (Corsica), MA, PT.

1. *Smyrnium perfoliatum*
2. *Levisticum officinale*
3. *Scandix pecten-veneris*

1

2

3

Bunium bulbocastanum GREAT PIGNUT A large, erect perennial to 50 cm (1 m) in which stems remain solid (not hollow) after flowering. Basal leaves 2–3-pinnately divided, to 15 cm long, withered when in flower. Umbels 35–60 mm across with 15–20 rays; petals 1–1.3 mm; styles bent in fruit. Fruit 3–4.5 mm, scarcely ridged. Predominantly eastern in distribution, but a casual naturalised elsewhere. ES, FR, IT.

Scandix

Slender annual herbs with thread-like, feathery leaves. Flowers with white petals and with minute sepals. *Fruits conspicuous, long, needle-like* and with cylindrical beaks, hairless.

S. pecten-veneris (leaf)

S. pecten-veneris (fruits)

Scandix pecten-veneris SHEPHERD'S NEEDLE A small, spreading, hairless annual to 40(50) cm with leaves 2–3(4)-pinnately divided, the segments widened towards the tips. Umbels simple or with 2–3 rays; *secondary bracts divided into linear lobes;* flowers small and white; petals 2–4.5 mm. Fruits highly distinctive: *elongated, 50–95 mm long*, at least ¹/₂ the length comprising the seedless beak which is *flattened and distinct*; borne in claw-like clusters. Common on disturbed and fallow ground. Throughout. *Scandix australis* is similar but with undivided (or bilobed) secondary bracts and fruit shorter, 11–35 mm, and *scarcely compressed*, and the beak *indistinct* from the rest of the fruit. Pine forests and shady places. Throughout.

Conopodium

Tuberous, hairless perennials with hollow stems. Flowers with deeply notched petals. Fruits ovoid, hairless and with low, rounded ridges. Rare in the Mediterranean.

Conopodium majus PIGNUT A slender, sparingly branched, hairless perennial to 40(75) cm with smooth, finely grooved stems carrying few stem leaves 2–3-pinnately divided into *lobes 3–10 mm long*; basal leaves soon withering. Umbels 40–70 mm across, white, nodding in bud, to 50 mm across, and with 6–12(14) rays; petals 1.5–1.8 mm; *secondary bracts 0–2(3)*. Fruits narrowly ovoid, 3–4.5 mm. Rare or absent in hot, dry areas. ES, FR, IT, PT. *Conopodium pyrenaeum* (Syn. *C. bourgaei*) is similar but with broader linear-lanceolate lobes on the stem leaves *10–40 mm long*, no bracts, and *0–1 secondary bracts*. Similar habitats; local outside of northern Iberian Peninsula. ES, PT.

C. subcarneum

Conopodium subcarneum (Syn. *C. capillifolium*) A small, tuberous plant 20–60(75) cm with widely stalked, hairless, lobed basal leaves and reduced, often entire uppermost leaves; leaf lobes 8–30 mm. Umbels 30–50 mm across with 11–15(18) rays and 0–2 bracts, and *5–10(12) secondary bracts*; petals 1–1.5 mm, white with a brown vein on the back. Fruit 2.8–4 mm, oblong, laterally flattened. Grassy places; absent from hot, dry regions and local outside of northern Iberian Peninsula. ES, PT.

Crithmum

Hairless, fleshy perennials with leaves 1–2-pinnately divided. Petals yellow-green. Fruit oblong, hairless, not compressed, and rather corky (spongy when fresh).

C. maritimum

Crithmum maritimum ROCK SAMPHIRE A short to medium, somewhat bushy, greyish, hairless perennial to 45 cm woody at the base. Leaves 1–2-pinnately divided with almost cylindrical, untoothed, *fleshy segments*; the base membranous and clasping the stem; *smelling of furniture polish when crushed*. Flowers yellow-green in umbels to 60 mm across with 10–32(36) rays, sepals absent; bracts and secondary bracts numerous (5–10), triangular. Fruit 3.5–5 mm, ovoid-oblong and corky, later purple. Common on cliff-tops, rocky shores and other maritime environments. Throughout.

Astydamia

Woody-based perennials. Umbels with up to 10 basal bracts; flowers yellowish, the *innermost opening first*. Seeds scattered with minute crystals on the surface and with a *long, forked* carpophore.

Astydamia latifolia A fleshy, aromatic, yellowish to greyish-green biennial to perennial 10–40 cm high with coarsely jaggedly toothed to pinnately lobed, succulent, shiny leaves. Umbels 60 mm–12 cm across, typically with 10–15 rays, and a similar number of broadly linear bracts at the umbel base and secondary bracts at the base of the flowers; flowers yellow. Fruits ovoid, 3-veined and corky when ripe. Coastal dunes and rocks in North Africa. DZ, MA, TN.

Oenanthe WATER-DROPWORT

Typically aquatic, hairless annuals to perennials, often with tuberous roots. Flowers with white petals. Fruits rounded on the back, with shallow, grooved ribs.

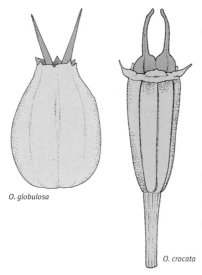

O. globulosa

O. crocata

Oenanthe globulosa MEDITERRANEAN WATER-DROPWORT A *short*, much-branched, perennial 15–60 cm with egg-shaped tubers and stems rather bluish-grey, hollow and distinctly *grooved*. Basal leaves 2-pinnately divided with oval to linear lobes, the upper leaves less divided and narrower. Umbels white with 2–8(15) lax, disparate rays; flowers male or cosexual, sepals persistent; *flower stalks thickened in fruit*. Fruit rounded, 3–5(7.5) mm with thickened lateral ridges. Marshes and riversides. ES, FR, IT, MA, PT. *Oenanthe crocata* is similar but *taller* and more robust to 1.5 m, with a terminal umbel with 12–30(40) rays; flower stalks *not* thickened in fruit, and the fruits *cylindrical*, 4–5.5 mm with styles $^1/_2$ as long. Similar habitats, and more frequent than the previous species. ES, IT, PT, FR.

1. *Crithmum maritimum*
2. *Astydamia latifolia*
3. *Oenanthe globulosa*
4. *Oenanthe crocata*
5. *Oenanthe pimpinelloides*

Oenanthe pimpinelloides CORKY-FRUITED WATER-DROPWORT An erect perennial with solid (pithy), ridged stems to 1 m with 1–3-pinnately divided basal leaves with linear to oval leaflets and 1-pinnately divided stem leaves with narrow, unlobed leaflets. Flowers white, borne in terminal umbels to 60 mm across with 6–15 rays which *thicken in fruit* (to >0.5 mm thick); bracts and secondary bracts bristle-like. Fruits 3–3.5 mm cylindrical and *corky and swollen at the base;* styles erect and equalling the fruit. Damp grassland; absent from the far south and east. ES, FR, IT.

Foeniculum FENNEL

Perennials with 3–4-pinnately divided leaves with slender lobes, *strongly aromatic.* Flowers with sepals absent, petals yellow. Fruit distinctly ridged and scarcely compressed.

F. vulgare

Foeniculum vulgare FENNEL A robust, tall, hairless, *strongly aromatic* perennial to 2.5 m, *smelling of aniseed.* Stems *bluish*, hollow when mature. Leaves feathery, 3–4-pinnately divided with *thread-like* segments 5–40 mm, light green and with sheathing bases. Umbels yellow with 5–44 rays without any bracts flowers borne in mid summer; petals 1.3–1.6 mm. Fruit oblong, 3–6(9) mm long, ridged. Very common and widespread in a range of dry habitats. Throughout.

Kundmannia

K. sicula

Erect perennial herbs with pinnately divided leaves with large, oval segments. Flowers with yellow petals. Fruits oblong, hairless and ribbed.

Kundmannia sicula KUNDMANNIA A medium, hairless perennial with erect stems 30 cm–1 m. Lower leaves 2-pinnately divided with an extra pair of lobes at the base of each main lobe; lobes 15–40 mm, oval, toothed, upper leaves 1-pinnately divided and coarsely toothed. Umbels yellow with 3–30 rays, bracts numerous and linear and keeled backwards. Fruit cylindrical, 6–10 mm long with slender ridges. Dry slopes and fallow ground; uncommon. Throughout.

Distichoselinum

Erect perennial herbs with leaves 4–5-pinnately divided. Flowers with yellow petals. Fruits winged.

D. tenuifolium

Distichoselinum tenuifolium (Syn. *Elaeoselinum tenuifolium*) A medium, hairless, *fennel-like* perennial to 1.3 m with distichous leaves (divisions in opposite rows) to 50 cm long, 4–5-pinnately divided into *short* terminal lobes just 1–3 mm long, slightly aromatic when crushed. Flowers yellow and borne in fennel-like umbels with 15–30 rays; bracts 4–12, all similar and to 25 mm long. Fruits 8–18 mm, winged. Maquis. Locally abundant in southern Iberian Peninsula. ES, PT.

Margotia

Very *strongly aromatic herbs* with rather flat basal rosettes of pinnately divided leaves. Flowers white. Fruits winged.

Margotia gummifera A herbaceous perennial to 1.8 m with a flat, basal rosette of leaves with purplish stalks. Leaves flat, 3–4-pinnately divided, the terminal divisions 1–4 mm, linear to linear-lanceolate; *dark shiny green or purplish, and very strongly aromatic* when crushed. Inflorescence an umbel with 20–35 rays and 3–8 linear to linear-lanceolate, drooping bracts; flowers white with 5 incurved petals. Fruits 7–15 mm, winged and papery. Coastal cliffs, dunes and garrigue in the southwest. ES, MA, PT.

Cachrys

Bushy perennials with leaves 2–4-pinnately divided into linear lobes. Flowers with yellow petals. Fruit with undulate wings and ridges. A difficult group for which keys are inconsistent between regional texts; further work is required to resolve the genus.

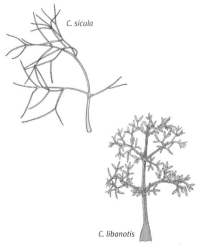

C. sicula

C. libanotis

Cachrys sicula A bushy, bluish-green perennial to 1.5 m with a solid, grooved stem. Leaves fennel-like; 4–6-pinnately divided, oval in outline with linear, somewhat flopping, *long lobes 15–50 mm* that are *slightly rough or smooth along the margins.* Flowers borne in umbels of up to 30 rays; yellow, and with sepals present. Bracts of the terminal umbel usually 1–2-pinnately lobed. Fruit 12–17 mm, *ovoid and with prominent ridges with protuberances.* Dry fields and scrub. Throughout. **Cachrys libanotis** is very similar to (and frequently confused with) *C. sicula,* but with leaves with rough to finely toothed margins and small, *smooth fruits* 6–11 mm (bracts of umbels usually entire, but this trait is not consistent across regions). Dry fields, garrigue and sea cliffs. Throughout.

---

1. *Foeniculum vulgare*
2. *Distichoselinum tenuifolium* (inset: fruits)
3. *Margotia gummifera*
4. *Cachrys sicula* (inset: fruits)

Prangos

Rounded, bushy, aromatic herbs with pinnately divided leaves with thread-like lobes. Flowers with yellow petals. Fruits large and corky.

Prangos trifida A hairless, herbaceous, very bushy, light yellow-green, fennel-like perennial to 1.2 m. Leaves roughly triangular in outline, *4–7-divided into long, linear* lobes 10–20 mm, giving the plant a feathery appearance; aromatic when crushed. Inflorescences stout with umbels of 10–20 rays of bright yellow flowers. *Fruits large, unwinged, 11–20 mm long, ovoid, purplish and rather corky when ripe.* Maquis and rocky pastures. ES, FR, IT, PT. A confused species with inconsistent records; herbarium specimens for *Cachrys trifida* (a synonym of *P. trifida*) from the Algarve appear to be distinct, but may better correspond with *C. sicula. Prangos ferulacea* is similar, robust, to 1.8 m. Umbels with 20–30 rays with linear-lanceolate bracts and linear secondary bracts. Fruit 10–25 mm, slightly laterally compressed with prominent crested or toothed wavy wings. IT (incl. Sicily).

Conium

Tall, erect, hairless biennials with compound leaves. Flowers with white petals. Fruits broadly ovoid with 5 prominent ribs.

Conium maculatum HEMLOCK A tall, erect, almost hairless biennial with *purple-spotted stems* to 2 m tall; strong-smelling and poisonous. Lower leaves large, to 50 cm long, and triangular in outline, 2–4-pinnately divided with fine leaflets. Umbels terminal and axillary, to 50 mm across with 8–10 rays and few, small, backwardly turned bracts and similar secondary bracts; flowers white. Fruit 2–3.9 mm, almost spherical with wavy ridges. Local in damp, grassy, and wasteland habitats, probably throughout. ES, FR, IT, MA, PT, TN.

Bupleurum

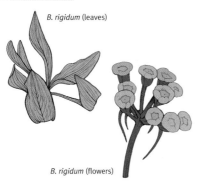

B. rigidum (leaves)

B. rigidum (flowers)

Hairless annuals or perennials with *simple leaves*. Flowers pale green or yellow, borne in small umbels surrounded by petal-like bracts. Sepals absent, petals not notched. Fruits prominently ridged.

Bupleurum rigidum A woody-based perennial to 1.5 m, with *large, leathery, broadly strap-like* leaves 10–45 cm which are stalked and clasping the stem, with 3–11(13) veins, and thickened margins. Umbels with 2–5(6) rays and short, slender bracts; petals yellow. Fruit 3–6(7) mm long, with prominent ridges. Common on dry rocky habitats and sea cliffs. ES, FR, IT, MA, PT.

Bupleurum rotundifolium THROW-WAX An erect, grey-green annual to 50(85) cm with hollow stems. Leaves elliptic to *rounded*, those below to 10 cm, tapering into the petiole, those above 10–60 mm, *clasping the stem entirely at the base*. Flowers borne in umbels to 30 mm across with few (3–8) rays and no bracts but conspicuously oval, fused secondary bracts (*Euphorbia*-like in appearance); petals yellow. Fruit 2.5–4 mm, with ridges, smooth in between. Arable and grassy habitats. ES, FR, IT, MA, PT.

1. *Cachrys libanotis* (inset: fruits)   4. *Bupleurum rigidum*
2. *Prangos trifida*   5. *Bupleurum spinosum*
3. *Bupleurum rigidum*

Bupleurum baldense SMALL HARE'S-EAR A slender, and *small*, erect annual to 10(30) cm high, often branched. Leaves linear to spoon-shaped, often curved, to 50 mm long and sharp-pointed. Umbels few-flowered, to 10 mm across, with *bracts longer than the longest rays* (inflorescence concealed); secondary bracts bristle-pointed. Fruit smooth, 1–3.5 mm. Dry, often coastal habitats. Southwest Europe. ES (incl. Balearic Islands), PT.

Bupleurum tenuissimum SLENDER HARE'S-EAR A slender, spreading to erect annual to 50 cm with wiry stems. Leaves 13–80 mm, linear-lanceolate to spoon-shaped, pointed. *Umbels tiny, <5 mm across,* with few (2–5) rays borne in the leaf axils, and very short-stalked. Fruit 1.5–3 mm. Saline grassy habitats. ES, FR, IT, MA, PT. *Bupleurum gerardii* is similar but has more sickle-shaped leaves and umbels with 3–10 rays. ES, FR, IT, PT.

Bupleurum falcatum SICKLE-LEAVED HARE'S-EAR An erect, hairless perennial with hollow stems to 80 cm. Leaves 20 mm–15 cm, herbaceous (not distinctly leathery), *narrowly spoon-shaped* and curved, those below stalked, those above clasping. Umbels stalked, to 40 mm across, with 3–15 rays and 1–4 unequal bracts, as well as 5 linear-lanceolate secondary bracts; petals yellow and minute. Fruit 2–4 mm, red-tipped. Dry, grassy habitats. ES, FR, IT, PT. *Bupleurum fruticescens* is similar but often *much-branched and bushy* to 1.2 m, with *woody* stems below and *leathery leaves* to 10(14) cm which are *linear to linear-lanceolate* and not clasping throughout. Umbels with 2–9 slender rays. Fruit 2–6 mm, with distinct stalks to 10 mm. Garrigue and scrub. Common in northern Spain. ES, MA. *Bupleurum spinosum* is sometimes treated as distinct, differing mainly in having much shorter leaves 5–40(60) mm and 2–6 stouter rays on the umbels. ES, MA. *Bupleurum acutifolium* is sparingly branched, with herbaceous (not woody) stems. Umbels with 3–9(14) rays. Seeds 3–5 mm, not winged. Southern Iberian Peninsula. ES, PT.

Bupleurum fruticosum SHRUBBY HARE'S-EAR A large, aromatic, evergreen, perennial *shrub* to 2.5(3) m. Leaves oblong with strong mid vein and prominent lateral veins, and *blue-grey beneath*. Stems reddish and much-branched. Umbels from previous years are persistent and woody, making the plant appear spiny. Leaves linear, 30 mm–13 cm. Flowers yellow, borne in open umbels with 3–20(25) rays; bracts and secondary bracts short and bristly; petals yellow. Fruit 5–7(8) mm. Stony gound and maquis; common in along the east coast of Spain, scattered elsewhere, probably throughout. ES.

Apium MARSHWORT

Aquatic, hairless perennials with pinnately divided leaves. Flowers white. Fruits laterally compressed.

Apium nodiflorum (Helosciadium nodiflorum) FOOL'S WATERCRESS A medium to tall, yellowish biennial to 1 m with a celery-like smell. Stems hollow, spreading and rooting at the lower nodes, then ascending. Leaves shiny, 1-pinnately divided with oval-lanceolate segments to 60 mm long. Umbels with 3–15 rays; flowers with whitish petals. Fruit 1.5–2.5 mm, ovoid and longer than wide and with thick ridges. Wet places, ditches and riverbanks. Throughout.

Apium inundatum (Syn. *Helosciadium inundatum*) LESSER MARSHWORT A straggling or prostrate, slender, often *submerged* perennial to 50(75) cm with 1-pinnately divided lower leaves with very narrow (almost hair-like) leaflets; leaves above pinnately divided with broader, often 3-lobed segments. Umbels arise from the leaf axils, with few (2–4) rays; flowers white. Fruit elliptic, 2.5–3 mm long, with thick ridges. Wet, muddy habitats. Probably throughout. *Apium crassipes* (Syn. *Helosciadium crassipes*) SARDINIAN CELERY is a similar, spreading aquatic perennial with pink-white flowers; a threatened Mediterranean endemic; fairly common in Tunisia. DZ, FR (Corsica), IT (incl. Sardinia and Sicily), TN.

Ridolfia

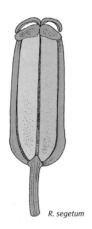

R. segetum

Tall, erect, aromatic, hairless annuals with 3–4(6)-pinnately divided leaves with thread-like segments. Flowers yellow. Fruits laterally compressed, longer than wide.

Ridolfia segetum RIDOLFIA A tall, fennel-like, bluish-green annual to 2.1 m, unpleasant-smelling when crushed (not like aniseed). Leaves 3–4(6)-pinnately divided with thread-like lobes 5–15 mm long and inflated petioles. Umbels yellow with many (8–56) slender, curved rays and no bracts or sepals; petals 1.9–2 mm, yellow. Fruit small, 1.2–2.7 mm long with slender ridges. Cultivated and waste land. Throughout; local though common in southeast Iberian Peninsula.

Ammi

Hairless annual or biennial herbs with erect, striated stems and leaves divided into linear or lanceolate segments. Flowers with white petals, borne in dense umbels. Fruits hairless and slightly laterally compressed.

A. visnaga

Ammi visnaga A robust, carrot-like annual or biennial to 1 m. Lower leaves 1-pinnately divided, upper leaves 1–2-pinnately divided, all somewhat feathery with narrow to *linear, thread-like lobes*. Inflorescence rather like those of *Daucus*; *flowers white, borne in dense umbels*; rays numerous (45–125), slender and spreading in flower but much *thickened and erect (bunching) in fruit*. Bracts divided, equalling or exceeding the rays. Fruit 2–2.8 mm. A weed of rocky and scrubby waste ground. Throughout, though only sporadic in many areas. *Ammi majus* is a similar erect annual to 1 m with broader, linear-lanceolate leaf segments and with 20–55 *rays slender and not bunching in fruit*. Fruit 1.5–2 mm. Common, sometimes abundant on disturbed ground. Throughout.

1. *Bupleurum fruticescens*
2. *Apium inundatum* PHOTOGRAPH: SERGE D. MULLER
3. *Bupleurum fruticescens*
4. *Apium crassipes* PHOTOGRAPH: SERGE D. MULLER

Angelica

Robust biennials and perennials with hollow stems and divided leaves, triangular in outline, with large lobes. Flowers borne in large umbels, with white or purlish petals. Fruits hairless with winged lateral ridges.

Angelica sylvestris A robust, erect biennial to 2.5 m, more or less hairless with thick, purple stems. Leaves green, not shiny, those below 2–3-pinnately divided into oblong-elliptic, finely toothed segments, those above variably divided into oval, toothed segments. *Umbels large, with up to 75, rather distant rays of white flowers*, often pink- or purple-flushed. Fruits 4–5 mm. Rare, in aquatic habitats, mainly in the north of the range. ES, FR, IT.

Ammoides

Slender annuals or biennials with divided leaves, cylindrical in outline. Flowers with white petals, borne in lax umbels. Fruits ridged and hairless.

Ammoides pusilla A *weak, very slender, hairless annual* to 50(75) cm. Leaves yarrow-like, 2-pinnately divided below with 7–12 pairs of *very short lobes*, the upper leaves with longer, *linear* segments; all leaves narrow in outline and yarrow-like. Umbels with 5–15(20) slender, *lax, unequal rays*; bracts absent or few; petals white. Fruit small, just 0.7–1.3 mm, with slender ridges; hairless. Dry habitats and field borders. Throughout.

Krubera (Capnophyllum)

Small, much-branched, spreading annuals with leaves 3–4-pinnately divided. Flowers white. Fruits prominently ridged.

Krubera peregrina (Syn. *Capnophyllum peregrinum*) A small, slender, hairless, spreading (not erect) annual 20–40 cm high with stems solid and grooved. Leaves broadly triangular in outline, 3-pinnately divided into linear segments. Flowers white, rays few (2–5) and outwardly-spreading; bracts absent, secondary bracts 4–6, shortly triangular. Fruit 4.3–6.5 mm, with *very prominent, rough ridges*, ovoid and laterally compressed. Mainly western. ES, FR, IT (incl. Sicily), MA, MT, PT.

Pastinaca PARSNIP

Slightly hairy, strong smelling biennials. Flowers with yellow petals with cut-off, incurved tips; sepals absent. Fruit ovoid and strongly compressed.

Pastinaca sativa WILD PARSNIP A medium to tall, erect, downy, branched biennial to 1(2) m with hollow, ridged and angled stems (sometimes solid and unridged). Basal leaves light green, *1-pinnately divided with broad, oval, lobed and toothed segments*. Umbels to 10 cm across with 5–20 rays with bracts and secondary bracts few or absent; *flowers yellow and fennel-like*. Fruit 4–7 mm, ovoid and flattened with narrowly winged edges. Damp fields and grassy habitats above sea level; rare in the Mediterranean and absent from the far south. ES, FR, IT. *Pastinaca lucida* is similar in form, but with leaves *undivided* (or infrequently, shallowly lobed). Field borders and rocky habitats. ES (Balearic Islands).

Ferula

Very tall, robust perennials with leaves 3–4-pinnately divided into linear lobes. Flowers with yellow petals. Fruits strongly compressed dorsally. DNA-based analysis shows that morphological traits are of limited value; reliable identification in the field may be impossible for closely related species (hence traditional keys and floras may not be reliable).

A. Leaves with long, flopping, thread-like lobes 10–40(80) mm long.

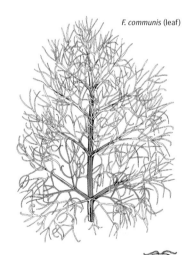
F. communis (leaf)

Ferula communis GIANT FENNEL A giant, robust perennial herb to 3.3 m, bushy at the base and a conspicuous feature of the landscape when in flower, and long-persisting after flowering. Leaves 3–4(6)-pinnately divided with bright green, *thread-like segments 10–40(80) mm long and just 0.5–1.3 mm wide*, and prominent sheathing bases; upper leaves reduced to sheathing bases only. *Inflorescences pyramidal and spreading*; umbels large and rounded, bright yellow-green, with 12–26(50) rays; bracts absent and secondary bracts few and soon-falling. Fruit large 10–16 mm long, oval, flattened with slender dorsal wings and numerous resin canals. Throughout. *Ferula arrigonii* is similar but with rather narrowly *cylindrical inflorescences*; umbels with fewer than 20 rays. Coastal slopes. FR (Corsica), IT (Sardinia). *Ferula glauca* is slender with alternate (not whorled) upper branches, leaf lobes *glaucous beneath,* narrower sheaths and elliptic fruits. Cliffs and sandy roadsides. IT (incl. Sicily).

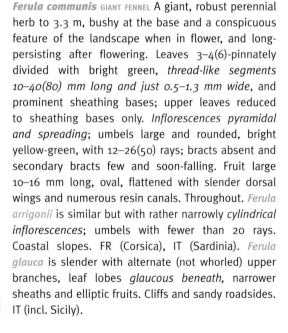

F. communis (fruit)

1. *Ammi majus*
2. *Angelica sylvestris*
3. *Pastinaca sativa*

F. tingitana (leaf)

F. tingitana (fruit)

B. Leaves with short, blunt, stiff lobes 5–10(14) mm long.

Ferula tingitana is similar to, and perhaps continuously variable with *F. communis* (levels of distinction between the species not clear), but generally with *shorter, narrowly wedge-shaped (not thread-like) leaf lobes 5–10(14) mm long*, and 1.3–6 mm wide, with or without down-turned margins. Umbels with 8–16(22) rays. Fruit 10–16 mm. Local, but under-recorded due to confusion. ES, PT. *Ferula loscosii* is smaller than the previous species, with leaves with *tiny segments just 0.5–2 mm long*. Umbels with 7–14 rays. Fruit 8–11 mm. Arid habitats. ES (south).

Opopanax

Robust perennials with 2-pinnately divided leaves and yellow flowers. Fruit oval, hairless and ribbed, sometimes shortly winged.

Opopanax chironium A tall perennial to 1.5(2) m superficially similar to *Ferula* in stature, though with rather shorter, broader and sparser inflorescences. Leaves long-stalked (18 cm) pinnately divided with *large, oval lobes* 12 x 10 cm, bristly hairy, withering during or soon after flowering. Flowers yellow, borne in rounded umbels with 7–10(20) rays. Fruit 8(10) mm, with a narrow, thick, white border. Rare and local, though frequent in southeast Italy. ES, FR, IT. *Opopanax hispidus* is similar but larger, to 3 m, with umbels with fewer (6–13) rays and fruits with a broad but thin white border. Strongly eastern. DZ, IT, TN.

Ferulago

Bushy perennial herbs with hairless stems, branched above and 2–4-pinnately divided leaves. Flowers yellow, borne in umbels on alternate branches. Fruits with narrow and expanded (wing-like) ridges.

Ferulago capillaris A bushy, light green perennial to 1.6 m with weakly grooved stems. Leaves triangular in outline, 2–4-pinnately divided with *sparse, short, linear, thread-like lobes* 5–30 mm. *Bracts and secondary bracts well-developed and conspicuous.* Flowers with yellow petals, the primary umbel with *many rays* (20–45). Fruit 10–19 mm, narrowly oval with well-developed lateral wings. South-central Iberian Peninsula. ES, PT. *Ferulago ternatifolia* is similar but with umbels with *fewer rays* (5–18). ES. *Ferulago brachyloba* has *zig-zagging stems* and umbels with very few, 5–9(11) rays. ES. *Ferulago nodosa* is characterised by its conspicuous, swollen stem nodes. A predominantly eastern Mediterranean species. IT (Sicily).

Elaeoselinum

Erect perennials with 3–4-pinnately divided leaves. Flowers with yellow petals. Fruits hairless and with prominent wings.

E. foetidum

Elaeoselinum foetidum A bright green, more or less hairless, rather fennel-like perennial to 2 m with leaves 3–4-pinnately divided with *triangular terminal lobes to 10 mm long*. Flowers with bright yellow petals, borne in umbels of 9–21(30) rays, 0–1 bracts, and numerous secondary bracts. Fruits 8–15 mm with *wide lateral wings to 3 mm that extend beyond the top of the fruit*. Dry sunny slopes and Garrigue. ES, MA, PT. *Elaeoselinum asclepium* is similar, but with feathery leaves with short, *linear, thread-like terminal segments* 1.5–4.5 mm long, arranged in a spiral. Fruit 6–12 mm, winged. ES (incl. Balearic Islands), IT (incl. Sicily), MA.

1. *Ferula communis* (inset top: fruits; bottom: leaf)
2. *Ferula tingitana* (inset top: leaf; bottom: flowers)
3. *Ferulago nodosa* (inset: stem node)
4. *Opopanax chironium* (fruits)

5. *Opopanax chironium*
6. *Elaeoselinum asclepium* (fruits)
7. *Elaeoselinum asclepium*

Thapsia DEADLY CARROT

Robust perennials with distinct basal rosettes of pinnately divided leaves. Flowers yellow, borne in rounded to flat umbels. Fruits broadly winged. High levels of variation have led to taxonomic confusion; it may not always be possible to identify species reliably in the field.

A. Leaves with *long, linear, untoothed segments.*

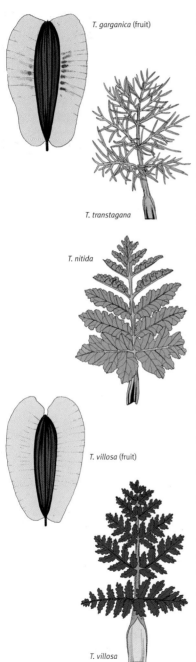

T. garganica (fruit)

T. transtagana

T. nitida

T. villosa (fruit)

T. villosa

Thapsia garganica A perennial herb to 1.4 m with a rosette of green, bushy leaves, triangular in outline, and 2-pinnately divided into linear to narrowly triangular, long, pointed lobes 50–80 mm long, somewhat flopping; hairless on the underside. Flowers yellow, borne in rounded umbels of 10–20(24) rays, carried on tall, erect, solitary stalks. *Fruits winged and large: 17–22 mm across;* wings 4–5(6) mm wide. DZ, ES (incl. Balearic Islands), IT (incl. Sardinia and Sicily), MA, TN. *Thapsia transtagana* is very similar, and often treated as a western form of *T. garganica,* differing mainly in having a hairy leaf undersurface and *larger fruits, 23–29 mm;* wings 3.5–7 mm wide. ES, MA, PT. *Thapsia gymnesica* is very similar to the previous species though with shorter, rather blunt terminal leaf segments just 4–11 mm long, and *smaller fruits to just 9–12 mm;* wings 1.5–2.5 mm. ES (Balearic Islands).

B. Leaves with narrowly oblong *toothed* segments, 1-pinnately divided.

Thapsia nitida (Syn. *T. maxima*) GREATER THAPSIA A perennial herb to 1.2 m with a rosette of *large, 1-pinnately divided triangular leaves with toothed lobes*; leaves rather shiny and dark green and purplish at the base. Flowers yellow, borne in rounded umbels of 10–20(24) rays carried on *very robust* and tall, erect, solitary stalks. Fruits winged, 9–13 mm; wings 2–3 mm. Common in the southwest in a range of habitats (and varying in form with the habitat). ES, MA, PT.

C. Leaves with narrowly-oblong *toothed* segments, 2–3-pinnately divided.

Thapsia villosa (Syn. *T. maxima*) HAIRY THAPSIA A variable, robust, dark green perennial to 1.9 m. Leaves hairy, flat, triangular, 12–40 cm long and *2–3-pinnately divided*, with whitish, sometimes slightly red-tinged sheathing bases; lobes toothed, and bristly hairy along the centre. Umbels yellow, borne on erect stems, with 9–29 rays, and very few bracts. Fruit 9–15 mm, elliptic with lateral wings 2–4(5) mm wide, and deeply notched

1. *Thapsia garganica* in fruit
2. *Thapsia nitida* (leaf)
3. *Thapsia villosa* (leaf)
4. *Thapsia nitida*
5. *Thapsia villosa*

at both ends. Var. *villosa* has terminal leaf segments 8–30 mm long. Various habitats, often alongside *T. nitida*. Var. *dissecta* has terminal leaf segments to just 3 mm long. Common and widespread in the west. ES, FR, MA, PT. *Thapsia minor* (Syn. *T. villosa* subsp. *minor*) LESSER THAPSIA is treated by some as a distinct species but differs only in having more *slender stalks with flat umbels of 4–10(12) rays*. Iberian Peninsula only. ES, PT.

Torilis HEDGE PARSLEY

Bristly annuals with adpressed hairs, solid stems, and pinnately divided leaves. Flowers white or purplish. Fruits with *prominent protuberances.*

Torilis nodosa KNOTTED HEDGE PARSLEY *A low to short, prostrate, rough-hairy annual* to 50 cm. Leaves 1–2-pinnately divided with deeply toothed segments. Flowers with pinkish-white petals, clustered on stalkless or short-stalked leaf-opposed umbels with 2–3 very short rays (<5 mm). Fruit egg-shaped, 2.5–3.5 mm long with *warts and straight bristles.* Common in open grassy areas and waste ground. Probably throughout.

Pseudorlaya

P. pumila

Small, prostrate, densely bristly annuals in maritime habitats. Flowers white or pinkish. Fruits with prominent bristles.

Pseudorlaya pumila PSEUDORLAYA A very *low to short, densely hairy,* rather fleshy annual 50 mm–20 cm, branched from the base. Leaves 2–3-pinnately divided with oval segments 2–5 mm. Umbels white to pale purple with 2–5(7) unequal rays of 8–12 flowers; petals more or less equal, some larger at the perimeter; bracts 2–5, linear. Fruit elliptic, 7.5–12 mm long, ridged and with hooked spines. Common in coastal habitats, especially on dunes. Throughout. *Pseudorlaya minuscula* is similar but smaller, and with *fruits 4.5– 7.5 mm long when mature.* Coastal dunes. ES, MA, PT.

Orlaya

Annuals with leaves 2–3-pinnately divided. Flowers with petals white or pink, deeply notched, often very unequal. Fruits oval or oblong with 5 *slender ribs* and 4 secondary ribs, *covered in spines.*

Orlaya grandiflora ORLAYA A branched or simple annual 10–50 cm, often hairy at the base. Leaves 2–3-pinnately divided into oval, toothed segments. Flowers white or pink, the outermost petals up to 8 x as long as the inner petals; umbels long-stalked, with 5–12 rays. Fruits 6–8 mm, egg-shaped with hooked bristles. A distinctive and common plant in Mediterranean grassland; absent from the far west and the islands. DZ, ES, FR, IT. *Orlaya daucoides* (Syn. *O. kochii*) is similar but with umbels with 2–4(5) rays. Throughout.

Daucus CARROT

Annual or biennials with leaves 2–3-pinnately divided. Umbels white, with pinnately lobed bracts; petals unequal, those of the outer flowers often larger. Fruit elliptic and spiny.

Daucus carota WILD CARROT A very variable, short to tall, hairy or hairless annual or biennial to 2.2 m with solid, often ridged stems. Leaves feathery with linear-lanceolate segments. Umbels with white flowers, many-rayed (9–130), often with a single purple flower in the centre, or purplish throughout; bracts pinnately lobed. Fruit oblong, 1.5–4 mm and shortly spiny; the spines not distinctly fused at the base; the umbel rays becoming conspicuously incurved when dry. Common in grassy places. Throughout. Subsp. *carota* grows to 1.1 m, with basal leaves 1–4-divided almost to the base (pinnatisect), and umbels up to 11 cm across when in bloom. Fruit 1.8–3.2 mm. Throughout. Subsp. *maximus* grows to 2 m, with basal leaves 1–3-pinnatisect, and large umbels, 12–23 cm across. Fruit 1.5–2.5 mm. Throughout. **Subsp.** *halophilus* (Syn. *D. halophilus*) is distinct, possibly warranting species rank, *short, robust, fleshy*, to 25 cm. Umbels 40 mm–12 cm across with 30–120 long rays; bracts shorter than the rays with 5–7 lanceolate lobes; secondary bracts 3-lobed at the tip; flowers white, pinkish at the centre of the umbel, with unequally large petals on the perimeter. Fruit small, 2–3.5 mm with yellowish spines joined at the base forming a crest. Rocky shores and sea-cliffs. Southwest Iberian Peninsula. PT.

Daucus muricatus is somewhat similar to *D. carota*, with bristly stems at the base and with umbels with markedly unequal rays; external petals 4–7 mm; fruits 5–8(10) mm, with slivery-white rows of long spines which are joined at the base. Throughout.

1. *Pseudorlaya minuscula*
2. *Daucus carota*
3. *Daucus carota* (fruiting head)
4. *Daucus carota* subsp. *halophilus*
5. *Orlaya grandiflora*

ARALIACEAE | IVY FAMILY

Woody climbers, trees and shrubs with alternate, simple, often lobed leaves. Flowers small, often borne in umbels and 5-parted; style(s) 1(5). Fruit a drupe or berry with 2–5 seeds.

Hedera IVY

Woody vines which climb by means of rootlets. Flowers borne in terminal umbels. Fruit berry-like. Species virtually identical and difficult to distinguish in the field.

Hedera helix IVY A vigorous evergreen climber or creeper with flexuous stems with adhesive roots. Leaves glossy, hairless and dark green, palmately lobed (>$^1/_2$ to the base), 13 cm across; leaves on flowering stems oval and unlobed. Flowers pale yellowish-geen and borne in terminal, dense spherical umbels. Berry globular and black when ripe, 7–8.3 mm across. Woods above sea level. Widespread but absent from arid areas and the far south. ES, FR, IT. *Hedera hibernica* (formerly treated as a subspecies of *H. helix*) differs very subtly in having pale yellow-brown (not whitish) hairs on the young leaves lying flat and all in 1 plane (not erect), leaves lobed <$^1/_2$ to the base, and with a more sweet and resinous-smelling sap. A predominantly Atlantic species, rare and patchy in Mediterranean Europe. ES. *Hedera maderensis* has young leaves with *reddish hairs with numerous (up to 24 rather than 13) branches*. Far southwest only. ES, PT.

PITTOSPORACEAE

Trees, shrubs and lianes with simple, alternate leaves. Flowers with 5 sepals, petals and stamens; style 1. Fruit a 2–4-parted capsule.

Pittosporum

Trees and shrubs with simple (rarely lobed) leaves. Flowers with 5 sepals and petals; style 1. Fruit a woody capsule with numerous seeds.

Pittosporum tobira A dense, dark green, evergreen shrub 2–6 m. Leaves 35 mm–10 cm, alternate, oval, deep green and leathery with inrolled margins. Flowers cream, with 5 petals 10–13 mm, fragrant, borne in dense, flat-topped clusters. Fruit 15–20 mm, splitting 3 ways. Widely cultivated in towns and on roadsides. Throughout. *Pittosporum undulatum* has broad, mid-green leaves 50 mm–15 cm that are narrowed at the base, with *undulate* (*not inrolled*) margins. Fruit 10–14 mm, splitting 2 ways. Occasionally planted in the region.

ADOXACEAE

A small family of perennials with opposite, toothed leaves. Inflorescences cymose; stamens 4–5; style 0 or 1. Fruit a 1–several seeded drupe (rarely an achene). This family now includes species that were previously described under the Sambucaceae or Caprifoliaceae.

Viburnum

Deciduous or evergreen shrubs with simple leaves. Flowers numerous, in compound clusters; stamens 5; style 0. Fruit a succulent 1-seeded drupe.

Viburnum tinus An evergreen shrub to 7 m with more or less hairless stems, and dark green, oval, short-stalked leaves which are sparsely hairy beneath. Flowers white, pinkish in bud, borne in dense, flattened heads to 90 mm across. Fruit a berry to 7 mm, *blue-black* when ripe. Locally common woods and hill scrub. Throughout.

Sambucus ELDER

Deciduous shrubs or perennials with pinnately divided leaves. Flowers borne in compound clusters; stamens 5. Fruit succulent, with 3–5 seeds.

Sambucus ebulus EUROPEAN DWARF ELDER A robust, vigorous, herbaceous perennial to 1.5(2) m with erect, grooved stems; unpleasant-smelling. Leaves pinnately divided with 7–13 narrow leaflets and conspicuous oval stipules at the base. Flowers borne in flat-topped inflorescences with 3 main rays; pinkish-white, with purple stamens. Fruit a black berry. Streamsides and field boundaries inland; frequent in southern France, scattered elsewhere. ES, FR, IT.

---

1. *Hedera helix*
2. *Pittosporum tobira*
3. *Viburnum tinus*
4. *Sambucus ebulus*

CAPRIFOLIACEAE | HONEYSUCKLE FAMILY

Woody shrubs and climbers with opposite, simple leaves. Flowers solitary, paired or in showy panicles on shrubs; calyx small; corolla regular or 2-lipped, fused below to form a tube; stamens mostly 5; ovary inferior. Fruit a berry or nutlet.

Lonicera HONEYSUCKLE

Deciduous or evergreen shrubs or climbers with simple, sometimes lobed leaves. Flowers stalkless, with a zygomorphic, 5-lobed corolla; stamens 5; style long. Fruit a several-seeded berry.

L. implexa

L. etrusca

Lonicera implexa HONEYSUCKLE An evergreen, much-branched, shrubby and woody climber to 3 m with bluish, hairless shoots. Leaves persistent, entire, 15–40 mm long, oval and pointed, stalkless and cone-shaped on the upper parts of the stems, hairless, dark green and shiny above, and bluish beneath. Inflorescence *stemless* with 5–6(8) flowers; corolla 32–40 mm, yellowish-white tinged red and fragrant, particularly at night; *style woolly, stamens slightly exserted*. Fruit a red berry 6–7.5 mm. Common on maquis, garrigue and among shrubs in thickets. Throughout. *Lonicera splendida* is a similar evergreen climber with *stalkless clusters of white-yellow flowers* in many-flowered clusters, and glandular corolla 35–45 mm; *styles hairless; stamens clearly exserted*. Mountains in southern Spain. ES. *Lonicera etrusca* is similar to the above species, but deciduous with leathery leaves, usually hairy beneath, and with inflorescences with clusters of flowers in groups of 3 *on stalks 30–40 mm long*. Fruit 5.2–6.2 mm, ovoid and red. Limestone woods and thickets. Throughout. *Lonicera biflora* is deciduous with oval leaves which are hairless and dark green above and finely hairy below. *Flowers paired in the axils*, terminally crowded; corolla 24–28 mm. Coastal. ES, IT (Sicily - naturalised), MA.

Scabiosa SCABIOUS

Annual or perennial herbs with simple or pinnately divided leaves, the lowermost often in a rosette. Flower-heads flat or domed and long-stalked; outer flowers longer than the inner; calyx with 5 long bristles; corolla 5-lobed.

Scabiosa atropurpurea MOURNFUL WIDOW A medium, hairy, bushy annual or biennial to 70 cm with erect, branched stems. Leaves oblong, untoothed and long-stalked below, pinnately lobed above. Flower-heads lilac, 15–35 mm across, becoming oblong in fruit, the outer florets 2 x the size as the centrals; involucral bracts *not* longer than the florets. Fruits with long calyx teeth. Common in a range of dry habitats, particularly on sandy, maritime waste ground. Throughout.

1. *Lonicera implexa*   3. *Scabiosa atropurpurea*
2. *Lonicera implexa* fruits   4. *Lomelosia stellata*

Lomelosia

Very similar to *Scabiosa* (and previously described in the same genus) but with a *pitted epicalyx*.

L. stellata

Lomelosia stellata (Syn. *Scabiosa stellata*) A shortly hairy annual to 40(70) cm with scattered longer hairs, erect, simple or branched. Leaves 10 mm–11 cm x 4–45 mm, oblong-elliptic, at least some leaves *toothed to pinnately lobed* with 4–6 pairs of linear segments, at least above. Flowers borne in distinctive heads 10–20 mm across (20–30 mm in fruit), conspicuous for the *cup-like and papery,* unlobed (involucral) bracts in which sit long-pointed, star-like (receptacular) bracts surrounding each floret. Local in disturbed and cultivated or developed areas near the coast. ES, FR, IT (Sardinia), MA, PT. *Lomelosia cretica* is similar but *only has entire leaves* 3–20 mm wide. ES (Balearic Islands), IT (incl. Sicily). *Lomelosia graminifolia* is similar to *L. cretica* but has *grass-like leaves* to just 1.5–3.5 mm wide. ES, MA.

Lomelosia divaricata (Syn. *Scabiosa sicula*) A short, erect or sprawling, hairy annual 10–50(70) cm. Leaves spoon-shaped to oval and toothed or not, the upper pinnately divided with narrow segments. Flower-heads mauve or reddish, 20 mm in flower, 7–19 mm in fruit, becoming spherical in fruit; *involucral bracts much exceeding the florets*. Dry, stony habitats. DZ, ES, IT (Sicily), MA, TN.

Pycnocomon

P. rutifolium

Perennials with entire to divided leaves. Flower-heads hemispherical with 7–12 bracts fused at the base; calyx with 8–14 bristles; corolla 4-lobed. Now established to belong within the *Lomelosia* group.

Pycnocomon rutifolium PYCNOCOMON Similar to *Scabiosa* species when in flower; a robust, hairy or hairless perennial to 1 m with erect, branched stems. Leaves slightly fleshy, 1–2-pinnately lobed, reduced above, sometimes unlobed below. Flower-heads *white or yellow*, 12–28 mm across, the outer florets slightly larger than the central; bracts below the flower-heads 7–8 fused together in the lower part forming a cup. Uncommon but widespread, on coastal sands. DZ, ES, IT, MA, PT. *Pycnocomon intermedium* is very similar but has pink to violet flower-heads 13–30 mm. Southwest Iberian Peninsula only. ES, PT.

Dipsacus

Robust, prickly biennials and perennials with entire or pinnately divided leaves. Flower-heads very dense; corolla 4-lobed; stamens 4; style 1.

Dipsacus fullonum WILD TEASEL A familiar, tall biennial to 2 m with robust, prickly, angled stems. Basal leaves forming a large rosette; oblong-elliptic, toothed or not *and prickly*; stem leaves in pairs and fused together on the stem forming a water-collecting cup. Flower-heads 40–80 mm, pink-purple, oblong-cylindrical with long, spiny bracts at the base. Local in damp places. Throughout. *Dipsacus comosus* is similar, but with all leaves pinnately divided or lobed almost to the midrib. ES, PT.

1

2

Knautia

Perennials with hairy stems. Flower-heads hemispherical with 1–2 rows of free bracts at the base (none below each flower); calyx with 8 long bristles; outer florets usually longer than the inner; corolla 4-lobed; stamens 4; style 1.

A. Plant an annual.

Knautia integrifolia A hairy annual with stems to 60 cm (1 m) arising from a basal leaf rosette; leaves deeply cut into oval segments, stem leaves divided into linear segments, or narrowly lance-shaped. Flower-heads 15–32 mm across, blue-violet, almost flat with spreading outer florets; calyx usually with 12–24 inconspicuous teeth. Fruits with numerous white hairs. Common in a range of habitats. Throughout.

B. Plant typically a perennial.

Knautia arvensis FIELD SCABIOUS A stout biennial or perennial with rough-hairy stems to 75 cm (1.2 m). Basal leaves roughly spoon-shaped, rough-hairy and undivided, stem leaves pinnately divided into segments with an oval-lanceolate end-segment. Flower-heads 25–40(50) mm across, blue-violet and hemispherical, borne on long hairy stalks; calyx with 8–14 teeth. Open woods and tracks, usually inland and on higher ground; absent from most islands. ES, FR, IT.

Cephalaria

Annual to perennial herbs or shrubs with leafy stems and entire to deeply lobed leaves. Flowers cosexual, borne in dense rounded or hemispherical heads; corolla 4-lobed; stamens 4; style 1. Fruit a capsule.

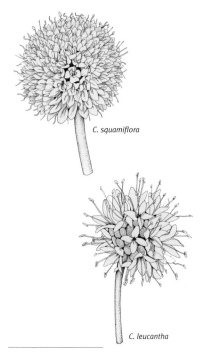

C. squamiflora

C. leucantha

A. Plant a small shrub with *leathery leaves*.

Cephalaria squamiflora A *shrub* to 90 cm with oval-lanceolate to elliptic leaves to 18 cm, *untoothed to shallowly lobed, leathery* and tapered at the base. Flower-heads 20–32 mm across, yellow-white with prominently exserted stamens; corolla 8–12 mm; bracts oval and hairy. Limestone crevices. ES (Balearic Islands), FR (Corsica), IT (Sardinia).

B. Plant a woody-based or herbaceous perennial, leaves *not leathery*.

Cephalaria leucantha A tall perennial herb to 1.3 m, woody at the base, with erect, branched stems. Leaves to 20 cm, *pinnately lobed* into narrow segments, hairy or hairless. Flower-heads 14–30 mm across, white-yellow with prominently exserted stamens; corolla 7–9(13) mm; bracts oval and hairy. Dry, stony habitats. Absent from the Balearic Islands. ES, FR, IT, MA. *Cephalaria syriaca* is similar but with blue or lilac flower-heads just 3–9 mm across; corolla 6–9 mm. Native to the eastern Mediterranean, occasionally naturalised in the region.

---

1. *Dipsacus fullonum*
2. *Knautia arvensis*

C. joppensis

Cephalaria transsylvanica A tall, rather hairy perennial to 1.5 m with erect, branched stems. Leaves to 20 cm, deeply pinnately lobed into elliptical to linear segments, toothed or not. Flower-heads 5–25 mm across, blue-lilac (or white); corolla 7–11 mm; bracts lanceolate and pointed, with ciliate margins. Fields and fallow land, absent from most islands. FR, IT. *Cephalaria joppensis* is a similar perennial to 2 m with leaves to 18 cm and whitish, bluish or pink flower-heads 6–18 mm across; corolla 5–9 mm; bracts pointed (mucronate). IT (south + Sicily).

Centranthus

Rhizomatous annual or perennial herbs with entire to deeply pinnately lobed, stalked or unstalked leaves. Flowers small, borne in dense panicles, pink or white; corolla tubular; stamen 1; stigma 1.

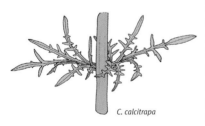

C. calcitrapa

Centranthus calcitrapa A small, hairless annual with simple or branched stems to 30 cm. Leaves rounded to oval, green flushed with purple, toothed or not, stalked below and stalkless above. Flowers *small,* with corolla tube 1–4.6 mm, white or pink, long, and short-spurred, borne in small terminal clusters. Fruit a feathery, persistent calyx. Common on sand dunes, low garrigue and bare waste places. Throughout.

Centranthus ruber RED VALERIAN A familiar, tufted, medium, and somewhat fleshy and waxy perennial to 80 cm. Leaves bluish, oval, pointed or blunt, clasping the stem above. Flowers with tube 5–12 mm, usually dark pink-red (sometimes white), spurred, borne in large, slightly fragrant, showy panicles. Fruit 1-seeded with a feathery, persistent calyx. Cultivated as an ornamental, and naturalised throughout.

Fedia

Small annual herbs with simple, stalked or unstalked leaves. Flowers small, pink-purple, zygomorphic, borne in small, dense clusters on opposite-spreading stems.

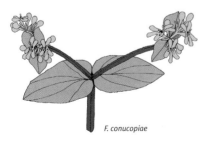

F. conucopiae

Fedia cornucopiae FEDIA A low to short, much-branched, hairless and slightly fleshy annual to 30(70) cm. Leaves spatula-shaped to elliptic, stalked and untoothed below; stalkless and toothed above, in opposite pairs. Flowers purplish-pink, 6.4–20 mm long, borne in stalkless clusters on *opposite-spreading stems*; corolla slender and 5-lobed, slightly 2-lipped; stamens 2–3, with 2 *fused together*. Very common in grassy places and fallow land. ES, MA, PT. *Fedia graciliflora* is similar but with corolla lobes flush with the tube (not at an oblique angle). An island endemic of Mallorca. ES (Balearic Islands).

1. *Centranthus calcitrapa*
2. *Centranthus ruber*
3. *Fedia cornucopiae*
4. *Fedia cornucopiae* (flowers)

Valerianella CORNSALAD

Small annuals with symmetrical branches. Flowers borne in the branch axils and in terminal clusters; stamens and stigmas 3. Species all rather similar; ripe fruits and calyx necessary for identification.

A. Calyx minute and inconspicuous.

Valerianella locusta COMMON CORNSALAD A small, variable, hairless annual to 15(40) cm with symmetrical, spreading branches, spoon-shaped, blunt, sometimes toothed leaves to 70 mm and dense terminal heads of pale lilac flowers; corolla 5-lobed; *calyx very small and 1-toothed.* Fruit hairless, 1.8–2.5 mm (1–1.5 mm wide), 2 x as thick as wide, scarcely longer than thick, and shallowly grooved. Locally frequent in grassy habitats and along tracks. Throughout.

B. Calyx conspicuous and often persistent.

Valerianella eriocarpa HAIRY-FRUITED CORNSALAD A small annual to 15(40) cm similar in appearance to *V. locusta* but with denser, compact flower clusters *and distinctive calyx nearly as broad as the fruit, strongly net-veined and deeply (2)5–6-toothed;* corolla 1–1.6 mm. Fruit 0.9–2 mm, with rigid hairs. Throughout. *Valerianella dentata* N
ARROW-FRUITED CORNSALAD is similar to *V. eriocarpa* but has laxer flower clusters and calyx *just ½ as broad as the fruit and scarcely veined.* Fruit 1.5–2 mm. ES, FR, IT, MA.

Valerianella discoidea A small annual similar to *V. locusta* with opposite-spreading stems and oval to narrowly spatula-shaped leaves which are blunt and toothed or untoothed. Flowers with corolla 1.5–2 mm, whitish-blue or mauve; *calyx well-developed, with many (6–12) spine-tipped lobes, hairy outside and in*. Fruit 1.5–2.1 mm. Fallow land, pastures and maquis clearings; common. Throughout. *Valerianella vesicaria* is distinguished by its *spherical inflated, bladder-like fruits*. IT (incl. Sardinia and Sicily). *Valerianella echinata* SPINY CORNSALAD is distinguished by its distinctive *fleshy, curved, spiny calyx lobes which are even visible in bud*, and thickened fruiting stalks. Corolla 2–2.5 mm. Fruit 7–9 mm. Throughout.

Valeriana VALERIAN

Perennials with small, funnel-shaped flowers with 5 lobes; stamens and stigmas 3. The most common species are described here, other species occur at higher altitudes.

A. Basal leaves pinnately divided or lobed.

Valeriana officinalis COMMON VALERIAN A tall, erect perennial to 1.5(2.3) m with single or clustered stems. Leaves pinnately divided, to 14 cm, borne in opposite pairs; the lowermost stalked, the upper almost stalkless; leaflets toothed. Flowers borne in terminal heads, with several lower clusters, corolla 4–7.8 mm, pink and funnel-shaped, 5-lobed with 3 protruding stamens. Fruit 2.6–4 mm. Aquatic habitats; absent from hot, dry areas and the far south and west. ES, FR, IT.

B. Basal leaves undivided.

Valeriana tuberosa TUBEROUS VALERIAN A rather fleshy, green, purple-flushed perennial to 70 cm with undivided, oblong basal leaves (sometimes shallowly lobed) and pinnately divided stem leaves with deeply dissected, linear lobes. Flowers borne in dense, *rounded* clusters; corolla 5–7.2 mm, pale pink-white. Fruits 2.7–3.9 mm, with silvery hairs. Frequent on maquis and scrub. Throughout.

Valeriana tripteris THREE-LEAVED VALERIAN A bushy, woody-based perennial to 50 cm with many grey-green, long-stalked, variable course-toothed leaves at the base; *middle stem-leaves trifoliate*. Flowers with corolla 4.7–8 mm, pinkish-white, borne in many-flowered, flat-topped clusters. Fruit 4–5.2 mm. Shady woods at high altitude and streamsides; local. ES, FR, IT.

CAMPANULACEAE | BELLFLOWER FAMILY

Annual or perennial herbs which exude a latex when cut, with alternate leaves. Flowers often large and showy, clustered or solitary; corolla bell-shaped, lobed, often blue or purple; stamens 5, fused or free; style 1. Fruit a capsule.

Campanula BELLFLOWER

Biennials to perennials. Flowers with 5-lobed calyx fused to the ovary, and funnel- or bell-shaped corolla. Fruit a capsule or berry. Many species mountain-dwelling.

A. Flowers (5)10–25(35) mm long.

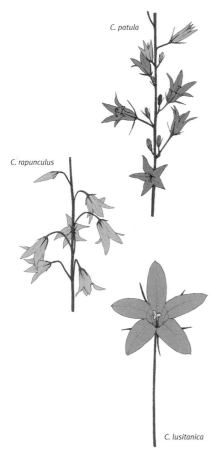

C. patula

C. rapunculus

C. lusitanica

Campanula patula SPREADING BELLFLOWER A medium, hairless or hairy biennial or perennial with erect or ascending, slender stems to 60 cm. Leaves gradually narrowed into their stalks. *Inflorescences wide-spreading*, with bell-shaped but open flowers with lobes as long as their tube, pale to purplish-blue (rarely white), 15–25(35) mm long, held on erect, slender stems with a bract in the middle; buds more or less nodding. Capsules 6–11 mm, erect with 10 veins. Grassy places on higher ground; absent from most islands, rare in the southwest. ES, FR, IT. *Campanula rapunculus* RAMPION is similar to *C. patula* and easily confused with this species, but with *simple, pyramidal inflorescences* with smaller, always pale blue-mauve flowers 10–22 mm, held on ascending to erect short stalks which have a bract at the base. Capsule 7–20 mm. Roadsides, meadows, cultivated land and other grassy places. Throughout.

Campanula lusitanica An annual with long spreading, virtually hairless stems to 40 cm. Leaves 21–40 mm, *round-oval* and slightly toothed above. Flowers borne on *long, slender stalks; corolla bright violet* or pale blue, 8–15 mm long, usually borne rather numerously in branched inflorescences. Capsules 2.8–10 mm, erect. Local on sandy soils. ES, MA, PT.

Campanula dichotoma A short, bristly annual to 15(35) cm tall with erect, regularly branched stems. Leaves ovate to elliptic and slightly toothed or untoothed, unstalked and ¹/₂-clasping the stem above, softly hairy. Flowers blue-lilac, 5–23 mm long, the corolla tube longer than its lobes, borne solitary in the leaf axils, forming a pyramidal inflorescence. Capsule 4.1–5.2 mm. Stony places on higher ground. ES (incl. Balearic Islands), IT (Sicily).

1. *Valerianella discoidea* (inset: fruiting calyx)
2. *Valeriana tuberosa*
3. *Campanula patula*

B. Flowers large and open, (25)30–50 mm long.

Campanula persicifolia NARROW-LEAVED BELLFLOWER A medium to tall, hairless perennial to 80 cm with non-flowering rosettes, and erect, unbranched stems with shiny, oval, blunt-toothed leaves below and toothed, linear-lanceolate leaves above. Flowers blue-violet, shallowly divided into broad, triangular lobes, *large and open, (25)30–50 mm.* Fruit 11–16 mm. Various habitats, and cultivated for ornament; absent from much of the south, west and Sardinia. ES, FR, IT.

C. Flowers very small, 2–5 mm long.

Campanula erinus ANNUAL BELLFLOWER A slender, rough-haired, spreading, branched annual to 35 cm with oval- to wedge-shaped leaves 10–20 mm, and *tiny, pale blue, reddish or white flowers with corolla just 2–5 mm,* borne in the axils of the branches; flowers very short-stalked. Capsule 2.5–3.6 mm. Throughout.

Asyneuma

Similar to *Campanula*, flowers borne in elongated spikes; *corolla lobes linear.*

Asyneuma limonifolium A slender, erect, often unbranched perennial 15–70 cm tall with mostly basal lanceolate, long-stalked leaves, toothed or not. Flowers blue-lilac, borne in elongated spikes, borne in groups of 1–3 in the axils of tiny triangular bracts, virtually stalkless; *corolla lobes linear.* Mountain slopes and garrigue, occasionally at sea level; strongly eastern. IT.

Jasione SHEEP'S-BIT

Hairless or hairy, annual or perennial herbs, usually with narrow leaves. Flowers lilac, blue or white, borne in small, compact, terminal, hemispherical heads. Capsule splitting by 2 apical short valves.

J. montana

J. corymbosa

A. Plant an annual or biennial.

Jasione montana SHEEP'S-BIT A low, *woolly-hairy annual* (or biennial) 5 –50 cm, erect or ascending, *leafless beneath the flowers,* simple or branched below. Leaves linear to oblong, undulate and weakly hairy at the margin, short-stalked. Involucral bracts entire, and *shorter* than the flowers. Flowers small, borne in *dense, blue heads* 5–3.5 mm across; corolla with 5 narrow lobes; stamens 5; stigmas 2, borne on styles which are hairy in the upper part. Dry, bare and rocky ground; local. ES, FR, IT, MA, PT. *Jasione corymbosa* is very similar (considered by some not to be distinct), differing in being sparsely hairy to hairless and in having stems that are leafy up to the inflorescence, the *leaf bases running down the stem* (decurrent) and involucral bracts *as long as or longer* than the flowers. Dry, sandy places. Records from south Portugal may refer to the previous species. ES, PT.

B. Plant a perennial.

Jasione crispa A more or less hairy *perennial with a stout woody stock* and short non-flowering shoots and longer erect or ascending flowering stems to 15(30) cm. Leaves leathery with thickened margins, linear-oblong to narrowly lanceolate, flat and entire or slightly toothed. Flowers borne in dense heads 10–25 mm across; corolla blue, 5-lobed to the base, with 5 stamens and 2 stigmas; calyx 5-toothed. Generally on higher ground in woods, screes and rocky habitats. ES, FR, MA, PT.

Solenopsis

Delicate annual or perennial herbs with simple, thin leaves. Flowers bell-shaped, zygomorphic, *2-lipped,* blue or lilac. Capsule 2-parted.

S. laurentia

Solenopsis laurentia (Syn. *Laurentia gasparrinii*) A small, slender, more or less hairless annual or perennial 60 mm–20 cm. Leaves 12–20 mm oblong to spatula-like, entire, often forming a basal rosette. Flowers solitary; small, 3–5(6) mm borne on stems with 1 or 2 secondary bracts; calyx 5-parted; corolla blue, lilac or white. Capsule 3.5 mm. Widespread but uncommon; damp habitats. Throughout. *Solenopsis balearica* is similar but a *perennial* with a densely leafy base. An island endemic. ES (Balearic Islands). **Solenopsis bicolor** is similar to both the previous species but with rather densely arranged, *toothed leaves.* A rare North African endemic. DZ, TN.

1. *Campanula rapunculus*
2. *Asyneuma limonifolium*
3. *Asyneuma limonifolium*
4. *Jasione crispa*
5. *Solenopsis bicolor*
   PHOTOGRAPH: SERGE D. MULLER
6. *Solenopsis bicolor*
   PHOTOGRAPH: SERGE D. MULLER

Legousia

Annual herbs. Flowers borne in panicles or racemes; calyx 5-lobed; corolla flattish, 5-lobed, fused at the base; stamens 5; style 1, stigmas 3. Fruit a cylindrical, 3-valved capsule.

Legousia speculum-veneris LARGE VENUS'S LOOKING GLASS A hairy, ascending to erect annual 10–25 cm high *much-branched from the base*. Leaves to 20 mm, alternate, oblong and slightly wavy, unstalked or scarcely stalked below. Flowers violet with a whitish centre, 15–23 mm across, star-like with 5 blunt lobes 8–9 mm, borne in lax panicles; calyx teeth 6–8 mm (almost as long as petal lobes); *filaments hairless*. Capsule 10–14 mm. Agricultural land and stony habitats. Throughout.

Legousia hybrida VENUS'S LOOKING GLASS An annual similar to the previous species to 30 cm but with *markedly wavy leaves* to 30 mm, and few, pink, violet or white flowers borne in terminal clusters with petal lobes just 2.5 mm, about ¹/₂ the length of the calyx lobes (5–6 mm); *calyx teeth erect in fruit*. Capsule 15–25 mm, constricted above. Similar habitats. Throughout.

Legousia falcata SPICATE VENUS'S LOOKING GLASS An erect to spreading annual, 15–50 cm, with hairless or bristly stems. Leaves 20–40 mm, oval, slightly wavy-margined and usually short-stalked below. Flowers violet to blue-purple, solitary or paired with lobes 7 mm, borne in a *spike-like inflorescence*; calyx teeth linear and curved, 4–8 mm (equal or $^1/_2$ length of tube), spreading, often exceeding the petals. Capsule 10–22 mm, narrowed at the top. Maquis and stony pastures. Throughout. *Legousia scabra* is a similar annual to 90 cm with oval to elliptic leaves and straight calyx teeth 8–10 mm ($^1/_2$–$^1/_3$ length of tube) and corolla lobes 6–7(12) (almost equal). Capsule 10–22 mm. ES, FR (south + Corsica), MA, PT.

Trachelium THROATWORT

Perennials with alternate leaves. Flowers borne in dense clusters; corolla with slender tube and 5 spreading lobes; stigmas 2–3. Fruit a capsule opening by 2–3 basal pores.

Trachelium caeruleum THROATWORT A tall, almost hairless perennial 30–70 cm, woody below with oval to lanceolate, toothed leaves 40 mm–10 cm, often with hairy margins. Flowers violet-purple (or white), 6–9 mm, borne in flat-topped clusters; corolla with 5 spreading lobes to 2 mm, exceeded by their tube (5–7 mm). Capsule 2 mm, pear-shaped. Damp, shady and rocky habitats. ES (incl. Balearic Islands), IT (incl. Sicily), PT.

Wahlenbergia

Herbaceous annuals or perennials with simple, alternate leaves. Flowers borne in lax cymes or panicles; corolla regular; stamens 3–5; style 1. Capsule opening by 3 apical valves.

Wahlenbergia lobelioides An erect, much-branched annual 20–30 cm. Leaves 20–70 mm, oblong, toothed or untoothed, tapered at the base into a winged stalk. Flowers solitary, pale blue, pink or white, funnel-shaped, 4–5 mm long with 4(5) lobes 2–2.5 mm, borne in spreading panicles; stamens 4(5). Capsule 5–12 mm, erect with 3–5 apical pores. Local, in various dry, sandy or rocky habitats. Subsp. *nutabunda* is the form in the region. DZ, ES, IT (incl. Sardinia and Sicily), MA, TN.

MENYANTHACEAE | BOGBEAN FAMILY

Aquatic or marsh-dwelling herbs with creeping stems bearing alternate lower leaves and floating or emergent upper leaves. Flowers with *5 fringed petals*; stamens 5; style 1. Fruit a capsule.

Nymphoides

Leaves simple and rounded, floating on the water surface. Flowers yellow with fringed petals, and long-stalked. Fruit an irregularly-splitting capsule.

Nymphoides peltata FRINGED WATER-LILY An aquatic perennial with creeping stems to 1.5 m, with floating leaves; resembling a water-lily in habit, but with very different flowers; leaves round to kidney-shaped, 30–90 mm across. Flowers solitary or few, bright yellow, 30–40 mm across with 5, distinctly *fringed lobes*. Local in the region and restricted to still and sluggish water. DZ, ES, FR, IT, MA.

---

1. *Nymphoides peltata*
2. *Bellis annua*
3. *Bellis sylvestris*
4. *Bellis prostrata* PHOTOGRAPH: SERGE D. MULLER
5. *Bellis perennis*

ASTERACEAE | DAISY FAMILY

Arguably the largest family of flowering plants (see also Orchidaceae). Herbs or perennials (sometimes shrubs) with alternate, opposite or rosette leaves, simple or compound. Flower-heads typically with an involucre of closely overlapping bracts around the base; flowers (florets) borne in congested heads (capitula); those in the centre (disk florets) often distinct from the edge (ray florets); stamens 5, fused around the style; ovary inferior. Fruit an achene, often (but not exclusively) with an appendage of bristles, hairs or scales (pappus).

Floral arrangement of an *Asteraceae* flower-head (capitulum)

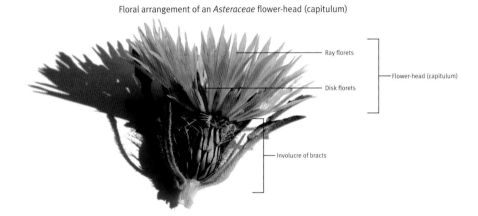

Ray florets

Flower-head (capitulum)

Disk florets

Involucre of bracts

Bellis DAISY

Annuals or perennials with basal leaf rosettes; leaves simple. Flower-heads solitary on long stalks; ray florets white to pink; disk flowers yellow. Pappus absent.

A. Stems branched and leafy below.

Bellis annua ANNUAL DAISY A low, normally *soft-bristly hairy annual* to 10 cm with short, erect, *leafy stems* (at least below). Leaves spatula-shaped, toothed or not, stalked below, *not borne in distinct rosettes*. Flower-heads white with a yellow disk, to 15 mm across, the rays sometimes tinged purple. Common in damp, grassy as well as dry open habitats. Throughout.

Bellis prostrata A low, *hairless* annual similar to *B. annua* but with *prostrate stems from which arise numerous large, long-stalked flower-heads*. Very rare, restricted to coastal wetlands of northeast Algeria, coastal reliefs in north Tunisia, and a single isolated population in Morocco. DZ, MA, TN.

B. Stems unbranched and leafless.

Bellis perennis DAISY A perennial to 12(20) cm with a *dense basal rosette*, leaves *abruptly* tapered at the base into a petiole, toothed. Flower-heads 12–25 mm across, borne on *unbranched, leafless stems*. Common on lawns, roadsides and other grassy habitats. Throughout. *Bellis sylvestris* SOUTHERN DAISY is similar to *B. perennis* but with *3-veined* leaves *gradually* tapered at the base into a petiole. Flower-heads *large*, to 40 mm across, borne on long, stout stalks to 30 cm; ray florets white but sometimes tinged purple on both sides. Achene compressed and bristly. Woods, thickets and shady roadsides. Throughout.

Filago CUDWEED

Downy or woolly annuals with alternate, untoothed leaves. Flower-heads inconspicuous, often numerous (8–40) per cluster. Pappus of simple hairs or absent.

A. Plant erect, with flower-heads (15)20–40 in rounded clusters *not exceeded by their leaves.*

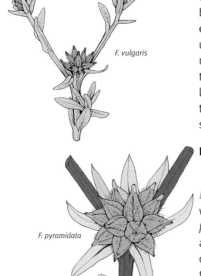

F. vulgaris

Filago vulgaris COMMON CUDWEED An erect, white-woolly annual to 30(40) cm, branched or unbranched below, always branched above in 2–3 forks. Leaves erect, lanceolate, to 20 mm long, wavy-edged and untoothed; widest in the basal $^1/_2$. Flower-heads 10–20 mm, borne as dense *rounded* clusters of (15)20–40 in the branch axils and terminally, *not* exceeded by the leaves immediately below; inner flower bracts with a transparent bristly tip. Common on bare disturbed, sandy ground. Throughout.

B. Plant erect, with flower-heads (5)10–20(25) per cluster, *exceeded by their leaves.*

Filago pyramidata BROAD-LEAVED CUDWEED An erect, white-woolly annual always branched and spreading *from the base*, often almost prostrate. Leaves oval and bristle-tipped. Flower-heads 10–20 mm, borne in dense clusters of (5)10–20(25) in the branch axils and terminally, *exceeded* by 2–4(5) leaves immediately below; outer flower bracts *with yellow points*, curving inwards in fruit. Open fallow and sandy waste places, and sandy cliff-tops; common. Throughout.

F. pyramidata

1. *Filago vulgaris*
2. *Filago pyramidata*

1

2

Filago lutescens RED-TIPPED CUDWEED An erect, often yellowish-woolly annual with irregularly branched *stems with yellow (not white) hairs,* and broader, not wavy, bristle-tipped leaves. Flower-heads borne in clusters of (5)10–20(25), over-topped by (0)1–2 leaves; *flower bracts yellowish*, with prominent, erect, *red-brown or purplish* bristle-tips. Local in sandy habitats; absent from most islands. ES, FR, IT, PT.

C. Plant erect, with flower-heads 3–8 per cluster, exceeded by their leaves.

Filago fuscescens An erect annual similar to *F. lutescens* but with flower-heads in groups of 3–8, with *reddish-brown flower bracts*. ES, MA.

D. Plant short-stalked or prostrate, often rosette or star-like (many species previously described under *Evax*).

Filago congesta CONGESTED CUDWEED A *low white-grey woolly annual with prostrate or ascending stems* (similar to *Evax*) with lanceolate to spatula-shaped leaves to 16 mm long. Flower-heads yellowish, borne in clusters of 3–6, with hairy, fine-pointed bracts. Sandy and rocky habitats. ES, FR, IT. *Filago mareotica* is similar, though often with short, erect stalks, simple below with lax, spreading branches above, each with a *solitary terminal, yellowish flower-head* 3–4 mm. Saline, bare habitats; rare. ES, MA, TN.

F. pygmaea

F. lusitanica

Filago pygmaea (Syn. *Evax pygmaea*) A very small, grey-felted annual to 40 mm high, branched at the base. Leaves oblong, *narrow* and blunt, to 15 mm long and 5 mm wide, and forming a rosette; upper leaves 2–3 x longer than flower-heads; all leaves without distinct stalks. Flower-heads borne in very compact clusters to 35 mm across; brownish-yellow. Fairly common on dry, bare, and stony places. Throughout. *Filago lusitanica* (Syn. *Evax lusitanica*) is very similar to *E. pygmaea*, but *virtually stemless*, with *broader* rosette leaves 12–25 mm long and 4–9 mm wide; greyish-green and rounded at the tip. ES, MA, PT. *Filago asterisciflora* (Syn. *Evax asterisciflora*) is similar to *F. pygmaea* but with well-developed stems (to 13 cm) and rigid, star-like leaf arrangements; leaves 15–40 mm long and 3–7 mm wide, exceeding the flower-heads x 4. Flower-heads numerous, borne in clusters 12–28 mm wide. Local and absent from many areas. ES, FR, IT, MA.

Logfia CUDWEED

Small woolly annuals. Previously treated either as distinct or under *Filago,* but recently resurrected based on DNA analysis; characterised by flower-heads solitary or in small clusters and *achenes always hairless.*

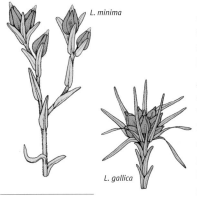

L. minima

L. gallica

Logfia minima (Syn. *Filago minima*) SMALL CUDWEED A low to short, grey silkily-haired annual to 25 cm. Stems very slender and branched above the middle. Leaves linear-lanceolate, 4–10 mm long. *Flower-heads in clusters of 3–7, with 5 marked angles,* 2–5 mm, ovoid to pyramidal, *not* over-topped by the leaves immediately below; outer flower bracts *woolly at the base*, but yellow and *hairless at the tip*. Open fallow and sandy waste places. Throughout, except for some islands. *Logfia gallica* (Syn. *Filago gallica*) NARROW-LEAVED CUDWEED is a similar annual to 33 cm with linear, *thread-like leaves (4)12–20(26) mm long*, the most apical *over-topping* the flower-heads which are borne 3–14, in groups of 3.5–5 mm, the inflorescence appearing very leafy; flower-bracts woolly and yellowish at the tip. Local, in similar habitats. Throughout.

1. *Filago lusitanica*
2. *Filago lutescens*
3. *Gnaphalium luteo-album*
4. *Gnaphalium uliginosum*

Gnaphalium CUDWEED

Inconspicuous annuals and perennials, often white-woolly-hairy. Flower-heads small, yellow-brown and bell-shaped with papery bracts, borne in clusters of (3)5–40. Pappus of simple hairs.

Gnaphalium sylvaticum (Syn. *Omalotheca sylvatica*) HEATH CUDWEED A grey-hairy perennial to 60 cm with non-flowering shoots and densely leafy stems. Leaves lanceolate to linear, *decreasing in size along the stem*, green above, woolly beneath, 1- or indistinctly 3-veined. *Inflorescence elongated* and lax and interrupted below, with numerous yellowish and small, narrowly bell-shaped flower-heads; florets compact; involucral *bracts dark blackish-brown*. Heaths and scrub; local and sporadic only in hot, dry areas. Throughout.

G. luteo-album

Gnaphalium luteo-album (Syn. *Laphangium luteoalbum*) JERSEY CUDWEED A short annual to 50 cm, *white-woolly*, unbranched or with branches spreading from the base then soon erect. Leaves broadly lanceolate, blunt, and running down the stem, *woolly on both sides*. Flower-heads terminal and densely ovoid, and *not overtopped* by their surrounding leaves (at least when mature); bracts elliptic, *shiny, straw-yellow*, only the outermost woolly below, and not bristle-tipped; florets yellowish, with red stigmas. Disturbed habitats. Throughout.

Gnaphalium uliginosum (Syn. *Filaginella uliginosa*) MARSH CUDWEED A short (often tiny) annual to 20 cm with *stems thickly white-woolly and branched throughout* with short lateral branches. Flower-heads in clusters of 3–10, *overtopped by the surrounding leaves;* bracts outward-spreading, flat and star-like after the seeds are shed. Disturbed habitats; common. ES, FR, IT, PT.

Helichrysum CURRY PLANT

Dwarf greyish aromatic shrubs with alternate, untoothed leaves. Flowers borne in dense clusters with papery bracts. Pappus absent.

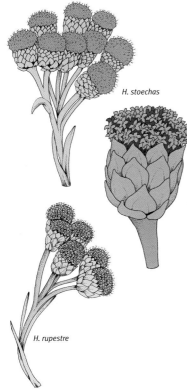

H. stoechas

H. rupestre

A. Flower-heads small (4–9 mm) and *roughly spherical*.

Helichrysum stoechas A variable, spreading, much-branched dwarf shrub to 50(70) cm. Leaves white-felted, narrowly linear, untoothed and slighlty aromatic (curry-scented) when crushed, 5–35(40) mm long, and 0.5–2.1 mm wide with down-turned margins. Flower-heads 4–7 mm, yellow, borne in clusters about 30 mm across; the series of bracts (involucre) bright shiny yellow and *spherical* in flower; the *outer flower bracts papery white*, overlapping and hairless or hairy at the base. Common on fixed sand dunes and garrigue. ES, FR, IT, MA, PT. *Helichrysum rupestre* is similar but with the largest leaves broader, 1–6 mm wide and *exceeding 30 mm in length*. Flower-heads 6–8(9) mm. ES, IT. *Helichrysum ambiguum* is distinguished by its broader leaves to 7 mm. Limestone cliffs. ES (Balearic Islands). *Helichrysum melitense* MALTESE EVERLASTING also has broad leaves to 45–55 mm long and 5–6 mm wide. A rare endemic of the Maltelse archipelago, restricted to just a few coastal cliff-tops. MT (Gozo).

B. Flower-heads small (4–5 mm) and narrowly *bell-shaped*.

Helichrysum italicum Similar to *H. stoechas*, a spreading, much-branched dwarf shrub to 40(50) cm. Leaves white-felted when young later hairless, linear, untoothed and strongly aromatic (curry-scented) when crushed, 60–40 mm long, 0.3–1.5 mm wide with down-turned margins. Flower-heads 4.5–5 mm, yellow, borne in *clusters of 15–25* about 80 mm across; the series of bracts (involucre) *dull mustard-yellow and oblong to narrowly bell-shaped*; the inner bracts at least 5 x longer than the outer. Common on sand dunes. Throughout. Subsp. *picardii* (treated by some authors at the species level) has inflorescences with fewer (*8–15*) *flower-heads* 3–5 mm. ES, MA, PT. *Helichrysum saxatile* is similar but with *longer leaves* to 70 mm long which are narrowly spatula-shaped. An island endemic of limestone cliffs. IT (Sardinia and Sicily).

H. italicum

1. *Helichrysum stoechas*
2. *Helichrysum stoechas*
3. *Helichrysum italicum*
4. *Phagnalon saxatile*

C. Flower-heads large (15–25 mm).

Helichrysum foetidum A robust, strong-smelling biennal to perennial to 1 m with oblong leaves to 12 cm at the base, 40–90 mm above and clasping the stem, white beneath, with slightly incurved margins. Flower-heads *large, 15–25 mm across* with an involucre of overlapping, outward-spreading, papery bracts; florets bright yellow. Native to South Africa but naturalised in southern Iberian Peninsula. ES, IT.

Phagnalon

Grey dwarf shrubs with alternate leaves and flower-heads borne solitary at the tips of the branches; flower-heads solitary, with densely overlapping bracts. Pappus of bristles.

P. rupestre

P. saxatile

Phagnalon rupestre A small shrub to 40 cm with erect to ascending, white-felted stems. Leaves *narrowly oval and more or less toothed*, small, 5–35 x 1.5–4.5 mm, green or whitish above and white-felted below, with down-turned margins. Flower-heads *solitary,* long-stalked (20–90 mm); yellowish; flower bracts brownish, membranous, somewhat hairy and closely overlapping. Fairly common on cliffs, sun-baked rocky ground and low coastal garrigue. Throughout. *Phagnalon saxatile* is a similar shrub to 30(50) cm with *linear* leaves 14–50 x 1.2–4 mm, sometimes broadest above the middle, *green* (not normally white-felted above) and white-felted below. Flower-heads *very long-stalked* (25 mm–13 cm); bracts with somewhat wavy margins, the tips often slightly recurved. Local in rocky and disturbed habitats. ES, FR, IT, MA, PT. *Phagnalon sordidum* is similar to the previous species but with long, narrow, linear leaves 7–42 x 0.5–1.5 mm that are white-woolly on both sides and *flower-heads borne in clusters of 2–4(6)* (not solitary). Rocky crevices. Common in southern France. ES, FR, IT.

Dittrichia

Annual or perennial herbs or small shrubs with sticky stems and simple, alternate leaves. Flower-heads borne in branched infloresences. Achenes abruptly contracted below the pappus.

D. viscosa

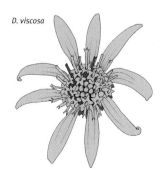

Dittrichia viscosa A densely glandular, sticky perennial 40 cm–1.3 m with stems woody at the base. Lower leaves bright green, linear, scarcely and sparsely toothed; upper leaves stalkless and semi-clasping the stem. Flower-heads 10–15(20) mm, 5–6 mm across, bright yellow, *with ray florets to 8–12 mm long*, much exceeding the flower bracts; bracts adpressed. Common to abundant in all disturbed and damp habitats. Throughout. Forms in Portugal with rigid, linear leaves, small flower-heads 3.5–4.5 mm across and a prostrate habit have been described as *D. maritima* and erect

1. *Phagnalon rupestre*
2. *Phagnalon sordidum*
3. *Dittrichia viscosa*
4. *Pulicaria dysenterica*
5. *Pulicaria odora*
6. *Pulicaria odora* (bracts)

perennials as *D. revoluta*; their distinction has not been analysed at the molecular level. *Dittrichia graveolens* is a more slender, erect annual 20–50 cm, with *flower-heads smaller, 3–5 mm across,* with outward-curving bracts *and with shorter rays* (3.5–4 mm), scarcely exceeding the involucre. Waste places. Throughout.

Pulicaria FLEABANE

Annual or perennial herbs with simple alternate leaves. Flower-heads yellow and daisy-like, borne terminally. Achenes with scales around the pappus.

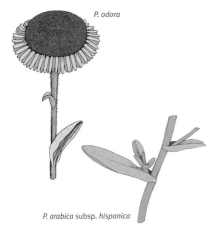
P. odora

P. arabica subsp. hispanica

Pulicaria dysenterica COMMON FLEABANE A rhizomatous, hairy perennial herb with downy or woolly, erect, branched stems to 1 m. Basal leaves oblong and withering by flowering time; stem leaves alternate, downy and clasping at the base. Flower-heads yellow, 15–30 mm across, borne in loose clusters; ray florets exceeding the disk florets; bracts sticky and hairy with long, fine tips. Damp, grassy habitats; widespread but local. Throughout. *Pulicaria odora* is similar but lacks creeping stolons and has basal leaves *not withered* at flowering time. Throughout. *Pulicaria arabica* subsp. *hispanica* (Syn. *P. paludosa*) is similar to the previous species but with narrow, *linear*, rigid leaves, smaller capitula, and achenes with erect to spreading hairs. ES, FR, PT.

Inula FLEABANE

Perennial herbs, sometimes woody below. Flower-heads yellow, often showy, usually with disk and ray florets, borne few, in branched, flat-topped clusters. Pappus 1 row of hairs.

A. Flower-heads with disk florets only, or ray florets very small.

Inula conyzae PLOUGHMAN'S SPIKENARD An erect, leafy, sparingly branched perennial to 1.3 m with hairy, oval, stalked lower leaves and narrowly oval-shaped upper leaves, all finely toothed and hairy beneath. Flower-heads small, 7–12 mm across, *numerous on each stem,* yellow; *ray florets absent*. Pappus reddish-white. ES, FR, IT, PT.

B. Flower-heads with disk and ray florets.

Inula verbascifolia A bushy, branched *white-downy* perennial to 50 cm with oval-elliptic, finely-toothed, slightly folded, pointed, grey leaves. Flower-heads bright golden yellow with disk and ray florets. Rocky habitats and cliffs. IT (south).

Inula salicina WILLOW-LEAVED INULA A slender, unbranched perennial to 70 cm with stiff, rough-margined leaves hairless above, the uppermost elliptic with heart-shaped bases $^1/_2$-clasping the stem and spreading horizontally; all leaves net-veined above. Flower-heads 25–45 mm, solitary or few, yellow, with long, slender rays. Moist habitats; absent from Corsica and Sardinia. ES, FR, IT.

Limbarda

L. crithmoides

Succulent, salt-tolerant perennials, previously included in the genus *Inula*. Pappus 1 row of hairs.

Limbarda crithmoides (Syn. *Inula crithmoides*) GOLDEN SAMPHIRE An erect, *succulent, maritime perennial* to 1 m, hairless with a woody base and *fleshy, crowded linear leaves* that are untoothed or 3-toothed at the apex. Flower-heads 15–25 mm, rather few, yellow with a golden disk, borne in flat-topped clusters. Salt-marshes, cliffs and other saline, coastal habitats. Throughout.

Rhanterium

Dwarf, desert-dwelling spiny subshrubs with branches covered in white, matted hairs. *Pappus bristles broad and flattened.*

Rhanterium suaveolens A small, much-branched, often domed, spiny subshrub 30–50 cm tall, *whitish-blue-grey-hairy throughout* with short, matted hairs. Leaves alternate, stalkless, small, linear and toothed. Flower-heads yellow with ray florets and tubular disk florets; *outer bracts distinctly recurved*. Achene with pappus of *bristles broad and flattened*. Mainly northern Sahara, dominant in some desert areas, extending locally into arid areas of the Mediterranean. TN.

1. *Limbarda crithmoides*
2. *Pallenis maritimus*
3. *Pallenis hierochuntica*
4. *Pallenis spinosa*

Pallenis

Annuals to subshrubs. Flower-heads yellow with bracts in 2–3 rows, leafy, the outermost spineless or spine-tipped. Pappus of scales. A genus now established to include species traditionally placed in *Asteriscus*.

P. spinosa

P. maritima

P. hierochuntica

Pallenis spinosa A slender, softly hairy annual to 60 cm with rigid stems, woody at the base with branches overtopping the main stem. Leaves elliptic, stalked below, unstalked and semi-clasping the stem above. Flower-heads daisy-like and bright yellow with a large disk to 20 mm across. Flower *bracts spine-tipped and 2 x the length of the ray florets*; inner flower bracts papery and not spine-tipped. Achene 2–2.5 mm. Common on coastal garrigue, dry roadsides and rocky places. Throughout.

Pallenis maritima (Syn. *Asteriscus maritimus*) YELLOW SEA ASTER A *short, compact, spreading subshrub* to 40 cm with stems woody at the base and much-branched; *not aromatic*. Leaves spatula-shaped, short-stalked and untoothed. Flower-heads numerous, bright yellow, to 40 mm across, surrounded by leafy flower bracts that are spine-tipped. Achene 1–1.5 mm. Common on coastal cliff-tops and rocks; rare inland. Throughout.

Pallenis hierochuntica (Syn. *Asteriscus pygmaeus*) RESURRECTION PLANT is a small, *virtually stemless* (main stem 0–7 mm) annual similar to *Asteriscus aquaticus* in general appearance (next page) but to just 15 cm high with small lateral branches. Flower-heads generally exceeded by their outer-most involucral bracts. Achene 1.2–1.9 mm. Bare, arid ground in North Africa. DZ, MA, TN.

1

2

3

4

Asteriscus

A. aquaticus

A genus very similar to *Pallenis* to which some species have recently been transferred, most notably *A. maritimus.*

Asteriscus aquaticus A similar species to the *Pallenis* group but an *aromatic annual* to 50 cm with erect to spreading stems, and flower bracts *greatly exceeding* the ray florets. Damp and sandy places, waste places and roadsides. Similar to *Pallenis spinosa* but lacking spine-tipped flower bracts. Achene 1.5–2 mm. Various damp habitats as well as garrigue. Throughout.

Bidens BUR-MARIGOLD

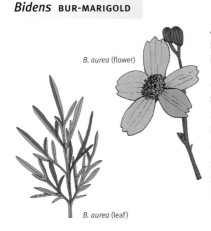

B. aurea (flower)

B. aurea (leaf)

Annuals with opposite leaves. Flower-heads rounded, often solitary, often without ray florets; bracts in 2 rows, the innermost papery. Pappus of 2–5 strong, barbed bristles.

Bidens aurea BIDENS A virtually hairless, medium to tall, slender perennial to 1 m. Leaves linear-lanceolate or 2-pinnately divided with linear lobes; toothed. Flower-heads held erect, daisy-like, yellow, white or mottled both, 35–50 mm across with only *5–6 large*, broad grooved ray florets; flower bracts all similar and much shorter than the ray florets. Achenes with 3–4 bristles. Native to Central America, naturalised in damp places. ES, FR, IT, PT.

Bidens tripartita TRIFID BUR-MARIGOLD A hairless to slightly hairy annual to 75 cm. Leaves with 1–2 pairs of deep lateral lobes and a short, winged stalk. Flower-heads yellow, erect, to 25 mm across, borne in branched clusters, *usually without ray florets but with leaf-like spreading bracts beneath.* Achenes with 2–4 bristles. Locally common on riversides and other damp habitats. Throughout.

Xanthium COCKLEBUR

Cosmopolitan annual weeds, native to the Americas. Leaves alternate. Flower-heads discoid. Fruiting head *conspicuously spiny*. Pappus absent.

X. strumarium

Xanthium strumarium ROUGH COCKLEBUR A stiffly branched, often aromatic, spineless annual to 1.2 m without spines. Leaves alternate and heart-shaped at the base, shallowly lobed and long-stalked. Flower-heads greenish and with male and female flowers borne separately in *lateral clusters. Fruiting heads covered in hooked spines; those at the apex not hooked.* Naturalised in damp habitats. Throughout. *Xanthium echinatum* STINKING COCKLEBUR is similar, *strongly aromatic* and with *incurved* (not straight) spines at the apex of the fruit. Throughout. *Xanthium spinosum* SPINY COCKLEBUR is similar but has *prominent beige spines* projecting from the leaf bases. Throughout.

Santolina

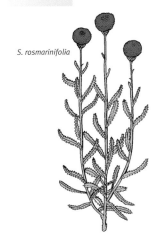

S. rosmarinifolia

Aromatic subshrubs with *pinnately divided leaves*. Flower-heads *long-stalked*, rounded without ray florets. Pappus absent.

Santolina chamaecyparissus LAVENDER COTTON A dense, aromatic, grey dwarf shrub to 60 cm with *crowded, crimped, narrow, sliver-felted leaves* and long-stalked, solitary yellow or cream flower-heads 6–10 mm across, lacking ray florets. Dry, rocky habitats and maquis. ES, FR, IT, PT. *Santolina rosmarinifolia* is a similar dense, aromatic, grey dwarf shrub to 60 cm. Leaves small, grey, narrowly oblong and lobed; lobes few and distant. Flower-heads yellow, borne terminally on numerous *leafless stalks thickened towards the top*. Dry, rocky habitats. ES, FR, PT.

Anthemis CHAMOMILE

Slightly hairy, often aromatic herbs or dwarf shrubs with *leaves cut into linear segments*. Flower-heads usually with white or yellow rays. Achenes obovate to obconical (not compressed), circular to square in cross-section, usually with about 10 smooth or rough *distinct* ribs. Pappus absent.

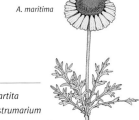

A. maritima

1. *Bidens tripartita*
2. *Xanthium strumarium*
3. *Santolina chamaecyparissus*
4. *Anthemis maritima* PHOTOGRAPH: GIANNIANTONIO DOMINA

A. Receptacle hemispherical in fruit.

Anthemis maritima A dwarf hairless or slightly hairy subshrub 20–40 cm tall with stout stems woody and rooting at the base; leafy above. Leaves pinnately lobed, *fleshy* and gland-dotted. Flower-heads daisy-like with white rays and a yellow disk, borne on *long, slender stalks* to 10 cm. Local on coastal sands; present on most islands. ES, FR, IT, MA, PT.

Anthemis tomentosa WOOLLY CHAMOMILE A low, ascending *woolly haired* annual with spreading to prostrate stems and 2-pinnately divided leaves. Flower-heads 20–30 mm across with a yellow disk and white rays; rays rather short; bracts grey-haired. IT (incl. Sicily). *Anthemis secundiramea* is similar but less hairy (not woolly) with lobed, slightly shiny leaves, and reddish stems. Flower-heads to 25 mm across. Coastal habitats. FR (Corsica), IT (Sardinia and Sicily).

B. Receptacle cone-shaped in fruit.

A. arvensis

Anthemis arvensis CORN CHAMOMILE A short to medium, hairy, aromatic herb with spreading or ascending branched, downy stems to 50 cm. Leaves to 50 mm long, oval in outline, 1–2-pinnately lobed with narrow, pointed segments, *woolly below*, especially when young. Flower-heads daisy-like with white rays and a yellow disk, to 40 mm across; flower bracts with brown, papery margins. Receptacle cylindrical-cone-shaped in fruit; achenes with 10 ridges. Disturbed, cultivated and fallow land; frequent. Throughout. *Anthemis cotula* STINKING MAYWEED is similar but *strong and unpleasant-smelling* with *almost hairless leaves* which are slightly fleshy and irregularly 2–3-pinnately lobed into linear segments. Rays reflexed on ageing. Achenes with rough ridges. Various habitats. Throughout.

Cota

C. tinctoria

Recently accepted as a new genus, differing from *Anthemis* in its obconical, flattened achenes with promiment lateral ribs or with 2–10 ribs on each face.

Cota tinctoria (Syn. *Anthemis tinctoria*) YELLOW CHAMOMILE An erect or ascending, branched, woolly haired perennial to 30(50) cm with leaves deeply twice cut into narrow, linear or narrowly oblong, toothed segments, woolly beneath. Flower-heads with 10–20 yellow ray florets 6–12 mm. Recectacle hemispherical in fruit; achenes 1.8–2.2 mm, angled. Dry, rocky and bare habitats. Throughout.

Chamaemelum CHAMOMILE

Annual or perennial herbs similar to *Anthemis* with deeply dissected leaves. Flower-heads with disk florets, and with the base of the corolla tube enlarged and inflated (pouch-like). Pappus absent.

Chamaemelum fuscatum A more or less *hairless* rather lax annual 60 mm–40(60) cm with erect to ascending stems. Leaves 2–3-pinnately divided below, 1-pinnately divided above into linear segments. Flower-heads somewhat *nodding* in bud; flower bracts with *blackish-brown* papery margins and tips. Achenes 1–1.4 mm, ovoid with about 30 ribs. Absent from the Balearic Islands. ES, FR, IT, MA, PT.

Chamaemelum nobile CHAMOMILE A creeping, *strongly, pleasantly aromatic* perennial with erect stems to 30 cm and gland-dotted leaves cut into linear segments with inrolled margins. Flower-heads long-stalked, with an involucre to 45 mm, with white rays and an orange-yellow disk. Achenes 1.4–1.6 mm, weakly ridged on 1 face. Fields, coastal and waste habitats. Native to Iberian Peninsula but now widely naturalised. Throughout.

Cladanthus

A genus closely related to *Anthemis* and *Chamaemelum* (and most species previously included in the latter genus).

C. mixtus

Cladanthus mixtus (Syn. *Chamaemelum mixtum*) A hairy, chamomile-like annual to 60 cm, normally with much-branched stems. Leaves oval in outline, 1–2-pinnately divided below, deeply toothed or 1-pinnately divided and *unstalked above*; leaf lobes linear-lanceolate and toothed or not. Flower-heads daisy-like with an involucre to 45 mm and spreading white rays and a yellow disk; rays 3-toothed at the tip; flower bracts with a wide, pale brown membranous margin. Achenes 1.2–1.6 mm, ovoid and weakly ridged. Cultivated, fallow and sandy waste ground and coastal sands; absent from the Balearic Islands. ES, FR, MA, PT.

1. *Achillea ageratum*
2. *Achillea santolina*
3. *Achillea millefolium*

1

2

3

Achillea

Perennial herbs with alternate, simple and wavy-edged to shallowly or deeply dissected leaves. Flower-heads congested into umbel-like flat-topped clusters; disk and ray florets white, pink or yellow. Pappus absent.

A. ageratum

A. Ray florets yellow.

Achillea ageratum An erect, simple or branched, hairy perennial with stems woody at the base. Leaves to 40 mm, lanceolate to linear in outline but pinnately divided; those of the non-flowering stems divided only in the upper part, *linear-toothed and entire below*. Flower-heads yellow. Local in damp habitats; absent from Sicily and much of the far south. ES, FR, IT, PT.

Achillea tomentosa YELLOW MILFOIL A dark green, grey-woolly perennial with spreading stems, and with linear leaves cut x 2 into linear, pointed segments. Flower-heads bright yellow, tiny, to 3 mm across, borne numerously in a dense, flat-topped cluster. Dry, grassy habitats and scrub. ES, FR, IT.

Achillea santolina SANTOLINA-LEAVED SNEEZEWORT A much-branched subshrub to 30 cm with woody, white-woolly stems and *pinnately lobed* leaves. Flower-heads yellow, each to 3 mm across, with very short ray florets (<1 mm) borne in domed inflorescences; flower-bracts with papery margins. A native of the eastern Mediterranean that is naturalised in Italy and possibly elsewhere. IT.

B. Ray florets white.

Achillea millefolium YARROW A strong-smelling, creeping perennial with clumped, erect, downy and furrowed, unbranched stems to 80 cm. Leaves much-divided 2–3-pinnately divided with lobes diverging in 3 dimensions (feathery), to 15 cm long, the upper leaves unstalked. Flower-heads 4–7 mm across, numerous (>25–50), aggregated into flattened, umbel-like inflorescences; ray florets white or pink; disk florets yellow. Cool, grassy habitats throughout but rather uncommon in the Mediterranean. Throughout, except for the Balearic Islands and Sicily. *Achillea ligustica* is similar but with shorter, wider, more finely divided leaves and smaller flower-heads (3(5) mm across). Absent from the Balearic Islands. ES, FR, IT (incl. Sicily).

A. maritima

C. Flower-heads *only with disk florets*; florets prolonged below into 2 enlarged spurs.

Achillea maritima (Syn. *Otanthus maritimus*) COTTONWEED A short, *densely white-woolly*, spreading perennial to 30 cm with robust, ascending stems. Leaves oblong-lanceolate, untoothed or blunt-toothed, fleshy and unstalked. Flower-heads few, 6–9 mm across, *yellow, rayless* and button-like, borne in lax, flat-topped clusters; flower bracts white-woolly. Fixed dunes and coastal sands; local. Throughout.

1. *Anacyclus clavatus*   3. *Glebionis segetum*
2. *Anacyclus valentinus* 4. *Glebionis coronaria*

Anacyclus

Chamomile-like annual or perennial herbs with alternate, 1–2-pinnately divided leaves. Outer achenes 2-*winged*, inner achenes unwinged; pappus absent.

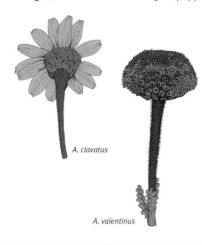

A. clavatus

A. valentinus

Anacyclus clavatus A short, widely branched, hairy annual 20–50 cm. Leaves alternate, 2–3-pinnately lobed with linear segments. Flower-heads daisy-like with *white recurved rays and a yellow disk*; flowers borne solitary on stalks *distinctly swollen* below the fruiting head; flower bracts with a narrow whitish or purplish margin. Sandy ground and coastal rocks. Throughout. *Anacyclus radiatus* is similar to *A. clavatus* but taller to 60 cm, with *flower-heads all yellow*, the rays purplish beneath. Sandy places. Throughout. **Anacyclus valentinus** is similar to *A. clavatus* but with flower-heads with *very short rays*, that do not exceed the involucre (*appearing rayless*). Similar habitats. DZ, ES, FR, MA, TN.

Glebionis (Chrysanthemum)

Annuals with simple leaves. Flower-heads with yellow, cream or white ray florets; receptacle without scales. Achene without a pappus. A genus now considered to be distinct from the ornamental chrysanthemum (*C. indicum*).

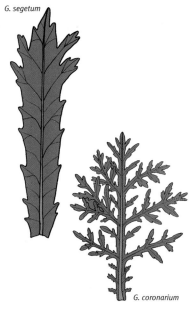

G. segetum

G. coronarium

Glebionis segetum (Syn. *Chrysanthemum segetum*) CORN MARIGOLD A short to tall, green, hairless, slightly fleshy annual to 60 cm with erect to ascending, branched or unbranched stems. Leaves glaucous, alternate, narrowly oval in outline, slightly toothed to *deeply toothed,* at least below. Flower-heads 30–70 mm across, bright yellow and daisy-like, with a flat disk. Achenes 2.5–3 mm, deeply ridged and unwinged. Common on disturbed fallow and cultivated land, roadsides, and coastal waste places. Throughout.

Glebionis coronaria (Syn. *Chrysanthemum coronarium*) CROWN DAISY A slightly hairy annual to 80 cm with leaves *2-pinnately lobed.* Flower-heads 40–80 mm across, yellow and daisy-like, with rays cream-white in the upper $^1/_2$ (or cream or yellow). Achenes 3–3.5 mm, deeply ridged and winged. Very common to abundant on disturbed land, roadsides, and coastal waste places. Throughout.

Ismelia

Closely related and similar to *Glebionis* but with keeled involucral bracts and flattened (not winged) disk achenes.

Ismelia carinata (Syn. *Chrysanthemum carinatum*) ANNUAL CHRYSANTHEMUM An annual similar to *Glebionis coronaria* but shorter (<50 cm). Flower-heads *brightly coloured with bands of yellow, maroon and white*; disk purple. Cultivated land, and in gardens. MA (naturalised elsewhere).

Coleostephus

Similar to *Glebionis* but with leaves *regularly finely toothed* (not deeply lobed or divided).

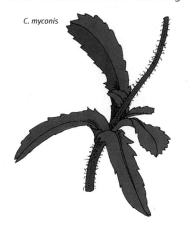

C. myconis

Coleostephus myconis A short to medium, sparingly branched, virtually hairless annual. Leaves oval, *regularly fine-toothed* (*not* lobed), the lower leaves tapered gradually into a stalk; the upper leaves unstalked and semi-clasping the stem. Flower-heads yellow and daisy-like, to 22 mm across with yellow or paler rays and a yellow disk. Cultivated, fallow and damp ground. Throughout, except the Balearic Islands. *Coleostephus paludosus* (Syn. *C. clausonis*) is similar but with spreading stems and *leaves irregularly toothed*. Similar habitats; absent from the Balearic Islands. ES, FR (Corsica), IT (incl. Sardinia and Sicily), PT, TN.

Lepidophorum

L. repandum

Similar to *Glebionis*, with *regular-coursely-toothed* (not deeply lobed or divided) leaves.

Lepidophorum repandum A *Glebionis*-like, hairless annual with sparingly branched stems to 50 cm. Leaves spatula-shaped to rectangular, *regular-coursely toothed* (not lobed), long-stalked below, stalkless above. Flower-head daisy-like and yellow, to 40 mm across. Fallow land and pine forests. ES, PT.

Leucanthemum   OX-EYE DAISY

Rhizomatous perennials. Flower-heads daisy-like, solitary or in groups of 2–3; ray florets white. Achenes with secretory canals.

Leucanthemum vulgare OX-EYE DAISY A variably hairy, clump-forming perennial to 75 cm with leafy stolons and erect, ridged stems. Leaves alternate, oblong, deeply toothed and stalked below, unstalked and clasping the stem above. Flower-heads large, 25–60(75) mm across and daisy-like, with long, white rays and a yellow disk. Locally frequent in undisturbed grassy places and meadows. Throughout, except the Balearic Islands.

Mauranthemum

A small genus recently separated from *Leucanthemum*.

Mauranthemum paludosum (Syn. *Leucanthemum paludosum*) An erect to ascending perennial to 40 cm similar to *L. vulgare* but without rhizomes and flower-heads just 20–30 mm across and with shorter rays. DZ, ES (incl. Balearic Islands), MA, PT, TN.

1. *Coleostephus myconis*
2. *Leucanthemum vulgare*
3. *Plagius maghrebinus*
PHOTOGRAPH: SERGE D. MULLER

Plagius

A small genus, virtually indistinguishable from *Leucanthemum*, with disk-like flower-heads and longitudinally ribbed achenes; pappus a short extension (corona) or absent.

Plagius maghrebinus A herbaceous annual with sparingly branched stems and alternate, toothed leaves. *Flower-heads small and button-like*, to <25 mm across, yellow, and with only disk florets. Achenes *without a pappus*. A North African endemic. DZ, MA, TN. *Plagius grandis* is similar but with flower-heads larger, 30–50 mm across. DZ, MA, TN. *Plagius flosculosus* is similar to *P. maghrebinus* but has achenes with *pappus present* (though inconspicuous, just 0.5–0.8 mm). FR (Corsica), IT (Sardinia).

Daveaua

Semi-aquatic, hairless annuals with leaves with *thread-like lobes* and white, daisy-like flowers. Achenes without ribs; pappus virtually absent.

Daveaua anthemoides A short, hairless annual 60–70 cm, branched above. Leaves alternate, pinnately divided with *very narrow, thread-like* linear segments. Flower-head 8–10 mm across, daisy like, with a yellow disk and white rays; solitary, to 35 mm across borne on stalks not thickened above; bracts triangular. Wet grasslands; rare and local. ES, MA, PT.

Erigeron FLEABANE

Annuals to perennials similar to *Conyza* with linear to oblong, entire to toothed, unstalked to shortly stalked leaves. Flower-heads with numerous ray florets in several rows and narrow and strap-like; bracts numerous and overlapping. Pappus 1 row of hairs (or with additional shorter hairs).

A. Ray florets erect and *little longer than the disk florets*.

Erigeron acris (Syn. *E. acer*) BLUE FLEABANE A variable, densely grey-hairy biennial to 60 cm with rigidly erect stems. Basal leaves narrow and entire, the uppermost unstalked. Flower-heads 10–15 mm across with 2 rows of erect, very *pale lilac ray florets scarcely longer than the yellowish disk florets*, borne on upward-spreading stalks, in spreading or flat-topped clusters. Throughout, except for the Balearic Islands, Sardinia and Sicily.

B. Ray florets spreading and *much longer than the disk florets*.

Erigeron annuus SWEET SCABIOUS An erect, branched, sparsely hairy biennial to 70 cm (1 m) with oval to lance-shaped, deeply lobed and coursely toothed, virtually hairless leaves. Flower-heads daisy like, with a yellow disk and *very slender pale lilac-white, outward-spreading rays*, borne in lax clusters. Native to North America, locally naturalised in the east. ES, FR, IT.

Conyza FLEABANES

Annual herbs with alternate, simple, narrow leaves borne densely along erect stems. Flower-heads inconspicuous but numerous; cream to pinkish, the central florets cosexual and the outermost female. Achenes with a pappus of hairs. Similar to *Erigeron* and possibly belonging to the same genus; treated as separate here in line with most floras.

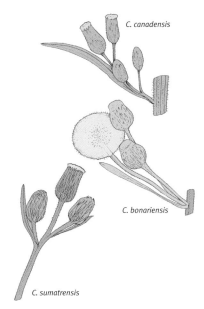

C. canadensis

C. bonariensis

C. sumatrensis

Conyza canadensis CANADIAN FLEABANE A tall, sparsely hairy annual to 1.5 m. Leaves alternate, narrowly oblong, stalked and yellow-green, often withered below and deciduous before flowering. Flower-heads very small, 3–5 mm, whitish, borne abundantly in a *cylindrical infloresence;* disk florets with 4-lobed corolla. Pappus cream. A weed native to North America very commonly naturalised on waste ground in towns. Throughout. *Conyza bonariensis* is a similar annual to 60 cm, and more densely brown to reddish-grey hairy, with larger flower-heads 6–10 mm borne in an *inflorescence with long lateral branches,* overtopping the main axis; disk flowers with 5-lobed corolla. Pappus dirty-white to reddish. Native to tropical America, commonly naturalised. Throughout. *Conyza sumatrensis* is similar to *C. bonariensis,* with flower-heads 5–7 mm borne in a *pyramidal inflorescence;* disk flowers with 5-lobed corolla. Pappus cream to grey. Native to the Americas, naturalised throughout.

Aster

Perennial herbs with simple alternate leaves. Flower-heads conspicuous, solitary or in panicles; ray florets blue, pink or white; bracts in 2–several rows. Pappus of 1–2 rows of hairs. A genus recently revised in light of DNA-analysis and found to comprise numerous genera (see *Tripolium* and *Galatella*).

A. Flower-heads conspicuous with lilac rays.

Aster sedifolius A medium to tall *rough* perennial with erect stems 15–25(60) cm with linear-elliptic, untoothed, unstalked, *gland-dotted leaves* with 1–3 veins and with clusters of smaller leaves in the axils. Flower-heads lilac, to 15 mm across and numerous, borne in crowded clusters; disk florets yellow, ray florets lilac. Dry hills and grassland; absent from the islands. ES, FR, IT.

---

1. *Erigeron acris*
2. *Erigeron annuus*
3. *Conyza canadensis*
4. *Conyza bonariensis*

B. Flower-heads inconspicuous with white-cream (to dull pink) rays.

Aster squamatus A lax, medium to tall, herbaceous annual or biennial to 1.8 m with linear to linear-lanceolate leaves; lateral branches short. Flower-heads borne in lax panicles, small with lanceolate, *purple-tipped involucral bracts with broad papery margins,* the outer reaching $^1/_2$ as high as the inner; rather inconspicuous and scarcely opening; florets dull white-cream, sometimes mauve or pink. Native of south America but naturalised in Spain and probably elsewhere in waste places on disturbed or saline soils. ES.

Tripolium

Previously grouped with *Aster. Succulent*, hairless annuals to biennials in saline habitats.

Tripolium pannonicum subsp. *tripolium* (Syn. *Aster tripolium*) SEA ASTER A hairless annual to biennial to 1 m with reddish, erect or ascending stems, branched from the base. Leaves *fleshy*, linear to linear-lanceolate, rounded in cross-section and clasping the stem; unstalked above. Flower-heads with bright blue-mauve rays and yellow disks, to 20 mm across, borne in large, flat-topped inflorescences. Local in maritime environments. Throughout.

Galatella

Previously grouped with *Aster.* Typically herbaceous perennials with prominently 1(3)-veined, glandular leaves, flower-heads with violet rays, bracts in 3–4 series and achenes with 1–2 ribs on each face.

Galatella linosyris (Syn. *Aster linosyris*) GOLDILOCKS ASTER A medium, hairless perennial with straggling or erect, very leafy stems. Leaves all 1-veined, linear, unstalked and rough-edged, gland-dotted, not fleshy, and pointing upwards. Flower-heads small, 12–18 mm across and *bright golden yellow, without rays*, borne in spreading clusters; bracts unequal and loosely adpressed. Dry habitats, absent from most islands. ES, FR, IT.

Cotula

C. coronopifolia

Small annual or perennial herbs with entire to deeply pinnately divided, alternate leaves. Flower-heads borne in leaf axils or terminally, without rays. Achene without a pappus.

Cotula coronopifolia BRASS BUTTONS A small, succulent, hairless, often patch-forming annual to 30 cm. Leaves rather fleshy; linear and entire or scarcely toothed, and clasping the stem at the base. Flower-heads all yellow, 8–12 mm across, without rays; long-stalked and often nodding. Achenes strongly compressed and winged (outer disk florets) or unwinged (inner disk florets). Native to South Africa, widely naturalised in damp and seasonally flooded habitats. Throughout.

Artemisia WORMWOOD

Annual or perennial herbs or low shrubs, often strongly aromatic. Leaves alternate, entire to divided. Flower-heads small, often nodding in a lax inflorescence. Achenes lacking (or virtually lacking) a pappus.

A. campestris

1. *Aster squamatus* (flower-heads)
2. *Aster squamatus* (in fruit)
3. *Cotula coronopifolia*
4. *Artemisia campestris* subsp. *maritima*
5. *Artemisia maritima*

Artemisia campestris FIELD SOUTHERNWOOD An *almost scentless*, low, mat-forming shrub to 75 cm with a stout, woody stock and numerous non-flowering shoots; stems ascending and slightly, but persistently, *silkily hairy*. Leaves silky when young, later hairless, 2–3-pinnately divided below, simple and stalkless above; leaf lobes *narrow, 0.3–1(1.5) mm wide*. Flower-heads 2–4 mm across, pale green, ovoid, short-stalked and erect or spreading; involucre hairless or virtually so; bracts with a wide, papery margin; corolla yellowish or reddish. Throughout. Subsp. *campestris* has very narrow linear leaf segments. Throughout. Subsp. *maritima* (Syn. *A. crithmifolia*) has less hairy stems and leaves, *fleshy, oblong-linear leaf lobes*, rounded on the upper sides, and broad panicles of flowers. Maritime sands and cliffs; common on Atlantic coasts. ES, PT.

Artemisia maritima A similar shrub to *A. campestris*, woody below, *strongly aromatic* with *markedly white-woolly* shoots and stems. Leaves 1–2-pinnately divided into linear segments. Flower-heads longer than wide, 1.5–3.5 mm across, yellowish or reddish. Predominantly coastal sands and dry salt marshes in the European Mediterranean northwards. ES, FR, IT.

Doronicum LEOPARD'S-BANE

Herbaceous, rhizomatous perennials. Leaves simple and alternate, the lowermost withered when in bloom. *Flower-heads large and yellow,* with a hairy receptacle. Pappus of short bristles, absent in outer florets.

Doronicum plantagineum LEOPARD'S-BANE A simple or sparingly branched, slender annual to 80 cm (1 m); hairless below, hairy above. Basal leaves pale green and *elliptic*, to 60 mm long, untoothed or weakly so and *gradually narrowed into long stalks*; upper leaves oval to triangular and clasping the stem. Flower-heads daisy-like with an orange-yellow disk, and long, slender, yellow rays, *borne solitary* (2–3), 45–80 mm across, on glandular-hairy stalks. Involucral bracts linear. Dry rocky places and woods. ES, FR, IT, PT. *Doronicum pardalianches* GREAT LEOPARD'S-BANE is a similar annual to 80 cm with broadly oval to *heart-shaped* leaves and *several (3–8) flower-heads* 30–45 mm across borne sparsely on branched, woolly-hairy stems. Rare in most of the west and south. ES, FR, IT.

Senecio

A large and diverse genus of annual or perennial herbs and shrubs with alternate, pinnately veined leaves. Flower-heads often numerous, borne in flat-topped clusters, usually yellow. Achene with a white or greyish, hairy pappus.

A. Some or all leaves deeply lobed; *ray florets absent*.

Senecio vulgaris GROUNDSEL A short, more or less hairy, rather succulent annual to 30 cm (usually less) with weak, sparingly and irregularly branched stems. Leaves coarsely lobed and toothed, oval in outline and short-stalked below, and semi-clasping the stem at the base. Flower-heads *small and without (or with few) rays*, to 5 mm across and yellow; involucre cylindrical with black-tipped bracts. Achenes <2.5 mm. A very common weed in a range of disturbed natural and urban habitats. Throughout.

B. Some or all leaves deeply lobed; *ray florets inconspicuous*.

S. viscosus

S. gallicus

S. lividus

Senecio viscosus STICKY GROUNDSEL A small, *very sticky*, rather unpleasant-smelling annual to 60 cm (usually less) with weak, freely or sparingly branched stems. Leaves dark grey-green, densely glandular-hairy and sticky, deeply and regularly lobed, and short-stalked below, and unstalked (but not clasping) above. Flower-heads to 12 mm, with 13 pale yellow, recurved ray florets, borne in a large, irregular corymb. Achenes hairless or hairy in the grooves. Open stony and sandy places; local and absent from hot, dry areas. ES, FR, IT, PT.

C. Some or all leaves deeply lobed; *ray florets conspicuous and spreading*.

Senecio gallicus A short, hairless or slightly hairy annual to 67 cm usually branched from the base. Leaves fleshy, pinnately divided with lobed segments; stalked below, unstalked above and *clasping* the stem. Flower-heads 5–8.5 (10) mm, bright yellow, borne in lax clusters. Achenes 2–2.5 mm. Common in the region on coastal sands; absent from some islands. Throughout. *Senecio lividus* is similar to *S. gallicus* but normally with fewer branches, sparsely glandular-hairy, with *smaller* flower-heads 4.5–6(7) mm with black-tipped bracts and short, *recurved* rays. Achenes 3.2–3.7(4.5) mm. Dunes and fallow land. ES, FR, IT, PT. *Senecio leucanthemifolius* is a similar ascending, branched annual to 25(30) cm, fleshy, hairless or slightly white hairy (sometimes purple-tinged). Flower-heads 7–8 mm with *black-tipped flower bracts*. Achenes 2–2.2 mm, hairless. Coastal habitats. Throughout.

D. Leaves *not deeply-lobed*.

Senecio angulatus A creeping, slightly *succulent* perennial with stems to 2 m long and *bluntly lobed, shiny green leaves*. Flowers yellow, typically *Senecio*-like and scented. A native of South Africa, commonly planted in the region. Throughout.

1. *Senecio vulgaris*
2. *Senecio viscosus*
3. *Senecio gallicus*
4. *Senecio lividus*
5. *Kleinia anteuphorbium*
   PHOTOGRAPH: SERGE D. MULLER

Kleinia

A predominantly tropical genus generally recognised as distinct from *Senecio*. Plants with a woody, succulent habit. Pappus hairy.

Kleinia anteuphorbium (Syn. *Senecio anteuphorbium*) A very distinctive, large, sprawling, *succulent* shrub with many thick, ascending, green to grey stems that are often eventually *leafless*; leaves soon-falling, when present, elliptic and with down-turned margins. Flower-heads without ray florets, whitish, and with purple bracts. Garrigue, maquis and desert fringe scrub. MA.

Jacobaea

A genus of species previously classified under *Senecio*, usually (thought not exclusively) characterised by perennials with pinnately lobed leaves. Pappus of simple hairs.

A. Plant a variably hairy, herbaceous perennial.

Jacobaea vulgaris (Syn. *Senecio jacobaea*) RAGWORT A large biennial or weak perennial to 1.5 m. Stems reddish, robust and erect, more or less hairless or slightly woolly (floccose) and unbranched below; *furrowed*. Basal leaves large to 20 cm long, *deeply pinnately lobed with a large blunt end-lobe* and stalked, often withered at flowering time; upper leaves semi-clasping the stem; all slightly hairy below. Flower-heads to 20 mm across, bright yellow and numerous, borne in *flat-topped clusters*. Involucre more or less hairless. Achenes of disk florets hairy. Pastures and grassy habitats; absent from hot, dry areas. Throughout.

B. Plant a densely white-hairy, shrubby perennial.

Jacobaea maritima (Syn. *Senecio bicolor*) A *white-felted* shrubby perennial to 50 cm, much-branched and leafy at the base, and often forming extensive patches. Leaves oval-lanceolate and toothed to pinnately lobed with narrow segments. Flower-heads to 15 mm across, borne in terminal clusters. Coastal habitats more or less throughout; common. DZ, ES, FR, IT, MA, TN. *Jacobaea ambigua* is similar but with fewer branches that are leafy throughout and with smaller flower-heads to just 12 mm across. Rocky habitats. IT (Sicily).

Calendula MARIGOLD

Annual or perennial, often aromatic herbs with alternate, undivided leaves. Flower-heads daisy-like with yellow or orange ray florets; bracts in 1–2 rows. *Achenes strongly curved*, without a pappus.

C. suffruticosa

C. arvensis

C. officinalis

Calendula suffruticosa A very variable (with numerous subspecies described), short to medium perennial, often with a woody stock. Leaves lanceolate, somewhat fleshy, pointed, with few, distant teeth; glandular. Flower-heads large, to 40 mm across, bright yellow, the rays sometimes red-tipped, to 20 mm long. Fruiting head with a *conspicuous outer row of spreading (weakly curved) achenes* 22 mm. Fairly common in rocky and sandy coastal habitats, especially sea-cliffs. Throughout.

Calendula arvensis FIELD MARIGOLD A short, slender, thinly hairy, ascending or spreading annual to 30 cm. Leaves oblong and finely toothed. Flower-heads daisy-like and *small*, 10–25(27) mm across; yellow-orange throughout or with a brownish disk. Fruiting head with an outer row of beaked and strongly incurved achenes. Common on fallow ground, dunes and open garrigue. Throughout. *Calendula officinalis* is similar but with larger flowers 40–70 mm across, yellow-orange with an orange or brownish disk. Widely cultivated as an ornamental, possibly naturalised throughout. *Calendula maritima* is a very rare Sicilian endemic with short, sticky hairs on the lower stems, and succulent, strong-smelling leaves. Rare, only in maritime habitats on the mainland Sicilian coast and neighbouring islets. IT (Sicily).

Arctotheca

Short, leafy annuals. Flower-heads yellow with long-spreading rays. Pappus of scales. Native to South Africa.

A. calendula

Arctotheca calendula An annual with spreading, white-woolly, leafy stems to 40 cm long. Leaves 70 mm–20 cm long, deeply lobed, roughly hairy above and softly white-hairy beneath. Flower-heads to 50 mm across, borne on long stalks; outer involucral bracts with a membranous margin, and often with a terminal appendage; inner bracts membranous; rays long, deep yellow and pale yellow at the tip, purplish beneath; disk florets blackish green. Sandy waste ground. Widely naturalised; common in southwest Iberian Peninsula. ES, PT.

1. *Jacobaea maritima*
2. *Calendula maritima* PHOTOGRAPH: GIANNIANTONIO DOMINA
3. *Calendula suffruticosa*
4. *Calendula arvensis* (fruits)

Gazania

Short, leafy perennials. Flower-heads conspicuous and brightly coloured. Pappus of scales. Native to South Africa.

Gazania rigens A bushy perennial to 50 cm with spreading stems, woody at the base. Leaves dark green, oblong, entire, and tapered gradually at the base into a long stalk, equalling the lamina; densely white-hairy below; lower leaves pinnately lobed. Flowers large and showy; to 80 mm across; rays bright orange or white with a basal black patch; disk orange. Cultivated as an ornamental in towns and gardens, probably throughout.

1. *Arctotheca calendula*
2. *Arctotheca calendula* (flower-heads)
3. *Carlina corymbosa*
4. *Carlina racemosa*
5. *Carlina vulgaris*
6. *Carlina gummifera*
7. *Carlina lanata*

Carlina CARLINE THISTLE

Spiny, thistle-like annuals or biennials with pinnately lobed leaves. *Flower-heads large, without ray florets, but bracts conspicuous and spreading.* Pappus of feathery hairs.

A. Flower-heads *large*, (35)60 mm–14 cm across, normally *solitary in a central leaf rosette*; yellowish, whitish or pink-purple.

Carlina acanthifolia A distinctive, stemless perennial to 10 cm with all leaves clustered into a basal rosette to 30 cm across. Leaves pinnately divided just >$^1/_2$-way to the midrib into spine-toothed segments, hairy-felted beneath. *Flower-heads large, usually >70 mm across, yellowish, and solitary*; ray florets absent. Stony pastures; absent from most islands. ES, FR, IT. *Carlina acaulis* is similar but with *flower-heads whitish*, due to bright silvery-white spreading hairs on the ray-like bracts, with dull purplish florets in the central disk. Mountains, south-centre of region. ES, FR.

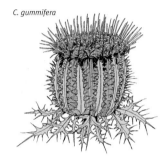

C. gummifera

Carlina gummifera (Syn. *Atractylis gummifera*) A thistle-like, low, robust, *stemless* perennial with leaves 15–40 cm in a lax rosette. Leaves virtually hairless, oblong in outline but deeply divided into narrow, spiny segments; stalks clasping at the base; *withered* when in bloom. Flower-heads large, 35–70 mm across and pink-purple without ray florets, with an involucre with outer bracts bearing 3 apical spines; inner bracts with a single brown spine. Achenes 10–12 mm. Local in dry fields, roadsides and fallow land; absent from much of the east and the Balearic Islands. ES, FR, IT, MA, PT.

B. Flower-heads 15–50 mm across, normally in clusters; yellowish or brownish.

C. corymbosa

Carlina corymbosa FLAT-TOPPED CARLINE THISTLE A pale, whitish or bluish short to tall, erect, spiny perennial; main stems usually solitary or few. Leaves linear-lanceolate, toothed and pinnately lobed, wavy, and spiny-margined; clasping the stem. *Flower-heads stalked*, borne terminally in flat-topped clusters, to 20(50) mm across with yellow florets and *bright golden-yellow ray-like bracts*. Common and widespread on dry stony slopes, sea-cliffs and garrigue. Throughout. *Carlina racemosa* is similar but with stems branched from the base and larger, unstalked flower-heads to 15 mm across that are *distinctly overtopped* by 1–2 surrounding flowering branches that arise from immediately below. Grassy and stony open habitats and fixed dunes. ES (south), IT (Sardinia), PT. *Carlina vulgaris* CARLINE THISTLE is similar to *C. corymbosa* but with pale silvery-yellow ray-like bracts. Widespread in the European Mediterranean, often inland. ES, FR, IT.

C. racemosa

C. Flower-heads 15–50 mm across, normally in *clusters; whitish or purplish.*

Carlina macrocephala A virtually hairless plant, or with some cobweb-like hairs, and stiff, erect stems to 40 cm, unbranched or sparingly branched. Leaves to 11 cm long, wavy-pinnately lobed with spiny teeth. Flower-heads to 30(50) mm across, the ray-like bracts whitish above and puplish beneath, 1.5–2 mm wide. Open woods. FR (Corsica), IT (centre + Sardinia and Sicily). *Carlina sicula* is similar but with silvery-white (or pink-flushed), broader ray-like bracts to 3 mm across. IT (south + Sicily).

Carlina lanata An erect plant with branched or unbranched flowering stems to 50 cm. Leaves white-woolly and lanceolate in outline, cut into spiny lobes, the uppermost clasping the stem. Flower-heads *with reddish-purple ray-like bracts and contrasting yellow disk florets.* Garrigue and rocky slopes. Throughout.

Atractylis

Similar to *Carlina*. Inflorescence with leafy bracts deeply-cut into spiny teeth, the innermost bracts papery-tipped, not brightly-coloured. Pappus silvery-haired.

A. cancellata

Atractylis cancellata A slender, thistle-like perennial with leaves in a lax rosette and with stems from 30 mm–30 cm. Leaves hairy, oblong in outline, toothed and shortly spine. Flower-heads large and pink-purple without ray florets, with an involucre to 20 mm, surrounded by upper leaves forming a *cage-like structure*; middle bracts with all spines similar. Common in dry fields, roadsides and fallow land. Throughout.

Atractylis humilis A slender, *low to short*, unbranched or scarcely branched, erect and hairy to hairless perennial. Leaves linear-oblong and spiny along the margins, the lowermost short-stalked and the upper leaves stalked and distinctly pinnately lobed. Inflorescence with *leaf-like outer bracts* encircling the flower-heads which are up to 25 mm across; inner bracts notched at the apex with spine-tips; florets purple-pink and all tubular. Often late-flowering, locally common on bare slopes and garrigue; absent from most islands. ES (incl. Balearic Islands), FR, MA.

Staehelina

Dwarf shrublets. Flower-heads narrow and cylindrical; ray florets absent; bracts not spiny. Pappus long and persistent.

Staehelina dubia A small, rounded shrub to 40 cm, *not typical of the daisy family, with the superficial appearance of a lavender*; stems white-felted, with crowded, linear to narrowly lanceolate leaves with slightly toothed, recurved margins, white-felted beneath. *Flower-heads narrow and cylindrical*, purple, to 30 mm long, borne solitary or in small clusters. Garrigue and rocky habitats. ES, FR, IT, PT.

1. *Atractylis cancellata*
2. *Atractylis humilis*
3. *Staehelina dubia*
4. *Echinops strigosus*
5. *Echinops spinosissimus*

Echinops

Thistle-like perennials with deeply pinnately lobed leaves. Flower-heads spiny, spherical and >25 mm across. Pappus of scale-like bristles.

Echinops sphaerocephalus PALE GLOBE THISTLE A stiffly erect thistle-like perennial to 2.5 m with deeply pinnately lobed spine-toothed leaves which are *green and slightly glandular above* and white-cottony beneath; bases winged and clasping. Flower-heads spherical, terminal and *pale grey to whitish-blue*, to 60 mm across; stamens with blue filaments. Achenes 6–10 mm. Most of the western European Mediterranean but absent from the islands. ES, FR, IT. ***Echinops strigosus*** is a similar, spiny, bushy and rigid perennial to 50 cm with leaves elliptic in outline and deeply divided with linear segments, *covered with dense, stiff hairs above*; softly white-hairy beneath. Flower-heads spherical, rather dull blue, to 70 mm across. Local in dry, scrubby places in the far southwest. ES, MA, PT. *Echinops ritro* GLOBE THISTLE is similar to the previous species but has leaves shiny-green above (white-cottony beneath) and *bright blue*. Rocky, uncultivated ground and cultivated as a garden plant. ES, FR, IT, MA, PT. ***Echinops spinosissimus*** is similar to the previous species but *markedly spiny*; densely woolly and glandular with downward-pointing leaves which are 2-pinnately divided. Flower-heads pale grey-blue *with sparse prominent spines protruding;* inner flower bracts fused to form a membranous tube. Stony pastures in the east of the region only. IT (south + Sicily).

Carduus THISTLE

Annuals to biennials with spiny-winged stems (at least in part), and alternate, spine-toothed leaves. Flower-heads rounded or cylindrical, often shaving brush-shaped; ray florets absent. *Pappus with many rows of simple hairs.*

A. Flower-heads ≤14 mm across (excluding flowers) and oblong-cylindrical.

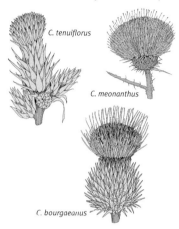

C. tenuiflorus

C. meonanthus

C. bourgaeanus

Carduus tenuiflorus SLENDER THISTLE An erect, narrowly branched biennial to 60(80) cm with stems *broadly winged* up to the flower-heads, grey to white-cottony. Flower-heads slender, 12–18 mm long and 5–10 mm across, borne in dense terminal clusters of 3–10; florets pale pink-red; bracts at least 1.5 mm wide. Fairly common on higher ground in dry open woods, and bare and disturbed habitats. ES, FR, IT, PT (possibly also North Africa). *Carduus meonanthus* is a maritime biennial similar to *C. tenuiflorus* but with *virtually hairless* stems and narrower bracts. ES, PT. *Carduus bourgaeanus* is an annual to 40 cm distinguished by its wide, *bell-shaped flower-heads* at least 10 mm wide. ES.

Carduus pycnocephalus A variable (with regional subspecies described), erect, narrowly branched biennial to 1 m tall with stems winged up to the flower-heads, similar to *C. tenuiflorus* but with leaves more densely white-cottony and stems not leafy but with interrupted spiny wings below the flower-heads; *flower-heads often solitary or in clusters 1–3; 14–20 mm long and 7–12(14) mm across*. Disturbed habitats. Widespread and common. DZ, ES, FR, IT, MA, TN. Subsp. *marmoratus* (Syn. *Carduus australis*) has stems with *broad wings 7–8 mm wide* and long apical spines on the leaf lobes to 30 mm long. IT. *Carduus cephalanthus* is distinguished by its leaves which are almost hairless beneath, and with conspicuously raised veins. *Flower-heads numerous, in clusters of 5–20*. Maritime habitats; mainland Iberian Peninsula and Italy, scattered across the islands except for the Balearic Islands. ES, IT. *Carduus fasciculiflorus* is very similar to *C. cephalanthus* but with *flower bracts with narrow membranous margins*. FR (Corsica), IT (south + Sardinia).

B. Flower-heads large, >14 mm across (excluding flowers) and spherical.

Carduus nutans NODDING THISTLE A variable (with numerous subspecies described), erect, robust, biennial thistle to 1 m with cottony, spiny-winged stems branched above and *not winged, and sparsely spiny to spine-free below the flower-heads*. Leaves pinnately lobed into spine-tipped segments, cottony below. *Flower-heads large* and nodding, usually solitary, 16–30 mm long and 20–60 mm across; florets bright red-purple, surrounded by robust, broad, reflexed spine-tipped bracts. Thickets and waste places, usually on higher ground. ES, FR, IT. Subsp. *macrocephalus* (Syn. *C. macrocephalus*) is the common form in the east of the region. IT.

Cirsium THISTLE

Biennials and perennials similar to *Carduus* with or without spiny wings. Flower-heads purple (or yellow). Pappus of many rows of *branched, feathery hairs* (not rough, unbranched hairs).

Cirsium vulgare SPEAR THISTLE An erect biennial to 1.5 m with cottony stems with *interrupted spiny wings*. Basal leaves to 30 cm long, deeply pinnately lobed with segments forked and spiny, and a *single long, pointed end-lobe*; upper leaves smaller, all leaves prickly. Flower-heads ovoid, 25–40 mm long and 20–50 mm across, with cottony bracts; the outer bracts with long spine-tips; florets reddish purple. Disturbed and cultivated habitats. Throughout.

Cirsium arvense CREEPING THISTLE A creeping perennial herb with leafy, mostly *unwinged stems* to 1.2 m and leaves not in a distinct basal rosette; leaves oblong and deeply divided with triangular, wavy, spiny lobes; upper leaves similar but clasping, sometimes forming a short wing along the stem. Flower-heads narrowly cylindrical, 10–22 mm long and 8–20 mm across, *pale pink-purple*, borne in loose clusters; *involucral bracts purple*. Grassland, usually not coastal. Rather uncommon except at higher altitudes. Throughout.

---

1. *Echinops sphaerocephalus*
2. *Carduus tenuiflorus*
3. *Carduus pycnocephalus*
4. *Carduus nutans*
5. *Cirsium vulgare*
6. *Cirsium arvense*

Notobasis

Similar to *Carduus* but with *flower-heads encircled by tough, spiny upper leaves.*

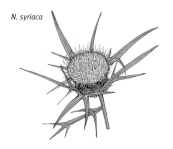

N. syriaca

Notobasis syriaca Syrian thistle A thistle-like annual 20 cm–2 m high with rigid stems *not* spiny-winged. Leaves alternate, dark green with paler veins; narrowly elliptic, pinnately lobed with spine-tipped, narrow triangular lobes, reduced and *clustered* around the stalkless flower-heads above. Flower-heads solitary, purple; bracts with spine-tips; involucre 18–23 mm long. Achenes 67 mm long. Field margins, fallow land and roadsides; common, sometimes abundant. Throughout.

Galactites GALACTITES

Thistle-like perennials with deeply dissected leaves. Flower-heads without ray florets, but with disk florets conspicuously spreading at the margins. Pappus long and feathery.

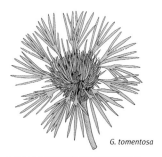

G. tomentosa

Galactites tomentosa GALACTITES A stiffly erect perennial 30–50 cm high with conspicuously *white-veined and variegated* dark green leaves; alternate, oblong, pinnately lobed with spiny lobes, *white-downy beneath.* Flower-heads pale purple, borne solitary or in clusters; the outer *ray florets long, even and spreading*; flower bracts tapered abruptly into spine tips; white-downy; 12–18 mm long. Achenes 3.5–5 mm long. Common in a range of dry and disturbed habitats. Throughout.

Onopordum SCOTCH THISTLE

Stout perennials with *spiny-winged stems*, often with cobweb-like hairs. Leaves spiny-margined and toothed to lobed. Flower-heads large and purple or white; all florets tubular and deeply 5-lobed. Pappus of many rows of simple hairs.

Onopordum acanthium SCOTCH THISTLE A large, spiny biennial to 2.5 m with broadly winged stems. Leaves oblong in outline, and pinnately lobed with broadly triangular, spiny-toothed segments, grey-white with cottony hairs. Flower-heads solitary, 20–60 mm across with tubular pink-purple florets 20 mm long; ray florets absent; bracts sepal-like, ending in yellow spines, woolly hairy at the base. Rather local, but widespread, on dry ground and hillsides. Throughout.

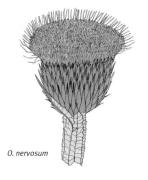

O. nervosum

Onopordum nervosum A large perennial to 2.7 m with yellowish hairy stems with *broad, veined wings to 20 mm* and long spines to 10 mm; branched at acute angles; plant cottony below, green above. Leaves large, to 50 cm, oblong-lanceolate and stalkless, green with paler veins; more or less hairless above and sparsely hairy and strongly veined below; divided with 6–8 pairs of lobes, each with a long spine-tip. Flower-heads pink, 30–50 mm with florets 30 mm long; bracts spine-tipped. South-central Portugal and southern and eastern Spain. ES, PT.

O. illyricum

Onopordum illyricum A stout perennial with spiny-winged stems and with grey or white-felted leaves, similar to *O. nervosum* but smaller, to 1.3 m with narrower wings *not veined*; remotely lobed leaves, and larger, purplish flower-heads to 60 mm across, with florets to 35 mm with conspicuous glands on the corolla. ES, FR, IT, PT. *Onopordum macracanthum* is similar but *cottony-white* throughout, with *sparsely* spiny stems. Flower-heads with a prominent series of very long, downward-pointing spines at the base. Rare and local in the south and west. DZ, ES, MA, PT.

1. *Notobasis syriaca*   4. *Onopordum acanthium*
2. *Galactites tomentosa* 5. *Onopordum illyricum*
3. *Galactites tomentosa*

Cynara ARTICHOKE

Stout perennials with leaves in a basal rosette or alternate; deeply divided into spiny segments. Flower-heads borne solitary or sparingly; purplish, blue or white. Pappus of many rows of feathery hairs.

C. humilis

A. Flower-heads borne on short to long stalks.

Cynara humilis A bushy, dark grey-green perennial to 80 cm with numerous basal leaves; stems white-hairy and unwinged. Leaves lanceolate in outline, *deeply 2-pinnately divided into linear segments with down-turned margins*, hairless and somewhat shiny above, white-hairy beneath; lower leaves short-stalked and upper leaves stalkless. Flower-heads large; involucre to 60 mm long, brownish when mature with purplish *bracts with prominent spine-tips*; florets purplish. Dry fallow ground on heavy soils. ES, PT.

C. cardunculus

C. algarbiensis

C. tournefortii

Cynara cardunculus CARDOON A large, bushy, *greyish or whitish* perennial to 1.8 m with numerous basal leaves; stems white-hairy and unwinged. Leaves thick, lanceolate in outline, deeply 1–2-pinnately lobed into *lanceolate* (not narrow and linear), *flat* segments; toothed, shortly and sparsely hairy and green above, white-hairy beneath; lower leaves short-stalked and upper leaves stalkless; leaves with *long spines to 35 mm clustered* at the base of each segment. Flower-heads large, 35–75 x 35–95 mm (excl. florets) with bracts narrowed into spreading spine-tips; florets violet-purple. Common on dry stony, fallow and waste places on heavy soils. Throughout. **Cynara algarbiensis** is similar to *C. cardunculus* but much smaller (30–50 cm), with leaves borne in flattish rosettes, markedly pale *grey-white* and hairy above, white-hairy and indistinctly veined beneath, with *variably short marginal spines* to 6 mm; flower-heads 20–30 x 40–50 mm, blue-purple. Disturbed and fallow ground or roadsides on clay soils; common in the Algarve. ES, PT. **Cynara scolymus** GLOBE ARTICHOKE is similar to *C. cardunculus* but with broad, oval, *blunt* (sometimes notched but *not spine-tipped*) involucral bracts. Of unknown origins but cultivated throughout.

B. Flower-heads stalkess.

Cynara tournefortii (Syn. *Arcyna tournefortii*) A *low, stemless*, robust, spiny, perennial. Leaves to 20 cm, oblong-lanceolate, virtually hairless above, grey-hairy below, divided into lobes that are toothed at the base (with yellow spine-tips), untoothed towards the tips. *Flower-heads borne large and solitary in the centre of the rosette*; involucre 80 mm; florets blue. Waste and fallow ground on heavy soils. ES, PT.

Silybum MILK THISTLE

Robust annuals to perennials with spineless stems. Flower-heads with *filaments fused at the base to form a tube*. Pappus with many rows of simple hairs.

Silybum marianum MILK THISTLE A robust, weakly spiny biennial to 1 m. Basal leaves oblong, pinnately lobed and prominently *white-veined and variegated*, virtually hairless and stalked beneath; stem leaves smaller, with fewer lobes and clasping the stem. Flower-heads purple, 25–40 x 50 mm (14 cm), borne solitary and terminally; bracts terminating in long, stout spines to 7 mm. Common and widespread on fallow, cultivated, waste ground and field margins. Throughout.

1. *Cynara algarbiensis*   4. *Cynara cardunculus*
2. *Cynara scolymus*   5. *Silybum marianum*
3. *Cynara humilis*

Centaurea CORNFLOWER, KNAPWEED

A large genus of annual to perennial herbs with alternate simple, entire to pinnate leaves. Flower-heads purple, pink, white or yellow; bracts with an apical appendage (bract characteristics important for identification). Pappus absent (or simple to toothed hairs or scales). A very large and complex genus (possibly comprising several genera) of which only a small subset are described.

A. Flower-heads yellow (except *C. horrida*); bracts ending with a single *prominent spine* and usually with smaller lateral spines.

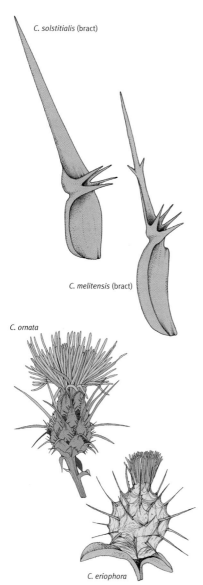

C. solstitialis (bract)

C. melitensis (bract)

C. ornata

C. eriophora

Centaurea solstitialis St. Barnaby's thistle A white-hairy, much-branched annual with *winged stems* to 60 cm, *grey woolly hairy or felted*. Lower leaves pinnately lobed or toothed, often withering by flowering-time; upper leaves linear-lanceolate and entire with spine-tips, the base decurrent down the stems to form a wing. Flower-heads usually solitary and *yellow*; involucre spherical to ovoid, to 12 mm across; bracts tipped with *stout, apical yellow spines 10–15 mm, with shorter lateral spines palmately arranged at the base.* Fallow ground and waste places, common. Throughout. *Centaurea melitensis* Maltese star thistle is a similar rough, grey-green (not densely grey-felted) annual with flower-heads in groups of 3–4 with surrounding leaves; *apical spines <10 mm with shorter lateral spines pinnately arranged along the lower ¹/₂.* Waste and bare places. Throughout. *Centaurea sicula* (Syn. *C. nicaeensis*) is similar to *C. solstitialis* but with *unwinged stems* and *larger flower-heads*, the stalks often swollen below; involucre 13–15 mm; spines on bracts 10–20 mm. DZ, ES, IT (incl. Sardinia and Sicily), MA, MT, TN.

Centaurea ornata A tall, slender and sparingly branched perennial 20–80 cm (1 m) tall. Leaves more or less hairless, undivided to 2-pinnately lobed with narrow segments to 3 mm wide. Flower-heads *dull yellow*, often tinged orange or brown; involucre spherical, 22–30 mm, bracts with *long, spreading apical spines* 15–35 mm long. Dry, scrubby places. ES, PT.

Centaurea eriophora An annual, easily identified by its rounded flower-heads with bracts *densely clothed in white, cobweb-like hairs* (arachnoid), with long spines projecting through; florets rather few, and yellow. DZ, ES, MA, PT.

1. *Centaurea solstitialis*   4. *Centaurea rupestris*
2. *Centaurea solstitialis*   5. *Centaurea calcitrapa*
3. *Centaurea melitensis*

Centaurea rupestris An erect-spreading perennial to 1 m with leaves *deeply pinnately divided with long, linear, almost thread-like segments.* Flower-heads yellow; bracts broad, green and leathery and brown at the tip, each with a deflexed-spreading, terminal yellowish spine, surrounded by small, white, filament-like appendages. Rocky slopes and garrigue. Strongly eastern in distribution, extending locally into southern France. IT, FR.

Centaurea horrida A woody, *domed, intricately branched, spiny, blue-grey subshrub* to 50 cm. Lower leaves pinnately divided, to 25 mm long, upper leaves simple or shallowly lobed. Flower-heads with white florets with purple tips; bracts terminating in a distinct brown spine, fringed with shorter spines, and woolly. A rare island endemic, restricted to just a few sea-cliffs at 4 sites. IT (Sardinia).

B. Flower-heads pink; bracts with *few to many*, prominent spines.

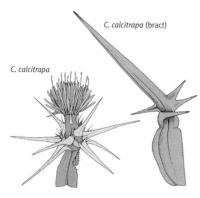

C. calcitrapa (bract)

C. calcitrapa

Centaurea calcitrapa RED STAR THISTLE A medium, much-branched, virtually hairless perennial 50 mm–50(80) cm with grooved stems. Leaves grey when young, glandular, remotely shallowly to deeply pinnately lobed with bristle-pointed lobes, often withered below when in flower; upper leaves smaller and narrower. Flower-heads *reddish-purple*, 12–17 mm with equal florets, surrounded by conspicuous *spreading, long and star-like spines 15–30 mm, >3 x as long as the longest laterals*; involucre ovoid-cylindrical. Pappus absent. Waste places, roadsides, bare, sandy ground and other disturbed habitats; common. Throughout.

C. sphearocephala

Centaurea sphaerocephala A variable, medium, bushy, branched annual 15–40(50) cm. Leaves cobweb-hairy; lyre-shaped to slightly lobed below, and pinnately lobed above; stalked below but clasping the stem above. Flower-heads mauve sometimes whitish in the centre, to 35 mm across, borne solitary; outer rays *spreading*; flower-bracts with *7–9 diverging yellowish apical spines (3)6–9 mm*. Common on coastal sands. More or less throughout, except southern France.

C. Flower-heads pink or purple; bracts ending in a papery fringed border or comb-like fringe.

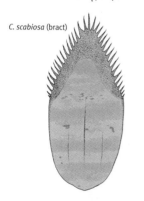

C. scabiosa (bract)

Centaurea scabiosa GREATER KNAPWEED An erect, downy perennial to 1.2 m with grooved stems, often branched above the middle. Basal leaves to 25 cm, stalked and usually *deeply pinnately divided into toothed lobes*; stem leaves similar but stalkless. Flower-heads large, to 60 mm across and solitary; florets red-purple (sometimes dull cream), *the outermost greatly enlarged and spreading, forming an outer ring*; bracts green, with fringed appendages. Mainly northern. ES, FR, IT.

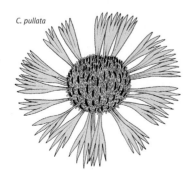

C. pullata

Centaurea pullata A short, branched or unbranched annual with rough-hairy leaves forming a basal rosette below; lyre-shaped to slightly lobed below, and pinnately lobed above. *Flower-heads mauve to bluish, or white, and large* to 50 mm across, borne solitary, often surrounded by leaves; outer florets greatly exceeding the inner, and widely spreading; flower-bracts hairless, with a conspicuous black margin and *comb-like* apex. Dry, rocky, stony and sandy habitats. Western in distribution, naturalised elsewhere. ES, MA, PT.

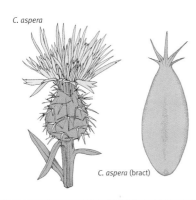

C. aspera

C. aspera (bract)

Centaurea aspera TOUGH STAR THISTLE A very variable, erect perennial to 50(80) cm, leaves pinnately lobed or toothed. *Flower-heads pink-purple* and with bracts with *appendages consisting of 3–5 chaffy, palmately arranged, recurved, spiny teeth*; the apical spine <5 mm, <1.5 x as long as the longest laterals; involucre broad at the base, narrowing at the apex. Common throughout.

1. *Centaurea sphaerocephala*
2. *Centaurea scabiosa*
3. *Centaurea pullata*
4. *Centaurea aspera*
5. *Centaurea jacea*
6. *Centaurea deusta*

Centaurea cineraria A variable, *white-grey-felted* spreading perennial with deeply pinnately lobed leaves. Flower-heads purple, rather small, to 32 mm and solitary; bracts green with a dark brown, spiny comb-like fringe. Coastal rocks. IT (incl. Sicily).

D. Flower-heads yellow; bracts ending in a papery fringed border or comb-like fringe.

Centaurea crocata A leafy annual with pinnately lobed, hairy, purplish leaves and ascending stems. *Flower-heads yellow* with *green, leathery bracts*, broad at the base, ending in a broad, deflexed, blackish border with regular, spine-like teeth. Southern Iberian Peninsula; local. ES, PT. *Centaurea prolongi* is very similar (and often confused) and has long-stalked leaves, flower-heads 15–18 mm across, and *deep golden to orange florets*. (Populations from southern Portugal are now believed to correspond with *C. crocata*). ES.

E. Flower-heads pink or purple; bracts distinctly enlarged into a broader, papery terminal section.

C. jacea (bract)

Centaurea jacea BROWN-RAYED KNAPWEED A downy perennial 20–60(80) cm with slender, rough stems, sometimes thickened below the flower-heads, unbranched, or with some branches above the middle. Leaves oval-lanceolate and unlobed or sometimes pinnately lobed. Flower-heads solitary, 15–20 mm across, with red-purple florets, the outermost larger; *bracts rather shiny, pale brown, rounded and irregularly jagged* (not regularly toothed). Absent from most islands. ES, FR, IT (incl. Sicily). *Centaurea deusta* is similar, but with bracts with *broad, silvery-papery margins*. IT.

1

F. Bracts entire (with no prominent fringe, spines or border).

Centaurea aeolica A bushy, *white-tomentose* perennial, often with arching or prostrate stems, with grey, pinnately divided leaves with linear-spatulate segments. Flower-heads mauve; involucral bracts compressed, oval-lanceolate and yellowish and entire. A rare Italian endemic in volcanic habitats. Two geographically isolated subspecies occur: Subsp. *aeolica* occurs on the Aeolian Islands of Sicily. Subsp. *pandataria* occurs on Ventotene near Naples. IT.

Cheirolophus

Very similar to *Centaurea* with leaves lobed or not; spineless. Flower-heads borne on long stalks swollen beneath the involucre. Outer achenes without a pappus.

Cheirolophus sempervirens A knapweed-like, roughly hairy perennial to 40(60) cm, somewhat woody at the base with *branches leafy throughout.* Leaves lanceolate, stalkless and long-pointed. Involucre 25 x 25 mm, globose, with hairless bracts; florets purple. Local in rocky places. Throughout.

Rhaponticum

Annuals and perennials similar to *Centaurea*. Flower-heads typically with conspicuously broad, papery bracts. Pappus hairy.

Rhaponticum coniferum (Syn. *Leuzea conifera*) PINECONE THISTLE A small, knapweed-like, white-hairy perennial to 30(45) cm with stems leafy at the top. Leaves alternate, oval-lanceolate, undivided or pinnately lobed below, divided into narrow segments above; white-felted beneath. Flower-heads borne solitary, pink, purple or whitish with a *prominent pine cone-like involucre* 32–40(50) x 32–37 mm, and rather inconspicuous florets. Local in pine woods and scrub. ES, FR, IT, MA, PT.

R. coniferum

Mantisalca

Very similar to *Centaurea*. Flower-heads with leathery, yellowish-green, spine-tipped bracts. Pappus of long, bristle-like scales. A controversial genus for which the number of species is disputed; the species below are similar, careful attention to the bracts and seeds is required.

M. salmantica

A. Plant a perennial.

Mantisalca salmantica A knapweed-like, erect, more or less hairy perennial to 1.5 m. Leaves crowded at the base in a rosette; oval in outline and pinnately lobed; upper leaves few, linear and toothed, to 35 cm. Flower-heads mauve, borne solitary or in lax branched clusters; involucre prominent and swollen in appearance; 13–19(21) x 6–11(15) mm, with oval, black-tipped bracts, each with a short *apical spine to 1 mm, often soon-falling.* Achene 3.2–4.7 mm with a *double pappus of bristles 2–4.5 mm.* Cultivated ground, roadsides and dry waste places. Common throughout (sporadic further east).

1. *Centaurea aeolica* PHOTOGRAPH: GIANNIANTONIO DOMINA

B. Plant an annual.

Mantisalca duriaei An annual to 1.5 m similar to *M. salmantica* but with involucral bracts with persistent or falling spines 2–2.6 mm. Outer achenes with a *short pappus 1–1.5 mm or absent*, the inner achene with a pappus 2–3.2 mm. Throughout (less common than *M. salmantica*). *Mantisalca spinulosa* is an annual to 60 cm. Involucral bracts with persistent spines 1.2–2.6 mm. *Achenes all with a pappus 2–3.5 mm.* ES (central, southeast). *Mantisalca delestrei* is very similar to *M. spinulosa* but is taller (>60 cm) with larger flower-heads with involucral bracts with *spines 2–6 mm.* DZ, MA. *Mantisalca cabezudoi* is an *annual to 40 cm, covered in long hairs at the base of the stems and leaves* and has involucral bracts with spines just 0.5–0.7 mm. ES (southeast).

1. *Carthamus lanatus*
2. *Klasea integrifolia* subsp. *monardii*
3. *Klasea baetica* subsp. *lusitanica*
4. *Carduncellus arborescens*

Klasea

Rhizomatous perennials with spineless, entire or deeply lobed leaves (often on the same plant). Flower-heads solitary, with oval-triangular bracts ending in a spine. Pappus of simple, persistent hairs.

Klasea baetica A rather short, clumped, rhizomatous perennial to 90 cm with *entire*, variably toothed (but not deeply lobed) and scarcely hairy leaves 35 cm long, ending in short spine-tips. Flower-heads borne on stalks sparsely leafy throughout, with an involucre 10–25 x 15–40 mm with narrowly oval involucral bracts which *gradually end in a pale brown, reflexed spine;* florets 20–35 mm, pink-purple. Garrigue and maquis. DZ, ES, MA, PT, TN. Subsp. *baetica* has involucral bracts ending in a spine-like appendage 2–7(9) mm. ES, MA, PT. Subsp. *alcalae* has involucral bracts ending in a spine-like appendage 3–6(9.5) mm. DZ, ES, MA, TN. Subsp. *lusitanica* has involucral bracts ending in *long spines 6–14 mm*. PT. *Klasea integrifolia* is a similar perennial to 50 cm with *shorter involucral spines, 0.5–4 mm long that arise apbruptly from the bract.* Flower-heads often short-stalked, and not leafy above. Subsp. *integrifolia* has corolla tube 20–40 mm and involucral bracts clearly concave and split along the edges. ES. Subsp. *monardii* has corolla tube, 28–33mm long and involucral bracts slightly concave and entire. ES, PT. *Klasea pinnatifida* has markedly lobed (rather than toothed) leaves with very *prominent pale lateral veins*, and involucral spines arising abruptly from the bracts. DZ, ES, MA, PT, TN. *Klasea flavescens* is similar to the species above but often has *white to yellow flower-heads* (sometimes pink-red like the above species) and involucral spines 3–8 mm long that are narrow and backwardly reflexed. Throughout (several regional subspecies occur).

Carthamus

Very spiny, often glandular *annuals.* Leaves pinnately lobed with spiny margins. Flower-heads solitary, surrounded by spiny, leaf-like bracts, yellow to orange. Pappus absent or *a series of narrow, pointed, persistent scales.*

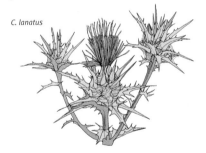
C. lanatus

Carthamus lanatus A very spiny, thistle-like annual 15–60 cm (1 m) with straw-coloured stems unbranched below, branched above, covered in white-woolly hairs when young. Leaves lanceolate, pinnately lobed with a spiny margin, withered below when in flower; clasping the stem above. Flower-heads 30 mm across, yellow; flower-bracts with a spine-toothed appendage. Pappus a series of scales as long as the achene. Common in bare, dry and sandy places. Throughout.

Carduncellus

Closely related to *Carthamus* but *woody perennials.* Flower-heads yellow or purple. *Pappus a series of persistent or shedding bristles.*

A. Flower-heads yellow.

Carduncellus arborescens (Syn. *Carthamus arborescens; Phonus arborescens*) A perennial superficially similar to *Carthamus lanatus*, but a more *robust perennial* to 2 m with broad involucral bracts, distinctly *long anthers to 10 mm*, and a shedding (not persistent) pappus. DZ, ES, MA.

B. Flower-heads blue-purple.

Carduncellus caeruleus A medium, greyish, hairy perennial 20–80 cm high with normally unbranched, erect, unwinged stems. Leaves rather shiny, grey-green, lyre-shaped and toothed or untoothed to pinnately lobed with bristle-tips; upper leaves semi-clasping the stem. Flower-heads blue-purple, borne solitary, surrounded by leafy bracts; involucre 16–30 x 20–30 mm; florets 14–22 mm, tubular and *deeply 5-lobed*. Fallow land and roadsides. Throughout.

Scolymus SPANISH OYSTER PLANT

Stout, spiny, thistle-like perennials. Flower-heads with outer bracts leaf-like and grading into the true upper leaves. Achenes flattened, not beaked, with pappus absent or few rigid hairs.

S. hispanicus

Scolymus hispanicus SPANISH OYSTER PLANT A robust, spiny, medium biennial or perennial to 80 cm with interrupted *spiny-winged* stems. Lower leaves yellowish green, oblong, pinnately lobed with sparse spines; upper leaves smaller and spinier; leaves with paler veins and border. Flower-heads golden-yellow, 25–40 mm across, *borne in long, narrow spike-like panicles; florets rayed* (not tubular); flower-bracts slightly hairy or hairless, with membranous margins, and narrowed into a sharp point. Frequent on fallow ground, roadsides and sandy waste ground. Throughout. *Scolymus maculatus* is very similar but with *broader wings 2–5 mm, at the narrowest, with prominent white margins*. Throughout. *Scolymus grandiflorus* is distinguished by its *markedly hairy* oval to linear flower-bracts. FR, IT.

Cichorium CHICORY

Annual or perennial herbs with a white latex when cut. Leaves toothed or lobed. Flower-heads numerous; florets all rayed and toothed at the tips. Achenes angular (not flattened); pappus a series of short scales.

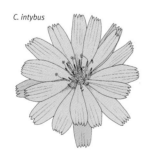

C. intybus

Cichorium intybus CHICORY A hairless to stiffly hairy, erect, branched perennial to 1 m. Basal leaves pinnately lobed and short-stalked below, lanceolate and clasping the stem above. Flower-heads bright sky-*blue*, 25–40 mm across, borne in narrow, leafy, branched spikes. Fairly common on fallow ground and roadsides. Throughout. *Cichorium endivia* is similar but a blue-grey annual with less deeply lobed leaves and *flower-stalks distinctly swollen* beneath the terminal flower-heads. Widely cultivated and naturalised. Throughout. *Cichorium spinosum* SPINY CHICORY is distinguished by its densely branched, mounded growth habit with spiny branches. ES (south + Balearic Islands), IT (incl. Sicily).

1. *Carduncellus caeruleus*
2. *Scolymus hispanicus*
3. *Cichorium intybus*
4. *Catananche caerulea*

Catananche

Rhizomatous perennials. Flower-heads solitary, each at the end of a long branch, with *shiny, papery bracts* loosely overlapping. Achenes 5–10-ribbed, not beaked; pappus of scales.

Catananche caerulea CUPIDONE A slender-stemmed perennial to 80 cm with short-adpressed hairs and linear to narrowly lanceolate, 3-veined leaves to 30 cm, often with lateral lobes at the base; stem leaves distant. Flower-heads blue, 25–35(40) mm across and solitary; all florets rayed and toothed at the tip; *bracts oval, papery and shiny, loosely overlapping,* each with a darker central stripe. Dry pastures. ES (incl. Balearic Islands), FR, IT, PT. *Catananche lutea* YELLOW CUPIDONE is similar, with narrower and smaller *yellow flower-heads*. IT (incl. Sardinia and Sicily).

Launaea

Shrubs and perennials, often intricately branched, with spineless leaves that exude a latex when cut. Flower-heads with ray florets only. Pappus of feathery bristles.

Launaea cervicornis A short, very dense, *mounded, cushion-like shrub* with *woody, spiny, zig-zgging stems* and basal, spineless leaves soon-withering to appear leafless. Flower-heads terminal and yellow with broad, flat, spreading, notched ray florets. Coastal cliffs and rocky garrigue on Mallorca and Menorca only. ES (Balearic Islands). *Launaea arborescens* is a similar, slightly mounded to open, *spreading shrub* with woody, spiny, zig-zagging stems branching at right angles. Leaves greenish, linear and alternate, soon withering and sparse. Flower-heads terminal and yellow. Coastal Garrigue and desert fringe habitats, mostly in North Africa. DZ, ES (southeast), MA.

Launaea fragilis (Syn. *L. resedifolia*) A short perennial with mostly basal leaves, and many, virtually leafless (but with bracts) long, straggling stems, often with a rather tangled appearance; leaves narrow and distantly pinnately lobed with spineless margins. Flower-heads solitary, yellow, to 25 mm across with all florets rayed. Achenes 5–7 mm. Coastal sands. ES, IT (Sicily), MA. *Launaea nudicaulis* is similar, with leaf margins with small white spines. Achenes 3–4 mm long. Dry saline or seasonally arid habitats; common in parts of North Africa. DZ, ES, MA, TN.

Tolpis

Annual to perennial herbs. Flower-heads with *very long, narrow, curved-spreading bracts*. Pappus typically with long, rough hairs.

T. barbata

Tolpis barbata ᴛᴏʟᴘɪs A variably sized, somewhat hairy annual to 90 cm with slender, spreading, branched stems. Leaves linear-lanceolate, toothed, lobed, or untoothed. Flower-heads *lemon yellow with a contrasting dark purple-brown centre*, to 30 mm across, borne on thickened stalks; florets all rayed; *thread-like bracts spread untidily beneath the flower-heads*. Pappus with rigid hairs. Common in disturbed sandy places, mostly in coastal areas. Throughout, except for some islands. *Tolpis virgata* has a straggling habit with ascending stems carrying *all yellow flower-heads* without a series of long, thread-like bracts below. FR, IT.

Hyoseris

Flower-heads borne solitary at the end of leafless stems, arising from a basal rosette of leaves. Pappus with *unequal yellowish hairs*.

Hyoseris scabra A low, dandelion-like annual with a rosette of basal leaves. Leaves pinnately divided with *backward-pointing* triangular lobes. Flower-heads borne on *ascending to prostrate (not erect), thickened stalks* 6–80 mm, swollen at, or above the middle and hollow above. Flower-heads yellow and dandelion-like, 8–10(30) mm across; florets all rayed. Achenes 6–7 mm. Common on grassy fallow land. Throughout. *Hyoseris radiata* is similar but a larger, leafier perennial with ascending stalks which exceed the leaves. ES, FR, IT, MA.

Hedypnois

Small annuals with rosette leaves. Flower-heads yellow, with *fleshy bracts in a single row, persisting and encircling the fruiting heads*. Achenes ribbed, not flattened, not beaked; pappus of scales.

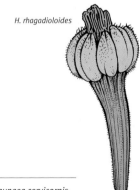

H. rhagadioloides

Hedypnois rhagadioloides (Syn. *H. cretica*) A variable, more or less hairy annual to 45 cm with mostly basal leaves, and *many, branched stems*. Leaves narrowly elliptic, entire to deeply lobed, with winged stalks below; stalkless above. Flower-heads 5–15 mm across, yellow and dandelion-like, borne on stalks thickened immediately below; involucre with narrow linear-lanceolate bracts *strongly incurved in fruit*. Fairly common in a range of dry and disturbed habitats across the region. *Hedypnois arenaria* is similar but with flower-stalks scarcely thickened, and involucral bracts *not*, or only slightly, incurved in fruit. Coastal sands. ES, MA, PT.

1. *Launaea cervicornis*
2. *Launaea arborescens*
3. *Launaea nudicaulis*
4. *Tolpis barbata*
5. *Tolpis barbata* (bracts)

Rhagadiolus

Hairy annuals. *Achenes few, lobe-like, borne in a star-like formation;* pappus absent.

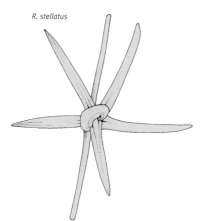

R. stellatus

Rhagadiolus stellatus STAR HAWKBIT A rather weedy, coarsely hairy annual with branched, spreading stems to 40 cm. Leaves 25 mm–14 cm, oblong, sparsely toothed to lobed and indistinctly stalked. Flower-heads yellow, small to 10 mm across, long-stalked in lax panicles. *Fruiting head a star-shaped* series of (5)7–8 long, slender, curved lobe-like achenes 10–16 mm long. Fallow land. Throughout. *Rhagadiolus edulis* is very similar but with fruiting heads with 5(6) shorter, straighter achenes 10–13 mm long. Throughout.

1. *Hedypnois rhagadioloides*
2. *Hedypnois rhagadioloides*
3. *Rhagadiolus stellatus*
4. *Urospermum picroides*
5. *Urospermum dalechampii*
6. *Hypochaeris radicata*

Urospermum

Flower-heads with 7–8 bracts *all in 1 row* and fused at the base; only ray florets present. Achene beaked; pappus of 2 rows of plumose hairs.

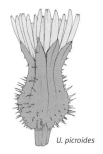

U. picroides

A. Plant an annual.

Urospermum picroides An *annual* with bristly stems 20–60 cm. Leaves bristly, rather large to 20 cm long, and sow thistle-like; toothed to lobed, oblong below and linear-lanceolate above. Flower-heads yellow and dandelion-like, borne on stalks thickened below; involucre 25 x 28 mm, cylindrical. Pappus white and fluffy (plumose). Waste ground. Throughout.

B. Plant a perennial.

Urospermum dalechampii A rather short, hairy perennial with rosettes of pinnately lobed basal leaves with backward-pointing lobes, the end segment enlarged; stem leaves oval-lanceolate and lobed or unlobed, the uppermost opposite. Flower-heads pale *sulphur-yellow coloured, often with a dark brown centre*; involucre 17–20 x 15–23 mm, cylindrical. Scrub and disturbed habitats. ES, FR, IT, MA.

Hypochaeris CAT'S-EAR

Annual to perennial herbs with rosettes of leaves and solitary or few, branched stems. Flower-heads yellow, with rayed florets only; bracts in several overlapping rows; *receptacle with scales between the florets*. Achenes finely ribbed, beaked or not; pappus 1–2 rows of brownish hairs.

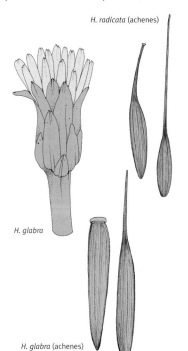

H. radicata (achenes)

H. glabra

H. glabra (achenes)

Hypochaeris radicata COMMON CAT'S-EAR A perennial to 60 cm with almost *hairless, branched stems*, and a basal rosette of bristly hairy, wavy-toothed, oblong, dark-tipped leaves to 25 cm. Flower-stalks *thickened* below the flower-heads; flower-heads yellow, 20–40 mm across, with a bell-shaped involucre; *bracts hairless,* dark-tipped. Central achenes 8–17 mm, beaked (marginal achenes usually also beaked). Common in grassy habitats above sea level. Throughout.

Hypochaeris glabra SMOOTH CAT'S-EAR is similar but with virtually hairless, glossy leaves, and *small, partially closed flower-heads* 10–15 mm across with ray florets scarcely exceeding the bracts. Central achenes 6–9(14) mm (*marginal achenes not beaked*). Throughout.

Hypochaeris achyrophorus A rough-hairy perennial 12–14 cm high with a basal rosette of leaves and, often branched, leafless stems thickened below the flower-heads. Leaves rather broad, spatula-shaped and lobed or unlobed. Flower-heads bright yellow, solitary, 10–15 mm across, and with *densely bristly, linear bracts*. Achenes 5–7 mm. Grassy habitats. Throughout.

Leontodon HAWKBIT

Rhizomatous perennials similar to *Hypochoeris* with unbranched stems without bracts and with forked leaves. Flower-heads solitary, with a *receptacle without scales between the florets*. Achenes finely ribbed, not or indistinctly beaked; pappus 2 rows of pale brown hairs (or scales).

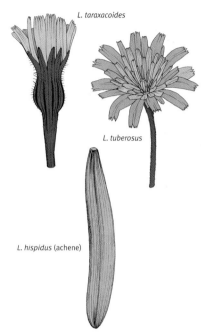

L. taraxacoides

L. tuberosus

L. hispidus (achene)

Leontodon taraxacoides LESSER HAWKBIT A bristly white-hairy perennial herb to 40 cm with leaves in a basal rosette, without tuberous roots. Leaves wavy-toothed to deeply pinnately lobed, sparsely bristly with *forked* bristles. Flower-heads borne solitary on *unbranched, leafless stalks* drooping in bud; flower-heads 12–25 mm across, the outer rays grey-violet underneath; bracts 7–11 mm, hairy or hairless except along the midribs. *Achenes with a ring of pappus of scales*. Common in grassy habitats. Throughout. *Leontodon tuberosus* is similar but with *tuberous* roots, and achenes with a *pappus of short hairs*. Similar habitats.Throughout.

Leontodon hispidus ROUGH HAWKBIT A perennial, rosette-forming herb to 60 cm high *covered in rough, white hairs*. Leaves oblong-lanceolate and narrow at the base, deeply wavy-toothed and *very hairy*. Flower-heads borne on stems to 40 cm long, *solitary*, with yellow florets longer than the bracts; bracts 11–13(15) mm. Achenes with a pappus of hairs. Grassy habitats, often above sea level. ES, FR, IT, MA.

Scorzoneroides

Perennials similar to *Leontodon*, without rhizomes, with simple leaves and (often) branched stems with scale-like bracts. Flower-heads often >1, with a receptacle without scales between the florets (but hairy); involucral bracts merging into stem scales. Achenes finely ribbed, not beaked; pappus 1 row of pale brown hairs.

S. autumnalis (achene)

Scorzoneroides salzmannii (Syn. *Leontodon salzmannii*) A small perennial with lobed leaves in a basal rosette, *completely hairless*, and with bright yellow flower-heads borne on stems branched above. Disturbed and sandy habitats; rare and local. ES, MA, PT. *Scorzoneroides autumnalis* (Syn. *Leontodon autumnalis*) AUTUMN HAWKBIT is a similar small perennial to 60 cm, *covered in white rough hairs*. Leaves lanceolate, wavy-toothed and narrowed at the base. Flower stems hairy throughout, with forked hairs; flower-heads solitary, with yellow florets much longer than the bracts; bracts 6–15 mm. Grassy places above sea level; local. ES, FR, IT.

1. *Leontodon hispidus*
2. *Picris hieracioides*
3. *Picris pauciflora*
4. *Helminthotheca echioides*
5. *Helminthotheca comosa* subsp. *lusitanica*

Picris OXTONGUES

Rough-hairy biennials to perennials with lax, branched stems. Flower-heads yellow, with rayed florets, the outer often with a reddish stripe; *outer bracts lanceolate, similar to the inner*. Achenes weakly ribbed, not, or shortly beaked; pappus 2 rows of off-white hairs.

Picris hieracioides HAWKWEED OXTONGUE A slender, *bristly* biennial with branched stems to 1 m. Leaves lanceolate to narrowly oblong, shallowly toothed or lobed, the upper leaves clasping the stem. Flower-heads long-stalked, yellow, to 35 mm across, the *involucral bracts lanceolate, spreading to recurved, simple hairs, dull green* (some blackish). Pappus cream-coloured. Various dry habitats; absent from the Balearic islands. ES, FR, IT.

Picris pauciflora A medium to tall annual or biennial with bristly untoothed to sparsely toothed leaves. Flower-heads borne on erect, branched, spreading panicles; rather few and lax; *involucral bracts all linear-lanceolate, rather concave and inwardly adpressed*, often reddish, and bristly; florets yellow. Common in southern France in a range of habitats; rare or absent elsewhere. FR.

Helminthotheca

Annuals to biennials similar to and previously grouped with *Picris*. Flower-heads with *oval outer bracts, much wider than the inner.* Achenes long-beaked; pappus white and feathery.

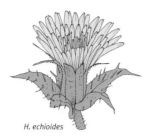

H. echioides

Helminthotheca echioides (Syn. *Picris echioides*) BRISTLY OXTONGUE A robust *very bristly* annual or biennial to 80 cm; *bristles arising from pimple-like bases.* Leaves elliptic to oblong, wavy-edged and pimply, the lower with winged stalks, the upper clasping the stem. Flower-heads yellow and numerous; *the outer-most involucral bracts 3–5, large and oval-heart-shaped, resembling an epicalyx,* 15–20 x 7–10 mm. Pappus white. A common weed of waste places. Throughout.

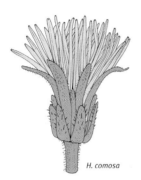

H. comosa

Helminthotheca comosa (Syn. *Picris comosa*) A very rough-hairy biennial to perennial with solitary, branched stems 30–90 cm tall. Leaves with scattered prominent spines with bulbous, pronounced bases. Flower-heads with yellow, rayed florets; involucre long, to 18 mm; *outer-most involucral bracts not heart-shaped at the base,* 4.5–7 x 2–2.5 mm. Achenes with a *long* beak, to 2 x as long as the body; pappus white. Maquis, garrigue, bare ground and rocky slopes. ES, MA, PT. Subsp. *comosa* has internal involucral bracts with spines 3–4 mm long. ES, MA, PT. Subsp. *lusitanica* (Syn. *Picris algarbiensis*) has internal involucral bracts with short spines, just 1–3 mm long. ES, PT.

Scorzonera VIPER'S-GRASS

Perennial herbs with solitary or several stems. Leaves often linear and entire. Flower-heads yellow, pink, purple or white; bracts borne in overlapping rows; florets all rayed and toothed at the tip. Achenes ribbed, with or without a tubular base, not beaked; pappus brownish-white, feathery. A taxonomically difficult group with conficting treatments among regional floras.

A. Flower-heads yellow.

Scorzonera hispanica A variable, leafy, *branched* perennial 30 cm–1.2 m. Leaves variably linear to spatula-shaped, narrowed at the base. Flower-heads yellow, typically 1–5, with florets 2 x as long as the involucre; involucre 20–25 mm with broad bracts 6–8 mm wide. Pastures, also an escape from cultivation. Probably throughout the European region. Unbranched forms and strictly linear leaves, and forms with basal rosettes and reduced stem leaves are recognised by some authors as *S. glastifolia* and *S. trachysperma*, respectively (IT).

1. *Scorzonera undulata*
   PHOTOGRAPH: GIANNIANTONIO DOMINA

2. *Tragopogon pratensis*

3. *Tragopogon dubius* (habit)

4. *Tragopogon dubius*

Scorzonera villosa VILLOUS VIPER'S-GRASS A low to short, hairy, tufted perennial, often with white, shaggy hairs; stems solitary or few and unbranched; *leaves linear and untoothed,* with whitish basal sheaths. Flower-heads yellow, the rays reddish on the reverse; flower-bracts often white-woolly and recurved. Grassy habitats. IT (incl. Sardinia).

B. Flower-heads pink-violet.

Scorzonera undulata A perennial 10–30 cm high with rosettes of keeled, markedly wavy, glaucous, short-hairy, broadly linear leaves; stems mostly simple and short. Flower-heads large and conspicuous, 30–50 mm across, pink-violet with darker centres, opening in the sun; bracts broad with recurved tips. Bare, arid, semi-desert and wasteland habitats. DZ, IT (Sicily), MA, TN.

Podospermum

A genus closely related to (and still often included in) *Scorzonera*. Annuals, biennials or perennials, often with pinnately lobed leaves. Achenes with a tubular base; pappus of hairs.

A. Flower-heads lilac.

Podospermum purpureum (Syn. *Scorzonera purpurea*) PURPLE VIPER'S-GRASS A hairless perennial with unbranched or sparingly branched stems 15–45 cm. Leaves linear and keeled along the centre. Flower-heads 30–50 mm across, *pale whitish purple to lilac, with long ray florets*; flower-bracts numerous, sepal-like and arranged in overlapping rows. Cool, grassy habitats in southern Europe. ES, FR, IT.

B. Flower-heads yellow.

Podospermum laciniatum (Syn. *Scorzonera laciniata*) CUT-LEAVED VIPER'S-GRASS A variable, short to medium annual or biennial with 1–2-pinnately lobed leaves with linear segments in rosettes; upper leaves clasping the stem. Flower-heads yellow, 15–25 mm across, solitary, borne on thick, unbranched stalks. Flower bracts lanceolate to oval, often with dark, recurved tips. Fruiting heads large 'dandelion clocks'. Cultivated land and waste places; common. Throughout. *Podospermum canum* (Syn. *Scorzonera jacquiniana*) is a similar *hairy perennial*, often with *branched stems*. Flower-heads yellow, the rays often grey-green, purple or red on the reverse; flower-bracts much shorter than the rays. Saline waste places. IT (incl. Sardinia and Sicily).

Tragopogon GOAT'S BEARD

Annual to perennial herbs with a white latex when cut. Stems usually solitary. Leaves linear, often rush-like. Flower-heads with only ray florets. Fruiting head a large 'dandelion clock'; achenes ribbed, beaked; pappus with feathery and simple hairs.

A. Flower-heads yellow.

Tragopogon pratensis GOAT'S BEARD An annual or perennial with erect stems to 75 cm, downy when young, hairless and greyish when mature. Lower leaves narrowly linear-lanceolate, to 30 cm with a distinct keel; stem leaves similar, ending in a fine point. Flower-heads solitary, long-stalked, to 50 mm across with yellow florets. Fruiting head a *large 'dandelion-clock'*. Outer achenes <30 mm. Cool, grassy habitats; absent from most islands. ES, FR, IT. ***Tragopogon dubius*** WESTERN SALSIFY is similar but with *stalks markedly swollen beneath flower-heads* and outer bracts much exceeding florets. Similar habitats and distribution. ES, FR, IT.

B. Flower-heads pink (species are similar and difficult to distinguish).

T. crocifolius

Tragopogon crocifolius is similar to the above species but with dark purple or brownish flower-heads with *yellow central florets*, and the ray florets less markedly exceeded by the flower bracts. Local in stony and grassy waste places. ES, FR, MA, PT. ***Tragopogon porrifolius*** SALSIFY is similar to the previous species but with *robust flower-heads of purplish florets borne on stalks markedly swollen beneath the heads.* ES, FR, MA, PT. Subsp. *australis* (Syn. *T. sinuatus*) is *minutely hairy*, with flower rays $^1/_2$ as long as the bracts. ES, FR, MA, PT.

Geropogon

Annuals similar to *Tragopogon* (still classified in this genus by many authors). Achenes ribbed, beaked; pappus of rigid (not plumose) hairs.

Geropogon hybridus (Syn. *Tragopogon hybridus*) A slender, hairless annual to 50 cm with branched stems. Leaves long-linear, hairless and rush-like. Flower-heads borne on inflated stalks; lilac-pink, with few, purplish central florets much-exceeded by the surrounding rays, which are in turn *greatly exceeded by the narrow, long-pointed flower bracts*. Stony fallow land, widespread. Throughout.

Reichardia

Annual to perennial herbs. Involucre pitcher-shaped with bracts in several rows, each with a white margin. Achenes 4–5-angled (at least the outermost); pappus of numerous rows of soft, simple hairs.

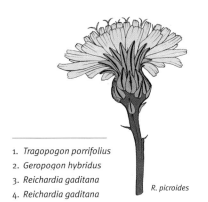

1. *Tragopogon porrifolius*
2. *Geropogon hybridus*
3. *Reichardia gaditana*
4. *Reichardia gaditana*

R. picroides

A. Plant a perennial, woody at the base.

Reichardia picroides A woody-based perennial with toothed to pinnately lobed leaves. Flower-heads yellow, borne on more or less leafless stalks scarcely thickened at the top; flower *bracts with narrow membranous margins* (<0.5 mm wide), the *florets all yellow* (*not purplish* at the base) though sometimes with a dark stripe on the outer surface. Outer achenes wrinkled, but *the inner smooth*, and appearing sterile. Cultivated and waste land. Throughout.

B. Plant typically an annual or biennial, not woody at the base.

Reichardia tingitana A variable, low to medium, hairless annual or biennial with oblong, toothed, pinnately lobed leaves, often with white pimples; basal leaves broadly winged at the base, the stem leaves few and linear, $^1/_2$-clasping the stem at the base. Flower-heads yellow, to 25 mm across, the *rays purplish at the base*, the outermost with a red stripe on the reverse; flower bracts to 15 mm long, with a membranous margin and hairless. Pappus *long, to 14 mm*. DZ, ES (south + Balearic Islands), MA. *Reichardia gaditana* is very similar, with leaves *smooth* (without pimples) with spine-toothed margins, narrowed at the base into a winged stalk. Involucral bracts to 22 mm long with wide, somewhat wavy, pale brown, membranous margins. All achenes wrinkled, and *pappus short, 7–9 mm long*. Dunes in western Iberian Peninsula. ES, PT. *Reichardia intermedia* is similar to *R. picroides* in habit and overall appearance, but an annual, and with *flower-bracts with membranous margins* to 1.25 mm wide, more similar to *R. tingitana*. DZ, ES, IT, MA, PT, TN.

Aetheorhiza

Herbaceous perennials *with long underground stolons* with white tubercles, from which arise leaves which are often $^1/_2$-buried in sand. Achenes with 4 grooves, not beaked; pappus of several rows of white, simple hairs.

Aetheorhiza bulbosa A small, green or purplish perennial to 30(55) cm with stolons and rhizomes and few, slender stems. Leaves hairless or hairy, all basal and *held prostrate on the surface of the sand;* elliptic and entire or lobed. Flower-heads 8–15 x 3–12 mm, borne terminally; yellow and dandelion-like; involucral bracts 14–15(16) mm, with blackish hairs. Achenes 3–4.5 mm. Common on coastal sands, rare elsewhere. Throughout.

Sonchus SOW-THISTLE

Annual or perennial herbs with stout, hollow stems exuding a white, sticky latex when cut. Leaves pinnately lobed or wavy with spiny margins; the stem leaves clasping at the base. Flower-heads yellow, all ray florets. Achenes flattened, ribbed, not beaked; pappus with 2 or more rows of white, simple hairs.

S. asper

Sonchus asper PRICKLY SOW-THISTLE A variably tall, erect, greyish or reddish, hairless (glandular above) annual to 1 m with simple or branched stems. Lower leaves spoon-shaped, sometimes pinnately lobed, with sharp, triangular toothed lobes, the lowest 2 *rounded* at the base, and *glossy* green on the upper surface; clasping the stem above. Flower-heads golden yellow, to 25 mm across. Achenes 2–3 mm. Very common in a wide range of habitats from towns to coastal dunes. Throughout. *Sonchus tenerrimus* is similar but with deeply pinnately divided leaves with all leaf lobes strongly constricted or linear at the base, and the terminal lobe equalling the laterals (not distinctly larger). Flower-stalks often white-hairy. Disturbed habitats; locally common. Throughout.

1. *Reichardia tingitana*   5. *Sonchus tenerrimus* (leaf)
2. *Aetheorhiza bulbosa*   6. *Sonchus oleraceus* (flowers)
3. *Sonchus asper*   7. *Sonchus oleraceus* (leaf bases)
4. *Sonchus tenerrimus*

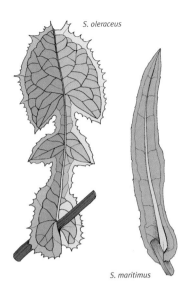

S. oleraceus

S. maritimus

Sonchus oleraceus SMOOTH SOW-THISTLE An erect annual to 1.5 m similar to *S. asper*; greyish or reddish, hairless (glandular above). Lower leaves spoon-shaped and variously, often deeply *pinnately lobed* with triangular toothed lobes, the *end-lobe distinctly larger* than the next pair down, and the lowest 2 *pointed* (not rounded) at the base, and *matte* (not glossy) green on the upper surface; leaves clasping the stem above. Flower-heads pale yellow, to 25 mm across. Achenes 2.5–3.8 mm. Very common in a range of habitats. Throughout.

Sonchus maritimus Similar to the above species but a *short, rhizomatous perennial* to 60 cm tall, with simple or sparingly branched stems, and leaves mostly at the base; linear and wavy-toothed or rarely lobed; the basal lobes rounded. Flower-stalks often *white-downy*. Local on damp coastal sands. DZ, ES, FR, IT, PT.

1 2 3
4 5 6 7

Lactuca LETTUCE

Annual, biennial and perennial herbs with a white latex when cut. Flower-heads with cylindrical involucre and all ray florets. Achenes flattened and ribbed; pappus with 2 rows of white simple hairs.

A. Flower-heads blue.

Lactuca perennis BLUE LETTUCE An erect, hairless perennial to 70 cm (1.2 m) with spineless, hairless, bluish leaves deeply pinnately cut into narrow, nearly entire lobes. Flower-heads blue or violet, borne in loose, spreading clusters, each with 12–20 flat, spreading rays (similar to flower-heads of *Chicorium* but borne on long stalks in loose, spreading clusters, not on short stalks in leaf axils). Achenes 5.5–8 mm, black and wrinkled. Absent from much of the far west, south and the islands. ES, FR, IT.

B. Flower-heads yellow; leaves prickly.

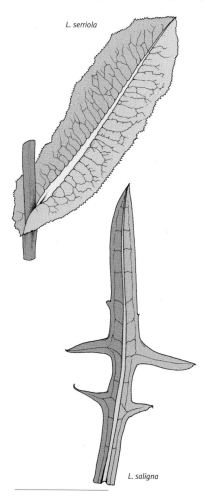

L. serriola

L. saligna

1. *Lactuca serriola*
2. *Lactuca perennis*
3. *Lactuca saligna*
4. *Chondrilla juncea*

Lactuca serriola PRICKLY LETTUCE A tall, greyish, stiffly erect annual or biennial to 1.5(2) m, unbranched below, and branched above with whitish stems. Leaves *held stiffly erect*, oblong-lanceolate, entire to pinnately divided with distant lobes below and sharply toothed, more or less hairless but spiny along the midrib beneath and along the margins; all leaves waxy and greyish. Flower-heads pale yellow and small, to 13 mm across, borne in a long, lax inflorescence. Achenes olive-grey, 3–4 mm (excl. beak). Local on waste ground. Throughout. *Lactuca virosa* is similar but taller, to 2 m, with *leaves spreading, not erect*, with rounded, clasping lobes at the base. Achenes 4.2–4.8 mm (excl. beak). Waste ground. Probably throughout.

C. Flower-heads yellow; leaves not prickly.

Lactuca saligna LEAST LETTUCE A slender annual to 75 cm (1 m) with *whitish stems* and greyish leaves, *not prickly,* with arrow-shaped bases clasping the stem, those below deeply pinnately lobed, those above linear-oblong and clasping. *Flower-heads borne in narrow, spike-like inflorescences with short branches;* florets pale yellow and reddish below. Achenes olive-grey, 2.8–3.5 mm with a white beak. Bare, grassy habitats. Throughout.

Lactuca viminea PLIANT LETTUCE A hairless biennial without prickles, often with numerous, erect stems 15–80 cm (1 m). Leaves dark grey-green, the lowermost pinnately lobed with linear-lanceolate toothed lobes, the stem leaves above lanceolate and more or less unlobed except at the base. Flower-heads pale yellow, borne in much-branched panicles; florets (4)5–8 per head. Achenes 6.5–12 mm. Throughout, except for some islands.

Chondrilla

C. juncea

Flower-heads cylindrical with 8–10 long, inner bracts and a row of very short, leafy outer bracts; disk florets absent. Achene ribbed, beaked or beakless, with scales at the base of the beak; pappus of simple hairs.

Chondrilla juncea A medium to tall greyish biennial, hairy, especially below. Stems normally solitary, to 1 m tall, stiff and *broom-like with few leaves*. Leaves oblong, lobed, withering below and linear and entire above. Flower-heads unstalked, in small clusters; florets all rayed. Sandy or scrubby open habitats; fairly common. Throughout.

Taraxacum DANDELION

Perennial herbs with tap roots, a basal rosette of leaves, and flower-heads borne on leafless, hollow stems that exude a milky latex. Achenes finely ribbed; pappus of white simple hairs. Only one, cosmopolitan species section is described here, however *Taraxacum* in its wider sense comprises a complex aggregate of numerous microspecies.

Taraxacum **sect.** *Ruderalia* (Syn. *T. officinale*) COMMON DANDELION A variable (group of) perennial(s) with leaves in a basal rosette and leafless, hollow stems. Leaves deeply pinnately lobed, with long, winged stalks at the base. Flower-heads bright yellow, the rays striped with purple beneath, on stalks to 40 cm tall; outer flower bracts 9–16 mm, backwardly curved. Fruiting heads borne as the familar 'dandelion clock'; achenes brown, 2.5–4 mm. Common in grassy and waste places across the region. Throughout.

1

2

3

4

Crepis HAWK'S-BEARD

Annual or perennial herbs with spirally arranged leaves with lobes pointing backwards and erect, branched stems. Flower-heads yellow with florets all rayed; flower bracts in 2 rows, the outer row often shorter and spreading. Achenes ribbed and flattened; pappus of white, brittle hairs.

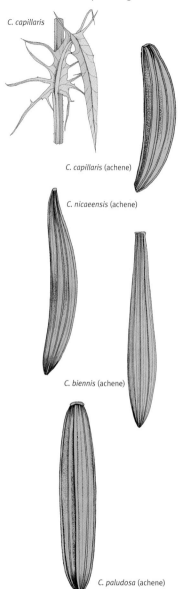

C. capillaris

C. capillaris (achene)

C. nicaeensis (achene)

C. biennis (achene)

C. paludosa (achene)

A. Achene without a beak; pappus white and silky.

Crepis capillaris SMOOTH HAWK'S-BEARD An erect, hairless annual to 75 cm, branched from the base, or above. Leaves sparsely hairy, glossy, entire to irregularly, narrowly pinnately lobed with a large triangular end-lobe, and narrower lateral lobes; upper leaves smaller and stalkless with *clasping, arrow-shaped bases*. Flower-heads erect in bud, yellow and dandelion-like, to 15 mm across, held erect in bud, in loose inflorescences; involucral bracts downy, often with blackish hairs, and adpressed. Achenes not beaked, 1.4–2.5 mm. Widespread in disturbed waste places. ES, FR, IT, PT. *Crepis nicaeensis* is similar but with outer bracts slightly spreading and achenes 2.5–3.8 mm, not beaked but narrowed above. Absent from most islands. ES, FR, IT.

Crepis biennis ROUGH HAWK'S-BEARD A rough-hairy biennial with stems to 1.2 m, often purple below. Basal leaves irregularly pinnately lobed with a large terminal lobe; stem leaves similar but *semi-clasping the stem and without markedly arrow-shaped bases*. Flower-heads to 30 mm, borne in loose clusters (crowded at first). Pappus equalling or slightly exceeding the involucre, and white and soft; achenes 4–7.5 mm. ES, FR, IT.

Crepis paludosa MARSH HAWK'S-BEARD A virtually hairless perennial with *hairless stems* to 80 cm, branched above. Basal and stem leaves oval-lanceolate, narrowed into short, winged stalks, the *uppermost leaves with arrow-shaped bases*; all leaves shiny, hairless and wavy-toothed. Flower-heads yellow, borne in few-flowered clusters; *bracts woolly with dense, blackish hairs*. Achenes 4–5.5 mm; pappus with *stiff, brittle, yellowish hairs*. Marshes and fens. Throughout.

1. *Taraxacum* sect. *Ruderalia*
2. *Andryala integrifolia*

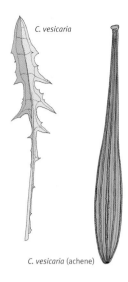

C. vesicaria

C. vesicaria (achene)

C. foetida (achene)

B. At least some achenes beaked, and ribbed.

Crepis vesicaria BEAKED HAWK'S-BEARD A robust *hairy perennial* with leaves with broad lobes, downy all over. Flower-heads to 15 mm across, with *orange-yellow* florets, the outer striped reddish externally; outer involucral bracts spreading; heads erect in bud. Achenes with *long, slender beaks,* 5–9 mm; fruiting pappus much exceeding the involucre. Meadows and waste places. Throughout. *Crepis foetida* STINKING HAWK'S-BEARD is similar but *strong-smelling and with flower-heads drooping in bud.* Central achenes long-beaked, 10–17 mm. Throughout. *Crepis rubra* PINK HAWK'S-BEARD is readily distinguished by its solitary or few *pink or whitish-pink flower-heads.* IT.

Andryala

Flower-heads many in a cluster; bracts in a single row with some additional bracts; only ray florets present. Pappus of soft hairs.

A. Flower-heads pale to mid-yellow, often with a darker centre; plant softly yellow-hairy.

A. integrifolia

Andryala integrifolia A variable, short to tall, white-hairy annual to perennial with sparingly branched, erect, leafy stems to 50 cm (1 m). Leaves oval-lanceolate, unlobed to lobed, semi-clasping the stem above, and densely covered in yellowish glandular hairs. *Flower-heads pale lemon yellow* (often with a darker centre), 20 mm across, borne abundantly in rather flat-topped clusters; flower bracts linear-lanceolate, and hairy; involucre 6–10 x 5–9 mm. Achenes 1 mm. Locally common in dry, sandy and grassy habitats. ES, FR, IT, MA, PT. *Andryala laxiflora* is very similar but with larger, fewer flower-heads (clusters flat-topped or elongated); involucre distinctly broader, to 15 mm wide (not 5–10 mm); pale to mid yellow, sometimes with a darker centre. ES, PT.

A. ragusina

B. Flower-heads uniformly bright yellow; plant densely white-felted.

Andryala ragusina A *densely white-felted* annual to perennial 30–50 cm with tufted basal, *pinnately lobed leaves.* Flower-heads *bright yellow* and with many, dense ray florets; involucre 7–13 x 5–18 mm, *without glandular hairs.* Achenes 2.4–2.7 mm. Local, mainly western. ES (inc. Balearic Islands), FR (incl. Corsica), PT. *Andryala agardhii* is also densely *white-felted* with folded, *narrowly spoon-shaped leaves* borne in dense basal clusters. Flower-heads *long-stalked and solitary*, yellow, with black-hairy bracts. Rocky screes and gorges. ES (south).

A. agardhii

Glossary

Achene, dry, 1-seeded, rather hard, indehiscent (non-splitting) fruit; there are often many achenes in a fruiting head, as in *Ranunculus*.

Actinomorphic, (of a flower) with a radially symmetrical (regular) shape that has multiple axes of symmetry, e.g. a lily flower; syn. **regular**.

Alien, not native, introduced to a region.

Alternate, arising at 1 axis per node; e.g. leaves arising at different heights along the stem (not opposite or whorled).

Angiosperm, flowering plant.

Annual, completing the life cycle in one year; typically without woody parts (herbaceous).

Anther, fertile, pollen-producing part of the stamen, typically on a terminal stalk (filament).

Apex, uppermost part of a structure.

Aril, succulent covering around the seed.

Ascending, arising upwards at an angle (curving).

Awl-shaped, long, pointed spike.

Awn, long, stiff bristle e.g. in florets of grasses (Poaceae).

Axil, point at which the leaf or leaf stalk joins the stem; adj. **axillary**.

Beak, elongated projection, usually on a fruit; adj. **beaked**.

Berry, succulent fruit, typically with >1 seed; seeds without stony coats.

Biennial, completing its life cycle in 2 years (often with leaves in a rosette in the first year and flowering in the second year).

Bifid, divided into 2 parts, typically deeply at the apex.

Blade, the main, often flattened, part of the leaf (or petal).

Bract, small, leaf-like structure, often subtending (beneath) a flower or inflorescence.

Bracteole, small secondary bract.

Bulb, underground storage organ composed of condensed stem and fleshy leaves.

Bulbil, small reproductive bulb borne in the leaf axil (sometimes among flowers e.g. in *Allium*).

Calyx, all the sepals of the flower (the outer whorls of the perianth if different from the inner).

Capitulum, a dense flower-head (inflorescence) composed of small stalkless flowers (often ray and disk florets) crowded together on a compound receptacle; typical of the daisy family (Asteraceae) and Dipsacaceae.

Carpel, female organ of the flower, comprising stigma, style and ovary.

Carpophore, a thin, sterile stalk above the pistil (typical in fruits of some Apiaceae and Caryophyllaceae).

Casual, introduced to an area and sporadic in its appearance, not persisting.

Cauliflory, flowers are produced directly from the primary branch or trunk.

Ciliate, fringed along the margin with hairs.

Cladode, flattened organs arising from the stem, typically resembling leaves.

Compound, made up of more than one similar part or segment (not simple), typically a leaf.

Cordate, heart-shaped (often the base of a leaf).

Corm, underground storage organ formed from a swollen stem base.

Cone, compact body of scales or bracts containing the reproductive structures in gymnosperms.

Corolla, all the petals of a flower (which may form a tube), the inner whorls of the perianth.

Corona, trumpet- or cup-shaped extension of the corolla in *Narcissus*, or fused filaments in some Apocynaceae.

Corymb, raceme in which the lower flowers have longer stalks, producing a flat-topped inflorescence; adj. **corymbose**.

Cosexual, flowers with both male and female reproductive organs, stamens and carpels, respectively.

Cupule, hardened, cup-like structure composed of bracts in Fagaceae.

Cyathium, specialised inflorescence of *Euphorbia* (Euphorbiaceae) with a cup-like structure containing a single carpellate (female) flower and several staminate (male) flowers.

Cyme, an inflorescence in which each flower terminates a branch; adj. **cymose**.

Deciduous, not persistent, for example leaves in autumn or petals after flowering.

Dehiscent, splitting.

Desiccation, drying out.

Dioecious, with male and female flowers on separate plants (individual plants of 1 sex).

Disk floret, small actinomorphic flower, forming part or all of a capitulum (in Asteraceae).

Divided, not entire (typically a leaf); divided into teeth, lobes or leaflets.

Drupe, succulent or spongy fruit, typically with 1 seed with a stony coat.

Elliptic, flat shape (typically a leaf), widest at the middle.

Endemic, restricted to a particular country, region or island.

Entire, whole; without distinct lobes, teeth or divisions.

Epicalyx, additional whorl of sepal-like bracts beneath the true calyx (e.g. in flowers of Malvaceae).

Epichile, distal part of the lip (labellum) in some orchid species (e.g. *Epipactis*) separated from the basal part (hypochile) by a joint.

Excerted, protruding (not included, such as anthers from a corolla tube).

Falcate, sickle-shaped.

Falls, outer perianth segments (tepals) of *Iris* flowers (Iridaceae).

Family, monophyletic group of related genera, the taxonomic group between the lower rank of genus and the higher rank of order.

Floccose, covered in soft, woolly hairs.

Floret, small individual flower making up part of a dense inflorescence, e.g. a component of the capitulum (in Asteraceae) or in grasses (Poaceae).

Flower-head, a group of flowers (inflorescence) such a capitulum (Asteraceae).

Follicle, dry, many-seeded fruit, dehiscent along 1 side; usually formed from 1 carpel.

Fruit, ripened ovary or ovaries of a flower containing 1 or more seed(s).

Genus (pl. **genera**), monophyletic group of related species, the taxonomic group between the lower rank of species and the higher rank of family; the generic name is the first part of the scientific binomial.

Garrigue, stunted sclerophyllous shrub-dominated vegetation at lower altitude, typically on coastal limestone (see also Maquis).

Geophyte, plant that survives the dry summer as a dormant underground bulb, corm or tuber.

Glabrous, not hairy.

Gland, organ of secretion, often in sticky plants; adj. **glandular** (often referring to hairs).

Glaucous, covered in a bluish, whitish or greyish waxy bloom (rather than green).

Globose, spherical.

Glume, (of grasses), bract below a spikelet

Gymnosperm, non-flowering, seed-bearing vascular plants such as Conifers.

Halophyte, salt-tolerant plant.

Head, group of flowers crowded together at the end of a stalk.

Hemiparasite, parasitic plant that gains some of its nutrition from another plant (host) but which also has chlorophyll and a root system; some hemiparasites can survive without a host (facultative hemiparasitism).

Herb, plant without woody parts; a soft and leafy annual, biennial or perennial in which aerial parts naturally die to ground level at the end of the growing season; adj. **herbaceous**.

Holoparasite, parasitic plant which gains all of its nutrition from another plant (the host) and which lacks chlorophyll and a true root system.

Hybrid, offspring of a cross between two different species, races, or varieties.

Hybridisation, the formation of hybrid offspring.

Hypochile, the basal part of the lip (labellum) of some orchid species (e.g. *Epipactis*); see also Epichile.

Inferior ovary, ovary situated beneath the point of insertion of other floral organs; syn. **epigynous ovary**.

Indehiscent, non-splitting.

Inflorescence, group of flowers with their floral stem (axis) and any associated bracts.

Internode, a part of the stem between two nodes.

Involucre, collection of involucral bracts (e.g. in Asteraceae).

Involucral bract, bracts surrounding a head of flowers (e.g. in Asteraceae).

Keel, boat-shaped structure formed by two lower petals in the pea family (Leguminosae), or a longitudinal ridge (typically on a leaf or petal).

Irregular, flower in which 1 or more members of the whorl, or several floral whorls, differ in form from the other members.

Labellum, lower-most petal of an orchid flower, often highly specialised, e.g. in bee orchids (*Ophrys*) and *Serapias*.

Lanceolate, narrowly ovate, spear- or lance-shaped.

Legume, term used to describe either the fruit or the plant itself in the pea family (Leguminosae).

Lemma, (of grasses) the lower of the pair of bracts (lemma and palea) that subtends the floret.

Ligule, (of grasses) a small membranous projection or ring of hairs at the junction of the leaf sheath and stem.

Limb, expanded portion of the calyx or corolla (distinct from the tube or throat).

Linear, long, narrow and parallel (typically the leaf of a grass).

Lip, region of the calyx or corolla sharply differentiated from the rest (see also Labellum).

Lobe, substantial division of a leaf, calyx or corolla.

Maquis, sclerophyllous shrub-dominated vegetation typical of the Mediterranean, often on deep, acidic substrates (see also Garrigue).

Membranous, paper-like or membrane-like in consistency.

Mericarp, 1-seeded portion formed by the splitting of a 2-many-seeded fruit.

Midrib, the central or main vein.

Monoecious, with separate male and female reproductive structures on the same individual plant; contrast dioecious.

Morphology, the appearance, form or structure.

Mucronate, with a short bristle-tip.

Mycoheterotrophy, process by which a non-photosynthetic plant obtains nutrition from a fungal symbiont (or sometimes from another plant via a shared fungal symbiont) living in its root system.

Native, naturally occurring in the area.

Naturalised, not native but well established.

Node, the position on the stem from which the leaves, branches or flowers arise.

Nut, dry, indehiscent, 1-seeded fruit with a woody, hard wall; often large in size.

Nutlet, a small, woody-walled nut (see also aAchene).

Obconical, an inverted cone, attached to the stalk by the pointed end.

Oblong, elongated but wide in shape, the middle part parallel-sided (usually describing a leaf).

Obovate, ovate (oval), and narrower at the base.

Obtuse, with a point >90°.

Opposite, of 2 organs arising from a common node (e.g. leaves from a stem).

Ovary, part of the carpel or pistil containing the ovules, and later, the seeds.

Ovate, oval in outline, or egg-shaped.

Ovoid, solid shape, oval/ovate in side view.

Ovule, structure containing the egg, becomes the seed after fertilisation.

Palea, (of grasses) the upper of the pair of bracts (lemma and palea) that subtends the floret.

Palmate, lobes or segments radiating from a common axis.

Panicle, branched, compound inflorescence.

Pappus, structure consisting of hairs, bristles or scales on the fruit (achene) of Asteraceae.

Parasite, plant obtaining nutrients from another plant (may be hemi- or holo-parasitic).

Pedicel, the stalk of an individual flower in an inflorescence or the stalk of a grass spikelet.

Peltate, leaf with stalk attached to the centre of the blade.

Perennial, living for more than two years, generally flowering every year, often woody at the base.

Perfoliate, (of a leaf or bract) with the base united (around the stem).

Perianth, all the non-sexual segments (i.e. calyx and corolla together) of a flower (see also **tepals**).

Petal, one of the segments of the inner whorl(s) of the perianth; **petaloid** is petal-like.

Petiole, the stalk of the leaf; adverb. **petiolate** (with a petiole).

Phyllode, a flattened, expanded petiole that resembles and functions as a leaf.

Phytogeographic, a geographic region defined by its flora (e.g. the *Mediterranean Floristic Region*).

Pinnate, (of a compound leaf) composed of leaflets arranged on opposite sides of a common axis (rachis); adj. **pinnately** (e.g. pinnately divided into leaflets).

Pinnatifid, (of a leaf) pinnately divided with the lobes cut nearly (but not quite) to the mid-vein.

Pinnatisect, (of a leaf) pinnately divided to the mid-vein, but lobes not contracted at the base to form discrete leaflets.

Plumose, with many fine filaments or branches, giving a feathery or 'fluffy' appearance (e.g. the pappus of some Asteraceae).

Pollinium (pl. **pollinia**), a mass of adhering pollen grains (e.g. of an orchid) that is shed and transported as a unit by a pollinator.

Prickle, a spiny outgrowth, broadened at the base.

Procumbent, trailing on the ground.

Pseudocopulation, the process by which a male insect (usually a bee or wasp) attempts to mate with the flower of a bee orchid (*Ophrys*) and in so doing, brings about cross fertilisation.

Receptacle, the thickened or expanded part of the stem from which the flowers or inflorescence arise (e.g. in Asteraceae).

Raceme, a simple unbranched inflorescence with stalked flowers borne on a single axis, the youngest flowers at the top; adj. **racemose**.

Rachis, stalk of a compound leaf or the central axis bearing the flowers.

Ray, a radiating branch of an umbel or cyathium (e.g. in Apiaceae or Euphorbiaceae).

Ray floret, a small zygomorphic flower often resembling a single petal in the inflorescence (flower-head) of Asteraceae. Contrast with disk floret.

Receptacle, the portion of the axis of a flower stalk on which the flower is borne.

Reflexed, bent downwards or backwards.

Regular, (of a flower) with a radially symmetrical (regular) shape that has multiple axes of symmetry, (see also Actinopmorphic). e.g. a *Lilium* flower.

Revolute, rolled back (e.g. inrolled margins of a petal).

Rhizome, horizontal underground stem; adj. **rhizomatous.**

Rosette, (of leaves) radiating from a central point on the ground.

Samara, dry, indehiscent (non-splitting) 1-seeded fruit with a membranous, wing-like extension for dispersal.

Scape, flowering stem without leaves (e.g. all leaves basal).

Schizocarp, a dry fruit splitting into 2 1-seeded portions (mericarps) e.g. fruits of Apiaceae (see also Mericarp).

Sclerophyll, hard leathery leaf containing a high proportion of thickened cells (sclereids); adj. **sclerophyllous.**

Sepal, typically leaf-like segments of the outer whorl(s) of the perianth (sometimes like a petal and then called a tepal).

Sessile, not stalked.

Simple, structure (e.g. a leaf) that is not divided into segments or lobes (not compound).

Spadix, spike-like organ bearing tiny male and female flowers at its base and surrounded by a spathe; characteristic of Araceae.

Spathe, large, leafy bract, sometimes brightly coloured; characteristic of Araceae.

Spike, an unbranched inflorescence (raceme) of stalkless flowers.

Spikelet, the basic unit of the inflorescence of grasses (Poaceae) and sedges (Cyperaceae); each flower cluster (e.g. in Plumbaginaceae).

Spur, hollow, tubular projection originating from the sepals or petals, often containing nectar.

Stamen, male reproductive organ of the flower consisting of a filament and anther (microsporangium).

Staminode, aborted or sterile stamens (not pollen-producing), sometimes resembling petals.

Standard, upper petals of flowers of species in the pea family (Leguminosae) and iris family (Iridaceae).

Stellate, star-shaped, e.g. a hair.

Steppic, referring to plant communities (used here in a Mediterranean context) typical of extreme, arid, continental environments (steppes).

Stigma, the part of the carpel or pistil that receives pollen and upon which the pollen germinates.

Stipule, small, leaf-like organ at the base of some leaf petioles (stalks), often in pairs; either simple or lobed.

Stolon, a spreading, above-ground shoot, or runner; adj. **stoloniferous**.

Style, stalk on an ovary bearing the stigma(s); sometimes absent.

Subshrub, small shrub or perennial with woody stems.

Subspecies, taxonomic subdivision of a species; usually a geographically, morphologically and or genetically distinct race.

Succulent, fleshy or juicy, e.g. stems of cacti and succulents.

Superior ovary, ovary positioned above the attachment points of other floral organs; syn. **hypogynous ovary**.

Taxon (pl. **taxa**), taxonomic unit of any rank, for example species, genus, subspecies or variety.

Teeth, divisions of a leaf, calyx or corolla; adj. **toothed**.

Tepals, the segments of the perianth, often used to describe segments which are not clearly petals or sepals (particularly in Liliaceae and related families).

Ternate, compound leaf with 3 leaflets.

Terminal, at the top/apex.

Throat, the opening where the tube joins the limb of the corolla or calyx.

Tree, a woody plant typically >5 m with a single trunk.

Trifoliate, compound leaf with 3 leaflets.

Tube, narrow, cylindrical part of the calyx or corolla, distinct from the limb, lobes or throat; adj. **tubular**.

Tufted, clustered together, e.g. a plant with numerous stems.

Umbel, flat-topped or convex 'umbrella-shaped' inflorescence consisting of a cluster of flowers with spreading stalks (pedicels) that arise from the apex of the peduncle; typical of Apiaceae.

Undulate, wavy at the margin.

Unisexual, flowers with either male or female reproductive organs only, contrast cosexual.

Variety, form of a species that is geographically, morphologically and or genetically distinct (but not distinct enough to warrant subspecies status).

Vascular, pertaining to the veins (conducting tissue) of an organ; all plants in this book are vascular.

Viscid, sticky.

Whorl, group of lateral organs borne >2 at each node (e.g. petals of a flower).

Woody, hard and wood-like, typically persistent.

Woolly, clothed with soft, shaggy hairs.

Xerophyte, drought-tolerant plant.

Zygomorphic, (of a flower) with a bilaterally symmetrical, irregular shape, e.g. an *Antirrhinum* flower.

Index of English Names

Main references are indicated by a bold page number, figures or drawings are indicated by a page number in italic text.

Index of Scientific Names

Accepted names are listed in italics, synonyms are listed in plain text. Main references are indicated by a bold page number, figures or drawings are indicated by a page number in italic text. (For synonyms, photographs or drawings will be captioned with the corresponding accepted name.)